The Microskills Hierarchy

A Pyramid for Building Cultural Intentionality

Determining Personal Style and Theory

Skills Integration
Sequencing skills in different theories

- Practice your own integration of skills.
- Various applications and theories use microskills differently.

Action Strategies for Change
Logical consequences, instruction/psychoeducation, stress management, therapeutic lifestyle changes and directives

Self-Disclosure and Feedback

Reflection of Meaning and Interpretation/Reframe

Empathic Confrontation

Focusing

The Five-Stage Interview Structure
Completing an interview using only the basic listening sequence and evaluating that interview for empathic understanding

Five Stages of the Counseling Session:

1. Empathic relationship
2. Story and strengths
3. Goals
4. Restory
5. Action

Reflection of Feeling

Encouraging, Paraphrasing, and Summarizing

Open and Closed Questions

Client Observation Skills

Basic Listening Sequence

Attending Behavior and Empathy
Culturally and individually appropriate visuals (eye contact), vocal qualities, verbal tracking skills, and body language

Ethics, Multicultural Competence, and Wellness

EIGHTH EDITION

Intentional Interviewing and Counseling

Facilitating Client Development in a Multicultural Society

Allen E. Ivey
University of Massachusetts, Amherst
Courtesy Professor, University of South Florida

Mary Bradford Ivey
Courtesy Professor, University of South Florida

Carlos P. Zalaquett
Professor, University of South Florida

BROOKS/COLE
CENGAGE Learning

Australia • Brazil • Japan • Korea • Mexico • Singapore • Spain • United Kingdom • United States

BROOKS/COLE
CENGAGE Learning

Intentional Interviewing and Counseling, Facilitating Client Development in a Multicultural Society, **Eighth Edition**

Allen E. Ivey, Mary Bradford Ivey, Carlos P. Zalaquett

Acquisitions Editor: Jon-David Hague

Developmental Editor: Julie Martinez

Assistant Editor: Mallory Ortberg

Editorial Assistant: Amelia Blevins

Managing Media Editor: Elizabeth Momb

Sr Brand Manager: Elisabeth Rhoden

Market Development Manager: Kara Parsons

Sr Content Project Manager: Rita Jaramillo

Art Director: Caryl Gorska

Manufacturing Planner: Judy Inouye

Rights Acquisitions Specialist: Don Schlotman

Production Service: Scratchgravel Publishing Services

Photo Researcher: PMG, Cheryl DuBois

Text Researcher: PMG, Punitha Rajamohan

Copy Editor: Peggy Tropp

Cover and Text Design: Baugher Design

Cover Image: Jrg Rse-oberreich

Compositor: MPS Limited

For product information and technology assistance, contact us at **Cengage Learning Customer & Sales Support, 1-800-354-9706**

For permission to use material from this text or product, submit all requests online at **www.cengage.com/permissions** Further permissions questions can be e-mailed to **permissionrequest@cengage.com**

Library of Congress Control Number: 2012951281

Student Edition:

ISBN-13: 978-1-285-06535-9

ISBN-10: 1-285-06535-2

Loose-leaf Edition:

ISBN-13: 978-1-285-17578-2

ISBN-10: 1-285-17578-6

Brooks/Cole
20 Davis Drive
Belmont, CA 94002-3098
USA

Cengage Learning is a leading provider of customized learning solutions with office locations around the globe, including Singapore, the United Kingdom, Australia, Mexico, Brazil, and Japan. Locate your local office at **www.cengage.com/global**

Cengage Learning products are represented in Canada by Nelson Education, Ltd.

To learn more about Brooks/Cole, visit **www.cengage.com/brookscole**

Purchase any of our products at your local college store or at our preferred online store **www.CengageBrain.com**

Printed in the United States of America
5 6 7   17 16 15 14

Love is listening.

Paul Tillich

To some of our groundbreaking colleagues:

Machiko Fukuhara
Professor, Tokiwa University, President of the Japanese Microcounseling Association

Thomas Daniels
Professor, Memorial University, Cornerbrook, Newfoundland

Weijun Zhang
Executive Coach, Shanghai, China

Philip Armstrong
Executive Director, Australian Counselling Association

Patricia Arredondo
Associate Vice Chancellor for Academic Affairs and Professor, Department of Educational Psychology, University of Wisconsin–Milwaukee

Nancy Baron
Director, Psycho-Social Services and Training Institute in Cairo and Global Psycho-Social Initiatives, Egypt

Fred Bemak
Professor and Director of the Diversity Research and Action Center, George Mason University

Kenneth Blanchard
CEO, Blanchard Training and Development

Michael D'Andrea
Executive Director, National Institute of Multicultural Competence, Professor, Seton Hall University

Óscar Gonçalves
Professor and Dean, School of Psychology— Neuropsychophysiology Lab, University of Minho, Portugal

David Goodrich
Executive Consultant, Sarasota, Florida

Ray Hosford
(Deceased) Professor, Department of Counseling Psychology, University of Wisconsin

Fran Howe
Private Practice, Hadley, Massachusetts

Maurice Howe
Professor and Former Chair, Department of Psychology, Swinburne University, Melbourne, Australia

Wha Myung
Director, Developmental Counseling and Therapy Institute, Daejeon, Korea

Jane Myers
Professor, Department of Counseling & Educational Development, University of North Carolina, Greensboro

Eugene Oetting
Professor Emeritus, Psychology Department, Colorado State University

Thomas A. Parham
Vice Chancellor for Student Affairs, University of California, Irvine

Paul Pedersen
Professor Emeritus, Department of Counseling and Human Services, Syracuse University

Sandra Rigazio-DiGilio
Professor, Department of Human Development & Family Studies, University of Connecticut

Lori Russell-Chapin
Professor and Director, Center for Collaborative Brain Research, Bradley University

Ernest Washington
Professor, Department of Teacher Education and Curriculum Studies, University of Massachusetts, Amherst

Derald Wing Sue
Professor, Department of Clinical and Counseling Psychology, Columbia University

ABOUT THE AUTHORS

Allen E. Ivey is Distinguished University Professor (Emeritus), University of Massachusetts, Amherst, and Courtesy Professor of Counseling at the University of South Florida, Tampa. He is the founder of Microtraining Associates, an educational publishing firm, and now serves as a consultant with Microtraining/Alexander Street Press. Allen is a Diplomate in Counseling Psychology and a Fellow of the American Counseling Association. He has received many awards, but is most proud of being named a Multicultural Elder at the National Multicultural Conference and Summit. Allen is author or coauthor of more than 40 books and 200 articles and chapters, translated into 22 languages. He is the originator of the microskills approach, which is fundamental to this book.

Mary Bradford Ivey is Courtesy Professor of Counseling, University of South Florida, Tampa. A former school counselor, she has served as visiting professor at the University of Massachusetts, Amherst; University of Hawai'i; and Flinders University, South Australia. Mary is the author or coauthor of 14 books, translated into multiple languages. She is a Nationally Certified Counselor (NCC) and a licensed mental health counselor (LMHC), and she has held a certificate in school counseling. She is also known for her work in promoting and explaining development counseling in the United States and internationally. Her elementary counseling program was named one of the ten best in the nation at the Christa McAuliffe Conference. She is one of the first 15 honored Fellows of the American Counseling Association.

Carlos P. Zalaquett is a Professor and Coordinator of Clinical Mental Health Counseling in the Department of Psychological and Social Foundations at the University of South Florida. He is also the Director of the USF Successful Latina/o Student Recognition Awards Program, Executive Secretary for the United States and Canada of the Society of Interamerican Psychology, and President of the Florida Mental Health Counseling Association and the Florida Behavioral Health Alliance. Carlos is the author or coauthor of more than 50 scholarly publications and four books, including the Spanish version of *Basic Attending Skills*. He has received many awards, including the USF Latinos Association's Faculty of the Year and Tampa Hispanic Heritage's Man of Education Award. He is an internationally recognized expert on mental health, psychotherapy, diversity, and education and has conducted workshops and lectures in seven countries.

Counseling CourseMate Website

Go to CengageBrain.com to access Counseling CourseMate, where you will find an interactive ebook, quizzes, videos, interactive counseling and psychotherapy exercises, the Portfolio of Competence, case studies, a practice test, and more. Students state that they have enjoyed this additional information and even report that they do better on course examinations.

We encourage you to use the CourseMate resources at the beginning and end of your work in each chapter. We have designed them to make your learning more meaningful and to help ensure competence in counseling and therapy practice.

If you did not receive a Printed Access Card for the CourseMate website packaged with your book, you can still purchase access to these resources at CengageBrain.com. Check with your professor to find out if the online resources are required.

CourseMate brings course concepts to life with interactive learning, study, and exam preparation tools that support the printed textbook. The CourseMate website provides a comprehensive study guide with many useful features to ensure competency:

- ▲ An integrated ebook
- ▲ More than 30 interactive case studies
- ▲ Flashcards to reinforce student learning and understanding
- ▲ Video clips illustrating counseling skills, including new examples of questioning, confrontation, and crisis counseling
- ▲ A new interactive mode of teaching empathic understanding, which includes a demonstration of varying empathic levels by Mary Bradford Ivey
- ▲ A chapter-by-chapter tutorial quiz in which incorrect answers link students to the section in the ebook where the correct answer is located
- ▲ Important forms and exercises to help students build portfolios they can present for field site placements and even for professional positions

CONTENTS

List of Boxes   *xiii*
Preface   *xv*

SECTION I
The Foundations of Counseling and Psychotherapy   1

CHAPTER 1   Toward Intentional Interviewing, Counseling, and Psychotherapy   3

Introduction: What Is the "Correct" Response to Offer a Client?   4
Interviewing, Counseling, and Psychotherapy   5
The Science and Art of Counseling and Psychotherapy   7
The Flexible Counselor and Client: Intentionality and Cultural Intentionality   7
Intentionality, Resilience, and Self-Actualization   8
The Core Skills of the Helping Process: The Microskills Hierarchy   10
Drawing Out Client Stories in a Well-Constructed Counseling and Psychotherapy
    Session   13
Empathic Relationship—Story and Strengths—Goals—Restory—Action   15
Our Multicultural World   17
RESPECTFUL Counseling and Psychotherapy   18
Neuroscience: Implications of This Cutting-Edge Science
    for the Future of Counseling and Psychotherapy   20
Your Natural Helping Style: An Important Audio or Video Exercise   23
Summary: Mastering the Skills and Strategies of Successful Relationships with Clients   25
Competency Practice Exercises and Portfolio of Competence   27

**CHAPTER 2   Ethics, Multicultural Competence, and the Positive Psychology
and Wellness Approach   29**

The Ethical Foundations of Counseling and Psychotherapy   30
Ethics in the Helping Process   31
Diversity and Multiculturalism   38
Multicultural Competence   41
Making Positive Psychology Work Through a Strengths and Wellness Approach   45
Positive Psychology: The Search for Strengths   46
Optimism, Resilience, and the Brain   46
The Brain, Stress, and the Wellness Approach   48
Assessing Client Wellness   51
Summary: Integrating Strength and Wellness, Ethics, and Multicultural Practice   57
Competency Practice Exercises and Portfolio of Competence   60
Determining Your Own Style and Theory: Critical Self-Reflection and Suggestions
    for Journal Assessment   62

CHAPTER 3   Attending Behavior and Empathy   63

Attending Behavior: The Foundation Skill of Listening   64
Attending Behavior: The Skills of Listening   65

Attending Behavior in Action: Getting Specific About Listening and Individual
and Multicultural Differences in Style   66
Visual/Eye Contact   66
Vocal Qualities: Tone and Speech Rate   67
Verbal Tracking: Following the Client or Changing the Topic   68
Body Language: Attentive and Authentic   71
Empathy   73
Example Counseling Session: I Didn't Get a Promotion—Is This Discrimination?   76
Training as Treatment: Social Skills, Psychoeducation, and Attending Behavior   79
Attending in Challenging Situations   80
Summary: The Samurai Effect, Neuroscience, and the Importance of Practice to Mastery   81
Competency Practice Exercises and Portfolio of Competence   83
Determining Your Own Style and Theory: Critical Self-Reflection on Attending and Empathy
and Suggestions for Journal Review   89

CHAPTER 4   Observation Skills   90

Are You a Good Observer?   91
Keeping Watch on the Counseling Session   92
Observe the Attending Patterns of Clients   92
Example Counseling Session: Is the Issue Difficulty in Studying or Racial Harassment?   95
Concepts Into Action: Three Organizing Principles   98
Summary: Observation Skills   106
Competency Practice Exercises and Portfolio of Competence   107
Determining Your Own Style and Theory: Critical Self-Reflection on Observation Skills   113

SECTION II
The Basic Listening Sequence: How to Organize a Session   115

CHAPTER 5   Questions: Opening Communication   117

Introduction: Defining and Questioning Questions   118
Empathy and Concreteness   121
Example Counseling Session: Conflict at Work   121
Concepts Into Action: Making Questions Work for You   125
Using Questions to Identify Strengths   129
Using Open and Closed Questions With Less Verbal Clients   130
Summary: Making Your Decision About Questions   131
Competency Practice Exercises and Portfolio of Competence   133
Determining Your Own Style and Theory: Critical Self-Reflection on Questioning   137
Our Thoughts About Benjamin   137

CHAPTER 6   Encouraging, Paraphrasing, and Summarizing: Key Skills
of Active Listening   138

Introduction: Active Listening   139
Empathy, Unconditional Positive Regard, and Active Listening Skills   140
Example Counseling Session: They Are Teasing Me About My Shoes   142
Counseling Children   145
Concepts Into Action: The Active Listening Skills of Encouraging, Paraphrasing,
and Summarizing   146
Diversity and the Listening Skills   149

Summary: Practice, Practice, and Practice   152
Competency Practice Exercises and Portfolio of Competence   154
Determining Your Own Style and Theory: Critical Self-Reflection
    on the Active Listening Skills   158
Our Thoughts About Jennifer   158

CHAPTER 7   Reflecting Feelings: A Foundation of Client Experience   160

Introduction: Reflection of Feeling   161
Comparing Paraphrasing and Reflection of Feeling   162
The Techniques of Reflecting Feelings   162
Empathy and Warmth   163
Example Counseling Session: My Mother Has Cancer, My Brothers Don't Help   164
Becoming Aware of and Skilled With Emotional Experience   167
The Language of Emotion   167
Helping Clients Increase or Decrease Emotional Expressiveness   170
The Place of Positive Emotions in Reflecting Feelings   172
Strategies for Positive Reflection   173
Summary: Reflection of Feelings Is a Key Skill, but Still Use It With Some Caution   175
Competency Practice Exercises and Portfolio of Competence   176
Determining Your Own Style and Theory: Critical Self-Reflection
    on Reflection of Feeling   181
List of Feeling Words   181

CHAPTER 8   How to Conduct a Five-Stage Counseling Session Using Only Listening Skills   183

The Basic Listening Sequence as a Foundation for Effective Communication
    in Many Areas of Life   184
Concepts Into Action: The Five-Stage Model for Structuring the Session
    With Many Theories of Counseling and Psychotherapy   187
Example Five-Stage Decisional Counseling Session: I Can't Get Along With My Boss   194
Taking Notes in the Session   200
Summary: The Well-Formed Session Using the Basic Attending Skills   201
Competency Practice Exercises and Portfolio of Competence   202
Determining Your Own Style and Theory: Critical Self-Reflection
    on Integrating Listening Skills   206

SECTION III
Focusing and Empathic Confrontation: Neuroscience, Memory,
and the Influencing Skills   207
How Memory Changes Are Enacted in the Session   208

CHAPTER 9   Focusing the Counseling Session: Exploring the Story From Multiple Perspectives   211

Introduction to Focusing   213
The Community Genogram   215
The Family Genogram   218
Debriefing a Community Genogram   218
Using Focusing to Examine Your Own Beliefs and Approach   223
Applying Focusing With a Challenging Issue   224
The Cultural/Environmental Context, Advocacy, and Social Justice   227
Summary: Being-in-Relation, Becoming a Person-in-Community   228

Competency Practice Exercises and Portfolio of Competence   230
Determining Your Own Style and Theory: Critical Self-Reflection on Focusing   234

CHAPTER 10 Empathic Confrontation and the Creative *New*: Identifying and Challenging Client Conflict   235

Defining Empathic Confrontation   236
Multicultural and Individual Issues in Confrontation   237
Empathy and Nonjudgmental Confrontation   237
The Skills of Empathic Confrontation: An Integrated Three-Step Process   238
The Client Change Scale (CCS)   243
Cultural Identity Development and the Confrontation Process   251
Cultural Identity Development and the Nelida/Allen Counseling Sessions   253
The CCS as a System for Assessing Change Over Several Sessions   255
Conflict Resolution and Mediation: A Psychoeducational Strategy
    for Creating the *New*   255
Summary: Supportive Challenge for the Creation of the *New*   256
Competency Practice Exercises and Portfolio of Competence   258
Determining Your Own Style and Theory: Critical Self-Reflection on Confrontation   264

SECTION IV
Interpersonal Influencing Skills for Creative Change   265

CHAPTER 11 Reflection of Meaning and Interpretation/Reframe: Helping Clients Restory Their Lives   267

Introduction: Defining the Skills of Reflection of Meaning and Interpretation/Reframe   268
Reflection of Meaning   270
Comparing Reflection of Meaning and Interpretation/Reframe   270
Example Counseling Session: Travis Explores the Meaning of a Recent Divorce   273
Concepts Into Action: The Specific Skills of Eliciting and Reflecting Meaning   276
Frankl's Logotherapy: Making Meaning Under Extreme Stress   280
The Skills of Interpretation/Reframing   283
Interpretation/Reframing and Other Microskills   284
Theories of Counseling and Interpretation/Reframing   286
A Cautionary Comment About Interpretation/Reframing   288
Summary: Facilitating Clients in Finding Their Meaning Core and Developing
    New Perspectives   288
Competency Practice Exercises and Portfolio of Competence   290
Determining Your Own Style and Theory: Critical Self-Reflection on Reflecting Meaning
    and Interpretation/Reframing   298
Our Thoughts About Charlis   298

CHAPTER 12 Self-Disclosure and Feedback: Immediacy and Genuineness in Counseling and Therapy   300

Defining Self-Disclosure   301
Empathic Self-Disclosure   303
Defining Feedback   306
Example Counseling Session: How Do I Deal With a Difficult Situation at Work?   308
Feedback and Neuroscience   310
Summary: Rewards and Risks of Self-Disclosure and Feedback   311

Competency Practice Exercises and Portfolio of Competence   312
Determining Your Own Style and Theory: Critical Self-Reflection on Self-Disclosure
    and Feedback Skills   315

**CHAPTER 13   Concrete Action Strategies for Client Change: Logical Consequences, Instruction/
Psychoeducation, Stress Management, and Therapeutic Lifestyle Changes   316**

Defining Logical Consequences   317
Defining Instruction and Psychoeducation Strategies   320
Stress and Stressors   323
Stress Management and Therapeutic Lifestyle Changes   326
Concepts Into Action: Psychoeducational Stress Management Strategies   327
Therapeutic Lifestyle Changes   331
Summary: Using the Concrete Action Strategies for Client Change   341
Competency Practice Exercises and Portfolio of Competence   342
Determining Your Own Style and Theory: Critical Self-Reflection
    on Influencing Skills   347

SECTION V
Skill Integration, Theory Into Practice, and Determining Personal Style   349

**CHAPTER 14   Skill Integration, Decisional Counseling, Treatment Planning,
and Relapse Prevention   351**

Decisional Counseling: Overview of a Practical Theory   352
The History of Pragmatic Decisional Counseling: Trait-and-Factor Theory   353
The Place of Decisional Counseling in Modern Practice   353
Decisional Counseling and Emotional Understanding   354
Importance of Case Conceptualization and Working Formulation   356
Planning the First Session and Using a Checklist   357
A Full Transcript of a Counseling Session: I'd Like to Find a New Career   360
Transcript Analysis   380
Skills and Their Impact on the Client   381
Additional Considerations: Referral, Treatment Planning, and Case Management   384
Maintaining Change: Relapse Prevention   388
Summary: Meeting the Clients Where They Are   390
Competency Practice Exercises and Portfolio of Competence   392
Determining Your Own Style and Theory: Critical Self-Reflection on Your Own Practice   393

**CHAPTER 15   How to Use Microskills and the Five Stages With Theories
of Counseling and Psychotherapy   394**

Part I: Microskills, Five Stages, and Theory   395
Part II: A Brief Summary of Theories Discussed in Earlier Chapters   397
Part III: Crisis Counseling and Cognitive Behavioral Therapy (CBT)   402
Crisis Counseling First Session Transcript   406
Cognitive Behavioral Session Transcript   411
Three Additional Theories   421
Summary: Practice and Integration Promote Personal Theory Development   421
Competency Practice Exercises and Portfolio of Competence   422
Determining Your Own Style and Theory: Critical Self-Reflection
    on Theoretical Orientations   425

CHAPTER 16 Determining Personal Style and Future Theoretical/ Practical Integration 426

Introduction: Identifying an Authentic Style That Relates to Clients 427
The Microskills Hierarchy: Assessing Your Competencies 427
Your Personal Style and Future Theoretical/Practical Integration 431
Summary—As We End: Thanks, Farewell, and Success in Your Future
    Growth and Professional Journey 432
Suggested Supplementary Readings for Follow-up on Microskills and
    Multicultural Issues 433

Appendix A The Ivey Taxonomy: Definitions and Anticipated Results 435

Appendix B The Family Genogram 441

Appendix C Counseling, Neuroscience, and Microskills 443

The Holistic Brain/Body and the Possibility of Change 444
Some Basic Brain Structures 445
Basics of Feeling and Emotion 447
The HPA and TAP and Other Structures 448
Neurons, Neural Networks, and Neurotransmitters 450
Microskills and Their Potential Impact on Change 454
The Brain's Default Mode Network: What's Happening When the Brain
    Is at Rest? 457
Social Justice, Stress Management, and Therapeutic Lifestyle Changes 458
Looking to the Future 460
Recommended for Further Study 460

References 462
Name Index 471
Subject Index 474

LIST OF BOXES

1.1 Research and Related Evidence That You Can Use: Microskills' Evidence Base   12

1.2 National and International Perspectives on Counseling Skills: Problems, Concerns, Issues, and Challenges—How Shall We Talk About Client Stories?   14

1.3 The Coming Paradigm Shift in Mental Health Counseling   20

1.4 Client Feedback Form (To Be Completed by the Volunteer Client)   24

2.1 National and International Perspectives on Counseling Skills: Professional Ethical Codes with Websites   31

2.2 Sample Practice Contract   36

2.3 National and International Perspectives on Counseling Skills: Multiculturalism Belongs to All of Us   44

2.4 A Six-Point Optimism Scale   47

2.5 Research and Related Evidence That You Can Use: Wellness Evidence Base   56

3.1 National and International Perspectives on Counseling Skills: Use With Care—Culturally Incorrect Attending Can Be Rude   71

3.2 Research and Related Evidence That You Can Use: Attending Behavior   72

3.3 Guidelines for Effective Feedback   85

3.4 Feedback Form: Attending Behavior   85

4.1 National and International Perspectives on Counseling Skills: Can I Trust What I See?   94

4.2 Research and Related Evidence That You Can Use: Observation   99

4.3 The Abstraction Ladder   102

4.4 Feedback Form: Observation   109

5.1 Research and Related Evidence That You Can Use: Questions   120

5.2 National and International Perspectives on Counseling Skills: Using Questions with Youth at Risk   128

5.3 Feedback Form: Questions   135

6.1 Listening Skills and Children   145

6.2 Accumulative Stress: When Do "Small" Events Become Traumatic?   146

6.3 Research and Related Evidence That You Can Use: Active Listening   150

6.4 National and International Perspectives on Counseling Skills: Developing Skills to Help the Bilingual Client   151

6.5 Feedback Form: Encouraging, Paraphrasing, and Summarizing   157

7.1 The Nonverbal Language of Emotion: Micro and Macro Feelings   169

7.2 National and International Perspectives on Counseling Skills: The Invisible Whiteness of Being and Feelings Toward Another Race   171

7.3 Research and Related Evidence That You Can Use: Reflection of Feeling   174

7.4 Feedback Form: Observing and Reflecting Feelings 179

8.1 National and International Perspectives on Counseling Skills: Demystifying the Helping Process 186

8.2 Feedback Form: Practice Session Using Only the Basic Listening Sequence 204

9.1 The Community Genogram: Three Visual Examples 216

9.2 National and International Perspectives on Counseling Skills: Where to Focus: Individual, Family, or Culture? 226

9.3 Research and Related Evidence That You Can Use: Focusing 227

9.4 Feedback Form: Focus 232

10.1 Research and Related Evidence That You Can Use: Confront, but Also Support 250

10.2 Cultural Identity Development 252

10.3 National and International Perspectives on Counseling Skills: A Practical Application of the Racial/Cultural Identity Development Model 254

10.4 Feedback Form: Confrontation Using the Client Change Scale 262

11.1 National and International Perspectives on Counseling Skills: What Can You Gain from Counseling Persons with AIDS and Serious Health Issues? 276

11.2 Questions Leading Toward Discernment of Life's Purpose and Meaning 279

11.3 Research and Related Evidence That You Can Use: Reflection of Meaning and Reframing 282

11.4 Feedback Form: Reflecting Meaning 294

11.5 Feedback Form: Interpretation/Reframe 296

12.1 The "1-2-3" Pattern of Listening, Influencing, and Observing Client Reaction 303

12.2 National and International Perspectives on Counseling Skills: When Is Self-Disclosure Appropriate? 305

12.3 Research and Related Evidence That You Can Use: Self-Disclosure 306

12.4 Feedback Form: Disclosure and Feedback 314

13.1 National and International Perspectives on Counseling Skills: Explorations Around the World: Learning New Approaches to Working With Cultural Strengths 322

13.2 Stress Management and TLC Instructional Strategies 327

13.3 Research and Related Evidence That You Can Use: The Benefits of Exercise 333

13.4 Feedback Form: Logical Consequences, Instruction/Psychoeducation, Stress Management, and TLCs 344

14.1 Checklist for the First Session 359

14.2 National and International Perspectives on Counseling Skills: What's Happening With Your Client While You Are Counseling? 383

14.3 Maintaining Change Worksheet: Self-Management Strategies for Skill Retention 389

14.4 Transcribing Sessions 393

15.1 Research and Related Evidence That You Can Use: Systematic Emergency Therapy for Sexual Assault, Personal Assault, and Accident Survivors 405

15.2 Research and Related Evidence That You Can Use: fMRI Predicts Response to CBT 420

15.3 Feedback Form: Counseling Theories 423

16.1 National and International Perspectives on Counseling Skills: Using Microskills Throughout My Professional Career 430

16.2 Your Natural Style of Counseling and Psychotherapy and the Future 432

 PREFACE

Welcome to the eighth edition of *Intentional Interviewing and Counseling: Facilitating Client Development in a Multicultural Society*, the original, most researched system in the basics of skilled counseling and psychotherapy. You will find a completely updated and rewritten revision, based on the latest research, and made even more user friendly. Empathy and the relationship/working alliance now take a more central place. There is also a new chapter. Several new transcripts of counseling interviews, as well as six additional video demonstrations, are on the accompanying optional CourseMate website.

The microskills approach has become the standard for counseling and psychotherapy skills training throughout the world. Based on more than 450 data-based studies, used in well over 1,000 universities and training programs through the world, the culturally sensitive microskills approach is now available in 23 translations. The emphasis is on clarity and providing the critical background for competence in virtually all counseling and psychotherapy theories.

Easy to teach and learn from, students will find that the optional CourseMate website helps ensure that they can immediately take the concepts presented in the textbook to the "real world."

An alternative version of this text is available. *Essentials of Intentional Interviewing* (2nd ed.) covers the crucial skills and strategies of interviewing, counseling, and psychotherapy in a briefer form, with less attention to theory, research, and supplementary concepts.

The Microskills Tradition and Basic Competencies

This book continues the original emphasis on competencies. What counts is what students *can do* to become competent and prepared for the next stages of their education and professional work. Students who work with this book will be able to

▲ Engage in the basic skills of the counseling or psychotherapy session: listening, influencing, and structuring an effective session with individual and multicultural sensitivity.
▲ Conduct a full session using only listening skills by the time they are halfway through this book.
▲ Master a basic structure of the session that can be applied to many different theories:

1. Develop an *empathic relationship* with the client.
2. Draw out the client's *story*, giving special attention to strengths and resources.
3. Set clear *goals* with the client.
4. Enable the client to *restory* and think differently about concerns, issues, and challenges.
5. Help the client move to *action* outside the session.

▲ Integrate ethics, multicultural issues, and positive psychology/wellness into counseling practice.
▲ Analyze with considerable precision their own natural style of helping and, equally or perhaps more important, how their counseling style is received by clients.

The Portfolio of Competence is emphasized in each chapter. Students have found that a well-organized portfolio is helpful in obtaining good practicum and internship sites and, at times, professional positions as well. Students may complain about the workload, but if they develop a solid portfolio of competencies, use the interactive website to reinforce learning, and engage in serious practice of skills and concepts, it will become clear how much they have learned. The portfolio concept and the interactive CourseMate website increase course satisfaction and ratings.

▶New Competency Features in This Eighth Edition

The coming decade will bring an increasing integration of mental and physical health services, as we move to new and more sophisticated and complete systems to help clients and patients. Innovations in team practice are bringing counselors and psychotherapists more closely together with physicians, nurses, and human service workers. Furthermore, neuroscience and brain research are leading to awareness that body and mind are one. Actions in the counseling session affect not only thoughts, feelings, and behaviors, but also what occurs in the brain and body. Many exciting new opportunities await both students and instructors.

This eighth edition of *Intentional Interviewing and Counseling* seeks to prepare students for culturally intentional and flexible counseling and psychotherapy. The following features have been added or strengthened as we prepare for this new future.

▲ *Empathy and empathic communication* have become even more central to the microskills framework. While they have always been there, they are now a centerpiece, associated with each and every skill. Students will be able to evaluate each intervention for its quality of empathic understanding. Every transcript in this text includes process discussions that illustrate the various levels of empathy. Students will be able to evaluate how their helping leads are received by clients.

▲ *Self-actualization, intentionality, and resilience* are clarified as goals for the counseling session. A new section focuses on what we would like to see for our clients as a result of the session. Of course, we want to facilitate their reaching their own desired ends, but we also seek to encourage the development of resilience skills to better cope with future stresses and challenges.

▲ *Crisis counseling and a new transcript of cognitive behavioral therapy* have been added to Chapter 15, where applications of microskills to counseling theories are demonstrated. The basics of crisis counseling and a transcript of what actually might happen in a session highlight the chapter. The CBT transcript shows the specifics of work with automatic thoughts and demonstrates clearly how students can use this strategy.

▲ *New chapter on self-disclosure and feedback* (Chapter 12). These two skills have increased space with special attention to their relationship to empathic understanding and immediacy.

▲ *Completely rewritten chapters on confrontation and focusing* (Chapters 9 and 10). To make these two chapters clearer and more practical for students, we have added a continuing transcript illustrating both skills over time. Students will see how empathic understanding is critical to both these skills. Students will also see how to evaluate change through the session using the Client Change Scale.

▲ *Five new session transcripts with process comments* illustrate how skills actually work in the session. While outlining the key aspects of counseling skills is essential, even more valuable is examining how they fit into the session, how to observe the results of empathic skill usage, and how the use of microskills affects client thoughts, feelings, and emotions. This better prepares the student for the most critical part of skill training—actual practice of the skills in role-played or real sessions.

▲ ***Increased integration of cutting-edge neuroscience with counseling skills.*** Counseling and psychotherapy change the brain and build new neural networks in both client and counselor through neural plasticity and neurogenesis. Special attention is paid to portions of the brain (with new illustrations) that are affected in the helping process. Neuroscience research stresses a positive wellness orientation to facilitate neural development, along with positive mental health. An appendix with additional practical implications is available. Students will find that virtually all of what we do in the helping fields is supported by neuroscience research.

▲ ***Stress management and therapeutic lifestyle changes (TLC)*** have become more central in this revised text. The critical issue of recognizing stress and its dangerous impact on the brain and body is emphasized throughout the text, while also recognizing that appropriate levels of stress can be positive and are necessary for learning and change. Research in wellness and neuroscience has revealed the importance of therapeutic lifestyle changes (TLC) as a supplement to stress management and all theoretical approaches.

▲ ***CourseMate, our optional online package,*** is a popular and effective interactive ancillary, authored by Allen and Mary Ivey and Carlos Zalaquett, that has been updated with new video demonstrations. The many case studies and interactive video-based exercises provide practice and further information leading to competence. Downloadable forms and feedback sheets make it easier for students to develop a Portfolio of Competence. Students who seriously use these resources report that they understand the session better and perform better on examinations. Transcripts and "how to's" for brief counseling, motivational interviewing, and coaching/counseling are major features of the online material.

▶For Instructors: Ancillary Materials

Intentional Interviewing and Counseling, eighth edition, has a full set of instructor and student ancillary materials that have been authored by Allen and Mary Ivey and Carlos Zalaquett to assist with teaching needs and learning comprehension.

Online Instructor Resource Guide   The online Instructor Resource Guide (IRG) has been revised in accordance with the changes in the book. There you will find chapter objectives, suggested class procedures, additional discussion of end-of-chapter exercises, and multiple-choice and essay questions. Possible supplementary videos are listed for each chapter. The IRG also includes basic information on developmental counseling and therapy (DCT) that many professors find useful in beginning skills courses. This material in the IRG may be duplicated for student use. The IRG is available electronically and can be ordered through your Cengage Learning sales representative.

Online Test Bank   An electronic Test Bank is available upon request from your Cengage Learning sales representative. Flexible quizzes and answers are also available on ExamView, which allows instructors to change items and create tests easily. If you have difficulties finding the Test Bank, contact your Cengage Learning sales representative at www.cengage.com or Allen Ivey at allenivey@gmail.com.

Two Sets of PowerPoint® Slides   These are available on the book's companion website at www.cengage.com/login. One set is quite detailed, covering all the concepts of each chapter. The second is abbreviated and covers the main concepts. You may download either or both sets and change and sort/reorder the slides according to your teaching preferences. You can then project them from your computer.

Acknowledgments

Our Thanks to Students   National and international students have been important over the years in the development of this book. We invite students to continue this collaboration. Weijun Zhang, a student of Allen, is now the leading coach and management consultant in China. He wrote many of the National and International Perspectives on Counseling Skills boxes, which enrich our understanding of multicultural issues. Penny John, another of Allen's students, is thanked for permission to use her class project session transcript as an example of brief counseling. Amanda Russo, a student at Western Kentucky University, allowed us to share some of her thoughts about the importance of practicing microskills. Michael Fitzsimmons, Temple University, provided comments on current literature in gender and sexual orientation studies.

We give special attention to two of Carlos's students, Nelida Zamora and SeriaShia Chatters. Nelida worked with us closely in the development of two sets of videos, *Basic Influencing Skills* (3rd ed.) and *Basic Stress Management Skills* for Alexander Street Press/ Microtraining Associates. She also gave permission to use a transcript of her demonstration session with Allen in Chapters 9 and 10. Dr. SeriaShia Chatters helped develop the DVD sets and book videos, important in making the nature of helping skills clear. She is now a faculty member at Zayed University in Dubai.

Our graduate students at the University of South Florida volunteered their time to participate in the videos that are on the supplemental website. They also assisted in updating the research boxes. We are especially appreciative of the quality work of Kerry Conca, Megan Hartnett, Jonathan Hopkins, Stephanie Konter, Floret Miller, Callie Nettles, and Krystal Snell.

Our Thanks to Our Colleagues   Owen Hargie, James Lanier, Courtland Lee, Robert Manthei, Mark Pope, Kathryn Quirk, Azara Santiago-Rivera, Sandra Rigazio-DiGilio, and Derald Wing Sue wrote the informative boxes on multicultural issues throughout the book. Robert Marx developed the Relapse Prevention form of Chapter 14. Discussions with Otto Payton and Viktor Frankl clarified the presentation of reflection of meaning. William Matthews was especially helpful in formulating the five-stage interview structure. Lia and Zig Kapelis of Flinders University and Adelaide University are thanked for their support and participation while Allen and Mary served twice as visiting professors in South Australia.

David Rathman, Chief Executive Officer of Aboriginal Affairs, South Australia, has constantly supported and challenged this book, and his influence shows in many ways. Matthew Rigney, also of Aboriginal Affairs, was instrumental in introducing us to new ways of thinking. These two people first showed us that Western individualistic ways of thinking are incomplete, and therefore they were critical in bringing us early to an understanding of multicultural issues.

The skills and concepts of this book rely on the work of many different individuals over the past 30 years, notably Eugene Oetting, Dean Miller, Cheryl Normington, Richard Haase, Max Uhlemann, and Weston Morrill at Colorado State University, who were there at the inception of the microtraining framework. The following people have been personally and professionally helpful in the growth of microcounseling and microtraining over the years: Bertil Bratt, Norma Gluckstern-Packard, Jeanne Phillips, John Moreland, Jerry Authier, David Evans, Margaret Hearn, Lynn Simek-Morgan, Dwight Allen, Paul and Anne Pedersen, Lanette Shizuru, Steve Rollin, Bruce Oldershaw, Óscar Gonçalves, Koji Tamase, and Elizabeth and Thad Robey.

The board of directors of the National Institute of Multicultural Competence—Michael D'Andrea, Judy Daniels, Don C. Locke, Beverly O'Bryant, Thomas Parham, and Derald Wing Sue—are now part of our family. Their support and guidance have become central to our lives.

Fran and Maurie Howe have reviewed seemingly endless revisions of this book over the years. Their swift and accurate feedback has been significant in our search for authenticity, rigor, and meaning in the theory and practice of counseling and psychotherapy.

Jenifer Zalaquett has been especially important throughout this process. She not only navigates the paperwork but is instrumental in holding the whole project together.

Julie Martinez has now worked with us as consulting editor over five editions of this book. At this point, we almost feel that she is a coauthor. It is always a pleasure to work with the rest of the group at Brooks/Cole, notably Jon-David Hague, Rita Jaramillo, Elizabeth Momb, Elisabeth Rhoden, Kara Parsons, and Caryl Gorksa. Our manuscript editor, Peggy Tropp, has become a valuable adviser to us. Anne and Greg Draus of Scratchgravel Publishing Services always do a more than terrific job. At this point, they too are beginning to feel like coauthors.

We are grateful to the following reviewers for their valuable suggestions and comments: Yvonne Barry, John Tyler Community College; Cynthia Cary, University of Hawaii Maui College; Danuta Chessor, University of Western Sydney, NSW Australia; Andres Consoli, San Francisco State University; Lorraine Guth, Indiana University of Pennsylvania; Jerri Montgomery, Wytheville Community College; Erin Olson, Dordt College; Derrick Paladino, Rollins College; and Alan Silliker, St. Bonaventure University. They shared ideas and encouraged changes that you see here, and they also pushed for more clarity and a practical action orientation.

Again, we ask you to send in reactions, suggestions, and ideas. Please use the form at the back of this book to send us your comments. Feel free to contact us also by email. We appreciate the time that you as a reader are willing to spend with us.

Allen E. Ivey, Ed.D., ABPP
alleniivey@gmail.com
Mary Bradford, Ed.D., NCC, LMHC
mary.b.ivey@gmail.com
Carlos Zalaquett, Ph.D., LMHC,
Licensiado en Psicología
carlosz@usf.edu

The Foundations of Counseling and Psychotherapy

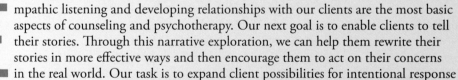

Empathic listening and developing relationships with our clients are the most basic aspects of counseling and psychotherapy. Our next goal is to enable clients to tell their stories. Through this narrative exploration, we can help them rewrite their stories in more effective ways and then encourage them to act on their concerns in the real world. Our task is to expand client possibilities for intentional response and action, which in turn lead to a self-actualization of potential, more effective decision making, and wellness. As counselors and therapists, we can make change happen whether we work in schools and colleges, community agencies, behavioral health organizations, or private practice.

Section I presents the skills of listening and empathic understanding; these skills are often enough to help clients look at themselves and make positive change. Building on this foundation, later chapters will discuss influencing skills and strategies, all designed to provide you with many possibilities to empower your clients to restory their lives, take charge, and move toward change and action.

Chapter 1. Toward Intentional Interviewing, Counseling, and Psychotherapy

This chapter offers an overview and a road map of what this book can do for you. We begin with the microskills hierarchy, with summaries of counseling skills and strategies, providing an outline of subsequent chapters. We also start by identifying your natural helping skills. You are not taking this course by chance; something has led you here, with unique abilities, oriented to helping others. You will be asked to reflect on what brings you to the helping field: What do you want to do as you examine the skills and strategies of counseling and therapy?

By the time you reach Chapter 8 of this book, you will be able to complete a full session using only listening skills. When you combine the listening microskills with the five-stage structure of counseling and psychotherapy, you are prepared for many types of sessions. The backbone of this book is the five-stage model: *empathic relationship—story and strengths—goals—restory—action*. The five stages represent a necessary and sufficient structure for interviewing, counseling, and psychotherapy that provides a key to the multiple theories and approaches of the helping field.

Chapter 2. Ethics, Multicultural Competence, and the Positive Psychology and Wellness Approach   This chapter presents three crucial aspects of all counseling and psychotherapy. Ethics—the professional standards that all major helping professions observe and practice—provides counselors and psychotherapists with guidelines on issues such as competence, informed consent, confidentiality, power, and social justice. Multicultural competence is about cultural awareness and sensitivity to the worldview of our clients. Positive psychology and wellness enable clients to identify their strengths and resources. This approach significantly facilitates resolving client life issues, focusing on what they "can do" rather than what they "can't do."

Chapter 3. Attending Behavior and Empathy   This chapter presents the most basic fundamentals of counseling and psychotherapy. Without attending skills, an empathic relationship cannot occur. Many beginning helpers inappropriately strive to solve the client's issues and challenges in the first 5 minutes of the session by giving premature advice and suggestions. Please set one early goal for yourself: Allow your clients to talk. Observe closely how they are behaving, verbally and nonverbally. Your clients may have spent several years developing their concerns, issues, and life challenges before consulting you. Listen first, last, and always.

Chapter 4. Observation Skills   This chapter builds on attending behavior and gives you the further opportunity to practice observing your clients' verbal and nonverbal behavior. You are also asked to observe your own nonverbal reactions in the session. Clients often come in with a "hangdog" and "down" body posture. Between your observation and listening skills, you can anticipate that they will later have more positive body language, as well as a new story and better view of self. You can help their body to stand up straight and their eyes to shine.

Begin this book with a commitment to yourself and your own natural communication expertise. Through the microskills approach, you can enhance your natural style with new skills and strategies that will expand your alternatives for facilitating client growth and development.

Toward Intentional Interviewing, Counseling, and Psychotherapy

We humans are social beings. We come into the world as the result of others' actions. We survive here in dependence on others. Whether we like it or not, there is hardly a moment of our lives when we do not benefit from others' activities. For this reason it is hardly surprising that most of our happiness arises in the context of our relationships with others.

—The Dalai Lama

Mission of "Toward Intentional Interviewing, Counseling, and Psychotherapy"

To present the major concepts underlying the skills and strategies of this book and provide an outline of what you can expect. The competency objectives listed below outline the goals of this chapter.

Chapter Goals and Competency Objectives Awareness, knowledge, skills, and actions developed through this chapter will enable you to

▲ Explore the session as both science and art. In this process, we ask you to reflect on yourself as a potential helper. While science undergirds what is said here, you as an independent artist will find your own integration of ideas, skills, and competencies.

▲ Define and discuss similarities and differences among interviewing, counseling, and psychotherapy, and review who actually conducts most of the helping sessions. This may be surprising and rewarding.

▲ Gain knowledge of the microskills step-by-step approach that provides an adaptable base on which to define your personal style and, later, your view of theories of counseling.

- ▲ Examine key goals of counseling and psychotherapy: self-actualization, resilience, and resolution of client issues.
- ▲ Develop increased awareness of what "multiculturalism" really is and the importance of being able to work with clients from widely varying backgrounds.
- ▲ Consider the place of cutting-edge neuroscience in your own work and for the future of the counseling and psychotherapy field.
- ▲ As your first practical exercise, record a counseling session demonstrating your natural style of communicating and helping. This first exercise concludes the chapter and provides a baseline so that later you can examine how your counseling style may have changed during your time with this book.

 Go to CengageBrain.com to access Counseling CourseMate, where you will find an interactive ebook, quizzes, videos, interactive counseling and psychotherapy exercises, the Portfolio of Competence, case studies, a practice test, and more.

▶Introduction: What Is the "Correct" Response to Offer a Client?

Sienna, 16 years old, is 8 months' pregnant with her first child. She says, "I wonder when I'll be able to see Freddy [baby's father] again. I mean, I want him involved; he wants to be with me, and the baby. But my mom wants me home. His mom said she's looking for a two-bedroom apartment so we could possibly live there, but I know my mom will never go for it. She wants me to stay with her until I graduate high school and, well, to be honest, so that this never happens again [she points to her belly]."

I listen carefully to her story and later respond, "I'm glad to hear that Freddy wants to be involved in the care of the child and maintain a relationship with you. What are your goals? How do you feel about talking through this with your mom?"

"I don't know. We don't really talk much anymore," she says as she slumps down in her chair and picks away at her purple nail polish. I reflect her sad feelings, but as I do so, she brightens up just a bit as she recalls that she does get along with her mother fairly well.

She then describes her life before Freddy, focusing mainly on the crowd she hung around, a group of girls whom she says were wild, mean, and tough. Her mood returns to melancholy, and she seems anxious and discouraged. At the same time, the session has gone smoothly and we seem to have a good relationship. I say, "I sense that you have a good picture of what you are facing. Well, it seems that there's a lot to talk about. How do you feel about continuing our conversation before sitting down with your mom?"

Surprisingly, she says, "No. Let's talk next week with her. The baby is coming and, well, it'll be harder then." As we close the session, I ask her, "As you look back on our talk together, what comes to mind?" Sienna responds, "Well, I feel a bit more hopeful and I guess you're going to help me talk about some important issues with my mom, and I didn't think I could do that."

This was the first step in a series of five sessions. As the story evolved, we invited Freddy for a session. He turned out to be employed and was anxious to meet his responsibilities, although finances remained a considerable challenge. This was followed by a meeting with both mothers in which a workable action plan for all families was generated. I helped Sienna find a school with a special program for pregnant teens.

This case exemplifies the reality of helping. We often face complex issues with no clear positive ending. If we can develop a relationship and listen to the story carefully, clearer goals develop, and solutions usually follow.

▶Interviewing, Counseling, and Psychotherapy

The terms *interviewing, counseling,* and *psychotherapy* are used interchangeably in this text. The overlap is considerable (see Figure 1.1), and at times interviewing will touch briefly on counseling and psychotherapy. Both counselors and psychotherapists typically draw on the interview in the early phases of their work. You cannot become a successful counselor or therapist unless you have solid interviewing skills.

Interviewing is the basic process used for gathering data, providing information and advice to clients, and suggesting workable alternatives for resolving concerns. Interviewers can be found in many settings, including employment offices, schools, and hospitals. Professionals in many areas also use these skills—for example, in medicine, business, law, community development, library work, and many government offices.

Counseling is a more intensive and personal process. Although interviewing to gain client information is critical, counseling is more about listening to and understanding a client's life challenges and, along with the client, developing strategies for change and growth. Counseling is most often associated with the professional fields of counseling, human relations, clinical and counseling psychology, pastoral counseling, and social work and is also part of the role of medical personnel and psychiatrists.

Psychotherapy focuses on more deep-seated difficulties, which often require more time for resolution. However, psychotherapists interview clients to obtain basic facts and

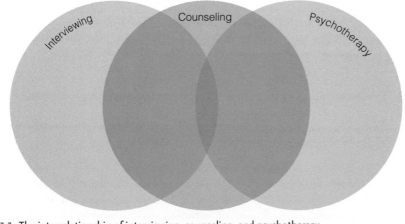

FIGURE 1.1 The interrelationship of interviewing, counseling, and psychotherapy.

TABLE 1.1 Numbers of Helping Professionals

School counselors	131,100
College counselors	52,370
Mental health counselors	114,180
Vocational rehabilitation counselors	110,690
Substance abuse and behavioral disorder counselors	76,600
Human service workers—social and human service assistants	384,200
Marriage and family counselors	33,990
Clinical, counseling, and school psychologists	100,850
Child, family, and school social workers	276,510
Medical and public health social workers	138,890
Psychiatrists	23,140
Professional coaches identified by the International Coaching Federation	30,000

Source: U.S. Department of Labor. (2012). *Occupational Outlook Handbook* 2010–2011. www.bls.gov/oco/ooh_index.htm. The Labor Department updates these data regularly.

information as they begin working with them, and often use counseling as well. The skills and concepts of intentional interviewing and counseling are equally effective for the successful conduct of longer-term psychotherapy. Historically, psychotherapy was the province of psychiatrists, but they are limited in number, and today they mostly offer short sessions and treat with medications. This means that most talking therapy will be conducted by professionals other than psychiatrists. Table 1.1 shows a total of 23,000 psychiatrists with more than 1 million other helping professionals.

Number of sessions.   Carlstedt (2011) has examined the number of sessions for 23,034 clients who underwent a total of 218,331 psychotherapy sessions. About 21% of the clients did not return after the first session, while 50% finished at four sessions. Another 25% completed their work in 5 to 10 sessions, and most had completed therapy by the 35th session. What are the implications for you? On the one hand, some clients have their needs met in one session—and that may be enough to resolve most issues. On the other hand, Sue and Sue (2013) note that close to 50% of non-White clients drop out after the first session. This suggests that although some clients find that one session provides the help and information they need, many others fail to find counseling and therapy valuable.

REFLECTION *Where Is Your Place in the Helping Field?*

You likely started this course with some vision of yourself in the helping field. Do you see yourself emphasizing interviewing, counseling, or psychotherapy? Given the many possibilities for service, which of the listed professions appeals to you most at this time? Would you rather work in the schools, a community mental health clinic, a hospital, or private practice?

Psychiatry, which has the smallest number of professionals, also has the greatest power to define the nature of the helping fields, particularly through the naming of client issues in their *Diagnostic and Statistical Manual of Mental Disorders*. Their diagnostic sessions take considerable time, up to 2 hours or more. At the same time, psychiatrists typically focus on medication in brief sessions and may not be able to provide time for needed talk therapy. What are your thoughts as to your responsibility in meeting the overall mental health needs of your country's total population?

▶The Science and Art of Counseling and Psychotherapy

This chapter introduces counseling as science and art. Counseling and psychotherapy now have a solid research and scientific base that enables us to identify many qualities and skills that lead to effectiveness. Science has demonstrated that the specifics of listening skills are identifiable and are central to competent helping. But an evidence-based approach by itself is not enough. You as counselor are similar to an artist whose skills and knowledge produce beautiful paintings out of color, canvas, and personal experience. You are the listener who will provide color and meaning to the interpersonal relationship we call helping.

Counselors who work with clients such as Sienna need to be competent in evidence-based counseling skills, as well as strategies drawn from theoretical systems of counseling and psychotherapy. But each client is unique, and scientific research cannot tell us precisely what will be most useful in helping each one. Thus, counseling and psychotherapy indeed have a scientific base, but applying that science is, in truth, an art form. Science provides a base, but you are the artist who makes it happen.

Like the artist, the musician, or the skilled athlete, you bring a natural talent to share with others, but you need to be ready for surprises and change direction when necessary. This is where your skill as an artist comes in. Which aspects of evidence-based psychotherapy are going to be useful in specific, often difficult situations? Although the theories, skills, and strategies of professional work remain essential, you are the one who puts them together and can uniquely facilitate the development and growth of others.

The following section extends science and art to the concepts of intentionality and cultural intentionality.

▶The Flexible Counselor and Client: Intentionality and Cultural Intentionality

If you don't like that idea, I've got another.

—Marshall McLuhan

There are many ways to facilitate client development. As you become increasingly competent, learn to blend what is natural for you with new skills and theory. *Intentionality* is a central goal of this text. We ask you to be yourself, but also to realize that to reach a wide variety of clients, you need to be flexible and constantly change behavior and learn new ways of being with the uniqueness of each client.

Beginning students are often eager to find the "right" answer for the client. In fact, they are so eager that they often give quick patch-up advice that is inappropriate. For example, your own personal issues or cultural factors such as ethnicity, race, gender, lifestyle, or religious orientation may have biased your response and session plan for Sienna. Intentional counseling and psychotherapy are concerned not with which single response is correct, but with how many potential responses may be helpful. We can define intentionality as follows:

> Intentionality, along with cultural intentionality, is acting with a sense of capability and deciding from among a range of alternative actions. The intentional individual has more than one action, thought, or behavior to choose from in responding to changing life situations. The culturally intentional individual can generate alternatives in a given situation and approach a problem from multiple vantage points, using a variety of skills and personal qualities, adapting styles to suit different individuals and cultures.
>
> The culturally intentional counselor or therapist remembers a basic rule of helping: *If a helping lead or skill doesn't work—try another approach!* A critical issue in counseling and psychotherapy is that the same comment may have different effects on individuals who have different personal life experiences and multicultural backgrounds, because everyone has a unique history with a unique pattern of communication. Watch out for stereotyping, because individuals, regardless of culture, vary widely.

REFLECTION *Developing Your Own Culturally Intentional Style*

What multiple paths has your life taken? How has each new experience or setting changed the way you think? Has this led to increasing flexibility and awareness of the many possibilities that are yours? Can you listen and learn from those who may sharply differ from you? What is your family and cultural background, and how does that affect the person you are?

▶Intentionality, Resilience, and Self-Actualization

People who feel good about themselves produce good results.

—Kenneth Blanchard

Many, even most, of our clients come to us feeling that they are not functioning effectively and are focused on what's wrong with them. They are *stressed*. Clients may feel *stuck, overwhelmed,* and *unable to act*. Frequently, they will unable to make a career or life decision. Often, they will have a *negative self-concept*, or they may be full of anger toward others. This focus on the negative is what we want to combat as we emphasize developing client intentionality, resilience, and self-actualization.

Intentionality

We cannot expect to solve all our clients' issues and challenges in a few sessions, but in the short time we have with them, we can make a difference. First think of what intentionality and flexibility mean for you as a counselor or therapist. Clients will benefit and become stronger as they feel heard and respected and become intentionally flexible with new ways to resolve their concerns. Resolving specific immediate issues, such as choice of a college major, a career change, whether to break up a long-term relationship, or handling mild depression after a significant loss, will help them feel empowered and facilitate further action.

Resilience and Intentionality

A longer-term goal of effective counseling and therapy is the development of client resilience.

> Resilience is a dynamic process whereby individuals exhibit positive behavioral adaptation when they encounter significant adversity, trauma, tragedy, threats, or even significant sources of stress. (Luthar, Cicchetti, & Becker, 2000)

The development of client intentionality is another way to talk about resilience. As counselors, we want to be flexible and move with changing and surprising events, but clients need the same abilities. Helping a client resolve an issue is our contribution to increasing client resilience. You have helped the client move from stuckness to action, from indecision to decision, or from muddling around to clarity of vision. Pointing out to clients who change that they are demonstrating resilience and ability facilitates longer-term success.

This book offers multiple skills and strategies, all designed to facilitate client resilience and positive development.

Self-Actualization

Carl Rogers and Abraham Maslow have focused on self-actualization as a goal of counseling and therapy. Closely related to both intentionality and resilience, self-actualization is defined as

> the curative force in psychotherapy—*man's tendency to actualize himself, to become his potentialities* . . . to express and activate all the capacities of the organism. (Rogers, 1961, 350–351)

> intrinsic growth of what is already in the organism, or more accurately of what is the organism itself. . . . Self-actualization is growth-motivated rather than deficiency-motivated. (Maslow, Frager, & Fadiman, 1987, p. 66)

Regardless of the situation in which our clients find themselves, we ultimately want them to feel good about themselves, in the hope of good results. This does not mean approval of inappropriate, unwise, and harmful actions. Empathy demands that we understand and be with the client, but we do not have to agree with or support actions that will harm the client or others.

Both Rogers and Maslow had immense faith in the ability of individuals to overcome challenges and take charge of their lives. You want to be there for the client in this process.

Counseling and psychotherapy sessions are indeed for the individual client, but let's not forget that the client exists in a multidimensional, multicultural, social context. Putting the client in context is all too often missing in counseling theory and practice.

REFLECTION *What Are the Goals of Counseling and Therapy?*

Self-actualization is a challenging concept. What does it mean to you? What experience and supports have led you to become more yourself, what you really are and want to be? How have you bounced back (resilience) from major challenges you have faced? What personal qualities or social supports helped you grow? What does this say to your own approach to counseling and psychotherapy?

Let us now turn to the skills and strategies that are aimed at developing intentionality, resilience, and self-actualization.

▶The Core Skills of the Helping Process: The Microskills Hierarchy

Counseling and psychotherapy require a relationship with the client; they seek to help clients work through issues by drawing out and listening to the client's story. This book presents key skills and strategies toward those ends.

Microskills are the foundation of intentional counseling and psychotherapy. They are the communication skill units of the counselor/client session that provide specific alternatives for you to use with many types of clients and all theories of counseling and therapy. You master these skills one by one and then learn to integrate them into a well-formed session.

When you are fully competent in the microskills, you are able to listen effectively and help clients change and grow. Effective use of microskills enables you to anticipate or predict how clients will respond to your interventions. And if clients do not do what you expect, you will be able to shift to skills and strategies that match their needs.

The **microskills hierarchy** (see Figure 1.2) summarizes the successive steps of intentional counseling and psychotherapy. The skills rest on a base of **ethics, multicultural competence**, and **wellness**. On this foundation rests the first microskill discussed in this text, **attending behavior**, followed by a set of basic listening skills. Attending and listening skills remain key to all the other skills and the five-stage structure of the counseling session. Through this book, you will have the opportunity to examine all the microskills in detail, read about further implications, and finally master each skill in practice and real sessions.

Once you have mastered attending behavior, you will move up the microskills pyramid to the empathic basic listening skills of **questioning**, **observation**, **encouraging**, **paraphrasing**, **summarizing**, and **reflecting feelings**. Higher is not necessarily better in this hierarchy. Unless you have developed skills of listening and respect, the upper reaches of the pyramid are meaningless. Develop your own style of being with clients, but always with respect for the importance of listening to client stories and issues. With a solid background in these central skills, you will be able to conduct a complete session using only listening skills.

The **five-stage** structure provides a framework for integrating the microskills into a complete counseling session. The **empathic relationship—story and strength—goals—restory—action** framework provides an overall system for you to use and serves as a checklist for all your meetings with clients.

Next you will encounter the influencing skills to help clients explore personal and interpersonal conflicts. **Focusing** will help you and the client to see cultural and contextual issues related to their concerns. **Empathic confrontation** is considered critical for client growth and change. **Interpretation/reframing**, **reflection of meaning**, **feedback**, and **self-disclosure** elaborate on possibilities for influencing clients more directly, particularly in the restorying phase of the conversation.

Concrete action strategies include an array of influencing skills in which you offer your own and tested ideas for change and development. The strategy of **logical consequences** is presented first, followed by specific examples of the best way to provide information and direction for the client. Emphasis will be given to **stress management**, **psychoeducation**, and **therapeutic lifestyle changes (TLCs)**, well-researched methods for building mental and physical health.

With a mastery of listening skills, the ability to conduct a session using only those skills, and a further command of the influencing skills and strategies, you are prepared to build competence in multiple theories of counseling and psychotherapy. You will find that microskills can be organized into different patterns utilized by different approaches. For example, if you have mastered the listening skills and the five-stage structure, you have a useful beginning to become fully competent in decisional, person-centered, and multicultural

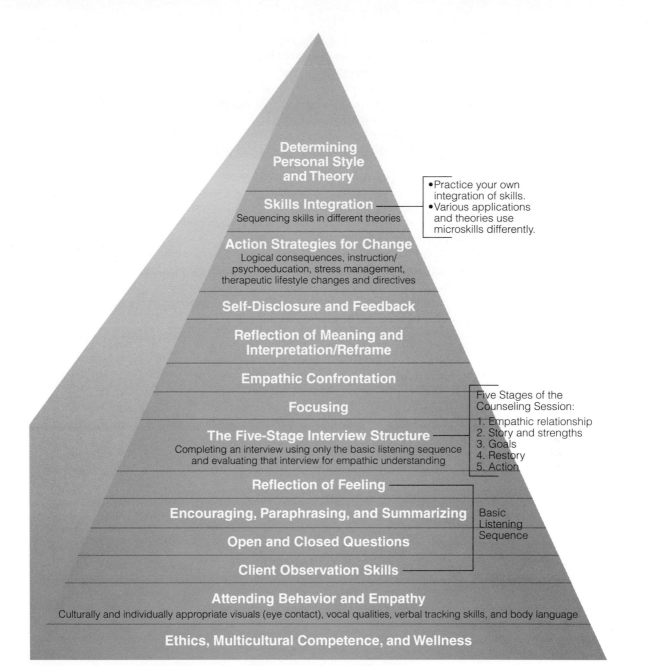

FIGURE 1.2 The microskills hierarchy: A pyramid for building cultural intentionality.

Copyright © 1982, 2003, 2012, 2014 Allen E. Ivey. Reprinted by permission.

counseling and therapy (MCT), as well as an introduction to logotherapy. In addition, crisis counseling and cognitive behavioral counseling are explored in Chapter 13.

At the apex of the microskills pyramid is integration of skills and determining your own personal style and theory or theories. Competence in skills, strategies, and the five stages are not sufficient; you will eventually have to determine your own approach to the practice of counseling and psychotherapy. Counselors and psychotherapists are an independent lot; the vast majority of helpers prefer to develop their own styles and, through eclecticism, move toward their own integrated blend of skills and theories.

As you gain a sense of your own expertise and power, you will learn that each client has a totally unique response to you and your natural style. While many clients may work well with you, others will require you to adapt to them and their style of being. Cultural intentionality will provide alternatives to help your varying clientele.

The model for learning microskills is practice oriented and follows a step-by-step progression, which will appear throughout the chapters of this text as a basic learning framework.

1. *Warm up.* Focus on a single skill or strategy and identify it as a vital part of the helping process.
2. *View.* See the skill in operation via a transcript and process analysis—or better yet, watch a live demonstration or view a video presentation on the Counseling CourseMate website.
3. *Read.* Read about the skill and/or hear a lecture on the main points of effective usage. Cognitive understanding is vital for skill development. However, understanding is not competence, nor does it show that you can actually do what you understand.
4. *Practice.* Ideally, use video or audio recording for skill practice; however, role-play practice with observers and feedback sheets is also effective.
5. *Generalize.* Complete a self-assessment. Integrate the skills and contract for action into the real world of counseling and therapy.

You can "go through" the skills quickly and understand them, but practicing them to full mastery is what makes for real expertise. We have seen many students "buzz" through the skills, but end with little in the way of actual competence. Teaching these skills to clients has also proven to be an effective counseling and therapeutic technique (Daniels, 2014).

The microskills of this book are key to developing an empathic relationship, drawing out the client's stories and issues, and ensuring that change and growth will be the result of your conversations with your clients. Box 1.1 summarizes more than 450 data-based studies on the microskills framework, now used in thousands of settings around the world.

BOX 1.1 Research and Related Evidence That You Can Use

Microskills' Evidence Base

More than 450 microskills research studies have been conducted (Daniels, 2014). The model has been tested nationally and internationally in more than 1,000 clinical and teaching programs. Microcounseling was the first systematic video-based counseling model to identify specific observable counseling skills. It was also the first skills training program that emphasized multicultural issues. Some of the most valuable research findings include the following:

▲ *You can expect results from microskills training.* Several critical reviews have found microtraining an effective framework for teaching skills to a wide variety of people, ranging from beginning interviewers and counselors to experienced professionals, who need to relate to clients more effectively. Teaching your clients many of the microskills will facilitate their personal growth and ability to communicate with their families or coworkers.

▲ *Practice is essential.* Practice the skills to mastery if the skills are to be maintained and used after training. *Use it or lose it!* Complete practice exercises and generalize what you learn to real life. Whenever possible, audio or video record your practice sessions.

▲ *Multicultural differences are real.* People from different cultural groups (e.g., ethnicity/race, gender) have different patterns of skill usage. Learn about people different from you, and use skills in a culturally appropriate manner.

▲ *Different counseling theories have varying patterns of skill usage.* Expect person-centered counselors to focus almost exclusively on listening skills whereas cognitive behaviorists use more influencing skills. Microskills expertise will help you define your own theory and integrate it with your natural style.

▲ *If you use a specific microskill, then you can expect a client to respond in anticipated ways.* You can predict how the client will respond to your use of each microskill, but each client is unique. Cultural intentionality prepares you for the unexpected and teaches you to flex with another way of responding or another microskill.

▲ *Neuroscience and brain research now support clinical and research experience with the microskills approach.* Throughout this book, we will provide data from neuroscience and brain research. This research explains and clarifies much of what counseling and psychotherapy have always done and, at the same time, increases the quality and precision of our practice.

You can view a comprehensive summary of microskills research on the Counseling CourseMate website at CengageBrain.com.

The importance of stories as a way to bring out information, thoughts, and feelings is discussed below. Microskills facilitate effective listening and bringing out the client's needs and wishes more clearly.

REFLECTION *What Does the Microskills Hierarchy Say to You?*

We are back to counseling as a science and an art. The hierarchy represents scientific observation and empirical research. That leaves you as the artist. How do you react to the metaphor of the microskills' serving as a palette of varying colors and hues? How would you, as counseling artist, choose from this palette? What makes most sense to you in terms of scientific and artistic balance?

▶Drawing Out Client Stories in a Well-Constructed Counseling and Psychotherapy Session

Counseling and psychotherapy are concerned with client stories. You will hear many different story lines—for example, stories of procrastination and the inability to take action, tales of depression and abuse, and most useful for client change, narratives of strength and courage. Your first task is to listen carefully to these stories and learn how clients come to think, feel, and act as they do. Sometimes, simply listening carefully with empathy and caring is enough to produce meaningful change.

You will also want to help clients think through new ways of approaching their stories. Through the conversation that is counseling and psychotherapy, you will help rewrite and rethink/restory old narratives into new, more positive and productive stories. The result can be deeper awareness of emotional experience, more useful ways of thinking, and new behavioral actions.

Development and growth are the aim of all that we do. Expect your clients to have enormous capacity for change. In the midst of negative and deeply troubling stories, one of your tasks is to search for strengths and resources that will empower the client. If you can develop exemplary models from the client's past and present, you are well on the way to combating even the most difficult client situation or story.

One brief example: Imagine that an 8-year-old child comes to you in tears, having been teased by friends. You listen and draw out the story. Through your warmth and caring interest you provide relationship, and the child calms down. The child has strengths, and you point out some of them. You talk about the wisdom of coming to talk about problems with a person who is there to help. You comment on a concrete example of the child's strengths, such as verbal or physical ability, a time you noticed the child helping someone else, or perhaps family members who support the child. You may read a short book or tell a story that metaphorically illustrates that problems can be overcome through internal strength. Next time, you notice that the child responds differently to friends' teasing.

How might your brief interaction with the child have effected this change? As a result of listening, developing positive assets and strengths, and gaining new perspectives through storytelling, you and the child rewrote the event and planned new narratives and actions for the future. The basic treatment structure used with the child can be expanded for counseling with adolescents, adults, and families: Listening to the story, finding positive strengths in that story or another life dimension, and rewriting a new narrative for action are what counseling and psychotherapy are about. In short: *empathic relationship—story and strengths—goals—restory—action.*

Of course, many times you will encounter complex issues. In such cases, as you listen, be careful not to minimize the story. Behind the tears of the child may be a history of abuse or other, more serious concerns. So, as you listen to stories, simultaneously search for more complex, unsaid stories that may lie behind the initial narrative.

Please take a moment to now review Box 1.2, which explores how traditional counseling too often focuses only on client problems. James Lanier suggests positive ways to draw out clients' stories and focus more on strengths.

BOX 1.2 National and International Perspectives on Counseling Skills

Problems, Concerns, Issues, and Challenges—How Shall We Talk About Client Stories?

James Lanier, University of Illinois, Springfield

There are different ways of listening to client stories. Counseling and therapy historically have tended to focus on client problems. The word *problem* implies difficulty and the necessity of eliminating or solving the problem. Problem may imply deficit. Traditional diagnosis such as that found in the *Diagnostic and Statistical Manual of Mental Disorders-IV-TR* (American Psychiatric Association, 2000) carries the idea of problem a bit further, using the word *disorder* with such terms as *panic disorder, conduct disorder, obsessive-compulsive disorder*, and many other highly specific disorders. The way we use these words often defines how clients see themselves. The forthcoming DSM-5 carries the idea of pathology even further with many new categories of "disorder" (www.dsm5.org/Pages/Default.aspx).

I'm not fond of problem-oriented language, particularly that word *disorder*. I often work with African American youth. If I asked them, "What's your problem?" they likely would reply, "I don't have a problem, but I do have a concern." The word concern suggests something we all have all the time. The word also suggests that we can deal with it—often from a more positive standpoint. Defining *concerns* as *problems* or *disorders* leads to placing the blame and responsibility for resolution almost solely on the individual.

Recently, there has been increasing and particular concern about that word *disorder*. More and more, professionals are realizing that the way people respond to their experiences is very often a *logical response to extremely challenging situations*. Thus, the concept of posttraumatic stress *disorder* (PTSD) is now often referred to as a *stress reaction*. Posttraumatic stress reaction (PTSR) has become an alternative name, thus normalizing the client's response. Still others prefer to *avoid naming* at all and seek to work with the thoughts, emotions, and behaviors of the stressed clients.

Finding a more positive way to discuss client concerns and stories is relevant to all your clients, regardless of their background. *Issue* is another term that can be used instead of *problem*. This further removes the pathology from the person and tends to put the person in a situational context. It may be a more empowering word for some clients. Carrying this idea further, *challenge* may be defined as a call to our strengths. All of these terms represent *an opportunity for change*.

As you work with clients, please consider that change, restorying, and action are more possible if we help clients maintain awareness of already existing personal strengths and external resources. Supporting positive stories helps clients realize the positive assets they already have, thus enabling them to resolve their issues more smoothly and effectively, and with more pride—specifically, they become more actualized. Then you can help them restory with a *can do* resilient self-image. Out of this will come action and generalization of new ideas and new behaviors to the real world.

▶Empathic Relationship—Story and Strengths—Goals—Restory—Action

Narrative theory is a relatively new model for understanding counseling and psychotherapy sessions (Holland, Neimeyer, & Currier, 2007; Monk, Winslade, Crocket, & Epston, 1997; White & Epston, 1990; Whiting, 2007). Narrative theory emphasizes storytelling and the generation of new meanings. The concepts of narration, storytelling, and conversation are useful frameworks as we examine skills, strategy, and theory in counseling and psychotherapy.

This book seeks to extend the narrative tradition of storytelling through the five stages of empathic relationship—story and strengths—goals—restory—action. These stages provide a framework within which you can view all counseling and therapy sessions. Theories of counseling as varied as person-centered, cognitive behavioral, brief counseling, and psychoanalytic/interpersonal approaches can be considered narratives or stories of the helping process. Each of these theoretical stories can be helpful to your clients at various times.

Drawing on the concepts of neuroscience (to be discussed later), restorying can also be described as "rewiring the brain." As the client writes new ways of thinking, new neural networks are being built and placed in long-term memory, changing or replacing old dysfunctional ways of thinking.

The narrative and microskills tradition will allow you to develop expertise in multiple theories and strategies. This approach will enable you to understand and become more competent in the multiple theories and strategies that you will encounter.

Empathic Relationship

No one wants to tell a story to someone who is not interested or who is not warm and welcoming. Unless you can develop rapport and trust with your client, expect little to happen. The relationship in every session will be different and will test your social skills and understanding. Basic to this relationship is being your own natural self and being open to others and to differences of all types. Your attending and empathic listening skills are key to understanding and will play a part throughout all sessions.

Another term for relationship is *working alliance*, which in turn is based on what is now called the *common factors approach*. Though not always cited, Carl Rogers and his person-centered theory have set the pace for our awareness of relationship and the working alliance. It has been estimated that 30% of successful counseling and therapy outcomes are due to relationship or common factors, consisting of caring, empathy, acceptance, affirmation, and encouragement (Imel & Wampold, 2008). Your ability to listen and be with the client is the starting point for the session.

Story and Strengths

The listening skills described in Section I are basic to learning how clients make sense of their world—the stories clients tell us about their lives, problems, challenges, and issues. We need to help them tell their stories in their own way. Attending and observation skills are critical, while encouraging, paraphrasing, summarizing, and reflection of feeling help fill out the story. Regardless of the theory used, these listening skills are central, although different counseling systems and theories may draw out different aspects of stories that lead in varying directions.

The listening and observational skills of Chapter 3 through integrative Chapter 8 will be key in drawing out clients' difficulties, concerns, and issues along with their strengths to solve these problems. At times, counseling and therapy can spiral down into a depressing repetition of negative stories—and even whining and complaining. Seek out and listen for times when clients have succeeded in overcoming obstacles. Listen for and be "curious about their competencies—the heroic stories that reflect their part in surmounting obstacles, initiating action, and maintaining positive change" (Duncan, Miller, & Sparks, 2004, p. 53).

Goals

If you don't know where you are going, you may end up somewhere else. Too many sessions wander and never have a focus. Once you have heard the story, and you and the client see the need for a new and more effective story, how would you and the client like the story to develop? What is an appropriate ending? If the client does not have a goal in mind, the new story may be irrelevant. This area is considered central in brief counseling, and counselors often start the session right here: "What do you want to happen today as a result of our conversation?"

Restory

If you understand client stories, strengths, and goals, you are prepared to help clients restory—generate new ways to talk about themselves. One valuable strategy for restorying is provided in Chapter 8, where you will demonstrate your ability to conduct a full session using only listening skills. Many times, effective listening is sufficient to provide clients with the strength and power to develop their own new narratives.

However, the restorying process is particularly well exemplified in the use of influencing skills such as confrontation, feedback, and logical consequences (Chapters 9–13). The five-stage framework with its many adaptations can be used to enable clients to find new ways of making meaning. The stages and skills are central in multiple theories of helping, but each counseling theory provides us with alternative ways to think and talk about client stories.

For example, you will find more emphasis on self-discovery, emotion, and meaning in person-centered counseling, whereas cognitive behavioral methods will actively seek to change clients' ways of thinking and behaving. The new story from a psychoanalytic perspective will differ greatly from the narrative of brief counseling. In this process of restorying, each theory will have a different language system. Yet all systems gain clarity when you view them as stories that offer clients new ways of being.

Awareness of theoretical diversity is central. All counseling theories seek to help clients find new ways to think about their concerns and problems, but each has its own story about the meaning and purpose of counseling. As a result, each theory's story may lead to different actions, different beliefs, and even different emotional experiences. As you define your own natural style, remain open to the multitude of possibilities offered by the professional helping field.

Action

Pay special attention to this final stage. All of the previous efforts will be of little use if the client does not go home with new actions. Contract with the client to *act and think* in new ways tomorrow and during the coming week. Homework has now become an essential part of the change process. You will know that you have made a difference if the client takes new thoughts, emotions, and behaviors into the "real world."

Develop your own theoretical/practical story of counseling and psychotherapy. If your work with this book is successful, in the end you will have developed a solid understanding of foundation skills and strategies, you will be able to conduct sessions from several perspectives, and—perhaps most essential—you will have begun the process of writing your own narrative, your own personal story/theory of counseling and psychotherapy. We hope that in your personal construction of theory and practice you will remain open to constant challenge and growth from your clients and from your professional colleagues.

REFLECTION *What Is Your Story of Counseling and Psychotherapy at This Point?*

What has been your past experience with those who may have offered you an empathic and caring relationship? Where might personal relationships have been damaging? How have positive relationships made a difference in your life?

▶Our Multicultural World

All interviewing and counseling is multicultural.

—Paul Pedersen

Psychology as a research practice and psychiatry . . . must start to take into account context, context, context. And that introduces a lot of different complications in terms of language, culture, socioeconomic status, etc.

—Jerome Kagan

Multiculturalism, also discussed as diversity or cross-cultural issues, is now defined quite broadly. Once it referred only to the major racial groups, but now the definition has expanded in multiple ways. The story is that we are all multicultural. If you are White, male, heterosexual, from Alabama, an Episcopalian, and able-bodied, you have a distinct cultural background. Just change Alabama to Connecticut or California, and you are very different culturally. Similarly, change the color, gender, sexual orientation, religion, or physical ability, and your cultural background changes significantly. Multiculturalism means just that— many cultures.

Whether we know it or not, we are all multicultural beings. Culture is like air: We breathe it without thinking about it, but it is essential for our being. Culture is not "out there"; rather, it is found inside everyone, markedly affecting our view of the world. Learn about and be ready to work with cultural difference, and seek to maintain that critical empathic relationship for more than one session unless that clearly is sufficient for the client.

Multicultural competence is imperative in counseling and psychotherapy. We live in a multicultural world, where every client you encounter will be different from the last and different from you. Without a basic understanding of and sensitivity to a client's uniqueness, the counselor will fail to establish a relationship and true grasp of a client's issues. Throughout, this book will examine the multicultural issues and opportunities we all experience.

▶RESPECTFUL Counseling and Psychotherapy

The RESPECTFUL model (D'Andrea & Daniels, 2001; Ivey, D'Andrea, & Ivey, 2012) makes it clear that multiculturalism refers to far more than race and ethnicity. As you review the list below, first identify your own multicultural background. Then examine your beliefs and attitudes toward those who are culturally similar to and different from you on each issue. How prepared are you to work with clients who are culturally different from you? And *all* clients are culturally different from you in some way.

R   Religion/spirituality. What is your religious and spiritual orientation? How does this affect your thoughts, feelings, and behaviors as a counselor?

E   Economic/class background. How will you work with those whose financial and social background differs from yours?

S   Sexual orientation and/or gender identity. How effective will you be with those whose gender and/or sexual orientation differs from yours?

P   Personal style and education. How will your personal style and educational level affect your practice?

E   Ethnic/racial identity. The color of a person's skin is one of the first things we notice. What is your reaction to different races and ethnicities?

C   Chronological/lifespan challenges. Children, adolescents, young adults, mature adults, and older persons all face different issues and challenges. Where are you in the developmental lifespan?

T   Trauma. It is estimated that 90% or more of all people experience serious trauma in their lives. Trauma underlies the issues faced by many of your clients. War, flood, rape, and assault are powerful examples, but a serious accident, divorce, loss of a parent, or being raised in an alcoholic family are more common sources of trauma. The constant repetition of racist, sexist, or heterosexist actions and comments can also be traumatic. What is your experience with life trauma? We now recognize it as a "normal" part of being alive for most of us.

F   Family background. We learn culture in our families. The old model of two parents with two children is challenged by the reality of single parents, gay families, and varying family structures. How has your life experience been influenced by your family history (both your immediate family and your intergenerational history)?

U   Unique physical characteristics. Become aware of disabilities, special challenges, and false cultural standards of beauty. In addition, physical health is a key part of mental health. Exercise, nutrition, yoga, and meditation are not traditionally included as part of formal helping theory and practice, but research and clinical experience have now changed our view and these are an essential part of modern counseling. How well do you understand the importance of the body in counseling and psychotherapy, and how will you work with others different from you in their physical characteristics and interests?

L   Location of residence and language differences. Whether in the United States, Great Britain, Turkey, Korea, or Australia, there are marked differences between south and north, east and west, urban and rural. Moreover, many of you reading this book, and certainly many of your clients, will come from a wide variety of nations. The small town of Amherst, Massachusetts, has 23 different languages represented in their schools. Remember that a person who is bilingual is advantaged and more skilled, not disadvantaged. What languages do you know, and what is your attitude toward those who use a different language from you?

Intersections among several multicultural factors are also critical. For example, consider the biracial family (e.g., Chinese and European, or African and Latino descent). Both children and parents are deeply affected, and categorizing an individual into just one multicultural category is inappropriate. Or think of the Catholic lesbian woman who may be economically advantaged (or disadvantaged). Or the South Asian male with a PhD who is gay. For many clients, sorting out the impact of their multiculturality may be a major issue in counseling.

REFLECTION *Culture Counts: Diversity, Multiculturalism, and Self-Assessment*

Consider the dimensions of the RESPECTFUL model and the intersecting multiple cultural identities we all have as part of our being:

R   Religion/spirituality

E   Economic/class background

S   Sexual orientation and/or gender identity

P   Personal style and education

E   Ethnic/racial identity

C   Chronological/lifespan challenges

T   Trauma

F   Family background

U   Unique physical characteristics

L   Location of residence and language differences

Using this list (and other dimensions that you might add), identify yourself as a multicultural being. Then examine yourself for personal preferences and biases. How much experience do you have with people who are different from you? How able are you to work with those who may be different from you? For example, if you are heterosexual, how able are you to work with clients from the gay, lesbian, bisexual, or questioning cultures? If any of these are part of your identity, how able are you to work with clients who are heterosexual or asexual? Now apply these questions to other dimensions of the RESPECTFUL model. What developmental steps do you need to take to increase your awareness and skills?

Box 1.3 reviews the paradigm shifts that are occurring in mental health counseling. Then we take a closer look at neuroscience and brain research, which will lead to major changes in the ways we think about counseling and psychotherapy in the next 10 years.

The Coming Paradigm Shift in Mental Health Counseling

Mental health services will no longer be routinely provided by solo-practice mental health practitioners. Instead, mental health care will be routinely provided as part of larger inter-professional group practices and institutional settings. Further, the mental health expert on the team will need a flexible armamentarium of interventions, and cannot rely solely on the 50-minute psychotherapy session. In addition, the mental health expert must be able to address a host of other behavioral issues, important to health and well being—medical regimen compliance, pain management, coping with disability or a life-threatening diagnosis, lifestyle behavioral change.

—Barbara Bennett Johnson, PhD, ABPP
President of American Psychological Association
(Johnson, 2012, p. 72)

The counseling and psychotherapy field is rapidly changing to a broadly based and integrated health-oriented profession. Individual counseling skills, strategies, and theories will remain important as a base for practice. Health care reform opens new opportunities for you, the beginning counselor or therapist—and our profession as a whole. Prevention of both mental and physical illness will become an ever-increasing part of our practice. The paradigm shift represented by Johnson's comment will be changing your view and our practice during the coming decades.

Stress management is becoming a central strategy for prevention of both mental and physical illness, regardless of our theoretical orientation to counseling and therapy. Therapeutic lifestyle changes (TLC) are a critical wellness part of managing both physical and mental health issues (Chapter 13 provides detailed discussions of both

stress management and TLCs). All these fall within the positive psychology and wellness framework outlined in Chapter 2, as foundational qualities of this book.

Evidence is clear that the body and mind are injured by the multiple stressors we and our clients face daily. Traumatic experiences, ranging from war and rape to death of a loved one, job loss, a complex divorce, bullying, and many others, cause unhealthy release of damaging hormones to the brain (and thus onward to the body). Abnormal stress can result in damage to structures in the brain, the heart, and other organs. Significant short- or long-term stressors often produce permanent injury.

At another level, you will find stress involved in virtually all of the issues clients face. Admission to college, career choice, financial issues, and coping with racial or gender harassment are what some would term "normal" concerns. But, these do not feel normal to our stressed and worried clients.

Neuroscience's research on the brain has become a practical influence on our understanding of physical and mental health. Neuroscientists and even genetics researchers emphasize the importance of prevention and the key role of stress management. In this new paradigm, counseling skills and strategies remain first-line necessities. We need to listen empathically to clients' stories, join them in their view of the world, and work with them on an egalitarian basis toward growth, development, and action. But we now can do this with a solid scientific base, as neuroscience findings reveal that many of the traditional approaches are effective and appropriate.

A holistic approach is replacing the traditional dualism of physical and mental health, as we move to a new and integrated paradigm.

▶Neuroscience: Implications of This Cutting-Edge Science for the Future of Counseling and Psychotherapy

Our interaction with clients changes their brain (and ours). In a not too distant future, counseling will be regarded as ideal for nurturing nature.

—Óscar Gonçalves

No longer can we separate the body from the mind or the individual from his or her environment and culture. Counseling and psychotherapy are moving closer to medicine, neurology, and brain science. Counselors once argued against the "medical model." Influenced by preventive medicine, accountability, and neuroscience research, however, physicians are increasingly aware that what happens in the body is deeply influenced by the mind. And we have helped lead this change in consciousness through effective helping skills and strategies.

The Brain and Stress

Some 80% of medical issues involve the brain and stress (Ratey, 2008a). You will find that, in one way or another, the vast majority of your work in counseling and psychotherapy includes stress as an underlying issue. Resolving stressors is critical in many styles of treatment. The evidence is clear that stress management and therapeutic lifestyle changes (see Chapter 13) are effective routes toward both mental and physical health and are necessary regardless of your counseling style or chosen theoretical approach.

Stress and stressful events leave a marked imprint on the brain. We need some stress for learning and for physical growth. Some people have compared the brain to a muscle: If it doesn't get exercise, it atrophies. But, like a muscle, it can be overstressed, which can result in damage and loss of neurons. Figure 1.3 shows the brain under severe stress (aversive condition) compared with the relative absence of stress (neutral condition).

The majority of clients who come to us have some form of stress underlying their issues and concerns. We can think of counseling and therapy as stress management. Listening to client stories is our first avenue to establishing an empathic relationship and understanding the client's world. Focusing on wellness and strengths enables clients to deal with stress better and to resolve their issues. The influencing skills of the second half of this book provide an array of additional skills to alleviate stress and enable clients to experience emotion and feelings more appropriately, think more constructively, and take behavioral change into action.

Neuroscience, the Brain, and Counseling

Whether in interviewing, counseling, or psychotherapy, the conversation changes the brain through the development of new neural networks. This is an example of *brain plasticity*.

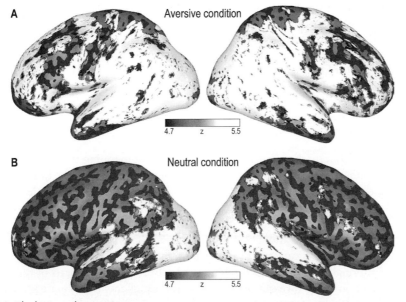

FIGURE 1.3 The brain under aversive stress.

Source: Hermans, E., van Marle, H., Ossewaarde, L., Henckens, A., Qin, S., Kesteren, M., Schoots, V., Cousijn, H., Rijpkema, M., Oostenveld, R., & Fernández, G. (2012). Stress-related noradrenergic activity prompts large-scale neural network configuration. *Science*, 334, 1151–1153. Reprinted with permission from AAAS.

Throughout our lives, we are adding and losing many millions of neurons, synapses, and neural connections. Effective counseling and therapy develop useful new neurons and neural connections in the brain. Both your own and your client's brain functioning can be measured through a variety of brain-imaging techniques, most notably functional magnetic resonance imaging (fMRI) (Hölzel et al., 2011; Logothetis, 2008).

"Neuroplasticity can result in the wholesale remodeling of neural networks . . . a brain can rewire itself" (Schwartz & Begley, 2003, p. 16). If we are indeed affecting the brain in all our sessions, then perhaps neuroscience can help us understand a bit more of what is happening between counselors and clients.

The idea of the brain being central in counseling practice is relatively new. In fact, the first convention presentation on the topic was when Allen spoke to the American Counseling Association in Hawai`i in 2006, and this counseling text is the first to give special attention to neuroscience. Thus, the idea of incorporating neuroscience knowledge into our practice is controversial and not accepted by all.

There are a number of other reasons why neuroscience and study of the brain will become more critical in future the years. Here are just a few:

1. The National Institute of Mental Health plans to institute a brain-based approach to counseling and therapy within the next 10 to 20 years, well within the time you will be practicing. This will replace the Diagnostic and Statistical Manual of Mental Disorders (DSM) with a totally new approach, one that is more amenable to the goals of the counseling process. It will be helpful if you take a few moments to visit www.nimh .nih.gov/research-funding/rdoc/nimh-research-domain-criteria-rdoc.shtml, or google "NIMH research domain criteria."

2. This new approach integrates neuroscience with medicine, counseling and therapy, developmental psychology, multicultural issues, and multiple sciences. Old clinical categories will be reevaluated and possibly eliminated. Diagnosis will be multidimensional and will lead to integrated treatment recommendations involving physicians, counselors and therapists, and the human services professions.

3. The media are full of new developments in neuroscience. Many of your clients will know this field through general reading and expect you to be knowledgeable as well.

4. You will find, like us, that a knowledge of neuroscience and the brain enables you to be a more skilled practitioner. Currently, all three authors of this book constantly think of how the client's brain (and our own) is influenced by the interaction. We find that, as a result, our work is more precise and successful. Neuroscience will improve your practice.

6. Neuroscience is stimulating and interesting, with constant new developments. Once one learns some basic vocabulary, it becomes fun and enjoyable.

Damasio (2003) remarked that the brain is constantly mapping the body while simultaneously being stimulated by body reactions. We know now that the heart's activity is engaged with the central nervous system, thus constituting a marker of high-level cognitive processes. As reported in a recent study by Oliveira-Silva and Gonçalves (2011), cardiac response is one of the best physiological markers of empathic response. Brain and heart . . . empathy and relationship . . . thought, feeling, behavior!

Neuroscience concepts can help us understand more precisely what is occurring in the client. Certain thoughts and feelings can be associated with different parts of the brain and body. With that knowledge, we can better work with our clients to facilitate growth. We can, and perhaps should, intentionally select culturally appropriate therapeutic strategies that meet the individual client's immediate and long-term needs, rather than relying on single theories or practices.

Some of you have taken biology or psychology courses (particularly social psychology) that speak with sophistication about the brain and its structures. For others, this will be a new topic with new vocabulary. Here are a few questions to consider: How do you react to the brief summary presented above? Do you agree that neuroscience and the brain are relevant to your work as a counselor or psychotherapist? What are your reasons? Given your position, how might what we share about neuroscience be relevant to you?

▶Your Natural Helping Style: An Important Audio or Video Exercise

At the beginning of this chapter, you were asked to give your own response to Sienna's multiple issues. Your response reflects you and your worldview. Your use of microskills and the five-stage structure must feel authentic. If you adopt a response simply because of recommendations, likely it will be ineffective for both you and your client. Not all parts of the microtraining framework are appropriate for everyone. You have a natural style of communicating, and these concepts should supplement your style and who you are. Learn these new skills, strategies, and concepts, but be yourself and make your own authentic decisions for practice.

Also, develop awareness of the natural style of the clients with whom you work, particularly if they are culturally different from you. What is this experience like for you? What do you notice about your client's style of communication? Use your observations to expand your competence and later add new methods and information to your natural style.

We believe the following is one of the most central exercises in the book:

Find someone who is willing to role-play a client with a decision that needs to be made, a concern, an issue, or an opportunity. Interview or practice counseling that "client" for at least 15 minutes, using your own natural style.

Read pages 31–37 of Chapter 2 and follow the ethical guidelines as you work with a volunteer client. Ask the client, "May I record this session?" Also inform the client that the video or audio recorder may be turned off at any time. Common sense demands ethical practice and respect for the client.

Video, with feedback from colleagues and/or clients, is the preferred way to examine your counseling style. Many of you will have either a small video recorder or a video-capable camera or phone. Video makes it possible for us to discover what we are really doing, not just what we think we are doing. Feedback from others helps us evaluate both our strengths and the areas where we might benefit from further development and growth.

The volunteer client can select almost any topic for the session. A friend or classmate discussing a school or job problem may be appropriate. A useful topic might be some type of interpersonal conflict, such as concern over family tensions, or a decision about a new job opportunity.

When you have finished, ask your client to fill out the Client Feedback Form (Box 1.4). In practice sessions, always seek immediate feedback from clients, classmates, and colleagues. We suggest that you use the Client Feedback Form for this purpose, with your own adaptations and changes, throughout your practice sessions.

BOX 1.4 Client Feedback Form (To Be Completed by the Volunteer Client)

You can download this form from the Counseling CourseMate website at CengageBrain.com.

_____ (Date)

_____
(Name of Counselor)

Instructions: Rate each statement on a 7-point scale, with 1 representing "strongly agree (SA)," 7 representing "strongly disagree (SD)," and the midpoint as "neutral (N)." You and your instructor may wish to change and adapt this form to meet the needs of varying clients, agencies, and situations.

	Strongly Agree SA			Neutral N		Strongly Disagree SD	
1. (Session beginning and introduction) The counselor explained to me the purpose of the session, what to expect, and discussed confidentiality and related issues.	1	2	3	4	5	6	7
2. (Awareness) The session helped me understand my decision, issue, concern, or opportunity for change more fully.	1	2	3	4	5	6	7
3. (Awareness) The counselor listened to me. I felt heard.	1	2	3	4	5	6	7
4. (Knowledge) I gained a better understanding of myself today.	1	2	3	4	5	6	7
5. (Knowledge) I learned about different ways to address my decision, issue, concern, or opportunity for change.	1	2	3	4	5	6	7
6. (Skills) This session helped me identify specific strengths and resources to help me work through concerns and issues.	1	2	3	4	5	6	7
7. (Action) I will take action and do something specific in terms of changing thinking, feeling, or behavior after this session.	1	2	3	4	5	6	7

What did you find helpful? What did the counselor do that was right? Think about what might have been said, how well he/she listened, and perhaps even his/her body language. Be specific; for example, not "You did great," but rather "You listened to me carefully when I talked about _____."

What, if anything, did the counselor miss that you would have liked to explore today or in another session? What might you have liked to happen that didn't?

Use this space or the other side for additional comments or suggestions.

You may also find it helpful to continue using this form, or some adaptation of it, in your work in the helping profession. Professional counselors and therapists seldom offer their clients an opportunity to provide them with feedback. In the interest of a more egalitarian session, consider this type of feedback from your clients as a regular part of your practice. We ourselves have learned valuable and surprising things through feedback, particularly when we may have missed something.

Please transcribe the audio or video for later study and analysis. You'll want to compare your first performance with practice sessions and, ideally, with another, more detailed analytic transcript and self-evaluation at the end of this course of study.

You can photocopy the Client Feedback Form here or download a copy from the CourseMate website. Downloading will enable you to add or change items if you so choose. Occasionally, adding specific items for individual clients may enable them to write things that they find difficult to put into words.

Self-Assessment

Review your audio or video recording and ask yourself and the volunteer client the following questions:

1. We build on strengths. What did you do right in this session? What did the client notice as helpful?
2. What stands out for you from the client feedback form?
3. What was the essence of the client's story? How did you help the client bring out his or her narrative/issues/concerns?
4. How did you demonstrate intentionality? When something you said did not go as anticipated, what did you do next?
5. Name just one thing on which you would like to improve in the next session you have. What actions will you take?

▶Summary: Mastering the Skills and Strategies of Successful Relationships with Clients

Welcome to the fascinating field of counseling and psychotherapy! You are being introduced to the basics of the individual counseling session, but the same skills are essential in group and family work. We now know that skills used in counseling and therapy affect brain development, thus resulting in longer-term change in our clients.

Physicians and nurses, managers in business settings, peer counselors, and many others have adopted the microskills training format as part of their profession and/or training. The original microskills format presented here has been translated into more than 22 languages and used in many varied settings, such as by AIDS and refugee counselors in Africa and Sri Lanka; top-line and middle managers in Sweden, Germany, and Japan; helpers working with trauma survivors from floods and hurricanes in Ceylon and Indonesia; Aboriginal social workers in Australia and Inuit community workers in the Canadian Arctic. The system works and is constantly changing and growing.

This first chapter frames the entire book. The following key points are the things we particularly want you to remember. The first competency practice exercise in this chapter asked you to examine yourself and identify your strengths as a helper. In the end, *you* are the person who counts, and we hope that you will develop your counseling skills based on your natural expertise and social skills. Good luck! We hope that you enjoy the journey.

▼▼▼▼	**Key Points**
Interviewing, counseling, and psychotherapy	These are interrelated processes that sometimes overlap. Interviewing may be considered the most basic; it is often associated with information gathering and providing necessary data to help clients resolve issues. Counseling focuses on normal developmental concerns, whereas psychotherapy emphasizes treatment of more deep-seated issues.
Counseling as both science and art	The field of counseling and therapy is now well supported by empirical research. More recently, neuroscience findings have added to our understanding, solidifying counseling as a scientific undertaking. Nonetheless, it is you, the counselor or therapist, who effectively integrates the many aspects of research and theory and creatively applies these findings to the client.
Intentionality	Achieving intentionality is the major goal of this book. Intentionality is acting with a sense of capability and deciding from among a range of alternative actions. The intentional individual has more than one action, thought, or behavior to choose from in responding to life situations.
Cultural intentionality	The culturally intentional individual can generate alternatives from different vantage points, using a variety of skills and personal qualities within a culturally appropriate framework.
Self-actualization and resilience	A major objective of counseling and psychotherapy is enabling clients to find their own direction and enhance their potential. Self-actualization requires resilience and the ability to rebound from the eventual stresses and challenges we all face.
Microskills	Microskills are single communication skill units (for example, questioning or reflection of feelings). They are taught one at a time to ensure mastery of basic counseling and therapy competencies.
Microskills hierarchy	The hierarchy organizes microskills into a systematic framework for the eventual integration of skills in a natural fashion. The microskills rest on a foundation of ethics, multicultural competence, and wellness. The attending and listening skills are followed by focusing, confrontation, influencing skills, and eventual skill integration.
Microskills learning model	Five steps are used to teach the single skills of counseling and psychotherapy: (1) warm up to the skill; (2) view the skill in action; (3) read and learn about broader uses of the skill; (4) practice; and (5) generalize learning from the session to daily life. The model is useful for teaching critical listening skills to clients.
Empathic relationship— story and strengths—goals —restory—action	Our first task is to help clients tell their stories. To facilitate development, we need to draw out narratives of their personal assets. With a positive foundation, clients may learn to write new stories with the possibility of new actions. James Lanier reminds us that language stressing a problem or disorder may get in the way of effective counseling and therapy.
RESPECTFUL counseling and psychotherapy	We need to be aware of social contextual issues that affect clients and include these dimensions in all our sessions: **R**eligion and spirituality **E**conomic/class background **S**exual orientation and/or gender identity **P**ersonal style and education **E**thnic/racial identity **C**hronological/lifespan challenges **T**rauma **F**amily background **U**nique physical characteristics **L**ocation of residence and language differences

Theory and microskills	All counseling theories use the microskills, but in varying patterns with differing goals. Mastery of the skills will facilitate your becoming able to work with many theoretical alternatives. The microskills framework can also be considered a theory in itself, in which counselor and client work together to enable the construction of new stories, accompanied by changes in thought and action.
Neuroplasticity	Counseling and psychotherapy will be increasingly informed by research in neuroscience, and you will want to keep abreast of new developments. Relevant studies and implications for practice will be presented throughout this book. Of particular importance is neuroplasticity, or "rewiring" of the brain. Successful therapy may be expected to help clients develop new neural connections.
You, microskills, and the helping process	Microskills are useful only if they harmonize with your own natural style. Earlier in this chapter, you were asked to audio or video record a session with a friend or classmate, and make a transcript. Later, as you learn more about session analysis in your counseling practice, continually examine and study your behavior. You'll want to compare it with your performance in another recording some months from now.

▶Competency Practice Exercises and Portfolio of Competence

Go to CengageBrain.com to access Counseling CourseMate, where you will find an interactive ebook, quizzes, videos, interactive counseling and psychotherapy exercises, the Portfolio of Competence, case studies, a practice test, and more.

Most of the chapters of this book end with both competency practice exercises and a Portfolio of Competence. We suggest that you stop after each chapter, do the practice exercises, and identify what you have learned from the chapter in terms of competencies. The Portfolio of Competence is structured in four levels of competence: (1) identification and classification, (2) basic competence, (3) intentional competence, and (4) psychoeducational teaching competence. *Identification and classification*, or *knowledge*, refers to your understanding of counseling concepts. *Basic competence* is your practice of the skills or strategies, showing that you know what they are and how to use them in a session. *Intentional* competence occurs when you can not only use a skill with predictable results, but also demonstrate the flexibility to change the skill or style in accordance with the client's immediate and long-term needs. *Psychoeducational teaching competence* is your ability to teach the skills to others. Many of the skills and strategies of this book can be taught to clients as part of counseling and therapy. In addition, you may be asked to conduct more formal presentations in which you teach other counselors, volunteer peer counselors, or others who may benefit from listening skills training, such as businesspeople, clergy, or community service workers.

Please take a moment now to start the process of competency assessment for this chapter using the following checklist.

Level 1: Identification and classification. Can you define and discuss the following concepts?

❑ Distinctions and similarities among interviewing, counseling, and psychotherapy
❑ Balance of science and art that makes sense to you as appropriate for counseling
❑ Meaning and importance of intentionality and cultural intentionality in counseling practice

- Definitions of resilience and self-actualization as potential goals for our clients
- Definition of the microskills hierarchy and its relevance to practice
- The potential value of neuroscience for the practice of counseling and psychotherapy

Level 2: Basic competence. We have asked you to take ideas from the chapter and actually try them out in your own life and/or the real world.

- Examining your own life experience as a multicultural being through the RESPECTFUL model
- Finding a volunteer client, conducting a session, obtaining client feedback, and evaluating your own natural style of helping

Intentional competence and *psychoeducational teaching competence* will be reviewed in later chapters.

We recommend that you keep a journal of your path through this course and your reflections on its meaning to you. Your first session will provide a critical foundation on which to build.

Ethics, Multicultural Competence, and the Positive Psychology and Wellness Approach

Ethics, Multicultural Competence, and Wellness

I am (and you also)

Derived from family

Embedded in a community

Not isolated from prevailing values

Though having unique experiences

In certain roles and statuses

Taught, socialized, gendered, and sanctioned

Yet with freedom to change myself and society.

—Ruth Jacobs[*]

Mission of "Ethics, Multicultural Competence, and the Positive Psychology and Wellness Approach"

To present three foundational dimensions, each different but related to competent practice. This chapter asks you to build your counseling sessions on professional ethics, multicultural sensitivity, and a strength-based wellness approach.

---

[*]R. Jacobs, *Be an Outrageous Older Woman*, 1991, p. 37. Reprinted by permission of Knowledge, Trends, and Ideas, Manchester, CT.

Chapter Goals and Competency Objectives Awareness, knowledge, skills, and actions in ethics, multicultural competence, and the strength and wellness approach will enable you to

▲ Apply key ethical principles in counseling and psychotherapy.

▲ Develop your own informed consent form.

▲ Define multicultural competence, including key aspects of awareness, knowledge, and skills.

▲ Apply wellness and positive psychology in your counseling and psychotherapy practice.

▲ Conduct a wellness assessment and assist the client to develop a wellness plan.

 Go to CengageBrain.com to access Counseling CourseMate, where you will find an interactive ebook, quizzes, videos, interactive counseling and psychotherapy exercises, the Portfolio of Competence, case studies, a practice test, and more.

▶ The Ethical Foundations of Counseling and Psychotherapy

If you behave ethically and with intentionality, you can anticipate that the relationship will proceed more smoothly and your client will be protected. A sense of ethics is part of trust building. Note the following brief description of ethical behavior and its implications for the session.

ETHICS	ANTICIPATED RESULT
Observe and follow professional standards, and practice ethically. Particularly important issues for beginning counselors are *competence, informed consent, confidentiality, power,* and *social justice.*	Client trust and understanding of the counseling process will increase. Clients will feel empowered in a more egalitarian session. When you work toward social justice, you contribute to problem prevention in addition to session healing work.

Kendra, age 25, enters your office and after some preliminaries in which you establish rapport, she expresses her concerns:

> I'm really upset. I've got a child at home with my mother, and I'm trying to work my way through community college, and my boss at the nursing home has been hitting on me. I want to leave, but I can't afford to stay in school without this job.

In working with all our clients, we need a sense of ethical practice, an awareness of their multicultural backgrounds, and an emphasis on their positive strengths. Each client we encounter is one of a kind. Kendra's uniqueness stems from her biological background and the way she has lived her life in connection to others. Family, community, and culture have deeply influenced Kendra's values and socialization. Our task, as counselors and therapists, is to facilitate her growth within the broad context in which she lives. Part of effective counseling is helping clients see strengths and resources around them. For example, we can help her recognize and appreciate strengths she has gained from her family and the community. Likely church and friends have been and perhaps still are important resources.

Even from her brief personal statement, you already have some ideas about Kendra. In all client sessions, ethical behavior needs to take center stage. Confidentiality and your own competence come to mind immediately.

Diversity is present in all sessions. First assume that Kendra is Irish American; then assume that she is African American. What if she is straight or lesbian? Considering the RESPECTFUL model will bring forth other issues.

In this chapter, we are also concerned with wellness strengths and resources. What do you already see in Kendra that indicates coping ability and potential positive assets for her growth? What are some family and community resources that Kendra might draw on for help?

You may wish to compare your thoughts with ours (see pages 57–58).

▶Ethics in the Helping Process

All major helping professions throughout the world have codes for ethical practice. It is essential that you read and thoroughly understand the ethical code of your profession.

Ethical codes promote empowerment for both counselors and their clients. They aid the helping process by (a) teaching and promoting the basics of ethical and appropriate practice, (b) protecting clients by providing accountability, and (c) serving as a mechanism to improve practice (Corey, Corey, & Callanan, 2011). Ethical codes can be summarized with the following statement: "Do no harm to your clients; treat them responsibly with full awareness of the social context of helping." We are responsible for our clients and for society as well. At times these responsibilities conflict, and you may need to seek detailed guidance from documented ethical codes, your supervisor, or other professionals.

Box 2.1 lists Internet sites for some key ethical codes in English-speaking areas of the globe. All codes contain information on competence, informed consent, confidentiality, and diversity. Issues of advocacy, power, and social justice are implicit in all codes, but most explicitly in counseling, social work, and human services. We strongly recommend that you review the complete text of the ethical code for your intended profession.

BOX 2.1	National and International Perspectives on Counseling Skills

Professional Ethical Codes With Websites

Listed below are some national and international ethical codes. Website addresses are correct at the time of printing but can change. For a keyword Web search, use the name of the professional association and the words *ethics* or *ethical code*.

American Academy of Child and Adolescent Psychiatry (AACAP)	www.aacap.org/galleries/AboutUs/AACAP_Code_of _Ethics.pdf.pdf
	www.aacap.org
American Association for Marriage and Family Therapy (AAMFT) Code of Ethics	www.aamft.org/imis15/content/legal_ethics/code_of _ethics.aspx
	www.aamft.org
American Counseling Association (ACA) Code of Ethics	www.counseling.org/files/fd.ashx?guid=ab7c1272-71c4 -46cf-848c-f98489937dda
	www.counseling.org
American Mental Health Counselors Association	http://www.amhca.org/assets/content/AMHCA_Code _of_Ethics_11_30_09b1.pdf
	www.amca.org
American Psychological Association (APA) Ethical Principles of Psychologists and Code of Conduct	www.apa.org/ethics/code/index.aspx
	www.apa.org

(continued)

BOX 2.1   (continued)

American School Counselor Association (ASCA) Code of Ethics	www.schoolcounselor.org/files/EthicalStandards2010.pdf www.schoolcounselor.org
Australian Psychological Society (APS) Code of Ethics	www.psychology.org.au/Assets/Files/Code_Ethics_2007.pdf www.psychology.org.au
British Association for Counselling and Psychotherapy (BACP) Ethical Framework	www.bacp.co.uk/admin/structure/files/pdf/566_ethical_framework_feb2010.pdf www.bacp.co.uk
Canadian Counselling Association (CCA) Code of Ethics	www.ccacc.ca/_documents/CodeofEthics_en_new.pdf www.ccacc.ca
Commission on Rehabilitation Counselor Certification (CRCC) Code of Professional Ethics for Rehabilitation Counselors	www.crccertification.com/filebin/pdf/CRCC_COE_1-1-10_Rev12-09.pdf www.crccertification.com
International Union of Psychological Science (IUPsyS) Universal Declaration of Ethical Principles for Psychologists	www.am.org/iupsys/resources/ethics/univdecl2008.pdf www.iupsys.net
National Association of School Nurses (NASN) Code of Ethics	www.nasn.org/RoleCareer/CodeofEthics www.nasn.org
National Association of School Psychologists (NASP) Principles for Professional Ethics	www.nasponline.org/standards/2010standards/1_%20Ethical%20Principles.pdf www.naspweb.org
National Association of Social Workers (NASW) Code of Ethics	www.naswdc.org/pubs/code/code.asp (English version); www.naswdc.org/pubs/code/code.asp?c=sp (Spanish version) www.naswdc.org
National Career Development Association (NCDA) Code of Ethics	associationdatabase.com/aws/NCDA/asset_manager/get_file/3395/code_of_ethicsmay-2007.pdf www.ncda.org
National Organization for Human Services (NOHS)	www.nationalhumanservices.org/index.php?option=com_content&view=article&id=43&Itemid=90 www.nationalhumanservices.org
New Zealand Association of Counsellors (NZAC) Code of Ethics	www.nzac.org.nz/NZAC%20CODE%20OF%20ETHICS.pdf www.nzac.org.nz
School Social Work Association of America (SSWAA) Ethical Guidelines	www.sswaa.org/index.asp?page=91 www.sswaa.org
Ethics Updates provides updates on current literature, both popular and professional, that relate to ethics.	ethics.sandiego.edu

Competence

Competent counselors and psychotherapists need awareness, knowledge, skills, and the ability to take appropriate action in the session. What is your awareness of the background of the unique client before you? What is your knowledge of ethics and helping theories and strategies? In the process of acting appropriately with the individual before you, are you aware of your own competence to handle the issues that are presented by the client?

The American Counseling Association's (2005) statement on professional competence includes diversity. Note the emphasis on continued learning and expanding one's qualifications over time.

> C.2.a. **Boundaries of Competence.** Counselors practice only within the boundaries of their competence, based on their education, training, supervised experience, state and national professional credentials, and appropriate professional experience. Counselors will gain knowledge, personal awareness, sensitivity, and skills pertinent to working with a diverse client population.

Regardless of the human services profession with which you identify, competence is key.

In working with a client you need to constantly monitor whether you are competent to counsel the individual on each issue presented. For example, you may be able to help the client work out difficulties occurring at work, but you discover a more complex problem that requires family counseling. You may need to refer the client to another counselor for family counseling while continuing to work with the job issues. If a client demonstrates severe distress or presents an issue with which you are uncomfortable, seek supervision.

Confidentiality

The American Counseling Association's (2005) ethical code states:

> Section B: Introduction. Counselors recognize that trust is the cornerstone of the counseling relationship. Counselors aspire to earn the trust of clients by creating an ongoing partnership, establishing and upholding appropriate boundaries, and maintaining **confidentiality**. Counselors communicate the parameters of confidentiality in a culturally competent manner.

As a student taking this course, you are a beginning professional; you usually do not have legal confidentiality. Nonetheless, you need to keep to yourself what you hear in class role-plays or practice sessions. Trust is built on your ability to keep confidences. Be aware that state laws on confidentiality vary.

Professionals encounter many challenges to confidentiality. Some states require you to inform parents before counseling a child, and information from sessions must be shared with them if they ask. If issues of abuse should appear, you must report this to the appropriate managerial and legal authorities. If the client is a danger to self or others, then rules of confidentiality change; the issue of reporting such information needs to be discussed with your supervisor. As a beginning counselor, you will likely have limited, if any, legal protection, so limits to confidentiality must be included in your approach to informed consent.

HIPAA Privacy

The Health Insurance Portability and Accountability Act of 1996 (HIPAA) is included here because, among other functions, it requires the protection and confidential handling of protected health information. Disclosure of private health information may have many consequences beyond physical health. Clients may lose their jobs, be rejected by friends or family, be denied insurance coverage, or suffer public humiliation, as illustrated by the following examples:

> A banker who also sat on a county health board gained access to patients' records and identified several people with cancer and called in their mortgages. . . . A candidate for Congress nearly saw her campaign derailed when newspapers published the fact that she had sought psychiatric treatment after a suicide attempt. . . . A 30-year FBI veteran was put on administrative leave when, without his permission, his pharmacy released information about his treatment for depression. (U.S. Department of Health and Human Services, 2000)

Situations such as these led to the development of national standards for the protection of health information.

Following is a summary of some key elements of the Privacy Rule, including who is covered, what information is protected, and how protected health information may be used and disclosed (U.S. Department of Health and Human Services, 2003). For a complete outline of HIPPA requirements, visit the website of the U.S. Department of Health and Human Services, Office for Civil Rights (www.hhs.gov/ocr/privacy/index.html).

1. *Who is covered by the Privacy Rule?* The Privacy Rule and its updates apply to health plans, to health care clearinghouses, and to any health care provider who transmits health information in electronic form. All these are called "covered entities." The following link helps professionals determine if they are covered: www.cms.gov /HIPAAGenInfo/06_AreYouaCoveredEntity.asp.

2. *Protected health information.* The Privacy Rule defines "protected health information (PHI)" as all individually identifiable health information held or transmitted by a covered entity or its business associate, in any form or media, whether electronic, paper, or oral. "Individually identifiable health information" is information, including demographic data, that identifies the individual, or could reasonably be used to identify the individual, and that relates to

   ▲ The individual's past, present, or future physical or mental health or condition
   ▲ The provision of health care to the individual
   ▲ The past, present, or future payment for the provision of health care to the individual

   Individually identifiable health information includes many common identifiers such as name, address, birth date, and Social Security number (Family Educational Rights and Privacy Act, 45 C.F.R. § 160.103).

   The Privacy Rule excludes from protected health information employment records that a covered entity maintains in its capacity as an employer as well as education and certain other records subject to, or defined in, the Family Educational Rights and Privacy Act, 20 U.S.C. § 1232g.

3. *De-identified health information.* There are no restrictions on the use or disclosure of de-identified health information. This is information that makes it impossible for others to identify a client. There are two ways to de-identify information: (1) a formal determination by a qualified statistician; or (2) the removal of specified identifiers of the individual and of the individual's relatives, household members, and employers. De-identification is adequate only if the covered entity has no actual knowledge that the remaining information could be used to identify the individual.

Recently the U.S. Department of Health and Human Services' Office for Civil Rights proposed to change the Privacy Rule to allow people to know how their health information has been used or disclosed (U.S. Department of Health and Human Services, 2011).

HIPAA currently requires health care organizations to track access to electronic protected health information, but they are not required to share this information with clients. The proposed rule represents another effort to promote accountability across the health care system and to ensure that providers protect private health information.

When you visit a physician, you are asked to sign a version of the privacy statement. Mental health agencies make their privacy statements clearly available to clients and often post them in the office.

Informed Consent

Counseling is an international profession. The Canadian Counselling Association (2007) approach to informed consent is particularly clear:

> *B4. Client's Rights and Informed Consent.* When counseling is initiated, and throughout the counseling process as necessary, counselors inform clients of the purposes, goals, techniques, procedures, limitations, potential risks and benefits of services to be performed, and other such pertinent information. Counselors make sure that clients understand the implications of diagnosis, fees and fee collection arrangements, record keeping, and limits of confidentiality. Clients have the right to participate in the ongoing counseling plans, to refuse any recommended services, and to be advised of the consequences of such refusal.

The American Psychological Association (2010) stresses that psychologists should inform clients if the counseling session is to be supervised:

> *Standard 10.01 Informed Consent to Therapy,* (c). When the therapist is a trainee and the legal responsibility for the treatment provided resides with the supervisor, the client/patient, as part of the informed consent procedure, is informed that the therapist is in training and is being supervised and is given the name of the supervisor.

In addition, the APA specifies:

> *Standard 4.03 Recording.* Before recording the voices or images of individuals to whom they provide services, psychologists obtain permission from all such persons or their legal representatives.

When you work with children, the ethical issues around informed consent become especially important. Depending on state laws and practices, it is often necessary to obtain written parental permission before counseling a child or before sharing information with others. The child and family should know exactly how the information is to be shared, and written records should be available to them for their comments and evaluation. A key part of informed consent is stating that both child and parents have the right to withdraw their permission at any point. Needless to say, these same principles apply to all clients—the main difference is parental awareness and consent.

When you enter into role-plays and practice sessions, inform your volunteer "clients" about their rights, your own competence, and what they can expect from the session. For example, you might say:

> I'm taking a counseling course, and I appreciate your being willing to help me. I am a beginner, so only talk about things that you want to talk about. I would like to record (or video) the interview, but I'll turn it off immediately if you become uncomfortable and erase it as soon as possible. I may share the recording in a practicum class or I may produce a written transcript of this session, removing anything that could identify you personally. I'll share any written material with you before passing it in to the instructor. Remember, we will stop any time you wish. Do you have any questions?

BOX 2.2 — Sample Practice Contract

BOX 2.2 Sample Practice Contract

The following is a sample contract for you to adapt for practice sessions with volunteer clients. (If you are working with a minor, add that the form must be signed by a parent as appropriate under HIPPA standards.)

Dear Friend,
I am a student in counseling skills at [insert name of college/university]. I am required to practice counseling skills with volunteers. I appreciate your willingness to work with me on my class assignments.

You may choose to talk about topics of real concern to you, or you may prefer to role-play an issue that does not necessarily relate to you. Please let me know before we start whether you are talking about yourself or role-playing.

Here are some important dimensions of our work together:

Confidentiality. As a student, I cannot offer any form of legal confidentiality. However, anything you say to me in the practice session will remain confidential, except for

certain exceptions that state law requires me to report. Even as a student, I must report (1) a serious issue of harm to yourself; (2) indications of child abuse or neglect; (3) other special conditions as required by our state [insert as appropriate].

Audio and/or Video Recording. I will be recording our sessions for my personal listening and learning. If you become uncomfortable at any time, we can turn off the recorder. The recording may be shared with my supervisor [insert name and phone number of professor or supervisor] and/or students in my class. You'll find that recording does not affect our practice session so long as you and I are comfortable. Without additional permission, recordings and any written transcripts are destroyed at the end of the course.

Boundaries of Competence. I am an inexperienced counselor; I cannot do formal counseling. This practice session helps me learn helping skills. I need feedback from you about my performance and what you find helpful. I may give you a form that asks you to evaluate how helpful I was.

_____   _____
Volunteer Client   Counselor

Date _____

You can use this statement as an ethical starting point and eventually develop your own approach to this critical issue. The sample practice contract in Box 2.2 may be helpful as you begin. You can download this form from CourseMate and adapt it to your own setting.

Power

The National Organization for Human Services (NOHS, 1996) comments on **power,** a significant ethical issue that often receives insufficient attention:

> *Statement 6.* Human service professionals are aware that in their relationships with clients power and status are unequal. Therefore they recognize that dual or multiple relationships may increase the risk of harm to, or exploitation of, clients, and may impair their professional judgment. However, in some communities and situations it may not be feasible to avoid social or other nonprofessional contact with clients. Human service professionals support the trust implicit in the helping relationship by avoiding dual relationships that may impair professional judgment, increase the risk of harm to clients or lead to exploitation.

Power differentials occur in society. The very act of helping has power implications. The client may begin counseling with perceived lesser power than the counselor. Awareness of and openness to talking about these issues help you work toward a more egalitarian relationship with the client. If your gender differs from that of your client, it can be helpful to bring up the gender difference. For example, "How comfortable are you discussing this issue with a man?" If your client is uncomfortable, discussing this issue further is wise. Referral may be necessary.

You will encounter many situations in which institutional or cultural **oppression** becomes part of the counseling relationship, even though you personally may not have been involved in that oppression. For example, a woman may have had bad experiences with men. An African American being counseled by a European American may perceive the counselor as potentially prejudiced, and a gay person may not feel safe with a heterosexual counselor. Those who have a disability may expect to be treated with a lack of real understanding by those more physically able. In each of these cases, discussion of differences in background and culture can be helpful early in the session.

Dual relationships occur when you have more than one relationship with a client. Another way to think of this is the concept of *conflict of interest*. If your client is a classmate or friend, you are engaged in a dual relationship. These situations may also occur when you counsel a member of your church or school community. Personal, economic, and other privacy issues can become complex issues. You can examine statements on dual relationships in more detail in the ethical codes.

In the past, the ethical ideal was to avoid all dual relationships; however, the term itself has multiple meanings. For this reason, some current codes of ethics (e.g., American Counseling Association, 2005) do not mention "dual relationship" but recognize three types of roles and relationships with clients: sexual/romantic relationships, nonprofessional relationships, and professional role change. The first type is banned because of its damaging effect on the clients. The other two may allow counselors to interact with a client in a nonprofessional activity as long as the interaction is potentially beneficial to the client and is not of a romantic or sexual nature. Of course you should always use caution.

Social Justice

The National Association of Social Workers (2008) suggests that action beyond the interview or session may be needed to address **social justice** issues. The code includes a major statement on social justice.

> *Ethical Principle: Social workers challenge social injustice.* Social workers pursue social change, particularly with and on behalf of vulnerable and oppressed individuals and groups of people. Social workers' social change efforts are focused primarily on issues of poverty, unemployment, discrimination, and other forms of social injustice. These activities seek to promote sensitivity to and knowledge about oppression and cultural and ethnic diversity. Social workers strive to ensure access to needed information, services, and resources; equality of opportunity; and meaningful participation in decision making for all people.

There are two major types of social justice action. The first and most commonly discussed is the importance of action in the community to work against the destructive influences of poverty, racism, and all forms of discrimination. These preventive strategies are now considered an essential dimension of the "complete" counselor or therapist. Getting out of the office and understanding societies' influence on client issues is central.

We now know that childhood poverty, adversity, and stress produce lifelong damage to the brain. These changes are visible in cells and neurons and include permanent changes in DNA (Marshall, 2010). Therefore, just treating children of poverty through supportive counseling is not enough. For significant change to occur, prevention and social justice action are critical.

The incidence of hypertension among African Americans is reported to be 4 to 7 times that of European Americans, and the long-term impact of racism is considered to be a major contributor (Hall, 2007). Here again, social justice action has both mental and physical implications.

Many professional associations focus on social justice issues. Among them are Counselors for Social Justice (counselorsforsocialjustice.com) and the International Association for the Advancement of Social Work with Groups (www.aaswg.org).

A second application of social justice action occurs within the session. When a female client discusses mistreatment and harassment by her supervisor, the issue of oppression of women should be named as such. The social justice perspective requires you to help her understand that the problem is not caused by her behavior or how she dresses. By naming the problem as sexism and harassment, you often free the client from self-blame and empower her for action. You can also support her in efforts to effect change in the workplace. On a broader scale, you can work in the larger community outside of the counseling office to promote fairer treatment for women in the workplace. Helping clients work through issues and concerns may not be enough. You also have a responsibility to promote community change through social action.

These same points hold true for any form of oppression that you encounter in the session, whether racism, ableism, heterosexism, classism, or other forms of prejudice. We need to remember that our clients live in relationship to the world. The microskill of focusing, discussed later in this book, provides specifics for bringing the cultural/environmental/social context into the session (Chapter 9).

REFLECTION *Ethical Professional Behavior*

Find the Code of Ethics for the profession in which you are interested, and respond to the following questions: What does the code says about *competence*, *confidentiality*, and *consent*? What are you planning to do to meet the ethical guidelines of your profession?

▶Diversity and Multiculturalism

All counseling and therapy are multicultural. The client brings many voices from the past and present to any counseling or therapy session.

—Paul Pedersen

The RESPECTFUL model, presented in Chapter 1, is central if we are to be truly empathic with the uniqueness of each client. Our individuality is developed in relation to others—the *self-in-relation*. Our self-in-relation is cultural and contextual. Thus, family, friends, community experience, the region in which we live, and contextual issues such as war or economic downturn are part of our own and our client's uniqueness.

The terms *diversity* and *multiculturalism* have come under serious attack in recent years, but as Paul Pedersen reminds us, we are all based in cultural experience. Though controversial, especially in the United States, multiculturalism, diversity, and pluralism are official policy in Australia, Canada, and New Zealand. These concepts of culture and context are central to the empathic counseling relationship.

Naming and Blaming   To take one example, Allen worked in a Veterans Hospital with Vietnam War survivors, who told incredible stories of battles, violence, and death. One of the central tasks the VA assigned to clinicians was to discover "malingerers," those veterans who might be seeking government benefits rather than treatment. At that point, the psychiatric establishment had not yet invented the diagnosis of posttraumatic stress disorder (PTSD).

Allen quickly learned that the veterans invariably had a story that was traumatic and a logical reason for their behavior. For example, one patient was diagnosed as bipolar after he had been arrested for demonstrating in protest against the Vietnam War. The veteran was always friendly and logical, until the war came up; then things would change and become more challenging. As the veteran explored his Vietnam experience, he came to realize, "I'm not crazy, the war is crazy. Locking me up in here is crazy." After this, Allen stopped looking for malingerers and started "normalizing" veterans' responses to traumatic experience.

PTSD did not really appear as a distinct diagnosis until veterans met together in consciousness-raising groups where they shared the commonality of their experiences. It was veterans, rather than "helping" professionals, who identified the typical behaviors now listed in the Diagnostic and Statistical Manual of Mental Disorders (DSM), including flashbacks of the trauma (day or night), troubling dreams and sleep difficulties, sudden crying or anger "out of nowhere," and various types of relationship and work difficulties.

In short, the DSM took logical behavior resulting from trauma and termed it pathological. But the veterans had discovered that the "problem" was in the war and the system, rather than their internal weakness. The word *disorder* places the problem in the client or patient; as a result, external causes tend to be ignored in treatment.

Recently, some members of the military have suggested the term "posttraumatic stress reaction" (PTSR), which places the problem or cause in the environment. This more respectful naming places the blame where it belongs, thus encouraging a more balanced approach to counseling and therapy. But while the new name makes sense, is it also a way to keep veterans from receiving the treatment they need? Even names with good intentions can have negative results and become a new form of "malingerer." Thus, a still more recent trend is to speak of "posttraumatic stress injury" (PTSI), mainly to ensure that the diagnosis is treatable. We hope that you will consider reducing the term to just "posttraumatic stress" (PTS) and not pathologize something that is almost always caused by environmental stress.

The RESPECTFUL model includes trauma as a cultural factor, and we now know that PTS occurs in response to many traumatic situations. Survivors of flood, earthquake, or famine, as well as survivors of violence, rape, severe accidents, cancer, and other medical difficulties, may all experience a variety of posttraumatic responses to the environmental insult. All of these are logical responses to an insane environment.

Part of effective treatment is normalizing clients' reactions and fears, helping them see that whatever they did or feel now is a normal and logical response to what they encountered. This "calming" and the processes of crisis management described in Chapter 15 are basic to any work with traumatized clients, even if you then engage in longer-term counseling.

Political Correctness   Another type of naming and blaming revolves around the idea of *political correctness (PC)*. This term is used, often negatively, to describe language that is deliberately intended to minimize any type of offense, particularly to racial, cultural, or other identity groups. The existence of PC has been alleged, denounced, and ridiculed by a wide range of commentators, both conservative and liberal.

Given this controversy, what is the appropriate way to name and discuss the various types of cultural diversity in the RESPECTFUL model? Because counselors and therapist should use language empathically, we urge that you use terms that the *client* prefers. Let clients define the name that is to be used for their cultural background. Respect is the issue here. The client's point of view is what counts in the issue of naming.

A woman is unlikely to enjoy being called a girl or a lady, but you may find some who use these terms. Some people in their 70s resent being called elderly or old, whereas others embrace and prefer this language. At the same time, you may find that the client is using language in a way that is self-deprecating. A woman struggling for her identity may use the word *girl* in a way that indicates a lack of self-confidence. The older person may benefit from a more positive view of the language of aging. A person struggling with sexual identity may find the words *gay* or *lesbian* difficult to deal with at first. You can help clients by exploring names and social identifiers in a more positive fashion.

Race and ethnicity are central issues. *African American* is considered the preferred term, but some clients prefer *Afro-Canadian* or *black*. Others will feel more comfortable being called Haitian, Puerto Rican, or Nigerian. A person from a Hispanic background may well prefer Chicano, Mexican, Mexican American, Cuban, Puerto Rican, Chilean, or Salvadoran. Some American Indians prefer Native American, but most prefer to be called by the name of their tribe or nation—Lakota, Navajo, Swinomish. Some people of European ancestry would rather be called British Australians, Irish Americans, Ukrainian Canadians, or Pakistani English. These people are racially Caucasian but also have an ethnic background.

The language of nationalism and region also needs to be considered. American, Irish, Brazilian, or New Zealander (or "Kiwi") may be the most salient self-identification. *Yankee* is a word of pride to those from New England and a term of derision for many Southerners. Midwesterners, those in Outback Australia, and Scots, Cornish, and Welsh in Great Britain often identify more with the values and beliefs of their region than with their nationality. Residents of many counties in Great Britain resent the more powerful Home Counties near London. Similarly, the Canadian culture of Alberta is very different from the cultures of Ontario, Quebec, and the Maritime provinces.

Following are some predictions that you can make when you take the broad array of multicultural issues into consideration. Remember the concept of intentionality. As you learn more about this area, you will increase your ability to respond appropriately to more and more clients who may be different from you. Be ready to flex intentionally as you learn more.

MULTICULTURAL ISSUES	ANTICIPATED RESULT
Base counselor and therapist behavior on an ethical approach with respect and an awareness of the many issues of diversity. Include the multiple dimensions described in this chapter. All of us have many intersecting multicultural identities.	Anticipate that both you and your clients will appreciate, gain respect, and learn from increasing knowledge in intersecting identities and multicultural competence. You, the counselor, will have a solid foundation for a lifetime of personal and professional growth.

Diversity and multiculturalism have become central to the helping professions throughout the world. If a client's needs are rooted in a multicultural issue in which you are not competent, you may need to refer the client. Over the long term, however, referral is inadequate. You also have the responsibility to build your multicultural competence through constant study and supervision and to minimize your need for referral.

▶Multicultural Competence

The American Counseling Association and the American Psychological Association have developed multicultural guidelines and specific competencies for practice (APA, 2002; Roysircar, Arredondo, Fuertes, Ponterotto, & Toporek, 2003; Sue & Sue, 2013). In these statements the words *multiculturalism* and *diversity* are defined broadly to include many dimensions.

The multicultural competencies traditionally include awareness, knowledge, and skills. Equally important is bringing these competencies into action. You need to become aware of specific issues, develop knowledge about multicultural issues, and master skills for daily practice in our multicultural world. Expect the issue of multicultural competence to become increasingly central to your professional helping career. Developing these competencies will take a lifetime of learning, as there is endless information to absorb.

Diversity and Ethics

The American Counseling Association (2005) focuses the Preamble to its Code of Ethics on diversity as a central ethical issue:

> The American Counseling Association is an educational, scientific, and professional organization whose members work in a variety of settings and serve in multiple capacities. ACA members are dedicated to the enhancement of human development throughout the life span. Association members recognize diversity and embrace a cross-cultural approach in support of the worth, dignity, potential, and uniqueness of each individual within their social and cultural contexts.

The Ethical Standards for Human Service Professionals (NOHS, 1996) include the following three assertions:

> *Statement 17* Human service professionals provide services without discrimination or preference based on age, ethnicity, culture, race, disability, gender, religion, sexual orientation or socioeconomic status.

> *Statement 18* Human service professionals are knowledgeable about the cultures and communities within which they practice. They are aware of multiculturalism in society and its impact on the community as well as individuals within the community. They respect individuals and groups, their cultures and beliefs.

> *Statement 19* Human service professionals are aware of their own cultural backgrounds, beliefs, and values, recognizing the potential for impact on their relationships with others.

Now, let us examine how you can implement these ethical imperatives.

Awareness: Be Aware of Your Own Assumptions, Values, and Biases

Awareness of yourself as a cultural being is a vital beginning. Unless you see yourself as a cultural being, you will have difficulty developing awareness of others. Understand your own multicultural background and the differences that may exist between you and those who come from other backgrounds. Learn about groups different from yours, and recognize your limitations and the need to refer clients on occasion.

The guidelines also speak of how contextual issues beyond a person's control affect the way a person discusses issues and problems. Oppression and discrimination, sexism, racism, and failure to recognize and take disability into account may deeply affect clients without their conscious awareness. Is the problem "in the individual" or "in the environment"? For example, you may need to help clients become aware that issues such as tension, headaches, and high blood pressure may result from the stress caused by harassment and oppression. Many issues are not just client problems but rather problems of a larger society.

Privilege is power given to people through cultural assumptions and stereotypes. McIntosh (1988) comments on the "invisibility of Whiteness." European Americans tend to be unaware of the advantages they have because of the color of their skin. The idea of special privilege has been extended to include men, those of middle- or upper-class economic status, and others in our society who have power and privilege.

Whites, males, heterosexuals, middle-class people, and others enjoy the convenience of not being aware of their privileged state. The physically able see themselves as "normal" with little awareness that they are only "temporarily able" until old age or a trauma occurs. Out of privilege comes stereotyping of the less dominant group, thus further reinforcing the privileged status.

You, the counselor or psychotherapist, face challenges. For example, if you are a middle-class European American heterosexual male and the client is a working-class female of a different race, she is less likely to trust you and rapport may be more difficult to establish. You must improve your awareness, knowledge, and skills to work with clients who are culturally different from you.

In summary, you first need to learn about yourself and whether you have a privileged status. Then your lifetime task is to avoid stereotyping any group or individual while continuing to learn as much as you can about various cultural groups. Recognize that individual differences within a cultural grouping often have greater impact than the cultural "label." Your client is a unique human being. Although cultural diversity influences development, always recognize the person before you as special and different from all others. Awareness of multicultural issues and diversity actually enhances individual differences and the ways in which each client is unique.

Knowledge: Understand the Worldview of the Culturally Different Client

Worldview is formally defined as the way you and your client interpret humanity and the world. Because of varying cultural backgrounds, we all view people and the larger world differently. Multicultural competence stresses the importance of being aware of our negative emotional reactions and biases toward those who are different from us. If you have learned to view certain groups through stereotypes, you especially need to listen and learn respect for the worldview of the client; be careful not to impose your own ideas.

All of us need to develop knowledge about various cultural groups, their history, and their present concerns. If you work with Spanish-speaking groups, learn the different history and issues faced by those from Mexico, Puerto Rico, the Caribbean, and Central and South America. What is the role of immigration? How do the experiences of Latinas/Latinos who have been in Colorado for several centuries differ from those of newly arrived immigrants? Note that diversity is endemic to the broad group we often refer to as "Hispanic."

The same holds true for European Americans and all other races and ethnicities. Old-time New England Yankees in Hadley, Massachusetts, once chained Polish immigrants in barns to keep hired hands from running away. The tables are now turned and it is people of Polish descent who control the town, but quiet tensions between the two groups still remain. Older gay males who once hid their identity are very different from young activists. Whether race or religion, ability or disability, we will constantly be required to learn more about our widely diverse populations.

Traditional approaches to counseling theory and skills may be inappropriate and/or ineffective with some groups. We also need to give special attention to how socioeconomic factors, racism, sexism, heterosexism, and other oppressive forces may influence a client's worldview.

Understanding various worldviews often comes first through academic study and reading. Another approach is to become actively involved in the client's community, attending community events, social and political functions, celebrations and festivals, and—most important—getting to know on a personal basis those who are culturally different from you.

Race, gender, sexual orientation, ability, and other multicultural dimensions do matter. You will need to continually study, learn, and experience more about this complex area throughout your career. Despite the election of President Barack Obama, racial disparities remain. Racial minorities still have more school dropouts and tend to be more dissatisfied with the educational system at all levels. While college attendance has nearly doubled, recent court decisions have resulted in fewer minorities at "top" state universities. Beyond the schools, minorities encounter more poverty and violence, income disparities, and a variety of discriminatory situations. Just getting a taxi in downtown New York is particularly difficult for an African American, even in business attire.

These large and small insults and slights, termed *microaggressions*, result in many physical and mental health problems (Sue, 2010a, 2010b). As you work with minority clients, remain aware of the impact that these external system pressures and oppression may be having on their concerns. One useful intervention in such situations is encouraging and helping these clients to examine how racism, sexism, heterosexism, and other forms of oppression may relate to their present headaches, stomach upsets, high blood pressure, and an array of psychological stressors.

In short, do not enter the helping field without awareness and knowledge that multicultural differences are always present. Clients from many backgrounds experience the harassment of insults and slights, which can be ultimately damaging. Be prepared to make this awareness part of your practice.

A final critical element in multicultural competence is to seek supervision and increase your own awareness, knowledge, and skills when you recognize that you are uncomfortable and perhaps even deficient in knowledge and skills.

Skills: Develop Appropriate Intervention Strategies and Techniques

A classic study found that 50% of minority clients did not return to counseling after the first session (cited in Sue & Sue, 2013). This book seeks to address cultural intentionality by providing you with ideas for multiple responses to your clients. *If your first response doesn't work, be ready with another.* Attend and use listening skills to understand and learn the worldview of others as they tell you their stories (Chapters 3–8). Focusing (Chapter 9) can help clients, who may be blaming themselves for a problem with a classmate or instructor, determine whether their issues are actually related to discrimination.

Traditional counseling strategies are being adapted for use in a more culturally respectful manner (Ivey, D'Andrea, & Ivey, 2012). Remain mindful of the history of cultural bias in assessment and testing instruments and the impact of discrimination on clients. Many clients from culturally diverse backgrounds simply don't trust testing due to a long history of historical racism, which has not yet been totally cleared. Over time, you will expand your knowledge and skills with traditional strategies and also with newer methods designed to be more sensitive to diversity. Box 2.3 depicts the ongoing process of becoming multiculturally aware.

Multiculturalism Belongs to All of Us

Mark Pope, Cherokee Nation and Past President of the American Counseling Association

Multiculturalism is a movement that has changed the soul of our profession. It represents a reintegration of our social work roots with our interests and work in individual psychology.

Now, I know that there are some of you out there who are tired of culture and discussions about culture. You are the more conservative elements of us, and you have just had it with multicultural this and multicultural that. And, further, you don't want to hear about the "truth" one more time.

There is another group of you that can't get enough of all this talk about culture, context, and environmental influences. You are part of the more progressive and liberal elements of the profession. You may be a member of a "minority group" or you have become a committed ally. You may see the world in terms of oppressor and oppressed.

Perhaps now you are saying, "good analysis" or alternatively, "he's pathetic" (especially if you disagree with me). I'll admit it is more complex than these brief paragraphs allow, but I think you get my point.

Here are some things that perhaps can join us together for the future:

1. We are all committed to the helping professions and the dignity and value of each individual.
2. The more we understand that we are part of *multiple cultures*, the more we can understand the multicultural frame of reference and enhance individuality.
3. *Multicultural* means just that—many cultures. Racial and ethnic issues have tended to

predominate, but diversity also includes gender, sexual orientation, age, geographic location, physical ability, religion/spirituality, socioeconomic status, and other factors.

4. Each of us is a multicultural being and thus all counseling and psychotherapy involve multicultural issues. This is not a competition as to which multicultural dimension is the most important. It is time to think of a "win/win" approach.
5. We need to address our own issues of prejudice—racism, sexism, ageism, heterosexism, ableism, classism, and others. Without looking at yourself, you cannot see and appreciate the multicultural differences you will encounter.
6. That said, we must always remember that the race issue in Western society is central. Yes, I know that we have made "great progress," but each progressive step we make reminds me how very far we have to go.

All of us have a legacy of prejudice that we need to work against for the liberation of all, including ourselves. This requires constantly examining yourself, honestly and painfully. You are going to make mistakes as you grow multiculturally; but see these errors as an opportunity to grow further.

Avoid saying, "Oh, I'm not prejudiced." We need a little discomfort to move on. If we realize that we have a joint goal in facilitating client development and continue to grow, our lifetime work will make a significant difference in the world.

Action: Awareness, Knowledge, and Skills Are Meaningless Unless You Act

Developing awareness, knowledge, and skills is essential. However, there is a big difference between achieving these and acting on them. The knowing–doing gap, a concept developed by Stanford professors Jeffrey Pfeffer and Robert Sutton, suggests that organizations and individuals frequently struggle to do what they know. Many people have learned about the importance of exercise, diet, and living meaningfully, yet they fail to translate this knowledge into action. Many students participate in basic skills and diversity courses but fail to implement what they have learned when working with clients. Clients usually learn positive coping strategies through the process of counseling and therapy but often fail to implement these in their lives.

You and your clients need to get what you know out of your head and into your actions. There is a difference between what you learn from reading a book and what you learn from experience. We really don't know something until we practice and use it. And we confirm or modify what we know when anticipated outcomes do or do not occur.

Awareness, knowledge, skills, and action are all essential to learning and mastering counseling. Our counseling and teaching model highlights the importance of action. The exercises in this and other chapters were included to help you actively engage in learning and applying the microskills.

Practice, practice, practice. Doing is the way!

REFLECTION *Multicultural Counseling*

Working with clients that are different from you is a given in today's society. How do you see that having a multicultural perspective will help you in working with different clients? What are some ways you can use to expand your multicultural awareness, knowledge, and skills?

▶Making Positive Psychology Work Through a Strengths and Wellness Approach

Traditional models of counseling and therapy treat counseling concerns and behavioral symptoms as indicators of underlying dysfunctional processes. The counselor or therapist points out client deficits, leading to a top-down professional relationship. This approach places the client in a passive (recipient) position, emphasizes the individual origin of symptoms, and prescribes what the client needs to do.

A strength-based wellness model, in contrast, treats concerns and behaviors as responses to life challenges, builds on the client's strengths and assets, and leads to a more egalitarian and empathic relationship in the counseling setting. This positive approach places the client in the role of an active and engaged agent and focuses attention on environmental and multicultural factors that may affect the client's situation. A wellness approach works in collaboration with clients (Zalaquett, Fuerth, Stein, Ivey, & Ivey, 2008).

If you help clients recognize their strengths and surrounding resources, you can expect them to use this awareness in positive ways. A positive approach needs to be the foundation for all counseling and psychotherapy. But we do need to work through client difficulties and life challenges. So, always listen to clients' troubling stories completely to ensure that they know that you have heard them fully—but always search inside their stories for examples of strength, competence, and resources.

A POSITIVE STRENGTHS AND WELLNESS APPROACH	ANTICIPATED RESULT
Help clients discover and rediscover their strengths by listening carefully for present strengths and resources. In addition, consider a wellness assessment. Find strengths and positive assets in clients themselves and in their support system. Identify multiple dimensions of wellness.	Clients who are aware of their strengths and resources can face their difficulties and discuss problem resolution from a positive foundation. Also, effective and positive counseling and psychotherapy can be anticipated to strengthen the frontal cortex and hippocampus, while potentially resulting in a smaller amygdala.

►Positive Psychology: The Search for Strengths

Recently, the field of counseling has developed an extensive body of knowledge and research supporting the importance of **positive psychology**, a strength-based approach. Psychology has overemphasized the disease model. Seligman (2009, p. 1) states, "We've become too preoccupied with repairing damage when our focus should be on building strength and resilience." Positive psychology brings together a long tradition of emphasis on positives within counseling, human services, psychology, and social work.

When clients come to us to discuss their issues and their concerns, they often use the word *problem*, indicating a negative approach to their life challenges. They talk with us about what is *wrong* with their lives and may even want us to *fix* things for them. Our role is to enable clients to live their lives more effectively and meaningfully. Central to this role is helping clients discover their own strengths.

Leona Tyler (1961), one of the first women to serve as president of the American Psychological Association, long ago anticipated the importance of wellness and positive psychology. She developed a practical system of counseling based on human strengths:

> The initial stages . . . include a process that might be called exploration of resources. The counselor pays little attention to personality weaknesses . . . [and] is most persistent in trying to locate . . . ways of coping with anxiety and stress, already existing resources that may be enlarged and strengthened once their existence is recognized.

Leona Tyler's positive ideas have been central to the microskills framework since its inception (Ivey, 1971; Ivey, D'Andrea, & Ivey, 2012). The strength- and resource-oriented model, *empathic relationship—story and strengths—goals—restory—action,* is an elaboration of these ideas. We now know that Tyler's emphasis on resources and strengths is basic to effective helping. Research and clinical practice using the concepts of optimism, positive psychology, wellness, and neuroscience all support her original and sometimes forgotten contributions.

Counseling with severely distressed clients will be most effective if we incorporate wellness concepts as a foundation for change. We need to listen to client difficulties and issues, but spend more time on positives and strengths. For a wellness, multiculturally sensitive approach to clinical and therapeutic work, we suggest that you consult *Theories of Counseling and Psychotherapy* (Ivey, D'Andrea, & Ivey, 2012). There you will find many specific competency-oriented strategies for therapy with a positive wellness orientation.

Basic to wellness are optimism and resilience.

►Optimism, Resilience, and the Brain

Optimists literally don't give up as easily and this links to greater success in life.

—Elaine Fox

Optimism is defined in various dictionaries with many positive words, among them hope, confidence, and cheerfulness. It also includes a trust in which we expect things to work out and to get better, a sense of personal power, and a belief in the future. Optimism is a key dimension of resilience and the ability to recover and learn from one's difficulties and challenges. A meta-analytic review of 50 studies on optimism found an increased ability to eliminate, reduce, or manage stressors and negative emotions. Furthermore, those with a more optimistic attitude were better able to approach and face their difficulties (Nes & Segerstrom, 2006). Optimists tend to live healthier lives, suffer less from physical illness, and, of course, feel better about themselves and their ability to cope (Kim, Park, & Peterson, 2011; Seligman, 2006).

(Please say how much you agree or disagree with the following statements: 1 = Strongly disagree, 2 = Somewhat disagree, 3 = Slightly disagree, 4 = Slightly agree, 5 = Somewhat agree, 6 = Strongly agree.)

1. _____ If something can go wrong for me it will.
2. _____ I'm always optimistic about my future.
3. _____ In uncertain times, I usually expect the best.
4. _____ Overall, I expect more good things to happen to me than bad.
5. _____ I hardly ever expect things to go my way.
6. _____ I rarely count on good things happening to me.

Copyright © 1994 by the American Psychological Association. Reproduced with permission.

Scheier, M., Carver, C., & Bridges, M. (1994). Distinguishing optimism from neuroticism (and trait anxiety, self-mastery, and self-esteem): A reevaluation of the Life Orientation Test. *Journal of Personality and Social Psychology, 67*(6),1063–1078. Copyright © 1994 by the American Psychological Association. Reproduced with permission.

Out of this research has come an effective, well-researched six-point scale to measure optimism (see Box 2.4). We suggest that you use this scale to assess your own level of optimism. At times you may want to share this scale with clients, or ask them questions to discover their level of optimism. Optimism leads to resilience and wellness. One of our goals in counseling and therapy is to increase resilience, and helping clients become more optimistic and hopeful is part of this process.

Using the scale in Box 2.4, you can develop an optimism score by adding items 2, 3, 4 and a pessimism score by adding 1, 5, and 6. These six items have proven both reliable and predictive. We suggest that you avoid scoring, but note that it is essential that you as a counselor have an optimistic orientation toward yourself, the world, and your client. Discuss the six individual items with clients as a way to understand more completely how they think about themselves and the world. For example, search out client optimistic and pessimistic stories. Using this as a counseling instrument can be useful to you and your client as you work together toward more positive thoughts, feelings, and behaviors.

Resilience is the actual ability to bounce back from difficulty. The child or adult who recovers quickly from a serious illness or accident is a clear example of resilience. A similar example is the person who grew up with abuse and poverty who overcomes all this to go to college and does well in work.

Rainy Brain, Sunny Brain: How to Retrain Your Brain to Overcome Pessimism and Achieve a More Positive Outlook is the title of a book by cognitive neuroscientist Elaine Fox (2012). She points out that we can change brain circuits to focus on the positive, but it takes effort on the part of both client and therapist. For clients with mildly pessimistic attitudes, she suggests the following:

1. Make a daily list of positive and negative events for several days—the little things that go right and wrong from the time you get up. You and your clients will be surprised at how many things go right, but we expect and ignore the good things.
2. Aim for three positive experiences for each negative one. Take care of yourself. What do you enjoy? What makes you happy or more relaxed? Become aware of positive events, and, if needed, plan at least three of these daily for every time something goes wrong.
3. Exercise every day. This is a central theme for counseling and therapy and stress management in this book. See Chapter 13 for a detailed discussion of exercise as central to mental and physical health.

4. Engage in mindfulness meditation, also emphasized in Chapter 13 as central in therapeutic lifestyle changes (TLCs).

In our work with clients, we want to encourage an attitude of optimism and the development of resilience. We do this by listening to their stories, searching out strengths and positives, and helping them commit to a wellness approach to life. We want to encourage positive thoughts and recall of memories of their own strengths, positive assets, and resources.

REFLECTION *Are You Rewiring Your Brain to Become More Positive?*

If you seek to help clients, obviously you as counselor or therapist need to be optimistic about their future and their ability to resolve their concerns. An optimistic helper has a better chance of producing an optimistic client. What are your thoughts on your own optimism and its implications for you as a profession helper?

A positive, optimistic approach to counseling and therapy can have a significant impact on the brain and the body. While the brain is immensely complex, we will initiate the discussion with just a few basics.

▶The Brain, Stress, and the Wellness Approach

In this section, we focus only on the most basic brain systems of immediate relevance to counseling and psychotherapy. For an expanded discussion of what some call the triune brain (cortex, limbic, and reptilian complex), see Appendix C. To summarize briefly, the reptilian complex includes the brain stem, the basal ganglia, and the cerebellum. These structures are concerned with physical management of the body—breathing, movement, circulation, and "fight or flight" when we face unusually stressful situations. Under severe stress, the reptilian complex fulfills it critical function of automatically saving us from danger. In simplest terms, the limbic area is the emotional center, while the cortex is the decision maker that enables us to think and act. Also explored in Appendix C is the relationship between the left and right hemispheres of the brain.

Let's begin with a summary of basic points that illustrate why understanding the brain and neuroscience will be increasingly important in your career.

1. Counseling can change the brain in both positive and negative ways, depending on your emphasis, respect, and ability to develop an empathic relationship. Virtually all clients come to you with some degree of harmful stress underlying their issues. Thus, understanding and managing stress is critical for effective counseling and therapy.
2. Stress has an impact on the brain. Although mild and appropriate stress is required for learning, too much stress is damaging, even to the point of destroying neurons and neural connections. The activating amygdala may enlarge, and critical memory systems in the hippocampus may shrink.
3. A wellness approach, plus stress management and competent counseling, will enable clients to cope more effectively with the issues that they face. Clients grow from strengths, not from their deficits.
4. Understanding the brain and its relationship to the helping process will become increasingly necessary in the next decade. There is a new vocabulary that is becoming part of counseling literature, research, and practice.

5. Neuroscience is bringing counseling and therapy together with neurology and medicine in a new approach to mental and physical health.

6. Drawing from neuroscience, we find that the positive psychology and wellness approach is basic to the future of counseling and psychotherapy

Neuroscience brings with it a new way of thinking about counseling and psychotherapy. Although our field has so far given little attention to this relationship, brain science research supports the change processes of counseling and therapy and provides new scientific foundations for what we have been doing. For example, empathy is no longer just a cognitive concept. Neuroscience defines the physical foundations of empathy, enables us to become more precise in our use of helping skills, and allows us to be with the client more fully.

Neuroscience also brings with it a vocabulary that has seldom been seen in our counseling practice and textbooks. For those who have taken certain introductory courses in biology and/or psychology, the terms will be familiar. For others, relax, the words will be repeated here several times, and you may already know many of them from articles and presentations in the media. Anticipate that terms such as *amygdala*, *hippocampus*, *prefrontal cortex*, and *thalamus* will soon be part of standard practice at all levels. Although in describing their major functions,we discuss various brain structures as if they were independent, it is important to remember that the brain is in fact complex, holistic, and interconnected.

The five senses of sight, hearing, touch, taste, and smell bring information to the brain. If the data are strong enough, information moves to the amygdala, often called the "energizer bunny" of the brain (see Figure 2.1). The amygdala then sends messages throughout the brain. In stressful situations, it focuses on the limbic system's **HPA axis** (the hypothalamus, the pituitary gland, and the adrenal glands), which controls our reactions to stress and many body processes such as mood, sexuality, immune system, and energy storage. Cortisol, produced in the adrenal glands, is critical for learning, but an overabundance of cortisol leads to stress, even traumatic stress, and can damage the brain. Thus, the HPA has both a protective and a learning function, but trauma and repetitive stress can be injurious.

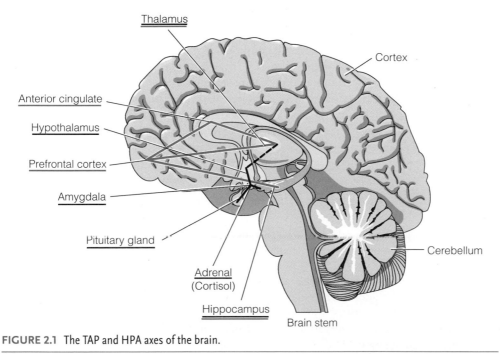

FIGURE 2.1 The TAP and HPA axes of the brain.

© Cengage Learning

When we explore client feelings and emotions, we seek to use positives to overcome the negative stressors.

Figure 2.1 also shows the **TAP axis** (thalamus, anterior cingulate cortex, and prefrontal cortex), which is sometimes described as the brain's executive CEO and decision maker. Counseling and psychotherapy seek to strengthen the TAP through a wide variety of therapeutic and wellness strategies aimed at working with positive strengths to counteract the damage caused by negative and stressful experiences. The left frontal cortex is the prime area for positive emotions such as gladness, joy, happiness, and even love. However, it is important to note that negative thoughts can become fixed in the TAP, leading to increased cortisol in the limbic system and increased awareness of negative memories in the hippocampus.

We cannot separate our thoughts from our emotions—the limbic HPA and executive TAP are always in communication. Sometimes the two are in a battle, and counseling and therapy's task is to strengthen the executive TAP, which is in charge of *emotional regulation*, so that our feelings do not overtake us inappropriately. In serious depression or addiction, the limbic HPA system takes over control of the prefrontal areas to the point that TAP virtually turns off. In such situations, counseling and therapy become more challenging and change may be quite slow.

The amygdala has been found to be the primary storehouse of the less positive feelings of sadness, anger, surprise, disgust, and fear—all in different ways protective to physical and mental systems and recognized throughout the world through cross-cultural research. Increasingly, however, fear is seen as the most basic, with the other feelings viewed as variations of the necessary protective feeling of fear, developed through evolution. This basic feeling quickly enables us to swerve to avoid an accident, duck when a baseball heads toward our head, or run when we see danger. However, these negative feelings can take control, and under severe stress, the amygdala can enlarge, while the hippocampus, central locus of memory, may become smaller. The frontal cortex, coming later in evolution, is the prime storehouse of positive feelings. Note that its physical distance from the amygdala indicates that the HPA fear response is likely to be first. Thus, psychotherapy, counseling, and wellness come late on the scene, and many of our clients need powerful, consistent, and sometimes long-term assistance if change is to occur.

Successful treatment encourages TAP emotional regulation and reduces the power of the negative feelings. Research even indicates that the amygdala may reduce in size and the memory center, the hippocampus, and the prefrontal cortex enlarge (Davidson & McEwen, 2012). Stress management (Chapter 13), cognitive behavioral therapy (Chapter 15), and wellness alternatives such as meditation and those described in this chapter are what make the difference. *Wellness and positive psychology are effective and a basis for growth and change in both mind and body.*

REFLECTION *Brain-Based Counseling and Psychotherapy*

As the National Institute of Mental Health plans to establish a brain-based approach to diagnosis and treatment, how do you plan now for coming change? The brain-based approach will include a blend of traditional diagnostic systems, greatly modified by continued new neuroscience, neurological, medical, and psychological discoveries. In addition, this new model includes extensive attention to environmental issues such as poverty and abuse. Developmental issues from childhood through old age will also be central. What do you think about this new future and your place in it? How will you prepare and keep up with our changing field?

The next section provides an organized and thoroughly researched approach to wellness, reinforcing the positive strengths of the executive TAP, facilitating emotional regulation, and thus enabling your clients to deal effectively with stress.

▶Assessing Client Wellness

The wellness orientation to counseling and psychotherapy has been most clearly defined and thoroughly researched by Jane Myers and Thomas Sweeney (2004, 2005). Their Indivisible Self model incorporates five research-identified dimensions. Each dimension has practical subcategories for assessing clients and facilitating their growth and development. The client who has a sense of wellness and its importance is well on the way toward strengthening the positive long circuit in the brain. In addition, explaining to clients how a positive approach promotes effective brain functioning, as well as physical health, often facilitates their willingness to follow a wellness plan.

The Indivisible Self holistic model stresses the importance of context (also see Chapter 9 on focusing). As appropriate to the individual client before you, it may be helpful to explore the multiple contexts of human development. For example, what is going on locally (family, neighborhood, community)? Problems here obviously affect the individual; at the same time, strengths and wellness assets can also be found here to counteract stressors and build resilience.

The Indivisible Self model points out that change in any part of the wellness system can be beneficial throughout the whole person—or it may damage many of the 17 dimensions of wellness. A problem or a positive change in one part of the system affects all the other parts. For example, a person may have all dimensions of wellness operating effectively but then encounter a difficult contextual issue, such as parents divorcing, a major flood or hurricane, or a major personal trauma. However, the individual may use wellness assets to surmount these challenges and come out of them stronger and more resilient.

Other contextual issues that may affect your clients' wellness include the institutions that define so much of their experience, such as education, religion, government, and business/industry. At an even broader level, politics, culture, environmental changes, global events, and the media can deeply affect clients. A change in social services, global warming, or a call-up for military service are three examples of contextual issues that can affect the individual.

A final contextual issue is lifespan development. Issues for a child entering the teenage years are very different from those for a teen entering the military, work, or college. Marriage or selection of a life partner, raising a family, and older maturity all present different contextual issues that need to be considered. Again, the Indivisible Self model reminds us that the individual is totally connected with the social context and with all parts of his or her developing personhood all the time.

The following exercise provides brief definitions of the 17 personal dimensions and some beginning wellness questions for you to explore with your clients. For a more detailed presentation of wellness research and the actual listing of factor analytic structures, we recommend consulting Myers and Sweeney (2004, 2005).

Your Own Wellness Assessment

Unless you do it, you won't remember it! This is a vital exercise that details the central importance and value of a wellness approach. You seldom will do all this with your clients, but including wellness as a central part of each session can be critical for success and effectiveness.

Again, this takes more than reading—it also takes thought and action. First focus on yourself to develop a solid understanding of the several wellness areas. Identify concrete examples and specifics available to you in each area. Write down your wellness strengths and personal assessment so that you are familiar with the process.

Then find a student colleague, friend, or family member and work through the wellness assessment with that person.

Dimension 1: The Essential Self Four aspects of the core self serve as a foundation for personal exploration of wellness. Each of these areas can provide resources and strengths for positive growth.

Spirituality. There is considerable evidence that those who have a spiritual or religious orientation have more positive attitudes and better mental health than those who lack such supports. Define spirituality and religion broadly, as a thoughtful agnostic or atheist often has many of the characteristics of a highly religious person. At times, the word values or *meaning* can be substituted for *spirituality* or *religion*. Chapter 12's focus on discernment and reflection of meaning will be helpful in integrating these issues into your counseling and therapy sessions.

▲ What strengths and supports do you gain from your spiritual/religious orientation? Be as specific as possible.
▲ How could you draw on this resource when faced with life challenges?
▲ Can you give a specific example of how spirituality has helped you in the past?

Gender identity. This area has two aspects—gender and sexual identity. Identifying men and women as positive role models and finding other positives about your own gender may help you develop unique strengths. Resilient family members or well-known heroes, meaningful to the client, who have surmounted major challenges can serve as role models. Sexual identity relates to one's identity as a heterosexual, gay, lesbian, bisexual, transsexual, transgender, or questioning person. You will find some clients who are unaware that heterosexuality is a sexual orientation. This lack of knowledge can lead to heterosexism. Seeking positive models and personal strengths can be a helpful route to wellness.

▲ What strengths can you draw on as a female or male?
▲ Who are some positive gender role models you have looked to during your life?
▲ What strengths do you draw from your sexual identity—as a heterosexual, a gay, a lesbian, a bisexual, or a transgendered person?
▲ Can you provide concrete examples of how your gender and sexual orientation have been significant in your development (including when you may not have been aware of their significance)?

Cultural identity—race and ethnicity. Research reveals that a positive attitude toward one's race and ethnicity is part of mental health and wellness. Being aware of the strengths of your race/ethnicity can be helpful in establishing who you are and your cultural history. Getting in touch with positive aspects of our ancestry, whether Aboriginal, African American, Italian, Korean, Maori, Navajo, or Swedish, can help us build strengths from our traditions and our families.

▲ What strengths do you draw from your race? Your ethnicity?
▲ What earned or unearned privileges do you enjoy because of your RESPECTFUL and ethic/cultural background? (Consider White privilege, religious privilege, economic privilege, etc.)
▲ Do you have family or role models that suggest ways of living effectively?
▲ Can you provide concrete examples of how your race and ethnicity have affected you and your development?

Self-care.   Virtually all the dimensions of wellness discussed here are part of effective self-care, whether it be cultural pride, good nutrition, or the ability to engage in a loving relationship. Factor analysis wellness issues added issues of personal safety, examples of which can range from fastening one's seat belt and avoidance of extreme risk behavior to stopping smoking and drugs, plus keeping alcohol in positive control.

Dimension 2: The Social Self   Connection with others is essential for wellness. We are selves-in-relation, and closeness to others is a central aspect of wellness. Counseling and psychotherapy often make this the major focus of treatment. However, we need to consider all dimensions of the wellness model in our conceptions of our clients. Two major components of the social self are identified here.

Friendship.   We are people in connection, not meant to be alone. It takes time to nurture relationships. This component focuses on your ability to be a friend and to have friends in healthy long-term relationships.

▲   Tell about your friends and what strengths they provide for you.
▲   Do you have a special friend, one with whom you have had a long-term relationship? What does that mean to you?
▲   Can you tell something specific about yourself as a friend, and what you have done to be a good friend to others?

Love.   Caring for special people, such as family members or a loved one, results in intimacy, trust, and mutual sharing. Sexual intimacy and sharing with a close partner are key areas of wellness.

▲   Describe some positive family stories. What are some positive memories about grandparents, parents, siblings, or your extended family?
▲   How does your family value you? As a grandparent/parent? Brother/sister? Child?
▲   If your immediate family relationships are not close, please share your experiences with your equivalent of family. (Examples: church/mosque/synagogue, community, cultural group, friendship groups.)
▲   Give an example of a positive love relationship and what this means to you.

Dimension 3: The Coping Self   To live effectively, we need to be able to cope with the situations around us. Four basic elements to help us have been identified for this dimension. Each of these is related to different issues in counseling and psychotherapy, and often different theoretical approaches will be useful for the various elements.

Leisure.   People who take time to enjoy themselves daily are better equipped to return to work or school the next day with more energy and less stress. This area is all too often forgotten in counseling's problem-solving approach. When you have time for fun, the many daily and long term challenges we face are more easily resolved.

▲   What leisure time activities do you enjoy?
▲   Equally important, do you take time to do them?
▲   When was the last time you did something fun, and how did it feel?
▲   Can you tell about a specific time when having fun and taking leisure time really benefited you?

Stress management.   Our approach to life, coupled with multiple commitments to family, career, religion, community, and even leisure activities, provides us with endless opportunities to be "stressed out." Data are accumulating that stress is perhaps the central issue in producing mental ill-health and that stress due to either short- or long-term trauma produces bodily changes and affects brain development. Can you help yourself build stress management resources? See Chapter 13 for elaboration of these issues.

- ▲ What do you do when you encounter stress?
- ▲ What specific skills and strategies do you use to cope with stress, and do you remember to use these strategies?
- ▲ Give at least one example of when you managed stress well.
- ▲ Exercise alleviates stress. Can you tell of a time when you exercised or did something else to help you calm down and relax?

Self-worth. This is a central issue for treatment. Self-esteem and feeling good about oneself are required for personal comfort and effective living. We need to accept our imperfections as well as acknowledge our strengths. This part of wellness is obviously central; unless we feel positive about ourselves, other aspects of wellness will be weak at best. It also illustrates the holistic and relatedness qualities found in the Indivisible Self model.

- ▲ What gives you a sense of self-worth and self-esteem?
- ▲ Can you tell about some specific times that you did something kind or helpful for others that you feel especially good about?
- ▲ How do you value your life contribution?
- ▲ What would you like to contribute to others and the world in the future?

Realistic beliefs. Life is obviously not all positives. We also need a clear grasp of reality, the ability to examine our own beliefs and those of others. We can get stuck with negative beliefs about ourselves and the world that undermine effective problem solving. You will find that cognitive behavioral theory and some of the strategies presented in the influencing skills section of this book are especially helpful here.

- ▲ How able are you to face up to difficult situations and see things as they really are?
- ▲ Do you have realistic beliefs and expectations about yourself and your abilities?
- ▲ Do you have realistic beliefs and expectations about others and their abilities?
- ▲ What has gone well for you in the past? The present? What positive anticipations do you have for the future?
- ▲ Is there a specific time you participated in a realistic assessment of yourself or others?

Dimension 4: The Creative Self   Research reveals five creative ways to have a positive impact on the world. Each of these can serve as a springboard for a wellness approach.

Thinking. This element includes the thoughts and thinking patterns that guide your life. Effective problem solving can lead to better personal adjustment. Avoid negative thoughts about yourself and others. An optimistic view is clearly helpful. Cognitive behavioral therapy (Chapter 13) makes this a major focus.

- ▲ What is the nature of your "inner speech"—words and ideas that you say to yourself "inside your head"? Are you encouraging to yourself? To others?
- ▲ How are you at problem solving? Tell about a time when you effectively resolved a difficult issue.
- ▲ Can you give an example or two of when positive thinking and optimism worked for you?

Emotions. Coupled with our thoughts are our feelings (e.g., glad, sad, mad, scared). The ability to experience emotion appropriate to the situation is vital to a healthy lifestyle.

- ▲ When have you felt and expressed emotion with a good result? Negative emotion? Positive emotion?

- ▲ Can you understand and support another's emotional experience and become attuned to the way this person experiences the world?
- ▲ How do you accept emotional support from others?

Control. People who feel in control of their lives see themselves as making a difference; they are in charge of their own "space." They do not seek to control others. Rather, they have a subjective feeling that they know what is expected to happen and that they can control present and future events. But we cannot always control the present or future. People who have realistic expectations when they face difficult life events tend to respond with more resilience that those who have no anticipation or those or who expect the worst.

- ▲ When have you been able to control difficult situations in a positive or realistic way, even though we cannot control all events?
- ▲ When have you had a positive sense of self-control? In relation to self? In relation to others?
- ▲ Provide specific examples of how you are in control of your own destiny. Again, provide concrete positive examples or those in which realism helped you cope more effectively.

Work. We need work to sustain ourselves; our work life takes as much of our time as sleep—or more. Much of our self-worth comes from our ability to contribute to the world through the work we do.

- ▲ What jobs have you most enjoyed or been most proud of?
- ▲ What kind of volunteer work do you do?
- ▲ What do you see as your major contributions or most supportive habits on the job? Write specific examples.

Positive humor. Laughing works! It opens the mind and refreshes the body. Humor is part of creativity and enjoying the moment. People with a sense of humor can often find something positive in the midst of real problems.

- ▲ What makes you laugh?
- ▲ Tell about your sense of humor.
- ▲ Is there some specific time when a sense of humor or laughing helped you deal with a difficult situation?

Dimension 5: The Physical Self The last two aspects of the research on wellness reveal an area that needs far more attention in counseling and psychotherapy. If a person is not doing well physically, even the best self-concept, ability to handle emotions, or ability to relate effectively to others is not enough. Special attention to the physical self as it relates to the body and mind is given in the discussion of therapeutic lifestyle changes in Chapter 13.

Nutrition. Eating a good diet is part of a wellness program. If a person is failing to eat well, referral to dietary counseling may be helpful. But focus here on strengths.

- ▲ How aware are you of the standards of good nutrition?
- ▲ How well do your present weight and eating habits reflect good nutritional standards?
- ▲ Can you provide concrete examples of how you have taken care of yourself in terms of nutrition in the past and present?

Exercise. New research, elaborated in Chapter 13, appears almost daily on the value of exercise and keeping the body moving. Recent evidence indicates that general health, memory, and cognitive functioning are all supported by regular exercise. Help your clients keep their bodies moving. One useful treatment for clients who may be depressed is exercise along with relaxation training. Self-evaluate your exercise. Make evaluation of exercise part of your counseling sessions, and help your clients plan for the future.

▲ What do you do for exercise?
▲ What types of exercise do you like best?
▲ How often do you exercise?
▲ Can you provide concrete examples of how exercise has been beneficial for you?
▲ How can you start a program of exercise?

Completing this exercise will familiarize you with the wellness assessment. Later, when you work with a volunteer or real clients, you will be more knowledgeable and competent. As part of the assessment, you will note weaknesses that can be addressed in counseling and through a wellness plan. But the focus is on finding strengths and positive assets for problem solving in the future.

Intentional Wellness Plan

Once you have a full or partial assessment, developing an **intentional wellness plan** with your clients will be helpful. This will take a minimum of half a session, though providing homework before the session will shorten the time. The first step is summarizing wellness strengths, as these will assist the client to move effectively in areas where change and further work will be useful. The second step is an honest appraisal of areas for improvement. Avoid overwhelming the client with too many immediate improvements for overall wellness—you could easily lose a discouraged client. Keep it simple. Together with the client, select one or two items from the wellness assessment and negotiate a contract for action. The influencing skills of the second half of this book will be helpful in establishing specifics. Check with your client regularly to see how the plan is working. As the client grows and develops, you can move to other dimensions.

We suggest developing an informal wellness plan as part of one session or as a dimension of a longer-term treatment plan. Clients work more effectively on their issues and challenges with a positive wellness approach. The growing interest in positive psychology supports wellness practices in a variety of settings and with clients of all ages. You may ask the client to complete a full wellness assessment as a homework assignment. This will give you a good picture of client strengths that you can draw on during difficult sessions.

| BOX 2.5 | Research and Related Evidence That You Can Use |

Wellness Evidence Base

Wellness has been studied in a wide variety of populations. This research has included many cross-cultural and cross-national studies involving participants of all ages, resulting in a database containing information on more than 12,000 individuals (Myers & Sweeney, 2005). Among the major findings has been that wellness measures are associated with general psychological well-being, a positive body consciousness as contrasted with body shame, healthy love styles, job satisfaction, ethnic identity, and acculturation.

Wellness study by Myers and Sweeney confirm that wellness is indivisible—that is, positive choices for living well in one area of life have implications for other areas of a person's physical, mental, emotional, and spiritual well-being. For example, if you recommend an exercise program as part of treatment, clients will likely be physically healthier, feel better emotionally, suffer less stress, and develop greater feelings of self-worth. If they understand their multicultural background, they can take more pride and develop a more solid self-concept. A wellness orientation over the lifespan is critical regardless of gender, culture, race, or geographic location. In terms of your practice, wellness is measurable in meaningful ways suitable for counseling and other educational interventions.

Chapter 13 will discuss stress management and therapeutic lifestyle changes (TLCs). TLCs are wellness strategies oriented to prevention. One example of a TLC in stress management is fitness training and exercise, which affect blood flow, cognitions, and neurogenesis. These, in turn, affect structures that support the brain's executive functions (Hillman, Erickson, & Kramer, 2008; Ratey, 2008b).

The left hemisphere is associated more with positive emotions, such as happiness and joy, while the right hemisphere and amygdala are more often associated with negative feelings (see Appendix C). In depression and deep sadness, brain scans reveal that the positive areas are less active (Davidson, Pizzagalli, Nitschke, & Putnam, 2002). Happiness involves physical pleasure, the absence of negative emotions, and positive meanings (Carter, 1999; Davidson, 2001). In effect, positive thoughts and action can help override fear, anger, and sadness. This finding is directly parallel to the wellness research of Myers and Sweeney cited above.

What, specifically, does this mean for your practice? When clients focus solely on problems and negative emotions, we can help them through a wellness approach that strengthens and nourishes the individual. In this way we can help clients "build a tolerance for negative emotions and gradually acquire a knack for generating positive ones" (Damasio, 2003, p. 275). So, the wellness approach is not just "window dressing." This strengthening can be both psychological (reminding clients of positive experiences and personal strengths) and physical (exercise and sports, nutrition, and adequate sleep). A base of strengths facilitates problem solving and working through the many complex issues we all face.

▶Summary: Integrating Strength and Wellness, Ethics, and Multicultural Practice

As we review this chapter, let us return to Kendra, the client presented at the beginning of this chapter. Kendra is a person-in-community, a self-in-relation to others. Her identity has been formed through multiple relations in her family, community, and broader society. These relationships have socialized her by placing her in certain roles and statuses. Yet she still has freedom to change herself and society, and our task is to facilitate that process.

Kendra presents her central concern as dealing with sexual harassment on the job, complicated by her need to finance her education. These are obviously serious stressors affecting Kendra and, at another level, affecting how her brain functions. A quick solution is to quit the job, but then the loss of income while searching for another job can be a serious issue. Beyond that, how are things going with her mother and child? Single parents lead complex lives.

Personal resources. While we need to address Kendra's major challenges, we also need to be fully aware of her strengths and ensure that Kendra is conscious of them as well. As part of getting to know Kendra, we'd discuss several of her positive assets, particularly her ability to balance work, parenting, and school. As we move into wellness assessment, we are drawing on positive memories from the past and the TAP, which will be helpful as a base for approaching her several challenges. Any client who manages as well as Kendra should be recognized as a person of strength, full of personal wellness assets. An empathic relationship will further strengthen Kendra's belief in herself. Further discussion will bring out other strengths. The skills she has in all these areas will combine to help her find a suitable solution to the current issue.

Community resources. The most obvious resource that Kendra has will likely be her family and her mother's support. Kendra's child is also an emotional support, even though likely challenging at times. Exploration of resources may also reveal friends and a church, mosque, or synagogue as a place of emotional and physical support. There may be scholarships and other support options available at the college. Finally, let us hope that there is someone in her work setting who can serve as a friend and advocate.

Social context and cultural background.   We believe that individual counseling is most effective when we are able to see the client in a social and cultural context. Pride in one's cultural background is part of a positive self-concept. There are two issues that we'd like you to consider at this time, although more could be commented on. First is our own awareness of multicultural issues and context that might affect Kendra. We need to be aware that Kendra is a woman, a single parent, and from a specific ethnic/racial background (which is unstated in this case). These three factors (plus others, of course) help us understand Kendra more completely. Our awareness and sensitivity to possible culture issues is vital, even if they are never discussed.

Second, as appropriate, we need to help Kendra understand the broader context of her issues. Most clearly, harassment tends to be a women's issue, and issues of sexual discrimination may need to be discussed. Other multicultural factors can be brought in as the sessions progress. What is a solution that is personally satisfying to Kendra? This may require a social justice approach and may involve helping her develop action plans to change the work setting rules. Or it may simply be recognizing the nature of the cultural context as Kendra makes a personal decision.

Key ethical issues include making sure that you are competent to work with her, obtaining appropriate informed consent, preserving confidentiality, and using counselor power responsibly. The possible sexual harassment needs to be explored, and Kendra will need your support as she moves to a decision here. At this moment in her life, Kendra appears to have real financial needs. People with lower incomes clearly do not have the same possibilities and privileges as those who are more economically stable. We may want to refer her for consideration for financial aid at her college.

▼▼▼▼ Key Points	
Identity	**I** am (and you also) **D**erived from family **E**mbedded in community **N**ot isolated from prevailing values **T**hough having unique experiences **I**n certain roles and statuses **T**aught socialized, gendered, and sanctioning **Y**et with freedom to change myself and society. (Jacobs, 1991)
Ethics and competence	You will practice only within the bounds of your competence, seek supervision when necessary, and refer appropriately, while supporting the client with a solid relationship as much as you can. You will continue to gain ethical knowledge and competence throughout your career.
Ethics and informed consent	You will obtain consent from role-playing and real clients and tell them the goals, procedures, benefits, and risks of counseling. Ensure that the client agrees to what has been outlined.
Ethics and confidentiality	You will keep confidences so far as legally possible in accordance with state law. As a beginning counselor in role-plays, you do not have legal confidentiality.
Ethics and power	You will maintain awareness of power differentials and seek to avoid dual or multiple relationships. Power differentials occur in many ways—economic status, gender, and other multicultural variables. The counselor is generally in a more powerful position than the client.
Ethics and social justice	You will maintain awareness that client concerns and issues may be the result of oppressive environments and, where possible, you will actively seek to enhance and protect the rights of your clients.

Multicultural competence	You will recognize that all individuals have dignity, and you will embrace an awareness of cultural differences of many types. You will not discriminate, and you will seek increasing knowledge of multicultural issues. You will become aware of your own multicultural background.
Action is essential	You will work to reduce the knowing–doing gap. You do not know something until you can do it. Practice afterwards, using achievement of expected outcome as a guide for further improvement.
Power and privilege	You will become aware that certain groups have more privileges and entitlements than others and consider these issues in your practice. Examples included in the text are White, male, and middle-class power and privilege. These three do not cover all forms of power and privilege, which are present and all countries and cultures.
Wellness and positive psychology	Clients come to us with many strengths and positive assets from their life experience and their own unique personal competencies. They have family and friends, cultural resources, and many other assets that need to be recognized. Once positive strengths have been clearly identified, problem solving and working through issues can be expected to work more smoothly.
Neuroscience and counseling	Neuroscience brings counseling and therapy together with neurology and medicine in a new approach to mental and physical health. Counseling can change the brain in both positive and negative ways, depending on your emphasis, respect, and ability to develop an empathic relationship. Understanding the brain and its relationship to the helping process is essential to your current and future work.
Stress and counseling	All your clients come to you experiencing some degree of stress. Continuous or excessive stress affects the brain and body in negative ways. Managing stress is key for effective counseling and therapy. Positive psychology and a wellness approach provide an effective framework for strength-based counseling.
Two brain axes	The brain is a holistic structure with two interrelated systems. The TAP axis (thalamus, anterior cingulate cortex, prefrontal cortex), known as the brain's CEO, controls our decision making. The limbic system's HPA axis (hypothalamus, pituitary gland, adrenal glands) controls our reactions to stress and many body processes such as mood, sexuality, immune system, and energy storage. In counseling, seek to strengthen the TAP through therapeutic and wellness strategies, explore client feelings and emotions, and use positives to overcome the negative effects of stressors.
The indivisible self	A wellness model developed by Sweeney and Myers includes the creative, coping, social, essential, and physical selves. Within these five categories are 17 specific dimensions of wellness.
Wellness assessment	Client strengths can be assessed in 17 dimensions: spirituality, gender identity, cultural identity, self-care, friendship, love, leisure, stress management, self-worth, realistic beliefs, thinking, emotions, control, work, positive humor, nutrition, and exercise. As needed, conduct a full wellness assessment with clients. Realistically, however, you may want to use a quick survey and out of that select with the client one or two key issues.
Wellness plan	The assessment of strengths and wellness assets can identify a balance of strengths and areas for growth. The client can then examine areas where more effort and planning might be helpful.

►Competency Practice Exercises and Portfolio of Competence

 Go to CengageBrain.com to access Counseling CourseMate, where you will find an interactive ebook, quizzes, videos, interactive counseling and psychotherapy exercises, the Portfolio of Competence, case studies, a practice test, and more.

Intentional counseling and psychotherapy are achieved through practice and experience. They will be enhanced by your own self-awareness, emotional competence, and ability to observe yourself, thus learning and growing in skills.

The competency practice exercises on the following pages are designed to provide you with learning opportunities in three areas:

1. *Individual practice.* A short series of exercises gives you an opportunity to practice the concepts.
2. *Group practice.* Practice alone can be helpful, but working with others in role-playing sessions or discussions is where the most useful learning occurs. Here you can obtain precise feedback on your counseling style. And if video or audio recordings are used with these practice sessions, you'll find that seeing yourself as others see you is a powerful experience.
3. *Self-assessment.* You are the person who will use the skills. We'd like you to look at yourself as a counselor or therapist through some additional exercises.

Individual Practice

Exercise 1. Review an Ethical Code   Select the ethical code from Box 2.1 that is most relevant to your interests and review it in more detail. Then visit the ethical code of another country or another helping profession and note similarities and differences on competence, informed consent, confidentiality, social justice, and diversity. What is your own position on these issues? Write your observations and comments in a journal.

Exercise 2a. Diversity: Experiential Project   This exercise will help you expand your awareness, knowledge, skills, and actions regarding different cultural groups. The exercise is adapted from Dr. Zalaquett's multicultural courses' syllabi.

Select a diverse group and setting that is different from your own (e.g., a predominantly African American church, a mosque, a GLBT event, a Silent Dinner for the Deaf community). Attend more than one event to gain a deeper cultural understanding of your chosen group. Then ask yourself the following questions:

▲ What did you see or hear that was consistent or inconsistent with your personal beliefs about this cultural/ethnic group?
▲ What did you learn about this group? What did you learn about yourself?
▲ What were your personal reactions in terms of your levels of comfort and feelings of acceptance and/or belonging?
▲ What experiences, beliefs, and values from your own upbringing do you think contributed to those personal reactions?
▲ How will this experience assist you in your professional development?

Exercise 2b. Diversity: Presentation   Conduct a 15- to 20-minute classroom presentation focusing on the group and the experiential activities chosen. To give a comprehensive presentation, please do the following:

▲ Provide handouts with basic information and a list of three peer-reviewed journal references to all class members.

▲ Provide the class with a sense of what that activity was like through an experiential activity, video, description, photos, art, interview, or discussion. Discuss how the experience expanded your awareness, knowledge, and skills; how you were challenged (comfort level); and what you learned about yourself.

▲ Discuss how this experience can be applied in counseling and therapy.

Please visit the following website for recent students' presentations: www.coedu.usf.edu /zalaquett/de/dep.htm. Past students' presentations are available at www.coedu.usf.edu /zalaquett/mcdp/m.htm.

Exercise 3. Personal Wellness Assessment Review the Sweeney-Myers wellness model contextual issues on page 51. What strengths and resources do you find in your own context?

Local	Institutional	Global (World Events)
Family	Education	Politics
Neighborhood	Religion	Culture
Community	Government	Global events
	Business/industry	Environment
	Media	

Review the individual personal strengths on pages 52–56. Then review your own strengths and resources.

Essential	Social	Coping	Creative	Physical
Spirituality	Friendship	Leisure	Thinking	Nutrition
Gender identity	Love	Stress management	Emotions	Exercise
Cultural identity		Self-worth	Control	
Self-care		Realistic beliefs	Work	
			Positive humor	

Group Practice

Exercise 4. Conduct a Wellness Assessment and Develop a Wellness Plan Now that you have engaged in a wellness assessment for yourself, meet with two other class members and engage in a wellness assessment with one of them. Conclude this practice with a discussion of a plan for the future. The third person will be an observer and provide comments and give feedback on the process. We recommend that your volunteer client fill out the Client Feedback Form from Chapter 1. Alternatively, do this as a homework assignment with a volunteer.

Exercise 5. Develop an Informed Consent Form Box 2.2 presents a sample informed consent form, or practice contract. This form can be downloaded from CourseMate. With your small group, develop your own informed consent form that is appropriate for your particular school situation and for your state or commonwealth.

Portfolio of Competence

Determining your own style and theory can be best accomplished on a base of competence. Each chapter closes with a reflective exercise asking your thoughts and feelings about what has been discussed. By the time you finish this book, you will have a substantial record of your competencies and a good written record as you move toward determining your own style and theory.

Use the following as a checklist to evaluate your present level of mastery. Check those dimensions that you currently feel able to do. Those that remain unchecked can serve as future goals. Do not expect to attain intentional competence on every dimension as you work through this book. You will find, however, that you will improve your competencies with repetition and practice.

Level 1: Identification and classification. You will need this minimal level of mastery for those coming examinations.

❑ Define and discuss the key aspects of ethics as they relate to counseling and psychotherapy: competence, informed consent, confidentiality, power, and social justice.
❑ Define and discuss the three dimensions of multicultural competence: becoming aware of your own assumptions, values, and biases; understanding the worldview of the culturally different client; developing appropriate intervention strategies and techniques.
❑ Define and discuss positive psychology and wellness.
❑ Define and discuss the contextual factors of the wellness model.
❑ Define and discuss the five personal dimensions of the wellness model: essential self, coping self, social self, creative self, and physical self.

Level 2: Basic competence. Here you are asked to perform the basic skills in a more practical context, such as an evaluation or an actual counseling session. This initial level of competence can be built on and improved throughout your use of this text.

❑ Write an informed consent form.
❑ Define yourself as a multicultural being.
❑ Evaluate your own wellness profile, both personal and contextual.
❑ Take another person through a wellness assessment.

Levels 3 and 4, intentional competence and teaching competence, will not be presented in this chapter. You will encounter them in Chapter 3 on attending behavior.

▶Determining Your Own Style and Theory: Critical Self-Reflection and Suggestions for Journal Assessment

Reflecting on yourself as a future counselor or psychotherapist in a written journal can be a helpful way to review what you have learned, evaluate your understanding, and think ahead to the future. Here are three questions that you may wish to consider.

1. What stood out for you personally in the section on ethics? What one thing did you consider most memorable for your practice? Some people consider ideas of social justice and action in the community a controversial topic. What are your thoughts?
2. How comfortable are you with ideas of diversity and working with people different from you? Can you recognize yourself as a multicultural person with many dimensions of diversity?
3. Wellness and positive psychology have been stressed as a useful part of the counseling and psychotherapy interview. At the same time, relatively little attention has been given so far to the very real problems that clients bring to us. While many difficult issues will be covered throughout this text, what are your personal thoughts at this moment on wellness and positive psychology? How comfortable are you with this approach?

CHAPTER 3

Attending Behavior and Empathy

Attending Behavior and Empathy

Ethics, Multicultural Competence, and Wellness

When someone really hears you without passing judgment on you, without taking responsibility for you, without trying to mold you, it feels good. When I have been listened to, when I have been heard, I am able to re-perceive my world in a new way and go on. It is astonishing how elements that seem insoluble become soluble when someone listens. How confusions that seem irremediable become relatively clear flowing streams when one is heard.

—Carl Rogers

Mission of "Attending Behavior and Empathy"

Attending and empathy are essential to establish a working relationship with your clients and a good understanding of their issues and concerns. Attending behavior, empathy, and observation skills are necessary (and sometimes sufficient) for effective, facilitative intentional counseling and psychotherapy. We consider them foundational skills.

This chapter and the next, Observation Skills, are complementary and may be read together. Attending focuses on the counselor's verbal and nonverbal behavior. Observation skills focus on the specifics of client nonverbal and verbal behaviors. Together, attending and observation form the foundation of empathic understanding, the relationship, and the working alliance.

Chapter Goals and Competency Objectives
Awareness, knowledge, skills, and actions in attending behavior and empathy will enable you to

▲ Establish an empathic relationship with your clients.

▲ Increase your skill in listening to clients and communicating that interest.

▲ Note your own patterns of attending, including selective attention. We all emphasize some issues while giving less attention to others, perhaps even avoiding certain concerns.

▲ Adapt your attending patterns to the needs of diverse individual and cultural styles of listening and talking.

▲ Develop recovery skills that you can use when you are lost or confused in the session. Even the most advanced professional doesn't always know what is happening. When you don't know what to do, attend!

▲ Understanding how the basics of neuroscience explain and expand the importance of attention and empathy.

▲ Learn how teaching the microskills of listening is a useful therapeutic strategy.

Go to CengageBrain.com to access Counseling CourseMate, where you will find an interactive ebook, quizzes, videos, interactive counseling and psychotherapy exercises, the Portfolio of Competence, case studies, a practice test, and more.

▶Attending Behavior: The Foundation Skill of Listening

Attending behavior, essential to an empathic relationship, is defined as supporting your client with individually and culturally appropriate verbal following, visuals, vocal quality, and body language. **Listening** is the central skill of attending behavior and is core to developing a relationship and making real contact with our clients.

Listening is more than hearing or seeing. You can have perfect vision and hearing, but be an ineffective listener. How can we define effective listening more precisely? The following exercise may help you to identify listening in terms of clearly observable behaviors.

Exercise: The Experience of Listening

One of the best ways to identify and define listening skills is to experience the opposite—poor listening. Think of a time when someone failed to listen to you. Perhaps a family member or friend failed to hear your concerns, a teacher or employer misunderstood your actions and treated you unfairly, or you called a computer helpline and never got someone who listened to your problem. These situations illustrate the importance of being heard and the frustration you feel when someone does not listen to you.

Please stop for just a moment and think back about what was going on when you felt that someone ignored you, distorted what you said, or just plain "didn't listen." How did you feel inside? What was the other person doing that showed he or she was not listening?

A more active, powerful way to define and clarify listening is to find a partner to role-play a session in which one of you plays the part of a poor listener. The poor listener should feel free to exaggerate in order to identify concrete behaviors of the ineffective counselor. The "client" ideally should continue to talk, even if the counselor appears not to listen. Then ask the "client" how he or she felt "inside" or emotionally when the counselor did not listen. Together, discuss the specific and observable *behaviors* that indicated lack of listening. Later, compare your thoughts with the ideas presented in this chapter.

An exaggerated role-play is often humorous. However, on reflection, your strongest memory of poor listening may be feelings of disappointment and even anger. It is the *observable behaviors* that affect the client immediately. Examples of poor listening and other ineffective interviewing behaviors are numerous—and instructive. If you are to be effective and competent, do the opposite of the ineffective counselor: Attend and listen!

This exercise demonstrates clearly that attending and listening behaviors make a significant difference in the session. Attending and listening are the ways in which you communicate empathy and understanding to the client. They are the behavioral roots of the working alliance and a good counseling relationship.

Brain imaging has demonstrated the importance of attending in another way. When a person attends to a stimulus such as the client's story, many areas of the brain of both counselor and client become involved (Posner, 2004). Specific areas of the brain show activity. In effect, attending and listening "light up" the brain in effective counseling and psychotherapy. Without attention, nothing will happen.

Now let us turn to further discussion of how skills and competence can "light up" a session.

▶Attending Behavior: The Skills of Listening

Attending behavior will have predictable results in client conversation. When you use each of the microskills, you can anticipate how the client is likely to respond. These predictions are never 100% perfect, but research has shown that the expected responses usually occur (Ivey, Ivey, & Daniels, 2014). If your first attempt at listening is not received, you can intentionally flex and change the focus of your attention or try another approach to show that you are hearing the client.

ATTENDING BEHAVIOR	ANTICIPATED RESULT
Support your client with individually and culturally appropriate visuals, vocal quality, verbal tracking, and body language.	Clients will talk more freely and respond openly, particularly about topics to which attention is given. Depending on the individual client and culture, anticipate fewer breaks in eye contact, a smoother vocal tone, a more complete story (with fewer topic jumps), and a more comfortable body language.

Attention is the connective force of conversations and of empathic understanding. We are deeply touched when it is present and usually know when someone is not attending to us. The way one attends deeply affects what is talked about in the session. Also, observe the client's reactions. Learning what to do and what not to do will help determine what might be better and more effective in helping that client.

Obviously, you can't learn all the possible qualities and skills of effective listening immediately. It is best to learn the behaviors and skills step by step. Attending and observation are the places to start. You will find that many advanced professionals, including highly paid psychiatrists and physicians need to go back and learn about listening and empathy.

Attending behavior is the first and most critical skill of listening. It is a necessary part of demystifying all counseling and psychotherapy. Sometimes listening carefully is enough to produce change.

To communicate that you are indeed listening or attending to the client, you need the following "three V's + B":*

1. **Visual/eye contact.** Look at people when you speak to them.
2. **Vocal qualities.** Communicate warmth and interest with your voice. Think of how many ways you can say, "I am really interested in what you have to say," just by altering your vocal tone and speech rate. Try that now and note the importance of changes in *behavior.*
3. **Verbal tracking.** Track the client's story. Don't change the subject; stay with the client's topic.
4. **Body language.** Be yourself—authenticity is essential to building trust. To show interest, face clients squarely, lean slightly forward with an expressive face, and use encouraging gestures. Especially critical, smile to show warmth and interest in the client.

Later, we will speak to individual and cultural differences in attending, but variations of three V's + B reduce counselor talk time and provide clients with an opportunity to tell their stories with as much detail as needed. As you listen, you will be able to observe your clients' verbal and nonverbal behavior. Note their patterns of eye contact, their changing vocal tone, their body language, and topics to which your clients attend and those that they avoid. Use your observation skills so that you can adapt your style to meet the needs of the unique person before you, who may come from a different community and cultural background than you.

These attending behavior concepts were first introduced to the helping field by Ivey, Normington, Miller, Morrill, and Haase (1968). Cultural variations in microskills usage were first identified as central to the model by Allen Ivey, when he worked with native Inuits in the central Canadian Arctic. He found that sitting side by side with them was more appropriate than direct eye contact (body language and visuals vary among cultures) and that developing a solid relationship was as important as staying on the verbal topic. Nonetheless, smiling, listening, and a respectful and understanding vocal tone are behaviors that "fit" virtually all cultures and individuals. As a result, Allen became much closer to the Inuits he was teaching. In short, attending behavior and listening are essential for human communication, but we need to be prepared for and expect individual and multicultural differences.

▶Attending Behavior in Action: Getting Specific About Listening and Individual and Multicultural Differences in Style

Listen before you leap! A common tendency of the beginning counselor is to try to solve the client's difficulties in the first 5 minutes. Think about it: Clients most likely developed their concerns over a period of time. It is critical that you slow down, relax, attend to client stories, and look for themes in their narratives. Use the three V's + B—visuals, vocals, verbals, and body language—to more fully understand the client's concerns and build rapport.

▶Visual/Eye Contact

Not only do you want to look at clients, but you also want to observe breaks in eye contact, both by yourself and by the client. Clients often tend to look away when thinking carefully or discussing topics that particularly distress them. You may find yourself avoiding eye

---

*We thank Norma Gluckstern Packard for the three V's acronym.

contact while discussing certain topics. There are counselors who say their clients talk about "nothing but sex" and others who say their clients never bring it up. Through breaks in eye contact or visual fixation, vocal tone, and body shifts, counselors indicate to their clients whether the current discussion topic is comfortable for them.

Cultural differences in eye contact abound. Direct eye contact is considered a sign of interest in European North American middle-class culture. However, even here people often maintain more eye contact while listening and less while talking. Furthermore, if a client from any cultural group is uncomfortable talking about a topic, it is probably better to avoid too much direct eye contact.

Research indicates that some traditional African Americans in the United States may have reverse patterns; that is, they may look more when talking and slightly less when listening. Among some traditional Native American and Latin groups, eye contact by the young is a sign of disrespect. Imagine the problems this may cause the teacher or counselor who says to a youth, "Look at me!" when this directly contradicts the individual's basic cultural values. Some cultural groups (for instance, certain traditional Native American, Inuit, and Aboriginal Australian groups) generally avoid eye contact, especially when talking about serious subjects. This is a sign of respect.

Persons with disabilities represent a cultural group that receives insufficient attention. They also represent the diversity that you will encounter in every session. Eye contact with persons with disabilities may vary, and we need to be careful not to label or counsel them in one way. Empathic understanding and effective listening require that we recognize uniqueness in each person. For example, in working with those who are blind, you can observe their behavior, but they cannot see you. However, your vocal tone communicates an immense amount of information. They often get much more from your vocal tone, speech hesitations, and conversational style than sighted clients. Working with deaf clients, on the other hand, makes your body language more important. If they can read lips, speak clearly and look at them directly. If they have a sign language interpreter available, then look at the client, *not* at the interpreter.

We suggest that those of you who do not face these challenges think of yourself as a member of the *temporarily able* culture. Age and life experience will bring most of you some variation of ability challenges. For older individuals, the issues discussed here may become the norm rather than the exception. Approach all clients with humility and respect.

▶Vocal Qualities: Tone and Speech Rate

Your voice is an instrument that communicates much of the feeling you have about yourself or about the client and what the client is talking about. Changes in pitch, volume, and speech rate, as well as breaks and hesitations, convey the same things as the nature of your eye contact. Throat clearing on your part or the client's may indicate that words are not coming easily. If clients are stressed, you'll observe that in their vocal tone as well as body movements. If the topic is uncomfortable for you or you pick up on the client's stress, your vocal tone or speech rate may change. Keep in mind that different people are likely to respond to your voice differently. Think of the radio and television voices that you like and dislike.

Verbal *underlining* is another useful concept. As you consider the way you tell a story, you may find yourself giving louder volume and increased vocal emphasis to certain words and short phrases. Clients do the same. The key words a person underlines by means of volume and emphasis are often concepts of particular importance. At the same time, expect some especially significant things to be said more softly. When talking about critical issues, especially those that are difficult to talk about, expect a lower speech volume. In these cases, seeking to match your vocal tone with the client's is usually appropriate.

Accent is a particularly good example of how people react differently to the same voice. What are your reactions to the following accents: Australian, BBC English, Canadian, French, Pakistani, New England, Southern United States? Obviously, we need to avoid stereotyping people because their accents are different from ours.

Exercise: Tone of Voice*

Try the following exercise with a group of three or more people.

Ask the members of the group to close their eyes while you speak to them. Talk in your normal tone of voice on any subject of interest to you. As you talk to the group, ask them to notice your vocal qualities. How do they react to your tone, your volume, your speech rate, and perhaps even your regional or ethnic accent? Continue talking for 2 or 3 minutes. Then ask the group to give you feedback on your voice. Summarize what you learn.

If you don't have a group easily available, spend some time noting the vocal tone/style of various people around you. What do you find most engaging? Do some types of speech cause you to move away from the speaker?

This exercise often reveals a point that is central to the entire concept of attending: People differ in their reactions to the same stimulus. Some people find one voice interesting; others find that same voice boring; still others may consider it warm and caring. This exercise and others like it reveal again and again that people differ, and that what is successful with one person or client may not work with another.

▶Verbal Tracking: Following the Client or Changing the Topic

Verbal tracking is staying with your client's topic to encourage full elaboration of the narrative. Just as people make sudden shifts in nonverbal communication, they change topics when they aren't comfortable. In middle-class U.S. communication, direct tracking is appropriate, but in some Asian cultures such direct verbal follow-up may be considered rude and intrusive.

Verbal tracking is especially helpful to both the beginning counselor and the experienced therapist who is lost or puzzled about what to say next in response to a client. *Relax*; you don't need to introduce a new topic. Ask a question or make a brief comment regarding whatever the client has said in the immediate or near past. Build on the client's topics, and you will come to know the client very well over time.

The Central Role of Selective Attention

The normal human brain is wired to attend to stimuli in a way that focuses on coping with the environment and what is near at hand. **Selective attention** is central to counseling and psychotherapy. Clients tend to talk about what counselors are willing to hear. In any session, your client will present multiple possibilities for discussion. Even though the topic is career choice, a sidetrack into family issues and personal relationships may be necessary before returning to the purpose of the counseling sessions. On the other hand, some counselors may not be as interested in career work; their career clients may end up talking about themselves and their personal history and end up in long-term therapy. How you selectively attend may determine the length of the session and whether or not the client returns.

---

*This exercise was developed by Robert Marx, School of Management, University of Massachusetts.

Observe the selective attention patterns of your clients. What do they focus on? What topics do they seem to avoid?

A famous training film (Shostrum, 1966) shows three eminent counselors (Albert Ellis, Fritz Perls, and Carl Rogers) all counseling the same client, Gloria. Gloria changes the way she talks and responds very differently as she works with each counselor. Research on verbal behavior in the film revealed that Gloria tended to match the language of the three different counselors (Meara, Pepinsky, Shannon, & Murray, 1981; Meara, Shannon, & Pepinsky, 1979). Each expert indicated, by his nonverbal and verbal behavior, what he wanted Gloria to talk about!

Should clients match your language and chosen topic for discussion, or should you, the counselor, learn to match your language and style to that of the client? Most likely, both approaches are relevant, but in the beginning, you want to draw out client stories from their own language perspective, not yours. What do you consider most central and meaningful in the session? Are there topics with which you are less comfortable? Some counselors are excellent at helping clients talk about vocational issues but shy away from interpersonal conflict and sexuality. Others may find their clients constantly talking about interpersonal issues, excluding critical practical issues such as getting a job.

REFLECTION   *Selective Attention*

Angelina: (speaks slowly, seems to be sad and mildly depressed) I'm so fouled up right now. The first term went well and I passed all my courses. But this term, I am really having trouble with chemistry. It's hard to get around the lab in my wheelchair and I still don't have a textbook yet. (An angry spark appears in her eyes, and she clenches her fist.) By the time I got to the bookstore, they were all gone. It takes a long time to get to that class because the elevator is on the wrong side of the building for me. (looks down at floor) Almost as bad, my car broke down and I missed two days of school because I couldn't get there. (The sad look returns to her eyes.) In high school, I had lots of friends, but somehow I just don't fit in here. It seems that I just sit and study, sit and study. Some days it just doesn't seem worth the effort.

There are several different directions a counselor could follow from this statement. Where would you go, given the multiple possible directions? Think of or list at least three possibilities for follow-up to this client statement. What is your overall reaction to this experience?

The multiple topics the client presented were chemistry, dealing with her wheelchair, the bookstore being out of books, the failure of the college to provide direct access, the car breaking down, missing school, friendship issues, and finally, that indication of depression, "Some days it just doesn't seem worth the effort."

Often the last thing a client says in a list of concerns needs to be a focus of the discussion, either now or later. In this case, we'd suggest keeping awareness of possible depression in mind and reflecting the main theme of the client's story. We recommend in these situations that you respond by helping the client decide where to start. "You must feel like you're being hit from all directions. What would you like to talk about first?" Angelina has a lot to talk about, but we can only talk about one thing at a time. At a later point in the session, return to the issues she mentioned at the beginning so that she knows you have heard her fully. It can be helpful to list the topics with Angelina and write a contract for how each will be addressed in future sessions, all the time observing signs of depression, which may require referral or longer-term counseling.

Some counselors consistently listen attentively to only a few key topics while ignoring other possibilities. Be alert to your own potential patterning of responses. Try to ensure that no issue gets lost, but avoid confusion by not seeking to solve everything at once.

The Value of Redirecting Attention

There are times when it may be inappropriate to attend to the here and now of client statements. For example, a client may talk insistently about the same topic over and over again. In such cases, intentional nonattending may be useful. Through failure to maintain eye contact, subtle shifts in body posture, vocal tone, and deliberate jumps to more positive topics, you can facilitate redirecting the session to other areas. Instead of actively changing the topic, you may want to ask for details from the repeating story. Remember that if clients have been traumatized (hospital, breakup of a long-term relationship, accident, burglary), they may need to tell their story several times.

A depressed client may want to give the most complete description of how and why the world is wrong and continue on with more negatives in their lives. We need to hear that client's story, but we also need to selectively attend and not pay attention to only the negative. Clients grow from strengths. Redirect the conversation, and when you observe a strength, a wellness habit (running, music), or a resource outside the individual that might be helpful, focus on that positive asset.

The most skilled counselors and psychotherapists use attending skills to open and close client talk, thus making the most effective use of limited time in the session.

The Usefulness of Silence

Sometimes the most useful thing you can do as a helper is to support your client silently. As a counselor, particularly as a beginner, you may find it hard to sit and wait for clients to think through what they want to say. Your client may be in tears, and you may want to give immediate support. However, sometimes the best support may be simply being with the person and not saying a word. Consider offering a tissue, as even this small gesture shows you care. In general, it's always good to have a box or two of tissues for clients to take even without asking or being offered. Of course, don't follow the silence too long, search for a natural break, and attend appropriately.

There is much more happening in the brain than just silence. It turns out that the auditory cortex remains active when you are attending or listening to silence. Your brain remains highly sensitive, as shown on an fMRI. Similarly, there is evidence that the brain's visual areas activate before the individual is consciously aware of seeing an object or person (Somers, 2006).

For a beginning counselor, silence can be frightening. After all, doesn't counseling mean talking about issues and solving problems verbally? When you feel uncomfortable with silence, look at your client. If the client appears comfortable, draw from her or his body language and join in the silence. If the client seems disquieted by the silence, rely on your attending skills. Ask a question or make a comment about something relevant mentioned earlier in the session.

Talk Time

Finally, remember the obvious: *Clients can't talk while you do.* Review your sessions for talk time. Who talks more, you or your client? With most adult clients, the percentage of client talk time should generally be more than that of the counselor. With less verbal clients or young children, the counselor may need to talk slightly more or tell stories to help the client talk. A 7-year-old child dealing with parental divorce may not say a word about the divorce initially. But when you read a children's book on feelings about divorce, he or she may start to ask questions and talk more freely.

▶Body Language: Attentive and Authentic

The anthropologist Edward Hall once examined film clips of Southwestern Native Americans and European North Americans and found more than 20 different variations in the way they walked. Just as cultural differences in eye contact exist, body language patterns also differ. Box 3.1 demonstrates the impact of our attending behavior on people from different cultures.

A comfortable conversational distance for many North Americans is slightly more than arm's length, and the English prefer even greater distances. Many Latin people prefer half that distance, and some people from the Middle East may talk practically eyeball to eyeball. As a result, the slightly forward lean we recommend for attending is not appropriate all the time.

What determines a comfortable interpersonal distance is influenced by multiple factors. Hargie, Dickson, and Tourish (2004, p. 45) point out the following:

Gender: Women tend to feel more comfortable with closer distances than men.
Personality: Introverts need more distance than extraverts.
Age: Children and the young tend to adopt closer distances.

BOX 3.1 National and International Perspectives on Counseling Skills

Use With Care—Culturally Incorrect Attending Can Be Rude

Weijun Zhang, Management Consultant, Shanghai, China

The visiting counselor from North America got his first exposure to cross-cultural counseling differences at one of the counseling centers in Shanghai. His client was a female college student. I was invited to serve as an interpreter. As the session went on, I noticed that the client seemed increasingly uncomfortable. What had happened? Since I was translating, I took the liberty of modifying what was said to fit each other's culture, and I had confidence in my ability to do so. I could not figure out what was wrong until the session was over and I reviewed the videotape with the counselor and some of my colleagues. The counselor had noticed the same problem and wanted to understand what was going on. What we found amazed us all.

First, the counselor's way of looking at the client—his eye contact—was improper. When two Chinese talk to one another, we use much less eye contact, especially when speaking with a person of the opposite sex. The counselor's gaze at the Chinese woman could have been considered rude or seductive in Chinese culture.

Although his nods were acceptable, they were too frequent by Chinese standards. The student client, probably believing one good nod deserved another, nodded in harmony with the counselor. That unusual head bobbing must have contributed to the student's discomfort. The counselor would mutter "uh-huh" when there was a pause in the woman's speech. While "uh-huh" is a good minimal encouragement in North America, it happens to convey a kind of arrogance in China. A self-respecting Chinese would say *er* (oh), or *shi* (yes) to show he or she is listening. How could the woman feel comfortable when she thought she was being slighted?

He shook her hand and touched her shoulder. I told our respected visiting counselor afterward, "If you don't care about the details, simply remember this rule of thumb: in China, a man is not supposed to touch any part of a woman's body unless she seems to be above 65 years old and displays difficulty in moving around."

"Though I have worked in the field for more than 20 years, I am still a lay person here in a different culture," the counselor commented as we finished our discussion.

Topic of conversation: Difficult topics such as sexual worries or personal misbehavior may lead a person to more distance.

Personal relationships: Harmonious friends or couples tend to be closer. When disagreements occur, observe how harmony disappears. (This is also a clue when you find a client suddenly crossing the arms, looking away, or fidgeting.)

Ability: Each person is unique. We cannot place people with physical disability in any one group. Consider the differences among the following: a person who uses a wheelchair, an individual with cerebral palsy, one who has Parkinson's disease, one who has lost a limb, or a client who is physically disfigured by a serious burn. They all may have the common problem of lack of societal understanding and support, but you must work with each individual from her or his own perspective. Their body language and speaking style will vary. Ensure that your working space makes any necessary physical accommodations, and attend to each client respectfully as a complete person.

A person may move forward when interested and away when bored or frightened. As you talk, notice people's movements in relation to you. How do you affect them? Note your own behavior patterns in the session. When do you markedly change body posture? A natural, authentic, relaxed body style is likely to be most effective, but be prepared to adapt and be flexible according to the individual client.

Your authentic personhood is a vital presence in the helping relationship. Whether you use visuals, vocal qualities, verbal tracking, or attentive body language, be a real person in a real relationship. Practice the skills, be aware, and be respectful of individual and cultural differences. Box 3.2 presents relevant research evidence for counseling skills.

BOX 3.2 | Research and Related Evidence That You Can Use

Attending Behavior

Empirical research on attending behavior has been extensive over the years, and we now have secure conclusions about its importance (see Ivey, Ivey, & Daniels, 2014, for a comprehensive report). In their review of the attending literature, Hill and O'Brien (2004) conclude that "smiling, a body orientation directly facing the client, a forward trunk lean, both vertical and horizontal arm movements, and a medium distance of about 55 inches between the helper and client are all generally helpful nonverbal behaviors."

Just because you think you are listening and being empathic does not mean that the client sees you that way. A particularly valuable research study found that European American counselors' perception of their expressed empathy and listening was not in accord with the perception of African Americans, who saw them as less effective (Steward, Neil, Jo, Hill, & Baden, 1998). The way a counselor can "be with" a client tends to vary among cultures. Nwachuku and Ivey (1992) tested a program of culture-specific training and found that variations to meet cultural differences were essential.

Researchers have documented the importance of communication skills training for physicians. Training improves physicians' communication, self-efficacy, confidence, and satisfaction with the training program. Furthermore, communication skills training has a positive effect on patient outcomes such as satisfaction and perception that the physician understood their disease (Ammentorp, Sabroe, Kofoed, & Mainz, 2007; Back et al., 2007; Bylund et al., 2008; Libert et al., 2007).

Bensing (1999a) provides a thought-provoking quote from her extensive review of the literature: "Simply looking at the patient has proven to be very important . . . and even silence can be very therapeutic, at least when it is used effectively" (p. 295). Among other findings, she notes that U.S. physicians are more detached and task-oriented than are Dutch physicians, who are rated as warmer and more involved. We can expect variations among both individuals and cultures in their style of attending.

Teaching the skills of attending has been successful with a variety of populations. Early work in treatment settings demonstrated the value of teaching attending behavior to hospitalized patients (Donk, 1972; Ivey, 1973). A study of adult schizophrenics showed that teaching social skills with special attention to attending behavior was successful and that patients maintained these skills over a 2-month period (Hunter, 1984). Shy, withdrawn people (avoidant personality) profited from social skills training in a well-controlled study by van der Molen (2006). Gearhart and Bodie (2011) have shown that teaching active listening and empathic skills builds closer relationships in many populations.

▶Empathy

Carl Rogers (1957, 1961) brought the importance of empathy to our attention. He made it clear that that we need to listen carefully, enter the world of the client, and communicate that we understand the client's world as the client sees and experiences it. Putting yourself "into another person's shoes" or viewing the world "through someone else's eyes and ears" is another way to describe empathy. The following quotation has been used by Rogers himself to define empathy.

> This is not laying trips on people. . . . You only listen and say back the other person's thing, step by step, just as that person seems to have it at that moment. You never mix into it any of your own things or ideas, never lay on the other person anything that the person did not express. . . . To show that you understand exactly, make a sentence or two which gets exactly at the personal meaning the person wanted to put across. This might be in your own words, usually, but use that person's own words for the touchy main things. (Gendlin & Hendricks, n.d.)

Again, please recall the importance of empathy to the relationship, the "working alliance"; empathy is considered 30% of *common factors* that make for successful counseling and psychotherapy (Miller, Duncan, & Hubble, 2005). When you provide an empathic response, you can predict how clients are likely to respond. Here is another description of empathy and the predictions that you can make.

EMPATHY	ANTICIPATED OUTCOME
Experiencing the client's world and story as if you were that client; understanding his or her key issues and saying them back accurately, without adding your own thoughts, feelings, or meanings. This requires attending and observation skills plus using the important key words of the client, but distilling and shortening the main ideas.	Clients will feel understood and engage in more depth in exploring their issues. Empathy is best assessed by the client's reaction to a statement and his or her ability to continue the discussion in more depth and, eventually, with better self-understanding.

Rogers's thinking led to extensive work by Charles Truax (1961), who is recognized as the first person to measure levels of empathic understanding. He developed a nine-point scale for rating level of empathic understanding (Truax, 1961). Robert Carkhuff (1969), who originally partnered with Truax, developed a five-point scale. These scales have been used widely in research and have practical applications for the session.

Many others have followed and elaborated on Rogers's influential definition of empathy (see Carkhuff, 2000; Egan, 2010; Ivey, D'Andrea,& Ivey, 2012). A common current practice is to describe three types of empathic understanding. This is the convention that we will use in this book. Chapter session transcripts will be evaluated on the following scale.

Subtractive empathy: Counselor responses give back to the client less than what the client stated, and perhaps even distort what has been said. In this case, the listening or influencing skills are used inappropriately.

Basic (interchangeable) empathy: Counselor responses are roughly interchangeable with those of the client. The counselor is able to say back accurately what the client has said. Skilled intentional competence with the basic listening sequence (see the later chapters of this book) demonstrates basic empathy. You will find this the most common counselor comment level in helping. Rogers pointed out that listening, by itself, is not only necessary but sufficient to produce client change.

Additive empathy: Counselor responses may add something beyond what the client has said. This may be adding a link to something the client has said earlier, or it may be a congruent idea or frame of reference that helps the client see a new perspective. Feedback and your own self-disclosure, used thoughtfully, can be additive.

These three anchor points are often expanded to classify and rate the quality of empathy shown in a session. You can use empathy rating in your practice with microskills. Later, in your professional work, keep checking whether you have maintained interest in your clients and are fully empathic.

Client: I don't know what to do. I've gone over this problem again and again. My husband just doesn't seem to understand that I don't really care any longer. He just keeps trying in the same boring way—but it doesn't seem worth bothering with him anymore.

Level 1 Empathy	(subtractive) That's not a very good way to talk. I think you ought to consider his feelings, too.
	(slightly subtractive) Seems like you've just about given up on him. You don't want to try anymore. (interpreting the negative)
Level 2 Empathy	(basic empathy or interchangeable response) You're discouraged and confused. You've worked over the issues with your husband, but he just doesn't seem to understand. At the moment, you feel he's not worth bothering with. You don't really care. (Hearing the client accurately is the place to start all empathic understanding. Level 2 is always central.)
Level 3 Empathy	(slightly additive) You've gone over the problem with him again and again to the point that you don't really care right now. You've tried hard. What does this mean to you? (The question adds the possibility of the client's thinking in new ways, but the client still is in charge of the conversation.)
	(additive and perhaps transformational) I sense your hurt and confusion and that right now you really don't care anymore. Given what you've told me, your thoughts and feelings make a lot of sense to me. At the same time, you've had a reason for trying so hard. You've

talked about some deep feelings of caring for him in the past. How do you put that together right now with what you are feeling? (A summary with a mild self-disclosure. The question helps the client develop her own integration and meanings of the issue at the moment.)

In the first half of this book, we recommend that you aim for interchangeable responses. What is most essential for empathic understanding is listening carefully and hearing the client accurately. This, by itself, often helps the client to clarify and resolve many issues. At the same time, be aware that slightly substractive empathy may be an opening to better understanding. You may see your helping lead as interchangeable, but the client may hear it differently. Use unpredicted and surprising client responses as an opportunity to understand the client more fully. *It's not the errors you make; it's your ability to repair them and move on that counts*!

Many other dimensions of empathic understanding will be explored throughout this book. For the moment, recall the following points as central.

1. Aim to understand your clients' experience and worldview in a nonjudgmental supportive fashion as they present their story, thoughts, and emotions to you.
2. Seek to communicate that understanding to the client, but avoid mixing "your own thing" in with what you say.

This is the surest route to reaching that critical Level 2 of interchangeable empathic responding.

Empathy and Mirror Neurons   Historically, counseling and therapy have advocated and demonstrated the importance of empathy, but empathy has always been a somewhat vague and sometimes controversial concept. Well-established findings from neuroscience have changed our thinking. Empathy is identifiable by means of functional magnetic resonance imaging (fMRI)and other key technologies. Key to this process are **mirror neurons**, which fire when a human or other primate acts *and* when they observe actions by another. Many psychologists believe that mirror neurons are one of the most significant discoveries in recent science.

One of the earliest studies of mirror neurons and empathy asked one of two closely attached partners to watch the other (through a one-way mirror) receive a mild shock. It was found that the brain of the shocked partner fired in two areas—one representing physical pain and the other emotional pain. At the same time, the observing partner's emotional pain center also fired when watching the shocked partner experience physical pain (Singer et al., 2004). Research consistently shows that the mirror neurons of children, adolescents, and adults diagnosed with antisocial or conduct disorder do *not* activate (Decety & Jackson, 2004). In fact, there is evidence that many people with this diagnosis show pleasure when observing others in pain.

These basic findings have been replicated many times in different ways. For example, Marci et al. (2007) found that skin conductance of patient–therapist pairs was high and parallel when they both indicated that they felt a communication of empathic understanding. This can work two ways in communication. Verbal communication is a joint activity, and an fMRI study found that "neural coupling" such as this disappears when story comprehension is not effective. When listening skills are not successfully implemented (e.g., subtractive), empathy falls apart.

Questions of interchangeable, subtractive, and additive empathy were explored by Oliveira and Gonçalves (2011). In this study, 40 participants watched actors respond to emotion-laden video stories. They found that participants who demonstrated higher levels of additive empathy had an increased heart rate, whereas those whose observations were interchangeable or subtractive had no "change of heart."

In short, your empathic being and your ability to listen and be with clients are a vital part of helping your clients grow and change. Listening and empathy are not just abstract concepts—they are clearly measureable, and they make a difference in other people's lives (Stephens, Silbert, & Hasson, 2010).

▶Example Counseling Session: I Didn't Get a Promotion—Is This Discrimination?

The following session example illustrates the importance both of empathic attending skills and of using these skills with awareness of cultural and gender differences.

Azara, a 45-year-old Puerto Rican manager, was not promoted, although she thinks her work is of high quality. She is weary of being passed over and seeing less competent individuals take the position she feels she deserves.

The first session segment is a negative example, designed to be particularly ineffective so that it provides a sharp contrast with the more positive effort that follows. In both examples, the counselor, Allen, has the task of developing a relationship and drawing out the client's story. Note how disruptive visual contact, vocal qualities, verbal tracking, and body language can lead to a poor session.

Negative Example

COUNSELOR AND CLIENT CONVERSATION	PROCESS COMMENTS
1. *Allen*: Hi, Azara, you wanted to talk about something today.	Allen fails to greet Azara warmly. He just starts and does nothing to develop rapport and a relationship, which are especially important in a cross-cultural session. He remains seated in his chair behind a desk. (The nonverbal situation is already subtractive.)
2. *Azara*: Yes, I do. I've come to you because there's been an incident at my job a couple of days ago. And I'm kind of upset about it.	Azara sits down and immediately moves ahead with her issues regardless of what Allen does. She is clearly ready to start the session.
3. *Allen*: What is your job?	Allen's voice is aggressive. He ignores Azara's upset feelings and asks a closed question. An appropriate vocal tone communicates warmth and is essential in any relationship.
4. *Azara*: Well, right now I'm an assistant manager for a company and I've worked at this company for 15 years.	Azara keeps trying. Allen looks down while she talks. Subtractive nonverbals.
5. *Allen*: So after 15 years you're still an assistant. When I was in business, I didn't take that long to get a promotion. Let me tell you about what I did to get ahead . . . (he goes on at length about himself).	The focus is taken away while Allen, the counselor, talks about himself. With this long response, he has more talk time than the client, Azara. The evaluative "putdown" is an example of how counselors inappropriately use their power and is, of course, totally subtractive.

COUNSELOR AND CLIENT CONVERSATION	PROCESS COMMENTS
6. *Azara*: Yeah I'm still an assistant after 15 years. But what I want to talk to you about is I was passed over for a promotion.	Is there an issue of discrimination here? By ignoring cultural issues, Allen will eventually lose this relationship. Allen has *no* idea what is going on. It might be cultural, it might not. This is definitely an important factor to consider, but he is not establishing an empathetic relationship. Regardless of who the client is, Allen would lose the relationship because of his inability to attend and listen!
7. *Allen*: Could you tell me a little bit more about some of the things you might have been doing wrong?	Still looking out the window, he returns to Azara with an open question, but he continues to ignore the main issue and topic jumps with an emphasis on the negative.
8. *Azara*: Well, I don't think I did anything wrong. I've gotten very good feedback from . . .	Azara starts defending herself here, but Allen interrupts. Changing topics and interruption are clear signs of the failure of empathic communication.
9. *Allen*: Well, they don't usually pass people up for promotions unless they're not performing up to standards.	The counselor supplies his interpretation, a subtractive negative evaluation without any data. He is not drawing out her story or really seeking to define her concerns.

Allen does not seem to listen to Azara. Furthermore, he confronts her inappropriately, and it is very unlikely that she will return for another session. Her European American male counselor just doesn't "get it."

But let's give Allen another chance. What differences do you note in this second session?

1. *Allen:* Hi, Azara. Nice to see you. Please come in and sit where it looks comfortable.	Allen stands up, smiles, and faces the client directly and shakes hands. First impressions are often key. Allen provided positive, facilitative nonverbals.
2. *Azara:* Thank you, nice to see you too.	She sits down and smiles in return, but appears tense.
3. *Allen:* Thanks for coming in.	The counselor likes to honor the client's willingness to come to the session. This often helps to equalize the power relationship that exists in counseling.
4. *Azara:* Thanks. I'm hopeful that you can help me.	Azara relaxes a little.
5. *Allen:* Azara, I looked at your file before you came in and I see that you'd like to talk a about a problem on the job. Is that right?	Looking at forms in the session is very likely to be subtractive. If you must look at files, share what you are looking at with your client. Prepare as appropriate to your setting.
6. *Azara:* Yes, that's right.	Her mouth is a little tense and she sits back.

Even in this brief period of time, Allen has conveyed to Azara a genuine warmth and readiness to hear her story. The session continues with Allen discussing with Azara information about the structure of the sessions. Allen also spends some time discussing cultural concerns with her, ensuring that she feels comfortable with Allen and that he understands some aspects of her culture. The session resumes with Azara describing the problem she is facing at work. Notice the difference in information gathered between the positive and negative examples.

COUNSELOR AND CLIENT CONVERSATION	PROCESS COMMENTS
13. *Allen:* So, there is a concern on the job. I'd like to hear about it.	Allen returns now to the job issue.
14. *Azara:* Okay, well, a few days ago I found out that I was passed over for a promotion at my job. And I've been with this company for 15 years. I was really pretty upset when I first found out, because the person who got the job, first of all is a male, he's only been with the company for 5 years. And you know I think I'm much better qualified than he is for this position. I've gotten really good evaluations from my supervisor, I have a great working relationship with my colleagues. . . . I was completely shocked to find out that I didn't get this promotion. 'Cause I was actually encouraged to apply for this job. And you know I didn't get it. This is . . . I'm just really, really angry.	Azara says a lot in this statement, and we as counselors sometimes have difficulty hearing it all. This is where the skills of paraphrasing and summarizing (Chapter 6) can be especially helpful. The task of these skills is to repeat what the client has said, but in a more succinct form.
15. *Allen:* 15 years compared to 5, and you are really, really angry. And what I've heard makes you angry is that you've had a good record, you were even asked to apply for this job, and finally this man who hasn't been there that long gets the job. Have I heard you correctly?	The counselor's summary of what has been said indicates that he has been engaging in verbal as well as nonverbal attending. "Have I heard you correctly?" is termed a *checkout* in the microskills framework. If you are accurate, the client will often say "yes" or even "exactly!" This represents a Level 2 interchangeable response.
16. *Azara:* Yes, you heard me. . . . Now the problem I'm having—I think it's discrimination, but now I have to decide what I'm going to do. If I'm going to file a complaint . . . will that upset my colleagues, will that get my boss, my supervisor, upset with me? I'm really worried about the consequences. I don't want to lose my job, but I think it's discrimination.	Having been heard, the client moves on.
17. *Allen:* Azara, it's a tough decision to make. If you file for discrimination, you set yourself up for a lot of hassles; if you don't file, then you're stuck with your anger and frustration. Could you tell me a bit more about that dilemma you are feeling?	Here Allen paraphrases the main ideas and reflects Azara's feelings as well. This is followed by an open question about the dilemma; Level 3 with some elements of additive empathy as he encourages her to tell more of her story.
18. *Azara:* Well, it's like I'm stuck, I don't know what to do. On the one hand, I think it's important to file the complaint because I think it will show the company that they really need to think about diversity in the workforce, and I'm kind of tired of being the only Latina working in this company for as long as I have, when you know they need to do something different. So I'm torn between that and being afraid of losing my job.	Azara summarizes key aspects of her conflict. The discrepancies or incongruity between herself and the company could be summarized this way: the responsibility to file a discrimination suit because it appears that the company is consistently being unfair versus the fear of losing her job if she takes this on.

COUNSELOR AND CLIENT CONVERSATION	PROCESS COMMENTS
19. *Allen:* So you're angry, afraid, frustrated. A lot of stuff comes together for you all at once.	Allen is sitting upright, forward trunk lean, supportive vocal tone while he reflects her emotions and her dilemma. Appropriate nonverbals are always central to maintaining an empathic relationship.
20. *Azara:* Yes, that's right. And I don't know what to do about that.	The client provides her own checkout and speaks of her puzzlement.
21. *Allen:* One thing I heard you saying that I'd like to understand a little bit more: You had good evaluations, . . . good relationships, success, a reasonable rate of promotion, at least raises along the line. I'd just like to hear at this point about examples of something specific that's gone right in the past—something you're proud of. Because when a person talks to me about difficulties, it kind of makes them feel a little embarrassed, and I'd like to understand some of your strengths. I've got a general understanding of your problem and we will come back to that. Could you tell me a little bit about some of your strengths too?	Now that the issues are clearer, Allen turns to the positive asset search. What are Azara's strengths that we can draw on as we work on these concerns? Note that Allen has avoided the use of the word "problem" as that is a self-defeatist negative view of client issues. You will find that most counseling training books use a problem-centered language. This is a clear example of upper level additive empathy.

The session continues from here, with Allen and Azara exploring her strengths.

Here we see a much stronger focus on Azara as a person with individual needs and feelings. A relationship has been established, and it is now possible to discuss multicultural issues as appropriate to the moment. Through attending and listening, we see her story and concerns more fully. A positive asset search for strengths has been initiated.

▶Training as Treatment: Social Skills, Psychoeducation, and Attending Behavior

Social skills training involves psychoeducational methods to teach clients an array of interpersonal skills and behaviors, such as listening, assertiveness, dating, drug refusal, mediation, and job interviewing procedures. Virtually all interpersonal actions can be taught through social skills training.

Training as treatment is a term that summarizes the method and goal of social skills training. Since its inception, the microtraining format of selecting specific skill dimensions for education has become basic to most psychoeducational social skills programs. As you extend the counseling and psychotherapy dimensions to skills training, think of the following steps:

1. Negotiate a skill area for learning with the client.
2. Discuss the specific and concrete behaviors involved in the skill, sometimes presenting them in written form as well.
3. Practice the skill with the client in a role-play in the individual or group counseling session.
4. Plan for generalization of the skill to daily life.

Shortly after the first work in identifying counseling and psychotherapy microskills, Allen Ivey was working with a first-year college student who suffered a mild depression and complained about lack of friends. Allen asked the student what he talked about with others in his dormitory. The student responded by continuing his list of complaints and worries.

With further probing, the student acknowledged that he spent most of his time with others talking about himself and his difficulties. It was easy to see that potential friends would avoid him. We all tend to move away from people who talk negatively and stay away from those who talk only about themselves and fail to listen to us.

On the spot, Allen talked to the student about attending behavior and its possible rewards. He suggested that the student might profit from actively listening to those around him rather than talking only about himself. Allen then presented the three V's + B and emphasized the importance of gaining trust and respect from others by listening. The student expressed interest in learning these skills, and a practice session was initiated there in the session. First, negative attending was practiced, and the student was able to see how his lack of listening might contribute to his isolation in the dormitory. Then positive attending was practiced, and the student discovered that he could listen.

Allen and the client discussed the specifics of selecting someone with whom to try these skills. When the student returned the following week, he wore a big smile and reported that he had found his first friend at the university. Moreover, he discovered an important side effect: "I feel less sad and depressed. First, I don't feel so alone and helpless. The second thing I noticed was that when I am attending to someone else, I am not thinking about myself and then I feel better." When you are attending to someone else, it becomes much more difficult to think negatively about yourself. Instruction in attending behavior is one of the foundations of social skills training.

Many types of clients can benefit from learning and practicing these skills. Allen Ivey found that teaching attending and other microskills to veterans at a VA hospital was sufficient to enable them to return to their families and communities (Ivey, 1973). Van der Molen (1984, 2006) used attending behavior and other microskills in a highly successful psychoeducational program in which he taught people who were shy (also known as the "avoidant personality") to become more socially outgoing. As just one other example, children diagnosed with attention deficit disorder (ADD) who receive skills training are less disruptive (Pfiffner & McBurnett, 1997). Effective psychoeducation can help children learn better in the classroom. Teaching couples to listen to each other can make a difference rather quickly! As you work through this book, think about how educating clients in the various microskills can be beneficial.

▶Attending in Challenging Situations

Don't be fooled by the apparent simplicity of the attending skills. Some beginning counselors and psychotherapists may think that these skills are obvious and come naturally. They may be anxious to move to the "hard stuff." The more we work with beginning and experienced counselors, the more we realize how difficult it is to master these skills. Yes, they can be learned, but it takes time, commitment, and intentional and deliberate practice.

You also may have wondered how attending behavior can be useful if you plan to work with challenging clients in schools, community mental health centers, or hospitals. The following examples from our personal experience illustrate the depth and breadth of attending.

Mary. Attending is natural to me and the basic listening sequence has always been central to my work with children, but even with all my experience with children, sometimes I am at a loss as to what to do next. After some analysis, I found that if I moved back to my foundation in attending skills and focused carefully on visuals, vocals, verbal following, and body language, I could regain contact with even the most troubled child. Similarly, in challenging situations with parents, I have at times found myself returning to a focus on attending behavior, later adding the basic listening sequence and other skills. Conscious attending has helped me many times in involved situations. Attending is not a simple set of skills.

I train older students to work as school peer mediators and another group to be peer tutors for younger children. I have found that using the exercise at the beginning of this chapter on poor attending and then contrasting it with good attending works well as an introductory exercise. I then teach attending skills and the basic listening sequence to my student groups.

Allen.   One of my most powerful experiences occurred when I first worked at the Veterans Administration with schizophrenic patients who talked in a stream of consciousness "word salad." I found that if I maintained good attending skills and focused on the exact words they were saying, they soon were able to talk in a more normal, linear fashion. I also found that teaching communication skills with video and video feedback to some troubled patients was effective. Sometimes attending was sufficient treatment by itself to move them out of the hospital. Depressed psychiatric inpatients, in particular, responded well to social skills training. However, I did find that highly distressed patients could learn only one of the four central dimensions at a time, as trying to grasp all four dimensions of attending was confusing for them. Thus, I would start with visual/eye contact and later move on to other attending skills.

Like Mary, when the going gets rough, I find that it helps me to return to basic attending skills and a very serious effort to follow what the client is saying as precisely as possible. In short, when in doubt, attend. It often works!

Practice, practice, practice—also known as "use it or lose it!"

▶Summary: The Samurai Effect, Neuroscience, and the Importance of Practice to Mastery

Japanese masters of the sword learn their skills through a complex set of highly detailed training exercises. The process of masterful sword work is broken down into specific components that are studied carefully, one at a time. Extensive and intensive practice is basic to a samurai. In this process of mastery, the naturally skilled person often suffers and finds handling the sword awkward at times. The skilled individual may even find performance worsening during the practice of single skills. Being aware of what one is doing can interfere with coordination and smoothness in the early stages.

Once the individual skills have been practiced and learned to perfection, the samurai retire to a mountaintop to meditate. They deliberately forget what they have learned. When they return, they find the distinct skills have been naturally integrated into their style or way of being. The samurai then seldom have to think about skills at all; they have become samurai masters.

What is samurai magic, you may ask? Intentional practice!

Once upon a time, it was believed that giftedness was inherited. Thus, many of us have been taught that Mozart and Beethoven had a magical gift. Baseball fans still believe that Ted Williams and Joe DiMaggio "had it in their genes." It is a bit different from that. The "magic" of a solely genetic predisposition to giftedness is now recognized as a scientific error, but that error is still promoted in the popular media. Natural talent is there, but it needs to be developed and nurtured with careful practice. Expertise across all fields depends on persistence, practice, and the search for excellence (Ericsson, Charness, Feltovich, & Hoffman, 2006).

The neuroscience of "giftedness" is discussed in detail by David Shenk in his book *The Genius in All of Us* (2010). He finds that whether one is a master musician or a superstar athlete, natural talent may be there, but the real test is many hours and often years of detailed practice. We now know that Mozart, with many natural talents, was bathed in music by his demanding father, who was one of the first to focus on a detailed study of techniques and skills. From the age of 3, Mozart received intensive instruction, and his greatness magnified over time. Ted Williams carried his bat to school and practiced until dark.

Intentional practice is the magic! This means that you need to recognize and enhance your natural talents, but greatness only happens with extensive practice. Practice is the breakfast of champions. Skipping practice means mediocre performance.

Here is what Shenk (2010, pp. 53–54) found that relates directly to you and your commitment to excellence in counseling and psychotherapy:

1. *Practice changes your body.* Both the brain and body change with practice.
2. *Skills are specific.* Each skill must be practiced completely before skills can be integrated into superior performance.
3. *The brain drives the brawn.* Changes in the brain are evident in scans. Areas of the brain relating to finger exercises or arm movements show brain growth in those areas. Expect the same in your brain as you truly master communication skills.
4. *Practice style is crucial.* One can understand attending behavior intellectually, but actually practicing the specific skills of attending is what makes the difference. One pass through is seldom enough.
5. *Short-term intensity cannot replace long-term commitment.* If Ted Williams did not continue to practice, his skills would have gradually been lost. You will want to take all the counseling skills you learn and use them regularly.
6. *Practice provides a continuous feedback loop*, which leads to even more improvement. Feedback from colleagues on your counseling style and skills is especially beneficial.

We are asking you to focus on your natural gifts in communication and then add to them through practice and sharpening of new skills. You may find a temporary and sometimes frustrating decrease in competence, just as can happen with samurai, athletes, and musicians. Some of you may experience some discomfort in practicing the skill of attending. Others may find attending so "easy" that you fail to become fully competent in this most basic of listening skills (many experienced professionals still can't listen effectively to their clients).

Learn the skills of this book, but allow yourself time for integrating these ideas into your own natural authentic being. It does not take magic to make a superstar, but it does require systematic and intentional practice to achieve full competence in counseling and psychotherapy. Make your own magic!

▼▼▼▼	Key Points
Central goals of listening	When we use attending behavior, we have empathic goals: to reduce counselor talk time while providing clients with an opportunity to examine issues and tell their stories. Selective attention may be used to facilitate more useful client conversation. But always use attending with individual and cultural sensitivity. Observation skills will enable you to stay more closely in tune with your clients.
Four aspects of attending	Attending behavior consists of four simple but critical dimensions (3 V's + B). All of these need to be modified to meet individual and cultural differences. 1. Visual/eye contact. 2. Vocal qualities. Your vocal tone and speech rate indicate much of how you feel about another person. 3. Verbal tracking. Don't change the subject. Keep to the topic initiated by the client. If you selectively attend to an aspect of the story or a different topic, realize the purpose of your change. 4. Body language: attentive and genuine. Face clients naturally, lean slightly forward, have an expressive face, and use facilitative, encouraging gestures.

Focusing your attention	Attending is easiest if you focus your attention *on the client rather than on yourself*. Again, your ability to be empathic and observe what is occurring in the client is central. Note what the client is talking about, ask questions, and make comments that relate to your client's topics. For example:
Client:	I'm so confused. I can't decide between a major in chemistry, psychology, or language.
Counselor:	(nonattending) Tell me about your hobbies. What do you like to do? *or* What are your grades?
Counselor:	Counselor: (attending) Tell me more. *or* You feel confused? *or* Could you tell me a little about how each subject interests you? *or* Opportunities in chemistry are promising now. Could you explore that field a bit more? *or* How would you like to go about making your decision?
Individual and cultural adaptation	Attending and empathy are vital in all human interactions, whether in counseling, in medical interviews, in business decision meetings, or with friends and family. Different individuals and cultural groups may have differing patterns of listening. For example, some people may find the direct gaze rude and intrusive, particularly if they are dealing with difficult material.
What is empathy?	Empathy is the ability to enter the world of the client and to communicate that we understand his or her world as the client sees and experiences it. Attending and listening behaviors are the ways in which you communicate empathy and understanding to the client.
What should you do when you don't know what to do next?	A simple but often helpful rule for counseling is to use attending skills when you become lost or confused about what to do. Simply ask the client to comment further on something just said or mentioned earlier in the session. If you are a basically empathic person, this likely will be enough to fill in those inevitable "holes" when you are not quite sure what is happening.

▶Competency Practice Exercises and Portfolio of Competence

 Go to CengageBrain.com to access Counseling CourseMate, where you will find an interactive ebook, quizzes, videos, interactive counseling and psychotherapy exercises, the Portfolio of Competence, case studies, a practice test, and more.

Intentional counseling and psychotherapy are achieved through practice and experience. Reading and understanding are at best a beginning. Some find the ideas here relatively easy and think that they can perform the skills. But what makes one competent in basic skills is practice, practice, practice.

The competency practice exercises on the following pages and on CourseMate are designed to help you develop competence in three ways:

1. *Individual practice.* Two take-home individual practice exercises offer basic practice of attending and observation. In addition, CourseMate provides a series of case studies and video examples.
2. *Group and/or individual practice with video or audio.* Virtually all of us now have access to cameras or cell phones with video capabilities. If you can record your practice sessions and review what happens, this provides the most useful feedback of all. There is no better way to improve than looking at yourself.
3. *Self-assessment and Portfolio of Competence.* Assessing your abilities in each area of this chapter will enable you to determine if you can take action using these skills, generalizing them to your daily life and to counseling and psychotherapy practice.

Individual Practice

Exercise 1. Deliberate Attending and Nonattending   During a conversation with an acquaintance, deliberately attend, listening more carefully than you usually do. Maintain appropriate eye contact with an open attentive posture. Use a supportive vocal tone, and focus carefully on what the other person is saying. Observe what happens and how conversations can change if you really seek to listen.

You may wish to contrast deliberate attending with nonattending. What happens when your eye contact wanders, your vocal tone shows disinterest, your body becomes more rigid, or you constantly change the topic?

What did you learn from this experience?

Exercise 2. Observation of Verbal and Nonverbal Patterns   Observe 10 minutes of a counseling session, a television interview, or any two people talking. Make a video recording so that repeated viewing is possible.

Visual/eye contact patterns.   Do people maintain eye contact more while talking or while listening? Does the "client" break eye contact more often while discussing certain subjects than others? Can you observe changes in pupil dilation as an expression of interest?

Vocal qualities.   Note speech rate and changes in intonation or volume. Give special attention to speech "hitches" or hesitations.

Verbal tracking.   Does the counselor stay on the topic, or does he or she topic jump? What are the patterns of selective attention?

Attentive body language.   Note gestures, shifts of posture, leaning, patterns of breathing, and use of space. Give special attention to facial expressions such as changes in skin color, flushing, and lip movements. Note appropriate and inappropriate smiling, furrowing of the brow, and so on.

Movement harmonics.   Note places where movement synchrony and echoing occurred. Did you observe examples of movement dissynchrony?

If possible, use your video recording to view the session several times.

Group and/or Individual Practice With Video or Audio

The following exercise can be adapted for individuals who do not have observers. Key to the mastery of skills is feedback. Thus, it is important to get and provide feedback about your practice. For group work, technology may not always be available; having others available to watch the session and provide feedback is a good substitute.

Many of you now have small cameras or smartphones that have both video and sound capability. We urge you to take the time to observe closely what you are doing. Practice with clear video feedback is the original and most effective route toward mastery and competence.

Exercise 3. Group Practice Using Attending Skills   The following instructions are designed for groups of four but may be adapted for use with pairs, trios, and groups up to five or six. Ideally, each group will have access to a video or audio recorder. However, the guidelines provided in Box 3.3 and the feedback form in Box 3.4 can provide enough structure for a successful practice session without the benefit of recording equipment.

Step 1: Divide into practice groups.   Get acquainted with each other informally before you go further.

BOX 3.3 Guidelines for Effective Feedback

To see ourselves as others see us.

To hear how others hear us.

And to be touched as we touch others . . .

These are the goals of effective feedback.

Feedback, one of the skill units of the basic attending and influencing skills developed in this book, is discussed in more detail in Chapter 12. But, even before we study this skill, here are some basic guidelines for feedback in practice sessions.

▲ *The person receiving the feedback is in charge.* Let the counselor in the practice sessions determine how much or little feedback is wanted.

▲ *Feedback includes strengths*, particularly in the early phases of the program. If criticism or negative feedback is requested by the counselor, add positive dimensions as well. People grow from strength, not from weakness.

▲ *Feedback is most helpful when it is concrete and specific.* Not "Your attending skills were good" but

"You maintained eye contact throughout except for breaking it once when the client seemed uncomfortable." Make your feedback factual, specific, and observable.

▲ *Corrective feedback should be relatively nonjudgmental.* Feedback often turns into evaluation. Stick to the facts and specifics, though the word *relatively* recognizes that judgment inevitably will appear in many different types of feedback. Avoid the words *good* and *bad* and their variations.

▲ *Feedback should be lean and precise.* It does little good to suggest that a person change 15 things. Select one to three things the counselor actually might be able to change in a short time. You'll have opportunities to make other suggestions later.

▲ *Check how your feedback was received.* The counselor's response indicates whether you were heard and how useful your feedback was. "How do you react to that?" "Does that sound close?" "What does that feedback mean to you?"

Step 2: Select a group leader. The leader's task is to ensure that the group follows the specific steps of the practice session. It often proves helpful if the least experienced group member serves as leader first. Group members then tend to be supportive rather than competitive.

BOX 3.4 Feedback Form: Attending Behavior

You can download this form from Counseling CourseMate at CengageBrain.com.

_____
DATE

_____   _____
NAME OF COUNSELOR   NAME OF PERSON COMPLETING FORM

Instructions: Provide written feedback that is specific and observable, nonjudgmental, and supportive.

1. *Visual/eye contact.* Facilitative? Staring? Avoiding? Sensitive to the individual client? At what points, if any, did the counselor break contact? Facilitatively? Disruptively?

(continued)

BOX 3.4 (continued)

2. *Vocal qualities*. Vocal tone? Speech rate? Volume? Accent? Points at which these changed in response to client actions? Number of major changes or speech hesitations?

3. *Verbal tracking and selective attention*. Was the client able to tell the story? Stay on topic? Number of major topic jumps? Did shifts seem to indicate counselor interest patterns? Did the counselor demonstrate selective attention in pursuing one issue rather than another? Did the client have the majority of the talk time?

4. *Attentive body language*. Leaning? Gestures? Facial expression? At what points, if any, did the counselor shift position or show a marked change in body language? Number of facilitative body language movements? Was the session authentic?

5. *Specific positive aspects of the session.*

6. *Empathic communication*. Rate the quality of counselor responses in the session as subtractive, interchangeable, or additive.

7. *Discussion question*: What areas of diversity do the counselor and client represent? How does this affect the session? Keep in mind that all clients have a RESPECTFUL cultural background.

Step 3: Assign roles for the first practice session.

▲ *Client.* The first role-play client will be cooperative and present a story, talk freely, and not give the counselor a difficult time.

▲ *Counselor.* The counselor will demonstrate a natural style of attending behavior, practicing the basic skills.

▲ *Observer 1.* The first observer will fill out the feedback form (Box 3.4) detailing the counselor's attending behavior. Observation of these practice sessions could also be called "microsupervision" in that you are helping the counselor understand his or her behavior in the brief session. Later, when you are working as a professional helper, continue to share your work with colleagues through verbal reports or audio or video recordings. Supervision is a vital part of effective counseling and psychotherapy, regardless of how long one has practiced.

▲ *Observer 2.* The second observer will time the session, start and stop any equipment, and fill out a second observation sheet as time permits.

Step 4: Plan and select topic. The counselor states her or his goals clearly, and the members of the group take time to plan their parts in the role-play. This seems obvious, but in the first few practice sessions, people can get confused about the process. The more concrete the plan, the greater is the likelihood of success.

The suggested topic for the attending practice session is "Why I want to be a counselor or therapist." The client talks about her or his desire to join the helping profession, or at least to consider it as a future career.

Other possible topics for the session include the following:

▲ A job that you liked and a job that you didn't (or don't) enjoy
▲ A positive experience that led you to learn something new about yourself

The topics and role-plays are most effective if you talk about something meaningful to you. You will also find it helpful if everyone in the group works on the same topic as roles are rotated. In that way, you can compare styles and learn from one another more easily.

While the counselor and the client plan, the two observers preview the feedback sheets.

Step 5: Record a 3-minute practice session using attending skills. The counselor practices the skills of attending, the client talks about the current work setting or other selected topic, and the two observers fill out the feedback sheets. Try not to go beyond 3 minutes, but find a comfortable place before stopping.

Step 6: Review the practice session and provide feedback to the counselor for 12 minutes. Feedback has been called the "breakfast of champions," so give special attention here. Note the suggestions for feedback in Box 3.3. As a first step, the role-play client gives her or his impressions of the session and completes the Client Feedback Form from Chapter 1 (Box 1.4). This is followed by counselor self-assessment and comments by the two observers.

Finally, as you review the audio or video recording of the session, start and stop it periodically. Replay key interactions. Only in this way can you fully profit from the recording media. Just sitting and watching television is not enough; use media actively.

Step 7: Rotate roles.   Everyone should have a chance to serve as counselor, client, and observer. Divide your time equally!

Some general reminders.   It is not necessary to compress a complete session into 3 minutes. Behave as if you expected the session to last a longer time, and the timer can break in after 3 minutes. The purpose of the role-play sessions is to observe skills in action. Thus, you should attempt to practice skills, not solve problems. Clients have often taken years to develop their interests and concerns, so do not expect to solve one of these issues or obtain the full story in a 3-minute role-play session. Written feedback, if carefully presented, is an invaluable part of a program of counseling skill development.

Exercise 4. Reflections on Your Natural Style of Listening and Attending   To build self-awareness, reflect on yourself as a listener and your natural empathic style as you move to action and generalization. Attending is fundamental to empathic understanding, one of the central dimensions of emotional intelligence.

Form a group of two or three students or practitioners and discuss the following questions (exercise and questions can be adapted for individual use). Some of these are thought questions to which you can respond in journal form. Others are action oriented.

▲   What led you to a course in counseling skills? Are you a "people person"? Have you had friends come to you to share their concerns and problems? Do you like to listen to others? What are your motivations?

▲   How comfortable are you with ideas of diversity and working with people different from you? Can you recognize yourself as a multicultural person with many dimensions of diversity?

▲   View yourself on video or listen to an audio recording of a session. One of the best ways to examine your natural style of listening is to replay the video or audio recording you made as you started this book. Again, if you have not made that first video, now is the time! What do you observe in an informal viewing of yourself?

Portfolio of Competence

Use the following checklist to evaluate your present level of mastery. Check those dimensions that you currently feel able to do. Those that remain unchecked can serve as future goals. Do not expect to attain intentional competence on every dimension as you work through this book. You will find, however, that you will improve your competencies with repetition and practice.

Level 1: Identification and classification.   You will need this minimal level of mastery for those coming examinations!

❑   Identify and count the three V's + B as your observe the session.
❑   Define subtractive, interchangeable, and addictive empathy.
❑   Observe movement harmonics and movement dissynchrony.
❑   Outline key elements of observation of nonverbal communication.

Level 2: Basic competence.   These are the fundamentals required before moving on to the next skill area in this book.

- ❑ Demonstrate culturally appropriate visuals/eye contact, vocal qualities, verbal following, and body language in a role-played session.
- ❑ Increase client talk time while reducing your own.
- ❑ Stay on a client's topic without introducing any new topics of your own.
- ❑ Hear a client accurately so that you demonstrate interchangeable empathy.
- ❑ Mirror nonverbal patterns of the client. The counselor mirrors body position, eye contact patterns, facial expression, and vocal qualities.

Level 3: Intentional competence.   In the early stages of this book, strive for basic competence and work toward intentional competence later. Experience with the microskills model is cumulative, and you will find yourself mastering intentional competencies with greater ease as you gain more practice. The following dimensions reflect intentional competence in attending and empathy.

- ❑ Understand and manage your own pattern of selective attention.
- ❑ Change your attending style to meet client individual and cultural differences.
- ❑ Note topics that clients particularly attend to and topics that they may be avoiding.
- ❑ Use attending skills with more challenging clients, while maintaining an empathic style.
- ❑ Through empathic attention and inattention, help clients move from negative, self-defeating conversation to more positive and useful topics. Conversely, help clients who are avoiding issues to talk about them in more depth.

▶Determining Your Own Style and Theory: Critical Self-Reflection on Attending and Empathy and Suggestions for Journal Review

This chapter has focused on the importance of empathic listening as the foundation of effective counseling practice. When in doubt as to what to do, listen, listen, listen! Individual and cultural differences are central—visual, vocal, verbal, and body language styles vary. Avoid stereotyping any group.

What single idea stands out for you among all those presented in this chapter, in class, or through informal learning? What stands out for you is likely to be a guide toward your next steps. How might you use ideas in this chapter to begin the process of establishing your own style and theory? Please turn to your journal and write your thoughts. What are your thoughts about using attending behavior in psychoeducational practice?

Observation Skills

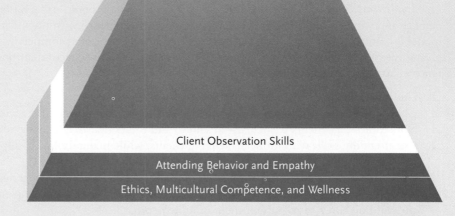

Client Observation Skills

Attending Behavior and Empathy

Ethics, Multicultural Competence, and Wellness

The scientific method may be summed up in the single word "observation."

—E. B. Titchener

Mission of "Observation Skills"

Sharpening your ability to observe what occurs between you and your clients verbally and nonverbally. Observation skills will guide you to key issues in the here and now of the counseling or therapy session. In addition, they will help you respond appropriately to both individual and multicultural differences.

Chapter Goals and Competency Objectives Awareness, knowledge, skills, and actions in observation will enable you to observe, understand, and make use of

▲ Nonverbal behavior. How do you and your clients behave nonverbally?

▲ Verbal behavior. How do you and your clients use language?

▲ Discrepancies and conflict. Much of counseling and psychotherapy is about working through conflict and coping with the inevitable stressful incongruities we all face.

▲ Varying individual and cultural ways of verbal and nonverbal expression. How can you flex intentionally and avoid stereotyping in your observation?

 Go to CengageBrain.com to access Counseling CourseMate, where you will find an interactive ebook, quizzes, videos, interactive counseling and psychotherapy exercises, the Portfolio of Competence, case studies, a practice test, and more.

▶Are You a Good Observer?

You can observe a lot by watching.

—Yogi Berra

Observation is the act of watching carefully and intentionally with the purpose of understanding behavior. In spite of what some professionals believe, mastering this skill is not easy.

What can you learn by observing, and why is it important? Through observation you get to know the client and what is conveyed by his or her verbal and nonverbal behavior. Client intentions, needs, meanings, and underlying emotions are often conveyed through nonverbals. Some authorities say that 85% or more of communication is nonverbal. How something is said can sometimes overrule the actual words used by you or your client. A keen observer discovers the many ways clients express their needs, emotions, and motivations.

Observation gives you key information to foster the relationship and facilitates empathic understanding of the client. Observation gives you information about when and what kind of intervention is needed and how the client responds to it. It also indicates what in the environment needs to change to facilitate client development and progress.

We suggest that you take time out and try the following exercise, as it will help you understand the challenges of effective observation in real life. Please stop for a moment and—before reading any further—view at least one of the following short videos:

www.youtube.com/watch?v=IGQmdoK_ZfY (or go to www.youtube.com and search for "The Monkey Business Illusion")

www.youtube.com/watch?v=kd2dQ26DdFQ (or go to www.youtube.com and search for "The Mentalist's Football Awareness Test")

These videos are both fun and interesting; more important, they illustrate that we can miss quite a bit in the process of observing.

These videos really show us what Yogi Berra is talking about. They are often used in neuroscience demonstrations because of their close relationship with attentional issues (Shabris & Simon, 2009). If we are carefully attending to a single event or client topic, we can easily fail to notice something that is even more important and miss critical nonverbal issues. Our attentional system, even the most vital and effective, can sometimes fool us. This can be an element in false memories, as well. The client might remember a negative event, but fail to recall that someone saved him or her at the last moment.

Observation can enable you to learn as much about yourself and your counseling skills as you do about your clients. By turning observation inward, you can tune into your own reactions and examine what lies within. This process of self-awareness may lead to growth and change. Whether the focus is on the client or on you, observation provides you with a compass to guide you through the session.

▶Keeping Watch on the Counseling Session

If you use observation skills as defined here, you can anticipate how the client is likely to respond.

OBSERVATION SKILLS	ANTICIPATED RESULT
Observe your own and the client's verbal and nonverbal behavior. Anticipate individual and multicultural differences in nonverbal and verbal behavior. Carefully and selectively feed back some here-and-now observations to the client as topics for exploration.	Observations provide specific data validating or invalidating what is happening in the session. Also, they provide guidance for the use of various microskills and strategies. The smoothly flowing session will often demonstrate movement symmetry or complementarity. Movement dissynchrony provides a clear clue that you are not "in tune" with the client.

What should you observe about client behavior in the here and now of the session? From your own life experience, you are already aware of many things that are important for a counselor or psychotherapist to notice. Brainstorm from what you already know and make a list.

But there are two people in the relationship. What about you? How are you affecting the client verbally and nonverbally? Looking at your way of being can be as critical as observing the client. Start by taking a brief inventory of your own nonverbal style. You might begin by thinking back to your natural style of attending (See Chapter 1, Your Natural Helping Style: An Important Audio or Video Exercise), but expand those self-observations. Better yet, make a video of you talking with someone on a topic of interest to you both, and then a topic that is less interesting or even boring. What is your interpersonal style, and how might it affect your relationship with others and in the session?

▶Observe the Attending Patterns of Clients

An ideal place to start practicing your observation skills is by noting your own and your client's style of attending.

Observe Nonverbal Behavior

> Facial expression, eye gaze, tone of voice, bodily motion, and the timing and intensity of response are all fundamental to emotional messages. . . . In every culture, we can identify characteristics of "basic" emotions—for example, as sadness, anger, or fear. In sadness, the face will show turned-down lips and squinted eyes, together with slower bodily motions. Anger will involve dilated pupils, widened orbital area, raised eyebrows, furrowed brow, and pursed lips. Fear combines raised eyebrows, flattened brow, and open mouth [plus often body movements away from the feared object]. (Siegel, 2012, p. 146, 153)

Clients may break eye contact, shift bodily movement, and change vocal qualities as their comfort level changes when they talk about various topics. You may observe clients crossing their arms or legs when they want to close off a topic, using rapid alterations of eye contact during periods of confusion, or exhibiting increased stammering or speech hesitations when topics are difficult. Jiggling legs, complete body shifts, or suddenly closing one's arms most often indicates discomfort. Hand and arm gestures may give you an indication of how the client is organizing things.

Random, discrepant gestures may indicate confusion. A person seeking to control or organize things may move hands and arms in straight lines and point fingers authoritatively. Smooth, flowing gestures, particularly those in harmony with the gestures of others, such as family members, friends, or the counselor, may suggest openness. If you watch yourself carefully on tape, you, as counselor, will exhibit many of these same behaviors.

However, be careful of identifying specific meanings with both nonverbal and verbal behavior because of individual and/or multicultural differences. For example, Siegel's statements about anger might be compared to what occurs in sexual attraction, that instant awareness of someone who really draws your attention. Here we often have involvement indicating desire—pupil dilation, widened eyes, perhaps even a glazed-over look or slight blinking, indicating excitement. Pupil dilation usually means involvement and interest, whereas contraction may mean disdain or disgust, and even be part of anger at times.

Nonverbal behavior may be an indication not only of feelings toward another person or situation, but also of immediate reactions to you as a person, including transference behavior indicating how the client is seeing you.

Observe Verbal Behavior

Language is basic to counseling and psychotherapy, and ways to consider verbal behavior range from detailed linguistic examination to the differing language systems of varying counseling and therapy theories. This chapter will consider four dimensions useful for direct verbal observation in the session: patterns of selective attention, client key words, abstract and concrete conversation, and "I" statements and "other" statements.

Observe Conflict, Incongruities, and Discrepancies

Whether you are helping clients work through problems, deal with issues, encounter challenges, or manage concerns, you will be facilitating the resolution of discrepancies, incongruity, and conflict in their lives. Out of your awareness of verbal and nonverbal behavior will come an increased ability to notice conflicts of many types. Stress comes from internal and external conflict. Examples of internal conflict and discrepancies include indecision, guilt, depression, and anxiety. Problems with interpersonal relations, cultural oppression, and work are three examples of external conflict. Of course, many of your clients will be dealing with both types of conflict.

Careful observation of multiple types of discrepancies give you a deeper understanding of where clients "really are" in terms of their issues. Conflict is literally the "stuff" of counseling and is often where you can help clients the most.

Individual and Multicultural Variations in Verbal and Nonverbal Behavior

As you engage in observation, recall that each culture has a different style of nonverbal communication. For example, in Turkey, "yes" is indicated by moving the head forward and "no" by raising the eyebrows. For a brief but helpful summary of such cultural differences, visit "The Importance of Nonverbal Communication" at www.expats-moving-and -relocation-guide.com/nonverbal-communication.html.

As another example, researchers have compared the touching behavior of different cultural groups. One study, which counted the average number of times friends touched each other in an hour while talking in a coffee shop, found that English friends did not touch each other at all, French friends touched 110 times, and Puerto Rican friends touched 180 times (cited in Asbell & Wynn, 1991). Croce (2003) cites a parallel study in which students at the University of Florida, Gainesville, touched twice, compared with 120 instances of touching in San Juan, Puerto Rico.

Smiling is a sign of warmth in most cultures, but in some situations in Japan, smiling may indicate discomfort. And, in all cultures, there can be a difference between a genuine smile and one that seeks to cover up real feelings. Eye contact may be inappropriate for the traditional Navajo, but highly appropriate and expected for a Navajo official who interacts commonly with White Arizona government staff. Similar issues with Native people can arise in other countries such as Australia, Canada, and New Zealand. In Nigeria, direct eye contact can be seen as intrusive, so looking at the shoulder is more appropriate.

Be careful not to assign your own ideas about what is "standard" and appropriate nonverbal communication. All helping professionals need to continue a lifetime study of nonverbal communication patterns and their variations. In counseling sessions, you will find that observing changes in nonverbal behavior may be as important as, or more important than, finding specific meanings in verbal communication style. Box 4.1 presents an interesting example of this. Edward Hall's *The Silent Language* (1959) remains a classic. Paul Ekman's work (2007) is the current standard reference for nonverbal communication.

| BOX 4.1 | National and International Perspectives on Counseling Skills |

Can I Trust What I See?

Weijun Zhang

James Harris, an African American professor of education, and I were invited by a national Native American youth leadership organization to give talks in Oklahoma. I attended Dr. Harris's first lecture with about 60 Native American children. Dynamic and humorous, Dr. Harris touched the heartstrings of everyone. But much to my surprise, when the lecture was over, he was very upset. "A complete failure," he said to me with a long face; "they are not interested." "No," I replied, "it was a great success. Don't you see how people loved your lecture?" "No, not at all." After some deliberation, I came to see why he could have such an erroneous impression.

"There were not many facial expressions," Professor Harris said. I said, "You may be right, using African American standards. However, that is not a sign that your audience was not interested. Native American people, in a way, are programmed to restrain their feelings, whether positive or negative, in public; as a result, their facial expressions would be hard to detect. Native American people have always valued restraint of emotion, considering this a sign of maturity and wisdom, as I know. Actually, in terms of emotional expressiveness, African American culture and Native American culture may represent two extremes on a continuum."

"They did not ask a single question, though I repeatedly asked them to," he said. I replied, "Well, they didn't because they respect you." "Come on, you are kidding me." I told him, "Many Native Americans are not accustomed to asking questions in public, probably for the following reasons. (a) If you ask an intelligent question, you will draw attention away from the teacher and onto yourself. That is not an act of modesty and may be seen as showing off. (b)

If your question is silly, you will be seen as a laughingstock and lose face. (c) Whether your question is good or bad, one thing is certain: You will disturb the instructor's teaching plan, or you may suggest the teacher is unclear. That goes against the Native American tradition of being respectful to the senior. So you can never expect a Native American audience to be as active as African Americans in asking questions.

"In today's situation, some kids probably did want to ask you questions, for you repeatedly encouraged them to. But, unfortunately, they still couldn't do it." He asked why. I said, "You waited only a few seconds for questions before you went on lecturing, which is far from enough. With Native Americans, you have to adopt a longer time frame. European Americans and African Americans may ask questions as soon as you invite them to; American Indians may wait for about 20 to 30 seconds to start to do so. That period of silence is a necessity for them. You might say that Native Americans are true believers in the saying 'Speech is silver, silence is golden.'" He asked, "Well, why didn't you tell me that on the spot, then?" "If it is respectful for those Native American kids not to ask you questions, Dr. Harris," I said, "how could you expect this humble Chinese to be so disrespectful as to come to the stage to correct you?"

The director of the Native American organization, who overheard my conversation with James Harris, approached me with the question "How do you possibly know all this about us Native Americans?" "Well, I don't think it is news for you that Native Americans migrated from Asia some thousands of years ago. You don't mind that your Asian cousins still share your ethnic traits, do you?" The three of us all laughed.

▶ Example Counseling Session: Is the Issue Difficulty in Studying or Racial Harassment?

Kyle Yellowhorse is a second semester junior business major at a large university. He was raised in a relatively traditional Lakota family on the Rosebud Reservation in South Dakota. Native American Indians are unlikely to come to counseling unless they are referred by others or the counselor has established herself or himself previously as a person who can be trusted.*

Kyle did well in his first two years at the university, but during the fall term, his B average dropped to barely passing. His professor of marketing has referred him to the campus counseling center. The counselor in this case is European American.

In this case, you will see a slow start, due at least partially to multicultural differences. See how the counselor uses observation skills to open Kyle to discussing his issues.

COUNSELOR AND CLIENT CONVERSATION	PROCESS COMMENTS
1. *Derek:* Kyle, come on in; I'm glad to see you.	Derek walks to the door; he smiles, shakes hands, and has direct eye contact.
2. *Kyle:* Thanks. (pause)	Kyle gives the counselor eye contact for only a brief moment. He sits down quietly.
3. *Derek:* You're from the Rosebud Reservation, I see.	Derek knows that contextual and family issues are often important to Native American Indian clients. Rather than focusing on the individual and seeking "I" statements, he realizes that more time may be needed to develop a relationship. Derek's office decorations include artwork from the Native American Indian, African American, and Mexican traditions, as well as symbols of his own Irish American heritage.
4. *Kyle:* Yeah. (pause)	Although his response is minimal, Kyle notes that Derek relaxes slightly in the chair.
5. *Derek:* There've been some hard times here on campus lately. (pause, but still no active response from Kyle) I'm wondering what I might do to help. But first, I know that coming to this office is not always easy. I know Professor Harris asked you to come in because of your grades dropping this last term.	During the fall term, the Native American Indian Association on campus had organized several protests against the school mascot—"the fighting Sioux." As a result, the campus has been in turmoil, with recurring instances of racial insults and several fights. Noting Kyle's lack of eye contact, Derek has reduced the amount of direct gaze and he also looks down. Among traditional people, a lack of eye contact generally indicates respect. At the same time, many clients who are depressed use little eye contact. (Interchangeable empathy)

continued on next page

---

*Many schools, elementary through university, have a small population of Native American Indians. For example, the Chicago public schools have slightly more than 1,000 Native Americans scattered throughout a large system. Most often, the Native American Indian population is invisible, and you may never know this cultural group exists unless you indicate, through your behavior and actions in the community, that you are a person who can be trusted and who wishes to know the community.

COUNSELOR AND CLIENT CONVERSATION	PROCESS COMMENTS
6. *Kyle:* Yeah, my grades aren't so good. It's hard to study.	Kyle continues to look down and talks slowly and carefully.
7. *Derek:* I feel honored that you are willing to come in and talk, given all that has happened here. Kyle, I've been upset with all the incidents on campus. I can imagine that they have affected you. But first, how do you feel about being here talking with me, a White counselor?	Derek self-discloses his feelings about campus events. He makes an educated guess as to why Kyle's grades have dropped. As he talks about the campus problems, Kyle looks up directly at him for the first time and Derek notes some fire of anger in his eyes. Kyle nods slightly when Derek says, "I feel honored." (Potentially additive)
8. *Kyle:* It's been hard. I simply can't study. (pause) Professor Harris asked me to come and see you. I wouldn't have come, but I heard from some friends that you were OK. I guess I'm willing to talk a bit and see what happens.	People who are culturally different from you may not come to your office setting easily. This is where your ability to get into the community can be so helpful in developing relationships. In this case, college counseling center staff have been active in the campus community, leading discussions and workshops seeking to promote racial understanding. As Kyle talks, Derek notices increased relaxation and senses that the beginning of trust and rapport has occurred. With some clients, reaching this point may take a full session. Kyle is bicultural in that he has had wide experience in European American culture as well as in his more traditional Lakota family. (The self-disclosure was additive.)
9. *Derek:* Thanks, maybe we could get started. What's happening?	We see that Kyle's words and nonverbal actions have changed in the short time that he has been in the session. Derek, for the first time, asks an open question. Questions, if used too early in the session, might have led Kyle to be guarded and say very little.
10. *Kyle:* I'm vice president of the Native American Indian Association—see—and that's been taking a lot of time. Sometimes there are more important things than studying.	Kyle starts slowly and as he gets to the words "more important things," the fire starts to rise in his eyes.
11. *Derek:* More important things?	The restatement encourages the client to elaborate on the critical issue that Derek has observed through Kyle's eyes.
12. *Kyle:* Yeah, like last night, we had a march and demonstration against the Indian mascot. It's so disrespectful and demeaning to have this little Indian cartoon with the big teeth. What does that have to do with education? They talk about "liberal education." I think it's far from liberal; it's constricting. But worse, when we got back to the dormitories, the car that belongs to one of our students had all the windows broken out. And inside was a brick with the words "You're next" painted on it.	Kyle is now sitting up and talking more rapidly. Anger and frustration show in his body—his fists are clenched and his face shows strain and tension. Women, gays, or other minorities who experience disrespect or harassment may feel the same way and demonstrate similar verbal and nonverbal behaviors. The car bashing incident clearly illustrates the harassing students' lack of "other" esteem and respect.

COUNSELOR AND CLIENT CONVERSATION	PROCESS COMMENTS
13. *Derek:* That's news to me. The situation on campus is getting worse. Your leadership of the association is really important and now you face even more challenges.	Derek shares his knowledge of the situation and paraphrases what Kyle has just said. He sits forward in his chair and leans toward Kyle. However, if Kyle were less fully acculturated in European American culture, the whole tone of this conversation would be quite different. It probably would have taken longer to establish a relationship, Kyle would have spoken more carefully, and his anger and frustration would likely not have been visible. The counselor, in turn, would be likely to spend more time on relationship development, use more personal self-disclosure, use less direct eye contact, and—especially—be comfortable with longer periods of silence. (Interchangeable)
14. *Kyle:* Yeah. (pause) But, we're going to manage it. We won't give up. (pause) But—I've been so involved in this campus work that my grades are suffering. I can't help the association if I flunk out.	Kyle feels heard and supported by the counselor. Being heard allows him to turn to the reasons he came to the counselor's office. He relaxes a bit more and looks to Derek, as if asking for help.
15. *Derek:* Kyle, what I've heard so far is that you've gotten caught up in the many difficulties on campus. As association vice president, it's taken a lot of time—and you are very angry about what's happened. I also understand that you intend to "hang in" and that you believe you can manage it. But now you'd like to talk about managing your academics as well. Have I heard you correctly?	Derek has used his observation skills so that he knows now that Kyle's major objective is to work on improving his grades. He wisely kept questioning to a minimum and used some personal sharing and listening skills to start the session. (Interchangeable)
16. *Kyle:* Right! I don't like what's going on, but I also know I have a responsibility to my people back home on the Rosebud Reservation, the Lakota people, and to myself to succeed here. I could talk forever on what's going on here on campus, but first, I've got to get my grades straightened out.	Here we see the importance of self-esteem and self-focus if Kyle is to succeed. But his respect for others is a central aspect of who he is as a person. He does not see himself as just an individual. He also sees himself as an extension of his group. Kyle has considerable energy and perhaps a need to discuss campus issues, but his vocal tone and body language make it clear that the first topic of importance for him today is staying at the university.
17. *Derek:* OK . . . If you'd like, later we might come back to what's going on. You could tell me a bit about what's happening here with the mascot and all the campus troubles. But now the issue is what's going on with your studies. Could you share what's happening?	Derek observes that Kyle's mission in the session right now is to work on staying in school. He asks an open question to change the focus of the session to academic issues, but he keeps open the possibility of discussing campus issues later in the session. (Potentially additive)

The session has now started. Kyle has obviously been observing and deciding if he can trust Derek. Fortunately, Derek has a good reputation on campus. He has regularly attended Native American powwows and other multicultural events. You will find if you work in a school or university setting that clients will be very aware of the effectiveness and trustworthiness of counseling staff.

Although focused on Native American Indian issues, this session has many parallels with other native people who have been dispossessed of their land and whose cultures have been belittled—Hawai'ians, Aboriginals in Australia, Dene and Inuit in Canada, Maori in New Zealand, and Celtic people in Great Britain. Moreover, the session in many ways illustrates what might happen in any type of cross-cultural counseling. A European American student meeting a counselor who is Latina/o or African American, for example (or vice versa if the roles of counselor and client are switched), might also have early difficulty in talking and establishing trust.

You will find that many sessions start slowly, regardless of the cultural background of the client. Your skill at observing nonverbal and verbal behavior in the here and now of the session will enable you to choose appropriate things to say. Some alternatives that may help you as you begin sessions include patience, a good sense of humor, and a willingness to disclose, share stories, and talk about neutral subjects such as sports or the weather. You will also find an early exploration of positive assets useful at times: "Before we start, I'd like to get to know a little bit more about you. Could you tell me specifically about something from your past that you feel particularly good about?" "What are some of the things you do well?" "What types of things do you like?" As time permits, consider conducting a full wellness review.

▶Concepts Into Action:
Three Organizing Principles

In working with patients [clients], if you miss those nuances—if you misread what they may be trying to communicate, if you misjudge their character, if you don't notice when their emotions, gestures, or tone of voice don't fit what they are saying, if you don't catch the fleeting sadness or anger that lingers on their face for only a few milliseconds as they mention someone or something you might otherwise not know was important—you will lose your patients [clients]. Or worse still, you don't.

—Drew Westen, 2007

Three organizing principles for understanding counseling interaction are stressed in this chapter: nonverbal behavior; verbal behavior; and conflict, discrepancies, and incongruities. Over time, you will gain considerable skill in drawing out and working with the client's view of the world and how you as a person relate to the client. The best way to learn observation skills may be to observe yourself and your client in a videotaped session.

Nonverbal Behavior

As you have seen, observing clients' attending behavior patterns is often central. Clients' body movements may express varying levels of comfort about certain topics or situations. Shifts in the tone of their voice may reveal the emotions felt in connection with specific events or relationships. Remember to observe your own nonverbals as therapist; your clients are clearly affected by them. "The voice of the therapist, regardless of what he or she says, should be warm, professional, competent—and free from fear" (Grawe, 2007, p. 411).

You can learn about nonverbal behavior research in Box 4.2. Periodically observe yourself on video, and carefully note your own nonverbal style. Continue this even long after you become a practicing professional to avoid falling into bad habits.

BOX 4.2 Research and Related Evidence That You Can Use

Observation

In the fields of observation chance favors only the prepared mind.

—Louis Pasteur

Research on nonverbal behavior has a long and distinguished history. Edward Hall's *The Silent Language* (1959) is a classic of anthropological and multicultural research and remains the place to start. Early work in nonverbal communication was completed by Paul Ekman, and his 1999 and 2007 summaries of his work are basic to the field.

A good place to learn about empathy and research on emotion is to view Paul Ekman's interview on YouTube (www .youtube.com/watch?v=3i1QFv_PtqM&feature=topics).

Sharpley (Sharpley & Guidara, 1993; Sharpley & Sagris, 1995) found that eye contact and forward trunk lean were highly correlated with ratings of empathy. Hill and O'Brien's review (2004) noted that clients used fewer head nods when they were reacting negatively. Hall and Schmid Mast (2007) found that participants shifted attention toward visual nonverbal cues and away from verbal cues when asked to infer feelings; when asked to infer thoughts, they did the reverse.

Miller (2007) studied communication in the workplace and found that "connecting," essential to the communication of compassion, included empathy and perspective taking. Haskard and coworkers (2008) report that physicians' nonverbal communication, particularly their tone of voice, plays a major role in their relationship with patients, affecting patient satisfaction and adherence to treatment. Health care providers need to be aware of the power of their tone of voice, which may inadvertently communicate their emotions and affect clients' satisfaction.

A classic study by Mayo and LaFrance (1973) remains useful and important today in highlighting issues of cultural change and acculturation. They observed,

> In interaction among Whites, a clear indication of attention by the listener is that he looks at the speaker. . . . Blacks do not look while listening, presumably using other cues to communicate attention. When Blacks and Whites interact, therefore, these differences may give rise to communication breakdowns. The White may feel he is not being listened to, while the Black may feel unduly scrutinized. Further, exchanges of listener-speaker roles may become

disjunctive, leading to generalized discomfort with the encounter. (p. 389)

Nearly 40 years later, another study found that when African Americans had a strong sense of their own racial identity, they showed more openness to Whites, whereas most Whites who had a strong identity with their own whiteness responded negatively to African Americans (Kaiser, Drury, Malahy, & King, 2010). This seems to be so even though verbal and nonverbal behaviors of African Americans have changed considerably over time (Greene & Stewart, 2011).

Especially relevant to this chapter is a study that examined women's attitudes toward real and imagined gender harassment. When women were asked to imagine being harassed in an interview, they said that they were angry. But those who were actually harassed felt *fear* (Woodzicka & LaFrance, 2002). As a partial explanation for what happened in this study, their earlier research found that women's nonverbal reactions to sexist humor were quite marked and different from those of men (LaFrance & Woodzicka, 1998).

Nonverbal Communication and Neuroscience

Neuroscientists report a number of fascinating multicultural findings that have immediate relevance to the counseling and therapy process. Among them are the following, reported by Eberhardt (2005):

▲ Japanese have been found to be more holistic thinkers than Westerners (Masuda & Nisbett, 2001). Expect the possibility of different cognitive/emotional styles when you work with people who are culturally different from you—but never stereotype!

▲ Brain systems can be modified by life experience (Draganski et al., 2004). As you learn to observe your client more effectively in the session, your brain is likely developing new connections. Expect your multicultural learning to become one of those new connections.

▲ Blacks and Whites both exhibit greater brain activation when they view same race faces and less when race is different (Golby, Gabrelli, Chiao, & Eberhardt, 2001). This could affect your work with a client whose race is different from yours, again suggesting that discussing racial and other cultural differences early in the session can be a helpful way to build trust.

Facial Expressions   For you, as the counselor, smiling is a good indicator of your warmth and caring. Your ability to develop a relationship will often carry you through difficult problems and situations. When it comes to observing the client, here are some things to notice. The brow may furrow, lips may tighten or loosen, flushing may occur, a client may smile at an inappropriate time. Even more careful observation will reveal subtle color changes in the face as blood flow reflects emotional reactions. Breathing may speed up or stop temporarily. The lips may swell, and pupils may dilate or contract. These seemingly small responses are clues to what a client is experiencing; to notice them takes work and practice. You may want to select one or two kinds of facial expressions and study them for a few days in your regular daily interactions and then move on to others as part of a systematic program to heighten your powers of observation.

Neuroscience offers critical data that we all need to consider. Grawe (2007) reviews key literature and points out that the amygdala, critical center of emotional experience, appears to be highly sensitive to "fearful, irritated, and angry faces . . . even when the faces have not been perceived consciously. . . . We can be certain that in psychotherapy the patient's amygdala will respond to even the tiniest sign of anger in the facial expressions of therapists" (p. 78). Self-awareness of your own being is obviously as valuable as awareness of client behavior.

Body Language   Discrepancies in nonverbal behavior may occur. For example, when a client is talking casually about a friend, one hand may be tightly clenched in a fist and the other relaxed and open. This could, and likely does, indicate mixed feelings toward the friend or something that the friend has just said about challenging issues.

Often people who are communicating well "mirror" each other's body language. Mirror neurons in the brain enable counselors to become empathic with their clients. When empathy is at a height, client and counselor may unconsciously sit in identical positions and make complex hand movements together as if in a ballet. This is termed *movement synchrony*. *Movement complementarity* refers to paired movements that may not be identical, but are still harmonious. For instance, one person talks and the other nods in agreement. You may observe a hand movement at the end of one person's statement that is answered by a related hand movement as the other takes the conversational "ball" and starts talking.

Movement dissynchrony is a good clue that your client has seen your counseling comment as subtractive. Lack of harmony in movement is common between people who disagree markedly or even between those who may not be aware they have subtle conflicts. You likely have seen this type of behavior in couples that you know have problems in communicating.

Some expert counselors and therapists deliberately "mirror" their clients. Experience shows that matching body language, breathing rates, and key words of the client can heighten counselor understanding of how the client perceives and experiences the world.

But be careful with deliberate mirroring. A practicum student reported difficulty with a client, noting that the client's nonverbal behavior seemed especially unusual. Near the end of the session, the client reported, "I know you guys; you try to mirror my nonverbal behavior. So I keep moving to make it difficult for you." You can expect that some clients will know as much about observation skills and nonverbal behavior as you do. What should you do in such situations? Use the skills and concepts in this book with *honesty* and *authenticity*. And talk with your client about their observations of you without being defensive. Openness works!

Acculturation Issues in Nonverbal Behavior: Avoid Stereotyping   We have stressed that there are many differences in individual and cultural styles. Acculturation is a fundamental concept of anthropology with significant relevance for the session.

Acculturation is the degree to which an individual has adopted the norms or standard way of behaving in a given culture. Because of the unique family, community, economic status, and part of the country in which a person is raised (and many other factors), no two people will be acculturated to general standards in the same way. In effect, "normative behavior" does not exist in any single individual. Thus, stereotyping individuals or groups needs to be avoided at all costs.

An African American client raised in a small town in upstate New York in a two-parent family has different acculturation experiences from those of an otherwise similar person raised in Los Angeles or East St. Louis. If the client is from a single-parent family, the acculturation experiences change further. If we alter the ethnic/racial background of this client to Italian American, Jewish American, or Arab American, the acculturation experience changes again. Many other factors, of course, influence acculturation—religion, economic bracket, and even being the first or second child in a family. Awareness of diversity in life experience is critical if we are to recognize uniqueness and specialness in each individual. If you define yourself as European American, Canadian, or Australian and you think of others as the only people who are multicultural, you need to rethink your awareness. All of us are multicultural beings with varying and singular acculturation experiences.

Finally, consider biculturality and multiculturality. Many of your clients have more than one significant community cultural experience. A Puerto Rican, Mexican, or Cuban American client is likely to be acculturated in both Hispanic and U.S. culture. A Vietnamese Canadian client in Quebec, a Ukrainian Canadian client in Alberta, and an Aboriginal client in Sydney, Australia, may also be expected to represent biculturality. All Native Americans and Hawai'ians in the United States, Dene and Inuit in Canada, Maori in New Zealand, and Aboriginals in Australia exist in at least two cultures. There is a culture among people who have experienced cancer, AIDS, war, abuse, or alcoholism. All of these issues and many others deeply affect acculturation.

In short, stereotyping any one individual is not only discriminatory; it is also naïve!

Verbal Behavior

As noted in the chapter introduction, counseling and psychotherapy theory and practice have an almost infinite array of verbal frameworks within which to examine the session. Four useful concepts for session analysis are presented here: key words, concreteness versus abstractions, "I" statements versus "other" statements, and identifying conflicts that clients show in their stories. We will also discuss some key multicultural issues connected with verbal behavior.

Key Words　If you listen carefully to clients, you will find that certain words appear again and again in their descriptions of situations. Noting their key words and helping them explore the facts, feelings, and meanings underlying those words may be useful. Key descriptive words are often the constructs by which a client organizes the world; these words may reveal underlying meanings. Verbal underlining through vocal emphasis is another helpful clue in determining what is most important to a client. Through intonation and volume, clients tend to stress the single words or phrases that are most salient to them.

Joining clients by using their key words facilitates your understanding and communication with them. If their words are negative and self-demeaning, reflect those perceptions early in the session but later help them use more positive descriptions of the same situations or events. Help the client change from "I can't" to "I can."

Many clients will demonstrate problems of verbal tracking and selective attention. They may either stay on a single topic to the exclusion of other issues or change the topic, either subtly or abruptly, when they want to avoid talking about a difficult issue. Perhaps the most difficult task of the beginning counselor or psychotherapist is to help the client stay on the topic without being overly controlling. Observing client topic changes is essential. At times it may be helpful to comment—for instance, "A few minutes ago we were talking about X." Another possibility is to follow up that observation by asking how the client might explain the shift in topic.

Abstract/Formal Operational

Clients high on the abstraction ladder tend to talk in a more reflective fashion, analyzing their thoughts and behaviors. They are often good at self-analysis, but may not easily provide concrete examples of their issues. They may prefer to analyze, rather than to act. Self-oriented, abstract theories, such as person-centered or psychodynamic, are often useful with this style.

Concrete/Situational

Clients who talk in a more concrete/situational style tend to provide specific examples and stories, often with considerable detail. You'll hear what they see, hear, and feel. Helping these clients to reflect on their situations and issues may be difficult. In general, they will look to the counselor for specific actions that they can follow. Concrete behavioral theories may be preferred.

More specific examples and extensions of these concepts can be found in *Developmental Counseling and Therapy: Promoting Wellness Over the Lifespan* (Ivey, Ivey, Myers, & Sweeney, 2005).

Concreteness Versus Abstraction Where is the client on the "abstraction ladder" (see Box 4.3)? Two major client communication styles—abstract and concrete—typically need different types of conversation (Ivey, Ivey, Myers, & Sweeney, 2005). Indeed, fMRI scans have shown that the prefrontal cortex actually fires differently with each style (Lane, 2008).

Clients who talk in a concrete/situational style are skilled at providing specifics and examples of their concerns and problems. The language of these clients forms the foundation at the "bottom" of the abstraction ladder. These clients may have difficulty reflecting on themselves and their situations and have difficulties seeing patterns in their lives.

Clients who are more abstract and formal operational, on the other hand, have strengths in self-analysis and are often skilled at reflecting on their issues. They are at the "top," but you will find that getting details from them on the concretes and specifics of what is actually going on may be difficult.

Of course, most adult and many adolescent clients will talk at both levels. Children, however, can be expected to be primarily concrete in their talk—and so are many adolescents and adults.

Clients with a concrete/situational style will provide you with plenty of specifics. The strength and value of these details is that you know relatively precisely what happened, at least from their point of view. However, they often have difficulties in seeing the point of view of others. Some clients with a concrete style may tell you, for example, what happened to them when they went to the hospital from start to finish, with every detail of the operation and how the hospital functions. Or ask a 10-year-old to tell you about a movie and you will practically get a complete script. Concretely oriented clients talking about a difficult interpersonal relationship may present the situation through a series of endless stories full of specific facts: "He said . . . and then I said. . . ." If asked to reflect on the meaning of their story or what they have said, they may appear puzzled.

Here are some examples of concrete/situational statements:

Child, age 5:	Jonnie hit me in the arm—right here!
Child, age 10:	He hit me when we were playing soccer. I had just scored a goal and it made him mad. He snuck up behind me, grabbed my leg and then punched me when I fell down! Do you know what else he did, well he . . .
Man, age 45:	I was down to Myrtle Beach—we drove there on 95 and was the traffic ever terrible. Well, we drove into town and the first thing we did was to

	find a motel to stay in, you know. But we found one for only $60 and it had a swimming pool. Well, we signed in and then . . .
Woman, age 27:	You asked for an example of how my ex-husband interferes with my life? Well, a friend and I were sitting quietly in the cafeteria, just drinking coffee. Suddenly, he came up behind me, he grabbed my arm (but didn't hurt me this time), then he smiled and walked out. I was scared to death. If he had said something, it might not have been so frightening.

The details are important, but clients who use a primarily concrete style in their conversation and thinking may have real difficulty in reflecting on themselves and seeing patterns in situations.

Abstract/formal operational clients are good at making sense of the world and reflecting on themselves and their situations. But some clients will talk in such broad generalities that it is hard to understand what they are really saying. They may be able to see patterns in their lives and be good at discussing and analyzing themselves, but you may have difficulty finding out specifically what is going on in their lives. They may prefer to reflect rather than to act on their issues.

Here are some examples of abstract/formal operational statements:

Child, age 12:	He does it to me all the time. It never stops. It's what he does to everyone all the time.
Man, age 20:	As I think about myself, I see a person who responds to others and cares deeply, but somehow I feel that they don't respond to me.
Woman, age 68:	As I reflect back on my life, I see a pattern of selfishness that makes me uncomfortable. I think a lot about myself.

Many counselors and psychotherapists tend to be more abstract/formal operational themselves and may be drawn to the analytical and self-reflective style. They may conduct entire sessions focusing totally on analysis, and an observer might wonder what the client and counselor are talking about.

In each of the conversational styles, the strength is also possibly a weakness. You will want to help abstract/formal clients to become more concrete ("Could you give me an example?"). If you persist, most of these clients will be able to provide the needed specifics.

You will also want to help concrete clients become more abstract and pattern oriented. This is best effected by a conscious effort to listen to their sometimes lengthy stories very carefully. Paraphrasing and summarizing what they have said (see Chapter 6) can be helpful. Just asking them to reflect on their story ("Could you tell me what the story means to you?" "Can you reflect on that story and what it says about you as a person?") may not work. More direct questions may be needed to help concrete clients step back and reflect on their stories. A series of questions such as these might help: "What one thing do you remember most about this story?" "What did you like best about what happened?" "What least?" "What could you have done differently to change the ending of the story?" Questions like these that narrow the focus can help children and clients with a concrete orientation move from self-report to self-examination.

Essentially, we all need to match our own style and language to the uniqueness of the client. If you have a concrete style, abstract clients may challenge or even puzzle you. If you are more abstract, you may not be able to understand and reach those with a concrete orientation (and they are the majority of our clients). Abstract counselors often are bored and impatient with concrete clients. If the client tends to be concrete, listen to the specifics and enter that client's world as he or she presents it. If the client is abstract, listen and join that client where he or she is. Consider the possibility of helping the client look at the concern from the other perspective.

"I" Statements and "Other" Statements Clients' ownership and responsibility for issues will often be shown in their "I" and "other" statements. Consider the following:

"I'm working hard to get along with my partner. I've tried to change and meet her/him halfway."

versus

"It's her/his fault. No change is happening."

"I'm not studying enough. I should work harder."

versus

"The racist insults we get on this campus make it nearly impossible to study."

"I feel terrible. If only I could do more to help. I try so hard."

versus

"Dad's an alcoholic. Everyone suffers."

"I'm at fault. I shouldn't have worn that dress. It may have been too sexy."

versus

"No, women should be free to wear whatever they wish."

"I believe in a personal God. God is central to my life."

versus

"Our church provides a lot of support and helps us understand spirituality more deeply."

We need to be aware of what is happening in and for our clients as well as learn what occurs in their relationships with others and in their families and communities. There is a need to balance internal and external responsibility for issues.

Review the five pairs of statements above. Some of them represent positive "I" and "other" statements; others are negative. Some clients attribute their difficulties solely to themselves; others see the outside world as the issue. A woman may be sexually harassed and see clearly that others and the environment are at fault; another woman will feel that somehow she provoked the incident. Counselors need to help individuals look at their issues but also to help them consider how these concerns relate to others and the surrounding environment.

The alcohol-related statements above may serve as an example. Some children of alcoholics see themselves as somehow responsible for a family member's drinking. Their "I" statements may be unrealistic and ultimately "enable" the alcoholic to drink even more. In such cases, the task of the counselor is to help the client learn to attribute family difficulties to alcohol and the alcoholic. In work with alcoholics themselves who may deny the problem, one goal is often to help move them to that critical "I" statement, "I am an alcoholic." Part of recovery from alcoholism, of course, is recognizing others and showing esteem for others. Thus, a balance of "I" and "other" statements is a useful goal.

You can also observe "I" statements as a person progresses through a series of sessions. For example, the client at the beginning of counseling may give many negative self-statements: "It's my fault." "I did a bad thing, I'm a bad person." "I don't respect myself." "I don't like myself."

If your sessions are effective, expect such statements to change to "I'm still responsible, but I now know that it wasn't all my fault." "Calling myself 'bad' is self-defeating. I now realize that I did my best." "I can respect myself more." "I'm beginning to like myself."

Multicultural differences in the use of "I" need to be recognized. We should remember that English is one of the very few languages that capitalizes the word "I." A Vietnamese immigrant comments:

> There is no such thing as "I" in Vietnamese. . . . We define ourselves in relationships. . . . If I talk to my mother, the "I" for me is "daughter," the "you" is for "mother." Our language speaks to relationships rather than to individuals.

Discrepancies, Mixed Messages, and Conflict

Life is full of contradictions that counseling can help resolve.

We all live with contradiction, conflict, incongruity, and discrepancies that all provide challenges. At the same time, they are opportunities opening our growth potential. The variety of discrepancies clients may manifest is perhaps best illustrated by the following statements:

"My son is perfect, but he just doesn't respect me."
"I deserve to pass the course." (from a student who has done no homework and just failed the final examination)
"That question doesn't bother me." (said with a flushed face and a closed fist)

Once the client is relatively comfortable and some beginning steps have been taken toward rapport and understanding, a major task of the counselor or psychotherapist is to identify basic discrepancies, mixed messages, conflicts, or incongruities in the client's behavior and life. A common goal in most counseling or therapy is to assist clients in working through discrepancies and conflict, but first these have to be identified clearly.

Examples of Conflict Internal to the Client

Discrepancies in verbal statements.   In a single sentence a client may express two completely contradictory ideas ("My son is perfect, but he just doesn't respect me" or "This is a lovely office you have; it's too bad that it's in this neighborhood"). Most of us have mixed feelings toward our loved ones, our work, and other situations. It is helpful to aid others in understanding their ambivalences.

Discrepancies between statements and nonverbal behavior.   Of central importance are discrepancies between what one says and what one does. A parent may talk of love for a child but be guilty of child abuse. A student may say that he or she deserved a higher grade than the actual time spent studying suggests. The client may "talk the talk" about support for multicultural, women's, or ecology causes, but fail to "walk the walk." The client may verbalize a desire to repair a troubled relationship while simultaneously picking at his or her clothes, or make small or large physical movements away from the counselor when confronted with a troubling issue.

Examples of Conflict Between the Client and the External World

Discrepancies and conflict between people.   "I cannot tolerate my neighbors." Noting interpersonal conflict is a key task of the counselor or psychotherapist.

Discrepancies between a client and a situation.   "I want to be admitted to medical school, but I didn't make it." "I can't find a job." In such situations the client's ideal world is often incongruent with what really is. Many people of color, gays, women, and people with disability find themselves in a contextual situation that makes life difficult for them. Discrimination, heterosexism, sexism, and ableism represent situational discrepancies.

Discrepancies in Goals   Goal setting is central in the *empathic relationship—story and strengths—goals—restory—action* model. In establishing the purposes of counseling or psychotherapy, you will often find that a client seeks incompatible goals. For example, the client may want the approval of his friends and his parents. But to win acceptance from peers, academic performance may suffer, and if academic performance pleases his parents, he has "sold out."

Discrepancies Between You and the Client   One of the more challenging issues occurs when you and the client are not in synchrony. This can occur on any of the dimensions above. Your nonverbal communication may be misread by the client. The client may avoid really facing issues. You may be saying one thing, the client another. A conflict in values or goals in counseling may be directly apparent or a quiet, unsaid thing, which still affects the client—and either can destroy your relationship. When you get too close to the truth, clients may wipe their nose. Does the client turn toward you or turn away?

In any of these situations, it is helpful to aid clients in understanding their ambivalences by summarizing the conflict, the client's thoughts, emotions, and behaviors, and/or the conflicting issues posed by someone else or a life situation. The summary of the conflict then can be followed by a variation of the basic challenge, such as "I hear you saying one side of the issue is (insert appropriate comment representing part of the conflict or discrepancy), but I also hear the other side as (insert the opposing side of the conflict)." Then, through further listening and observation, the client may come up with her or his own unique solution.

The issues of conflict, discrepancies, and contradiction will be explored in much more detail in Chapter 10 on confrontation as a constructive challenge.

▶Summary: Observation Skills

The counselor seeks to observe client verbal and nonverbal behavior with an eye to identifying discrepancies, mixed messages, incongruity, and conflict. Counseling and therapy frequently focus on problems and their resolution. A discrepancy is often a problem. At the same time, discrepancies in many forms are part of life and may even be enjoyed. Humor, for example, is based on conflict and discrepancies. In addition, observation will help you develop recovery skills that you can use when you are lost or confused in the session. Even the most advanced professional doesn't always know what is happening. When you don't know what to do, attend!

▼▼▼▼	Key Points
Importance of observation	The self-aware counselor is constantly aware of the client and of the here-and-now interaction in the session. Clients tell us about their world by nonverbal and verbal means. Observation skills are a critical tool in determining how the client interprets the world.
Three key items to observe	**Observation skills focus on three areas:** 1. *Nonverbal behavior.* Your own and the client's eye contact patterns, body language, and vocal qualities are basic. Shifts and changes in these may indicate client interest or discomfort. A client may lean forward, indicating excitement about an idea, or cross his or her arms to close it off. Facial clues (brow furrowing, lip tightening or loosening, flushing, pulse rate visible at temples) are especially noteworthy. Large-scale body movements may indicate shifts in reactions, thoughts, or the topic.

	2. *Verbal behavior.* Noting patterns of verbal tracking for both you and the client is of major concern. At what point does the topic change, and who initiates the change? Where is the client on the abstraction ladder? If the client is concrete, are you matching her or his language? Is the client making "I" statements or "other" statements? Do the client's negative statements become more positive as counseling progresses? Clients tend to use certain key words to describe their behavior and situations; noting these descriptive words and repetitive themes is helpful.
	3. *Conflict and discrepancies.* Incongruities, mixed messages, and contradiction are manifest in many and perhaps all sessions. The effective counselor is able to identify these discrepancies, to name them appropriately, and sometimes to feed them back to the client. These discrepancies may be between nonverbal behaviors, between two statements, between what clients say and what they do, between incompatible goals, or between statements and nonverbal behavior. They may also represent a conflict between people or between a client and a situation. In addition, your own behaviors may be positively or negatively discrepant.
	Simple, careful observation of the session is basic. What can you see, hear, and feel from the client's world? Note your impact on the client: How does what you say change or relate to the client's behavior? Use these data to adjust your observation skills.
Multicultural issues	Observation skills are essential with all clients. Note individual and cultural differences in verbal and nonverbal behavior. Always remember that some individuals and some cultures may have a different meaning for a movement or use of language from your own personal meaning. Use caution in your interpretation of nonverbal behavior.
Movement harmonics	Movement harmonics are particularly interesting and provide a basic concept that explains much verbal and nonverbal communication. When two people are talking together and communicating well, they often exhibit movement synchrony or movement complementarity in that their bodies move in a harmonious fashion. When people are not communicating clearly, movement dissynchrony will appear; body shifts, jerks, and pulling away are readily apparent.

▶Competency Practice Exercises and Portfolio of Competence

 Go to CengageBrain.com to access Counseling CourseMate, where you will find an interactive ebook, quizzes, videos, interactive counseling and psychotherapy exercises, the Portfolio of Competence, case studies, a practice test, and more.

Many concepts have been presented in this chapter; it will take time to master them and make them a useful part of your counseling and psychotherapy. Therefore, the exercises here should be considered introductory. Further, we suggest that you continue to work on these concepts throughout the time that you read this book. If you keep practicing the concepts in this chapter throughout the book, material that might now seem confusing will gradually be clarified and become part of your natural style.

Many of you now have small cameras or smartphones that have both video and sound capability. We urge you to take the time to observe closely what you are doing. Practice with clear video feedback is the original and most effective route toward mastery and competence.

Individual Practice

Exercise 1. Observation of Nonverbal Patterns   Observe 10 minutes of a counseling session, a television interview, or any two people talking. Make a video recording so that repeated viewing is possible.

Visual/eye contact patterns.   Do people maintain eye contact more while talking or while listening? Does the "client" break eye contact more often while discussing certain subjects than others? Can you observe changes in pupil dilation as an expression of interest?

Vocal qualities.   Note speech rate and changes in intonation or volume. Give special attention to speech "hitches" or hesitations.

Attentive body language.   Note gestures, shifts of posture, leaning, patterns of breathing, and use of space. Give special attention to facial expressions such as changes in skin color, flushing, and lip movements. Note appropriate and inappropriate smiling, furrowing of the brow, and so on.

Movement harmonics.   Note places where movement synchrony and echoing occurred. Did you observe examples of movement dissynchrony?

   If possible, watch your video recording so that you can view the session, or part of it, several times. Be sure to separate behavioral observations from impressions on the Feedback Form (Box 4.4).

a.   Present the context of your observation. Briefly summarize what is happening verbally at the time of the observation. Number each observation.
b.   Observe the session for the following and describe what you see as precisely and concretely as possible: visual/eye contact patterns, vocal qualities, attentive body language, and movement harmonics.
c.   Record your impressions. What interpretations of the observation do you make? How do you make sense of each observation unit? And—most essential—are you cautious in drawing conclusions from what you have seen and noticed?

Exercise 2. Observation of Verbal Behavior and Discrepancies   Observe the same session again, but this time pay special attention to verbal dimensions and discrepancies (which, of course, will include nonverbal dimensions). Consider the following issues, and provide concrete evidence for each of your decisions as to what is occurring in the session. Again, separate observations from your interpretation and impressions.

Verbal tracking and selective attention.   Pay special attention to topic jumps or shifts. Who initiates them? Do you see any pattern of special topic interest and/or avoidance? What does the listener seem to want to hear?

Key words.   What are the key words of each person in the communication?

Abstract or concrete conversation.   Is this conversation about patterns or about specifics? Are the people involved approaching this issue in a similar fashion?

Conflict and discrepancies.   What incongruities do you note in the behavior of either person? Do you locate any discrepancies between the two? What issues of conflict might be important?

Exercise 3. Examining Your Own Verbal and Nonverbal Styles   Make a video recording of yourself with another person in a real conversation or session for at least 20 minutes. Do not make this a role-play. Then view your own verbal and nonverbal behavior and that of the person you are talking with in the same detail as in Exercises 1 and 2. What do you learn about yourself?

BOX 4.4 Feedback Form: Observation

You can download this form from the Counseling CourseMate website at CengageBrain.com.

_____
DATE

NAME OF COUNSELOR   NAME OF PERSON COMPLETING FORM

Instructions: Observe the client or counselor carefully during the role-play session and immediately afterward complete the nonverbal feedback portion of the form. As you view the video or listen to the audio, give special attention to verbal behavior and note discrepancies. If no recording equipment is available, one observer should note nonverbal behavior and the other verbal behavior.

Nonverbal Behavior Checklist

1. *Visuals.* At what points did eye contact breaks occur? Staring? Did the individual maintain eye contact more when talking or when listening? Changes in pupil dilation?

2. *Vocals.* When did speech hesitations occur? Changes in tone and volume? What single words or short phrases were emphasized?

3. *Body language.* General style and changes in position of hands and arms, trunk, legs? Open or closed gestures? Tight fist? Playing with hands or objects? Physical tension: relaxed or tight? Body oriented toward or away from the other? Sudden body shifts? Twitching? Distance? Breathing changes? At what points did changes in facial expression occur? Changes in skin color, flushing, swelling or contracting of lips? Appropriate or inappropriate smiling? Head nods? Brow furrowing?

4. *Movement harmonics.* Examples of movement complementarity, synchrony, or dissynchrony? Echoing? At what times did these occur?

(continued)

CHAPTER 4 *Observation Skills* 109

BOX 4.4 (continued)

5. *Nonverbal discrepancies.* Did one part of the body say something different from another? With what topics did this occur?

Verbal Behavior Checklist

1. *Verbal tracking and selective attention.* At what points did the client or counselor fail to stay on the topic? To what topics did each give most attention? List here the most central key words used by the client; these are important for deeper analysis.

2. *Abstract or concrete?* Which word represents the client? How did the counselor work with this dimension? Was the counselor abstract or concrete?

3. *Verbal discrepancies.* Write here observations of verbal discrepancies in either client or counselor.

Exercise 4. Classifying Statements as Concrete or Abstract Following are examples of client statements. Classify each statement as primarily concrete or primarily abstract. You will gain considerably more practice and thus have more suggestions for interventions in later chapters. (Answers to this exercise are at the end of the chapter.)

For each statement, circle C (concrete) or A (abstract):

C A 1. I cry all day long. I didn't sleep last night. I can't eat.

C A 2. I feel rotten about myself lately.

C A 3. I feel very guilty.

C A 4. Sorry I'm late for the session. Traffic was very heavy.

C A 5. I feel really awkward on dates. I'm a social dud.

C   A   6. Last night my date said that I wasn't much fun. Then I started to cry.

C   A   7. My father is tall, has red hair, and yells a lot.

C   A   8. My father is very hard to get along with. He's difficult.

C   A   9. My family is very loving. We have a pattern of sharing.

C   A   10. My mom just sent me a box of cookies.

Group Practice

Exercise 5. Practicing Observation in Groups   Many observation concepts have been discussed in this chapter. Obviously you cannot observe all these behaviors in one single role-play session. However, practice can serve as a foundation for elaboration at a later time. This exercise has been selected to summarize the central ideas of the chapter.

Step 1: Divide into practice groups.   Triads or groups of four are most appropriate.

Step 2: Select a group leader.

Step 3: Assign roles for the first practice session.

▲   Client, who responds naturally and is talkative.
▲   Counselor, who will seek to demonstrate a natural, authentic style.
▲   Observer 1, who observes client communication, using the Feedback Form: Observation (Box 4.4).
▲   Observer 2, who observes counselor communication, using the Feedback Form. Here the consultative microsupervision process usually focuses on helping the counselor understand and utilize nonverbal communication more effectively. Ideally you have a video available for precise feedback.

Step 4: Plan.   State the goals of the session. As the central task is observation, the counselor should give primary attention to attending and open questions. Use other skills as you wish. After the role-play is over, the counselor should report personal observation of the client made during that time and demonstrate basic or active mastery skills. The client will report on observations of the counselor.

The suggested topic for the practice role-play is "Something or someone with whom I have a present conflict or have had a past conflict." Alternative topics include the following:

My positive and negative feelings toward my parents or other significant persons
The mixed blessings of my work, home community, or present living situation

The two observers may use this session as an opportunity both for providing feedback to the counselor and for sharpening their own observation skills.

Step 5: Conduct a 6-minute practice session.   As much as possible, both the counselor and the client will behave as naturally as possible discussing a real situation.

Step 6: Review the practice session and provide feedback for 14 minutes.   Remember to stop the audio or video recording periodically and listen to or view key items several times for increased clarity. Observers should give special attention to careful completion of the feedback sheet throughout the session, and the client can give useful feedback via the Client Feedback Form of Chapter 1.

Step 7: Rotate roles.

Portfolio of Competence

Use the following as a checklist to evaluate your present level of mastery. Check those dimensions that you currently feel able to do. Those that remain unchecked can serve as future goals. Do not expect to attain intentional competence on every dimension as you work through this book. You will find, however, that you will improve your competencies with repetition and practice.

Level 1: Identification and classification

❑ Ability to note attending nonverbal behaviors, particularly changes in behavior in visuals—eye contact, vocal tone, and body language
❑ Ability to note movement harmonics and echoing
❑ Ability to note verbal tracking and selective attention
❑ Ability to note key words used by the client and yourself
❑ Ability to note distinctions between concrete/situational and abstract/formal operational conversation
❑ Ability to note discrepancies in verbal and nonverbal behavior
❑ Ability to note discrepancies in the client
❑ Ability to note discrepancies in yourself
❑ Ability to note discrepancies between yourself and the client

Level 2: Basic competence. Nonverbal and verbal observation skills are things that you can work on and improve over a lifetime. Therefore, use the intentional competence list below for your self-assessments.

Level 3: Intentional competence. You will be able to note client verbal and nonverbal behaviors in the session and use these observations at times to facilitate session conversation. You will be able to match your behavior to the client's. When necessary, you will be able to mismatch behaviors to promote client movement. For example, if you first join the negative body language of a depressed client and then take a more positive position, the client may follow and adopt a more assertive posture. You will be able to note your own verbal and nonverbal responses to the client. You will be able to note discrepancies between yourself and the client and work to resolve those discrepancies.

Developing mastery of the following areas will take time. Come back to this list later as you practice other skills in this book. For the first stages of basic and intentional mastery, the following competencies are suggested:

❑ Ability to mirror nonverbal patterns of the client. The counselor mirrors body position, eye contact patterns, facial expression, and vocal qualities.
❑ Ability to identify client patterns of selective attention and use those patterns either to bring talk back to the original topic or to move knowingly to the new topic provided by the client.
❑ Ability to match clients' concrete/situation or abstract/formal operational language and help them to expand their stories in their own style.
❑ Ability to identify key client "I" and "other" statements and feed them back to the client accurately, thus enabling the client to describe and define what is meant more fully.
❑ Ability to note discrepancies and feed them back to the client accurately. (Note that this is an important part of the skill of confrontation and conflict management, discussed in detail in Chapter 10.) The client in turn will be able to accept and use the feedback for further effective self-exploration.
❑ Ability to note discrepancies in yourself and act to change them appropriately.

Level 4: Psychoeducational teaching competence. You will demonstrate your ability to teach others observation skills. Your achievement of this level can be determined by how well your students can be rated on the basic competencies of this self-assessment form. Certain of your clients in counseling may be quite insensitive to obvious patterns of nonverbal and verbal communication. Teaching them beginning methods of observing others can be most helpful to them. Do not introduce more than one or two concepts to a client per session, however!

❑ Ability to teach clients in a helping session the social skills of nonverbal and verbal observation and the ability to note discrepancies
❑ Ability to teach the skills to small groups

▶Determining Your Own Style and Theory: Critical Self-Reflection on Observation Skills

This chapter has focused on the importance of verbal and nonverbal observation skills, and you have experienced a variety of exercises designed to enhance your awareness in this area.

Again, what single idea stands out for you among all those presented in this chapter, in class, or through informal learning? What stands out for you is likely to be useful as a guide toward your next steps. What are your thoughts on multicultural differences? What other points in this chapter struck you as useful in your practice? How might you use ideas in this chapter to begin the process of establishing your own style and theory?

Correct Responses for Exercise 4

1 — C	5 — A	8 — A
2 — A	6 — C	9 — A
3 — A	7 — C	10 — C
4 — C		

The Basic Listening Sequence: How to Organize a Session

A ttending behavior and observation skills are basic to all the communication skills of the microskills hierarchy. Without individually and culturally appropriate attending behavior, there can be no counseling or psychotherapy.

This section adds to attending skills by presenting the basic listening sequence, which will enable you to elicit the major facts and feelings pertinent to a client's concern. Through the skills of questioning, encouraging, paraphrasing, reflecting feelings, and summarizing, you will learn how to draw out your clients and understand the way they think about their stories.

Chapter 5. Questions: Opening Communication   We encounter questions every day. Most theories of counseling now use questions rather extensively. Examples include cognitive behavioral therapy (CBT), brief counseling, and motivational interviewing. This chapter explains open and closed questions and their place in your communication. However, the use of questions is sometimes controversial. Several experts would argue that this chapter belongs *after* the critical and central skills of accurate and reflective listening.

Chapter 6. Encouraging, Paraphrasing, and Summarizing: Key Skills of Active Listening   Here you will explore the clarifying skills of paraphrasing, encouraging, and summarizing, which are foundational to developing a relationship and working alliance with your client. They are central also for drawing out the story.

Chapter 7. Reflecting Feelings: A Foundation of Client Experience   This skill gets at the heart of the issue and truly personalizes the session. You will learn how to bring out the rich emotional world of your clients. Reflecting feelings is a challenging skill to master fully and requires special attention, but many believe that real change is founded on emotions and feelings.

Chapter 8. How to Conduct a Five-Stage Counseling Session Using Only Listening Skills Once you have mastered observation skills and the basic listening sequence, you are prepared to conduct a full, well-formed session comprising five stages. You will be able to conduct this session using only skills of the basic listening sequence. Furthermore, you will be able to evaluate your skills and those of others for level of empathic understanding. Not only do we need to listen; we need to listen empathetically. Some instructors will want to include the material on empathy as reading with Chapters 6 and 7.

This section, then, has ambitious goals. By the time you have completed Chapter 8, you will have attained several major objectives, enabling you to move on to the influencing skills of interpersonal change, growth, and development. At an intentional level of competence, you may aim to accomplish the following in this section:

1. Develop competence with the basic listening sequence, enabling the client to tell the story. In addition, draw out key facts and feelings related to client issues.
2. Observe clients' reactions to your skill usage and modify your skills and attending behaviors to complement each client's uniqueness.
3. Conduct a complete session using only listening and observing skills.
4. Evaluate your counseling session for its level of empathy; in effect, examine yourself and your ability to communicate warmth, positive regard, and other, more subjective dimensions of counseling and psychotherapy.

When you have accomplished these tasks, you may find that your clients have a surprising ability to solve their own problems, issues, concerns, or challenges. You may also gain a sense of confidence in your own ability as a counselor or psychotherapist.

"When in doubt, listen!" This is the motto of this section and the entire microskills framework.

Questions
Opening
Communication

Open and Closed Questions

Client Observation Skills

Attending Behavior and Empathy

Ethics, Multicultural Competence, and Wellness

Asking questions of your client can be helpful in establishing a basis for effective communication. Effective questions open the door to knowledge and understanding. The art of questioning lies in knowing which questions to ask when. Address your first question to yourself: if you could press a magic button and get every piece of information you want, what would you want to know? The answer will immediately help you compose the right questions.

—Robert Heller and Tim Hindle

Mission of "Questions"

Questions can be used to enhance the session and draw out the client's story. Two types of questions, open and closed, are central. Open questions elicit more information and give the client more room to respond. Closed questions elicit shorter responses and provide specific information. Be aware that, like attending behavior, questions can encourage or discourage client talk. With questions, however, the counselor generally takes the lead. The client is often talking within the counselor's frame of reference. Questions potentially can take the client away from self-direction.

Chapter Goals and Competency Objectives Awareness, knowledge, skills, and actions in questioning will enable you to

▲ Draw out and enrich client stories by bringing out a more complete description, including background information and needed details.

▲ Choose the question style that is most likely to achieve a useful anticipated result. For example, *what* questions often lead to talk about facts, *how* questions to feelings or process, and *why* questions to reasons.

▲ Open or close client talk, intentionally, according to the individual needs of the client.

▲ Use questions in a culturally sensitive and respectful way.

 Go to CengageBrain.com to access Counseling CourseMate, where you will find an interactive ebook, quizzes, videos, interactive counseling and psychotherapy exercises, the Portfolio of Competence, case studies, a practice test, and more.

Benjamin is in his junior year of high school, in the middle third of his class. In this school, each student must be interviewed about plans after graduation—work, the armed forces, or college. You are the high school counselor and have called Benjamin in to check on his plans after graduation. His grades are average, he is not particularly verbal and talkative, but is known as a "nice boy."

Reread the quotation that introduced this chapter. What are some questions that you could use to draw Benjamin out and help him think ahead to the future? And what might happen if you ask too many questions?

To compare your thoughts with ours, see page 137 at the end of this chapter.

▶Introduction: Defining and Questioning Questions

Skilled attending behavior is the foundation of the microskills hierarchy; questioning provides a useful framework for focusing the session. Questions help a session begin and move along smoothly. They assist in pinpointing and clarifying issues, open up new areas for discussion, and aid in clients' self-exploration.

Questions are an essential component in many theories and styles of helping, particularly cognitive behavioral therapy (CBT), brief counseling, and much of career decision making. The employment counselor facilitating a job search, the social worker conducting an assessment interview, and the high school guidance counselor helping a student work on college admissions all need to use questions. Moreover, the diagnostic process, while not counseling, uses many questions.

Types of Questions

This chapter focuses on two key styles of questioning: open and closed questions.

Open questions are those that can't be answered in a few words. They tend to facilitate deeper exploration of client issues. They encourage others to talk and provide you with maximum information. Typically, open questions begin with *what, how, why,* or *could*. For example, "Could you tell me what brings you here today?"

Closed questions enable you to obtain specifics and can usually be answered in very few words. They may provide useful information, but the burden of guiding the talk remains on the counselor. Closed questions often begin with *is, are,* or *do*. For example, "Are you living with your family?"

If you use open questions effectively, the client may talk more freely and openly. Closed questions elicit shorter responses and may provide you with specific information. Following are some of the results you can anticipate when using questions.

OPEN QUESTIONS	ANTICIPATED RESULT
Begin open questions with the often useful *who, what, when, where,* and *why. Could, can,* or *would* questions are considered open but have the additional advantage of being somewhat closed, thus giving more power to the client, who can more easily say that he or she doesn't want to respond.	Clients will give more detail and talk more in response to open questions. *Could, would,* and *can* questions are often the most open of all, because they give clients the choice to respond briefly ("No, I can't") or, much more likely, explore their issues in an open fashion.
CLOSED QUESTIONS	**ANTICIPATED RESULT**
Closed questions may start with *do, is,* or *are.*	Closed questions may provide specific information but may close off client talk.
Effective questions encourage more focused client conversations with more pertinent detail and less wandering.	

Questioning Questions

Some theorists and practitioners are concerned about the use of questions and believe that they are best studied and learned *after* expertise is developed in the reflective listening skills of Chapters 6 and 7. They point out that once questions are presented to those beginning counseling, the listening skills of paraphrasing, reflecting feelings, and summarizing may receive insufficient attention. Certainly, excessive use of questions takes the focus from the client and can give too much power to the counselor. Your central task in this chapter is to find an appropriate balance in using questions with clients.

REFLECTION *Why Do Some People Object to Questions?*

Take a minute to recall and explore some of your own experiences with questions in the past. Perhaps you had a teacher or a parent who used questions in a manner that resulted in your feeling uncomfortable or even attacked. Write here one of your negative experiences with questions and the feelings and thoughts the questioning process produced in you.

How was your difficult personal experience with questions?
How did it make you feel?
What thoughts did you have?
What do you think about it now, and how does it make you feel about questions?

People often respond to this exercise by describing situations in which they were put on the spot or grilled by someone. They may associate questions with anger and guilt. Many of us have had negative experiences with questions. Furthermore, questions may be used to direct and control client talk. School discipline and legal disputes typically use questions to control the person being interviewed. If your objective is to enable clients to find their own way, questions may inhibit your reaching that goal, particularly if they are used ineffectively. For these reasons, some helping authorities—particularly those who are humanistically oriented—object to questions.

Additionally, in many non-Western cultures, questions are inappropriate and may be considered offensive or overly intrusive. Nevertheless, questions remain a fact of life in our culture. We encounter them everywhere. The physician or nurse, the salesperson, the government official, and many others find questioning clients basic to their profession. Most counseling theories espouse using questions extensively. Cognitive-behavioral therapy and brief counseling, for example, use many questions. The issue, then, is how to question wisely and intentionally.

The goal of this section is to explore some aspects of questions and, eventually, to determine their place in your communication skills repertoire. Used carefully, questioning is a valuable skill. Box 5.1 presents some interesting research regarding the use of questions.

Sometimes Questions Are Essential: "What Else?"

Clients do not always spontaneously provide you with all necessary information, and sometimes the only way to get at missing data is by asking questions. For example, the client may talk about being depressed and unable to act. As a helper, you could listen to

---

BOX 5.1 Research and Related Evidence That You Can Use

Questions

Research results can be a guide or a confirmation for what types of questions work best in certain situations or with particular theoretical orientations. Not too surprising is confirmation that open questions produce longer client responses than closed ones (Daniels & Ivey, 2007; Tamase, Torisu, & Ikawa, 1991).

Different theoretical orientations to helping vary widely in their use of questions. Person-centered leaders tend to use very few questions, whereas 40% of the leads of a problem-solving group leader were questions (Ivey, Pedersen, & Ivey, 2001; Sherrard, 1973). You will find that cognitive behavioral therapy, motivational interviewing, and brief counseling use questions extensively (Chapter 15). Your decision about which theory of helping you use most frequently will in part determine your use of questioning skills.

Understanding emotions is central to the helping process. Clients may most easily talk about feelings when they are asked about them directly (Hill, 2009; Tamase, 1991). If you are to reflect feelings and explore background emotions (Chapter 7), you will often have to ask questions.

Questions and Neuroscience: Memories, False or True?
Questions are often a good way to help a client discuss issues from the past residing in long-term memory, lodged primarily in the executive prefrontal cortex and hippocampus (Kolb & Wishaw, 2009). The goal of questions is to obtain information that will enhance client growth and ultimately generate new, more positive and accurate, neural networks.

However, questions that lead clients too much can result in their constructing stories of things that never happened. In a classic study, Loftus (1997, 2011) found that false memories could be brought out simply by reminding people of things that never happened. In other studies, brain scans have revealed that false memories activate different patterns than those that are true (Abe et al., 2008). Obviously, you as a counselor or therapist don't have that information available, and you may not know whether the memories a client reports are true or false. Be careful of putting your ideas into the client's head by means of probing questions.

If you use too many questions, you may have the best of intentions, but the possibility is that clients will end up taking your point of view rather than developing their own. We want clients to find their own direction. And directive intrusive questions may even lead to new false client memories.

the story carefully but still miss underlying issues relating to the client's depression. The open question "What is happening in your life right now or with your family?" might bring out information about an impending separation or divorce, a lost job, or some other issue underlying the concern. What you first interpreted as a classical clinical depression becomes modified by what is occurring in the client's life, and treatment takes a different direction.

An incident in Allen's life illustrates the importance of questions. His father became blind after open heart surgery. Was that a result of the surgery? No, it was because the physicians failed to ask the basic open question "Is anything else happening physically or emotionally in your life at this time?" If that question had been asked, the physicians would have discovered that Allen's father had developed severe and unusual headaches the week before surgery was scheduled, and they could have diagnosed an eye infection that was easily treatable with medication.

In counseling, a client may speak of tension, anxiety, and sleeplessness. You listen carefully and believe the problem can be resolved by helping the client relax and plan changes in her work schedule. However, you ask the client, "What else is going on in your life?" Having developed trust in you because of your careful listening and interest, the client finally opens up and shares a story of sexual harassment. At this point, the goals of the session change.

Useful questions from the helper that can provide more complete data include the following:

> What else is going on in your life?
> Looking back at what we've been talking about, what else might be added? You may even
>   have thought about something and not said it.
> Could you tell me a bit about whatever occurs to you at this moment? (This question often
>   provides surprising and helpful new information.)
> What else might a friend or family member add to what you've said? From a _____
>   perspective (insert ethnicity, race, sexual preference, religious, or other dimension), how
>   could your situation be viewed? (These questions change the focus and help clients see
>   their issues in a broader, network-based context of friends, family, and culture.)
> Have we missed anything?

▶ Empathy and Concreteness

To be empathic with a client requires that you understand specifically what the client is saying to you. Concreteness is valuable in empathic understanding. Seek specifics rather than vague generalities. As counselors, we are most often interested in specific feelings, specific thoughts, and specific examples of actions. One of the most useful of all open questions here is "Could you give me a specific example of . . . ?" Concreteness helps the session come alive and clarifies what the client is saying. Likewise, communication from the counselor — the directive, the feedback skill, and interpretation—needs to be concrete and understandable to the client.

▶ Example Counseling Session: Conflict at Work

Virtually all of us have experienced conflict on the job: angry, difficult customers, insensitive supervisors, lazy colleagues, or challenges from those whom we may supervise. In the following set of transcripts, we see an employee assistance counselor, Jamila, meeting with Kelly, a junior manager who has a conflict with Peter.

Closed Question Example

The first session illustrates how closed questions can bring out specific facts but can sometimes end in leading the client, even to the point of putting the counselor's ideas into the client's mind.

COUNSELOR AND CLIENT CONVERSATION	PROCESS COMMENTS
1. *Jamila:* Hi, Kelly. What's happening with you today?	Jamila has talked with Kelly once in the past about difficulties she was having in her early experiences supervising others for the first time. She begins with an open question that could also be seen as a standard social greeting.
2. *Kelly:* Well, I'm having problems with Peter again.	Jamila and Kelly have a good relationship. Not all clients are so ready to discuss their issues. More time for developing rapport and trust will be necessary for many clients, even on return visits.
3. *Jamila:* Is he arguing with you?	Jamila appears interested, is listening and demonstrating good attending skills. However, she asks a closed question; she is already defining the issue without discovering Kelly's thoughts and feelings. (Subtractive empathy)
4. *Kelly:* (hesitates) Not really, he's so difficult to work with.	Kelly sits back in her chair and waits for the counselor to take the lead.
5. *Jamila:* Is he getting his work in on time?	Jamila tries to diagnose the problem with Peter by asking a series of closed questions. It is much too early in the session for a diagnosis. (Subtractive; note that the counselor is supplying concreteness, not the client.)
6. *Kelly:* No, that's not the issue. He's even early.	
7. *Jamila:* Is his work decent? Does he do a good job?	Jamila is starting to grill Kelly. (Subtractive)
8. *Kelly:* That's one of the problems; his work is excellent, always on time. I can't criticize what he does.	
9. *Jamila:* (hesitates) Is he getting along with others on your team?	Jamila frowns and her body tenses as she thinks of what to ask next. Counselors who rely on closed questions suddenly find themselves having run out of questions to ask. They continue searching for another closed question usually further off the mark.
10. *Kelly:* Well, he likes to go off with Daniel, and they laugh in the corner. It makes me nervous. He ignores the rest of the staff—it isn't just me.	
11. *Jamila:* So, it's you we need to work on. Is that right?	Jamila has been searching for an individual to blame. Jamila relaxes a little as she thinks she is on to something. Kelly sits back in discouragement. (Very subtractive and off the mark)

COUNSELOR AND CLIENT CONVERSATION	PROCESS COMMENTS
12. *Kelly:* (hesitates and stammers) . . . Well, I suppose so . . . I . . . I . . . really hope you can help me work it out.	Kelly looks to Jamila as the expert. While she dislikes taking blame for the situation, she is also anxious to please and too readily accepts the counselor's diagnosis.

Closed questions can overwhelm clients and can be used as evidence to force them to agree with the counselor's ideas. While the session above seems extreme, encounters like this are common in daily life and even occur in counseling and psychotherapy sessions. There is a power differential between clients and counselors. It is possible that a counselor who fails to listen can impose inappropriate decisions on a client.

Open Question Example

The session is for the client, not the counselor. Using open questions, Jamila learns Kelly's story rather than the one she imposed with closed questions in the first example. Again, this interview is in the employee assistance office.

COUNSELOR AND CLIENT CONVERSATION	PROCESS COMMENTS
1. *Jamila:* Hi, Kelly. What's happening with you today?	Jamila uses the same easy beginning as in the closed question example. She has excellent attending skills and is good at relationship building.
2. *Kelly:* Well, I'm having problems with Peter again.	Kelly responds in the same way as in the first demonstration.
3. *Jamila:* More problems? Could you share more with me about what's been happening lately?	Open questions beginning with "could" provide some control to the client. Potentially a "could" question may be responded to as a closed question and answered with "yes" or "no." But in the United States, Canada, and other English-speaking countries, it usually functions as an open question. (Aiming toward concreteness, this is interchangeable empathy.)
4. *Kelly:* This last week Peter has been going off in the corner with Daniel, and the two of them start laughing. He's ignoring most of our staff, and he's been getting under my skin even more lately. In the middle of all this, his work is fine, on time and near perfect. But he is so impossible to deal with.	We are hearing Kelly's story. The anticipated result from open questions is that Kelly will respond with information. She provides an overview of the situation and shares how it is affecting her.
5. *Jamila:* I hear you. Peter is getting even more difficult and seems to be affecting your team as well. It's really stressing you out and you look upset. Is that pretty much how you are feeling about things?	When clients provide lots of information, we need to ensure that we hear them accurately. Jamila summarizes what has been said and acknowledges Kelly's emotions. The closed question at the end is a perception check, or checkout. Periodically checking with your client can help you in two ways: (a) It communicates to clients that you are listening and encourages them to continue; (b) it allows the client to correct any wrong assumptions you may have. (Interchangeable)

continued on next page

COUNSELOR AND CLIENT CONVERSATION	PROCESS COMMENTS
6. *Kelly:* That's right. I really need to calm down.	
7. *Jamila:* Let's change the pace a bit. Could you give me a specific example of an exchange you had with Peter last week that didn't work well?	Jamila asks for a concrete example. Specific illustrations of client issues are often helpful in understanding what is really occurring.
8. *Kelly:* Last week, I asked him to review a bookkeeping report prepared by Anne. It's pretty important that our team understand what's going on. He looked at me like, "Who are you to tell *me* what to do?" But he sat down and did it that day. Friday, at the staff meeting, I asked him to summarize the report for everyone. In front of the whole group, he said he had to review this report for me and joked about me not understanding numbers. Daniel laughed, but the rest of the staff just sat there. He even put Anne down and presented her report as not very interesting and poorly written. He was obviously trying to get me. I just ignored it. But that's typical of what he does.	Specific and concrete examples can be representative of recurring problems. The concrete specifics from one or two detailed stories can lead to a better a understanding of what is really happening. Now that Jamila has heard the specifics, she is better prepared to be helpful.
9. *Jamila:* Underneath it all, you're furious. Kelly, why do you imagine he is doing that to you?	Will the "why" question lead to the discovery of reasons? (Subtractive, note Kelly's response)
10. *Kelly:* (hesitates) Really, I don't know why. I've tried to be helpful to him.	The microskill did not result in the expected response. This is, of course, not unusual. Likely this is too soon for Kelly to know why. This illustrates a common problem with "why" questions.
11. *Jamila:* Gender can be an issue; men do put women down at times. Would you be willing to consider that possibility?	Jamila carefully presents her own hunch. But instead of expressing her own ideas as truth, she offers them tentatively with a "would" question and reframes the situation as "possibility." (Potentially additive)
12. *Kelly:* Jamila, it makes sense. I've halfway thought of it, but I didn't really want to acknowledge the possibility. But it is clear that Peter has taken Daniel away from the team. Until Peter came aboard, we worked together beautifully. (pause) Yes, it makes sense for me. I think he's out to take care of himself. I see Peter going up to my supervisor all the time. He talks to the female staff members in a demeaning way. Somehow, I'd like to keep his great talent on the team, but how when he is so difficult?	With Jamila's help, Kelly is beginning to obtain a broader perspective. She thinks of several situations indicating that Peter's ambition and sexist behavior are issues that need to be addressed. (Here we see Kelly adding new thoughts to help her look at the situation. Jamila's previous comment was additive.)
13. *Jamila:* So, the problem is becoming clearer. You want a working team and you want Peter to be part of it. We can explore the possibility of assertiveness training as a way to deal with Peter. But, before that, what do you bring to this situation that will help you deal with him?	Jamila provides support for Kelly's new frame of reference and ideas for where the session can go next. She suggests that time needs to be spent on finding positive assets and wellness strengths. Kelly can best resolve these issues if she works from a base of resources and capabilities. (Potentially additive; looks for concrete specifics of what Kelly can do rather than what she can't do)

COUNSELOR AND CLIENT CONVERSATION	PROCESS COMMENTS
14. *Kelly:* I need to remind myself that I really do know more about our work than Peter. I worked through a similar issue with Jonathan two years ago. He kept hassling me until I had it out with him. He was fine after that. I know my team respects me; they come to me for advice.	Kelly smiles for the first time. She has sufficient support from Jamila to readily come up with her strengths. However, don't expect it always to be that easy. Clients may return to their weaknesses and ignore their assets.
15. *Jamila:* Could you tell me specifically what happened when you sat down and faced Jon's challenge directly?	This "could" question searches for concrete specifics when Kelly handled a difficult situation effectively. Jamila can identify specific skills that Kelly can later apply to Peter. At this point, the session can move from problem definition to problem solution.

In this excerpt, we see that Kelly has been given more talk time and room to explore what is happening. The questions focused on specific examples clarify what is happening. We also see that question stems such as *why, how,* and *could* have some predictability in expected client responses. The positive asset search is a particularly relevant part of successful questioning. Issues are best resolved by emphasizing strengths.

You are very likely to work with clients who have similar interpersonal issues wherever you may practice. The previous case examples focus on the single skill of questioning as a way to bring out client stories. Questioning is an extremely helpful skill, but do not forget the dangers of using too many questions.

▶Concepts Into Action: Making Questions Work for You

Questions make the session work for me. I searched through many questions and found the ones that I thought most helpful in my own practice. I then memorized them and now I always draw on them as needed. Being prepared makes a difference.

—Norma Gluckstern Packard

Questions can be facilitative, or they can be so intrusive that clients want to say nothing. Use the ideas presented here to help you define your own questioning techniques and strategies and how questioning fits with your natural counseling style.

Questions Can Help Begin the Session

With verbal clients and a comfortable relationship, the open question facilitates free discussion and leaves plenty of room to talk. Here are some examples:

"What would you like to talk about today?"
"Could you tell me what prompted you to see me?"
"How have things been since we last talked together?"
"The last time we talked, you planned to talk with your partner about your sexual difficulties. How did it go this week?"

The first three open questions provide room for the client to talk about virtually anything. The last question is open but provides some focus for the session, building on material from the preceding week. These types of questions will work well for a highly verbal client. However, such open questions may be more than a nontalkative client can handle. It may be best to start the session with more informal conversation—focusing on the weather, a positive part of last week's session, or a current event of interest to the client. You can turn to the issues for this session as the client becomes more comfortable.

The First Word of Open Questions May Determine Client Response

Question stems often, but not always, result in anticipated outcomes. Use the following guidelines and you'll be surprised how effective these simple questions can be in gathering information.

What questions most often lead to facts.

"What happened?"
"What are you going to do?"

How questions may lead to an exploration of process or feeling and emotion.

"How could that be explained?"
"How do you feel about that?"

Why questions can lead to a discussion of reasons. Use *why* questions with care. While understanding reasons may have value, a discussion of reasons can also lead to sidetracks. In addition, many clients may not respond well because they associate *why* with a past experience of being grilled.

"Why is that meaningful to you?"
"Why do you think that happened?"

Could, can, or *would* questions are considered maximally open and also contain some advantages of closed questions. Clients are free to say, "No, I don't want to talk about that." *Could* questions suggest less counselor control.

"Could you tell me more about your situation?"
"Would you give me a specific example?"
"Can you tell me what you'd like to talk about today?"

Give it a try and you'll be surprised to see how effective these simple guidelines can be.

Open Questions Help Clients Elaborate and Enrich Their Story

A beginning counselor often asks one or two questions and then wonders what to do next. Even more experienced therapists can find themselves hard-pressed to know what to do next. To help the session start again and keep it moving, ask an open question on a topic the client presented earlier in the session.

"Could you tell me more about that?"
"How did you feel when that happened?"
"Given what you've said, what would be your ideal solution to the problem?"
"What might we have missed so far?"
"What else comes to your mind?"

Questions Can Reveal Concrete Specifics From the Client's World

The model question "Could you give me a specific example?" is one of the most useful open questions available. Many clients tend to talk in vague generalities; specific, concrete examples enrich the session and provide data for understanding action. Suppose, for example, that a client says, "Ricardo makes me so mad!" Some open questions that aim for concreteness and specifics might be

"Could you give me a specific example of what Ricardo does?"
"What does Ricardo do, specifically, that brings out your anger?"
"What do you mean by 'makes me mad'?"
"Could you specify what you do before and after Ricardo makes you mad?"

Closed questions can bring out specifics as well, but even well-directed closed questions may take the initiative away from the client. However, at the discretion of the counselor, closed questions may prove invaluable:

"Did Ricardo show his anger by striking you?"
"Does Ricardo tease you often?"
"Is Ricardo on drugs?"

Questions like these may encourage clients to say out loud what they have only hinted at before.

Questions Have Potential Problems

Questions can have immense value, but we must not forget their potential problems.

Bombardment/grilling.   Too many questions may give too much control to the counselor and tend to put many clients on the defensive.

Multiple questions.   Another form of bombardment, throwing out too many questions at once may confuse clients. However, it may enable clients to select which question they prefer to answer.

Questions as statements.   Some counselors may use questions to lead clients to answers that the counselor wants to hear. They can also be judgmental—for example, "Don't you think it would be helpful if you studied more?" This question clearly puts the client on the spot. On the other hand, "What do you think of trying relaxation exercises when you are tense?" might be helpful to get some clients thinking in new ways. Consider alternative and more direct routes of reaching the client. A useful standard is this: If you are going to make a statement, do not frame it as a question.

Why questions.   *Why* questions can put clients on the defensive and cause discomfort. As children, most of us experienced some form of "Why did you do that?" Any question that evokes a sense of being attacked can create client discomfort and defensiveness. Many experts suggest not using the *why* question at all.

However, the *why* question often becomes valuable as you help clients explore deeper issues of understanding. In Chapter 11, the microskills of interpretation/reframing and eliciting and reflecting meaning are frequently searching for underlying issues; they are explicitly or implicitly searching for *whys*. The developmental questions of discernment of life vision and meaning and the work of Viktor Frankl are particularly concerned with the *whys* of living.

In Cross-Cultural Situations, Questions Can Promote Distrust

If your life background and experience are similar to your client's, you may be able to use questions immediately and freely. If you come from a significantly different cultural background, your questions may be met with distrust and given only grudging answers. Questions place power with the counselor. A poor client who is clearly in financial jeopardy may not come back for another session after receiving a barrage of questions from a clearly middle-class counselor. If you are African American or European American and working with an Asian American or a Latino/a, an extreme questioning style can produce mistrust. If the ethnicities are reversed, the same problem can occur.

Allen was conducting research and teaching in South Australia with Aboriginal social workers. He was seeking to understand their culture and their special needs for training. Allen is naturally inquisitive and sometimes asks many questions. Nonetheless, the relationship between him and the group seemed to be going well. But one day, Matt Rigney,

---

BOX 5.2 · National and International Perspectives on Counseling Skills

Using Questions With Youth at Risk

Courtland Lee, Past President, American Counseling Association, University of Maryland

Malik is a 13-year-old African American male who is in the seventh grade at an urban junior high school. He lives in an apartment complex in a lower-middle-class (working-class) neighborhood with his mother and 7-year-old sister. Malik's parents have been divorced since he was 6, and he sees his father very infrequently. His mother works two jobs to hold the family together, and she is not able to be there when he and his sister come home from school.

Throughout his elementary school years, Malik was an honor roll student. However, since starting junior high school, his grades have dropped dramatically, and he expresses no interest in doing well academically. He spends his days at school in the company of a group of seventh- and eighth-grade boys who are frequently in trouble with school officials.

This case is one that is repeated among many African American early teens. But this problem also occurs among other racial/ethnic groups, particularly those who are struggling economically. And the same pattern occurs frequently even in well-off homes. There are many teens at risk for getting in trouble or using drugs.

While still a boy, Malik has been asked to shoulder a man's responsibilities as he must pick up things his mother can't do. Simultaneously, his peer group discounts the importance of academic success and wants to challenge traditional authority. And Malik is making the difficult transition from childhood to manhood without a positive male model.

I've developed a counseling program designed to empower adolescent Black males that focuses on personal and cultural pride. The full program is outlined in my book *Empowering Black Males* (1992) and focuses on the central question, "What is a strong Black man?" Although this question is designed for group discussion, it is also one for adolescent males in general, who might be engaging in individual counseling. The idea is to use this question to help the youth redefine in a more positive sense what it means to be strong and powerful. Some related questions that I find helpful are

▲ What makes a man strong?
▲ Who are some strong Black men that you know personally? What makes these men strong?
▲ Do you think that you are strong? Why?
▲ What makes a strong body?
▲ Is abuse of your body a sign of strength?
▲ Who are some African heroes or elders that are important to you? What did they do that made them strong?
▲ How is education strength?
▲ What is a strong Black man?
▲ What does a strong Black man do that makes a difference for his people?
▲ What can you do to make a difference?

Needless to say, you can't ask an African American adolescent or a youth of any color these questions unless you and he are in a positive and open relationship. Developing sufficient trust so that you can ask these challenging questions may take time. You may have to get out of your office and into the school and community to become a person of trust.

My hope for you as a professional counselor is that you will have a positive attitude when you encounter challenging adolescents. They are seeking models for a successful life, and you may become one of those models yourself. I hope you think about establishing group programs to facilitate development and that you'll use some of these ideas with adolescents to help move them toward a more positive track.

whom Allen felt particularly close to, took him aside and gave some very useful corrective feedback:

You White fellas! . . . Always asking questions! Let me tell you what goes on in my mind when a White person asks me a question. First, my culture considers many questions rude. But I know you, and that's what you do. But this is what goes on in my mind when you ask me a question. First, I wonder if I can trust you enough to give you an honest answer. Then, I realize that the question you asked is too complex to be answered in a few words. But I know you want an answer. So I chew on the question in my mind. Then, you know what? Before I can answer the first question, you've moved on to the next question!

Allen was lucky he had developed enough trust and rapport that Matt was willing to share his perceptions. Many people of color have said that this kind of feedback represents how they feel about interactions with White people. People with disabilities, people who may be gay, lesbian, bisexual, transgender, questioning, spiritually conservative persons, and many others—anyone, in fact—may be distrustful of the counselor who uses too many questions.

On the other hand, questions can be useful in group discussions to help at-risk youth redefine themselves in a more positive way, as suggested in Box 5.2.

▶Using Questions to Identify Strengths

Personal Strength Inventory

Clients tend to talk about their problems and what they can't do. This puts them "off balance." A structured questioning format can help clients identify their previous successes and strengths. Basically, we can help them center and feel better about themselves through a strength inventory. What is the client doing right?

As part of any session, I like to do a strength inventory. Let's spend some time right now identifying some of the positive experiences and strengths that you either have now or have had in the past.

▲ Could you tell me a story about a success you have had sometime in the past? I'd like to hear the concrete details.

▲ Tell me about a time in the past when someone supported you and what he or she did. What are your currently available support systems?

▲ What are some things you have been proud of in the past? Now?

▲ What do you do well, or what do others say you do well?

Cultural/Gender/Family Strength Inventory

Here we move outside the individual and look at context for positive strengths.

▲ Looking at your ethnic/racial/spiritual history, can you identify some positive strengths, visual images, and experiences that you have now or have had in the past?

▲ Can you recall a friend or family member of your own gender who represents some type of hero in the way he or she dealt with adversity? What did that person do? Can you develop an image of her or him?

▲ We all have family strengths despite frequent family concerns. Family can include our extended family, our stepfamilies, and even those who have been special to us over time. For example, some people talk about a special teacher, a school custodian, or an older person who was helpful. Could you tell me concretely about them and what they mean to you?

Positive Exceptions to the Concern

Searching for times when the problem doesn't occur is often useful. This is another approach that is common in brief counseling. With this information, you can determine what is being done right and encourage more of the same.

▲ Let's focus on the exceptions. When is the problem or concern absent or a little less difficult? Please give me an example of one of those times.

▲ Few problems happen all the time. Could you tell me about a time when it didn't happen? That may give us an idea for a solution.

▲ What is different about this example from the usual?

▲ How did the more positive result occur?

▲ How is that different from the way you usually handle the concern?

Many clients will be hesitant to say good things about themselves. Your observations and feedback can be helpful to them in developing a new view of themselves. The "What else?" question provides an opportunity for the client to add more strengths and resources to your feedback.

You obviously will not have time for all these possible strength and positive asset searches. But when you focus only on the negative story, you place your clients in a very vulnerable position. Do not use the positive asset search to cover up or hide basic issues. Rather, wellness strengths are resources for clients to use in resolving their concerns.

▶Using Open and Closed Questions With Less Verbal Clients

Generally, open questions are preferred over closed questions. Yet it must be recognized that open questions require a verbal client, one who is willing to share with you. Here are some suggestions to encourage clients to talk with you more freely.

Build Trust at the Client's Pace

A central issue with hesitant clients is trust. Extensive questioning too early can make trust building a slow process with some clients. If the client is required to meet with you or is culturally different from you, he or she may be less willing to talk. Trust building and rapport need to come first, and your own natural openness and social skills are necessary. With some clients, trust building may take a full session or more.

Search for Concrete Specifics

Some counselors and many clients talk in vague generalities. We call this "talking high on the abstraction ladder." This may be contrasted with concrete and specific language, where what is said immediately makes sense. If your client is talking in very general terms and is hard to understand, it often helps to ask, "Could you give me a *specific concrete* example?"

As the examples become clearer, ask even more specific questions: "You said that you are not getting along with your teacher. What specifically did your teacher say (or do)?" Your chances for helping the client talk will be greatly enhanced when you focus on concrete events in a nonjudgmental fashion, avoiding evaluation and opinion.

Following are some examples of concrete questions focusing on specifics.

▲ Draw out the linear sequence of the story: "What happened first? What happened next? What was the result?"

▲ Focus on observable concrete actions: "What did the other person say? What did he or she do? What did you say or do?"

▲ Help clients see the result of an event: "What happened afterward? What did you do afterward? What did he or she do afterward?" Sometimes clients are so focused on the event that they don't yet realize it is over.

▲ Focus on emotions: "What did you feel or think just before it happened? During? After? What do you think the other person felt?"

Note that each of these questions requires a relatively short answer. These types of open questions are more focused and can be balanced with some closed questions. Do not expect your less verbal client to give you full answers to these questions. You may need to ask closed questions to fill in the details and obtain specific information. "Did he say anything?" "Where was she?" "Is your family angry?" "Did they say 'yes' or 'no'?"

A *leading* closed question is dangerous, particularly with children. In earlier examples, you have seen that a long series of closed questions can bring out the story, but may provide only the client's limited responses to *your* questions rather than what the client really thought or felt. Worse, the client may end up adopting your way of thinking or may simply stop coming to see you.

▶Summary: Making Your Decision About Questions

We began this chapter by asking you to think carefully about your personal experience with questions. Clearly, their overuse can damage the relationship with the client. On the other hand, questions do facilitate conversation and help ensure that a complete picture is obtained. Questions can help the client bring in missing information. Among such questions are "What else?" "What have we missed so far?" and "Can you think of something important that is occurring in your life right now that you haven't shared with me yet?"

Person-centered theorists and many professionals sincerely argue against the use of any questions at all. They strongly object to the control implications of questions. They point out that careful attending and use of the listening skills can usually bring out major client issues. If you work with someone culturally different from you, a questioning style may develop distrust. In such cases, questions need to be balanced with self-disclosure and listening.

Our position on questions is clear: We believe in questions, but we also fear overuse and the fact that they can reduce equality in the session. We are impressed by the brief solution-focused counselors who seem to use questions more than any other skill, but are still able to respect their clients and help them change. On the other hand, we have seen students who have demonstrated excellent attending skills regress to using only questions. Questions can be an easy "fix," but they require listening to the client if they are to be meaningful.

The positive asset search has been a foundation of the microskills program since its beginnings. We believe that Carl Rogers was correct when he focused on positive regard and unconditional acceptance. We have noted again and again that therapy all too often ends in a self-defeating repetition of problems. Questions that bring out strengths and resources often lead clients to specific assets that they can use to help resolve issues and problems.

The most useful chapter summary will be your impressions and decisions. Where do you personally stand on the use of questions?

▼▼▼▼	**Key Points**
Value of questions	Questions help begin the session, open new areas for discussion, assist in pinpointing and clarifying issues, and assist the client in self-exploration.
Open questions	Questions can be described as open or closed. *Open questions* are those that can't be answered in a few words. They encourage others to talk and provide you with maximum information. Typically, open questions begin with *what*, *how*, *why*, or *could*. One of the most helpful of all open questions is "Could you give a specific example of . . . ?"
Closed questions	*Closed questions* are those that can be answered in a few words or sentences. They have the advantage of focusing the session and bringing out specifics, but they place the prime responsibility for talk on the counselor. Closed questions often begin with *is*, *are*, or *do*. An example is "Where do you live?" Note that a question, open or closed, on a topic of deep interest to the client will often result in extensive talk time if interesting or important enough. If a session is flowing well, the distinction between open and closed questions becomes less relevant.
Newspaper questions for context	Key to your understanding after reading the chapter is the discovery of the general framework for diagnosis and question asking provided by the newspaper reporter framework of *who*, *what*, *when*, *where*, *how*, and *why*. ▲ *Who* is the client? What are key personal background factors? Who else is involved? ▲ *What* is the issue? What are the specific details of the situation? ▲ *When* does the problem occur? What immediately preceded and followed the situation? ▲ *Where* does the issue occur? In what environments and situations? ▲ *How* does the client react? How does he or she feel about it? ▲ *Why* does the problem or concern happen?
"What else?" questions	*What else* is there to add to the story? Have we missed anything? "What else?" questions bring out missing data. These are maximally open and allow the client considerable control.
The need for a positive approach	Counseling and psychotherapy are typically seen as focused on life challenges and problems. But this focus needs to be balanced with questions that bring out client strengths, supports in the family or friendship group, and past and present accomplishments. Counseling session training can overemphasize concerns and difficulties. A positive approach is needed for balance.
Multicultural issues	All these questions may turn off some clients. Some cultural groups find North American rapid-fire questions rude and intrusive, particularly if asked before trust is developed. Yet questions are very much a part of Western culture and provide a way to obtain information that many clients find helpful. Questions help us find the client's personal, family, and cultural/contextual resources. If properly structured and your clients know the real purpose is to help them reach their own goals, questions may be used more easily.
Be positive	Emphasizing only negative issues results in a downward cycle of depression and discouragement. The positive asset search, strength emphasis, positive psychology, and wellness need to balance discussion of client issues and concerns. What is the client doing right? What are the exceptions to the problem? What are the client's new options? How would these options enrich the client's life?

▶Competency Practice Exercises and Portfolio of Competence

 Go to CengageBrain.com to access Counseling CourseMate, where you will find an interactive ebook, quizzes, videos, interactive counseling and psychotherapy exercises, the Portfolio of Competence, case studies, a practice test, and more.

Take time to master the many concepts and skills presented in this chapter and make them a useful part of your counseling or psychotherapy. These exercises will help you achieve this goal, but you should continue to work on these concepts throughout this book and beyond. With practice, all these materials will become clearer and, most important, will become a part of your natural style.

Individual Practice

Exercise 1. Writing Closed and Open Questions   Select one or more of the following client stories and then write open and closed questions to elicit further information. Can you ask closed questions designed to bring out specifics of the situation? Can you use open questions to facilitate further elaboration of the topic, including the facts, feelings, and possible reasons? What special considerations might be beneficial with each person as you consider age-related multicultural issues?

> *Jordan* (age 15, African American): I was walking down the hall and three guys came up to me and called me "queer" and pushed me against the wall. They started hitting me, but then a teacher came up.
>
> *Alicja* (age 35, Polish American): I've been passed over for a promotion three times now. Each time, it's been a man who has been picked for the next level. I'm getting very angry and suspicious.
>
> *Dominique* (age 78, French Canadian): I feel so badly. No one pays any attention to me in this "home." The food is terrible. Everyone is so rude. Sometimes I feel frightened.

Write open questions for one or more of the above. The questions should be designed to bring out broad information, facts, feelings and emotions, and reasons.

> Could . . . ?
> What . . . ?
> How . . . ?
> Why . . . ?

Now generate three closed questions that might bring out useful specifics of the situation.

> Do . . . ?
> Are . . . ?
> Where . . . ?

Finally, write a question designed to obtain concrete examples and details that might make the problem more specific and understandable.

Exercise 2. Observation of Questions in Your Daily Interactions   This chapter has talked about the basic question stems *what, how, why,* and *could,* and how clients respond differently to each. During a conversation with a friend or acquaintance, try these five basic question stems sequentially:

> *Could* you tell me generally what happened?
> *What* are the critical facts?

How do you feel about the situation?
Why do you think it happened?
What else is important? What have we missed?

Record your observations here. Were the anticipated results or outcomes fulfilled? Did the person provide you, in order, with (a) a general picture of the situation, (b) the relevant facts, (c) personal feelings about the situation, and (d) background reasons that might be causing the situation?

Group Practice

The following exercise is suggested for practice with questions. The objective is to use both open and closed questions. The instructional steps for practice are abbreviated from those described in Chapter 3, on attending behavior. As necessary, refer to those instructions for more detail on the steps for systematic practice.

Exercise 3. Systematic Group Practice on Open and Closed Questions

Step 1: Divide into practice groups.

Step 2: Select a group leader.

Step 3: Assign roles for the first practice session.

▲ Client
▲ Counselor
▲ Observer 1, who uses the Feedback Form: Questions (Box 5.3) and leads the microsupervision process. Remember to focus on counselor strengths as well as areas for improvement.
▲ Observer 2, who runs equipment, keeps time, and also completes the form.

Step 4: Plan. The counselor should plan to use both open and closed questions. Include in your practice session the key *what*, *how*, *why*, and *could* questions. Add *what else* for enrichment.

Discuss a work challenge. The client may share a present or past interpersonal job conflict. The counselor first draws out the conflict, then searches for positive assets and strengths.

Suggested alternative topics might include the following:

▲ A friend or family member in conflict
▲ A positive addiction (such as jogging, health food, biking, team sports)
▲ Strengths from spirituality or ethnic/racial background

Step 5: Conduct a 3- to 6-minute practice session using only questions. The counselor practices open and closed questions and may wish to have handy a list of suggested question stems. The client seeks to be relatively cooperative and talkative but should not respond at such length that the counselor has only a limited opportunity to ask questions. More time will be needed if you decide on a more challenging topic.

Step 6: Review the practice session and provide feedback to the counselor for 12 minutes. Remember to stop the audio or video recording periodically and listen to or view key happenings several times for increased clarity. Generally speaking, it is wise to provide some feedback before reviewing the recording, but this sometimes results in a failure to view or listen to the recording at all.

Step 7: Rotate roles.

Feedback Form: Questions

You can download this form from Counseling CourseMate at CengageBrain.com.

_____ (DATE)

_____   _____
(NAME OF COUNSELOR)   (NAME OF PERSON COMPLETING FORM)

Instructions: On the lines below, list as completely as possible the questions asked by the counselor. At a minimum, indicate the first key words of the question (*what, why, how, do, are*, and so on). Indicate whether each question was open (O) or closed (C). Use additional paper as needed. Does the session focus on strengths and goal attainment? How well did the counselor listen as well as ask questions?

1. _____

2. _____

3. _____

4. _____

5. _____

6. _____

7. _____

8. _____

9. _____

10. _____

1. Which questions seemed to provide the most useful client information?

2. Provide specific feedback on the attending skills of the counselor.

3. Discuss the use of the positive asset search and wellness, as well as the use of questions.

Portfolio of Competence

Determining your own style and theory, the apex of the microskills hierarchy can be best accomplished on a base of competence. Each chapter closes with a reflective exercise asking your thoughts and feelings about what has been discussed. By the time you finish this book, you will have a substantial record of your competencies and a good written record as you move toward determining your own style and theory.

Use the following checklist to evaluate your present level of mastery. Check those dimensions that you currently feel able to do. Those that remain unchecked can serve as future goals. Do not expect to attain intentional competence on every dimension as you work through this book. You will find, however, that you will improve your competencies with repetition and practice.

Level 1: Identification and classification

❑ Ability to identify and classify open and closed questions
❑ Ability to discuss, in a preliminary fashion, issues in diversity that occur in relation to questioning
❑ Ability to write open and closed questions that might anticipate what a client will say next

Level 2: Basic competence.   Aim for this level of competence before moving on to the next skill area.

❑ Ability to ask both open and closed questions in a role-played session
❑ Ability to obtain longer responses to open questions and shorter responses to closed questions

Level 3: Intentional competence.   Work toward intentional competence throughout this book. All of us can improve our skills, regardless of where we start.

❑ Ability to use closed questions to obtain necessary facts without disturbing the client's natural conversation
❑ Ability to use open questions to help clients elaborate their stories
❑ Ability to use *could* questions and, as anticipated, obtain a general client story ("Could you tell me generally what happened?" "Could you tell me more?")
❑ Ability to use *what* questions to facilitate discussion of facts
❑ Ability to use *how* questions to bring out feelings ("How do you feel about that?") and information about process or sequence ("How did that happen?")
❑ Ability to use *why* questions to bring out client reasons ("Why do think your spouse/lover responds coldly?")
❑ Ability to bring out concrete information and specifics ("Could you give me a specific example?")

Level 4: Psychoeducational teaching competence.   As stated earlier, do not expect to become skilled in teaching groups or peer counselors skills at this point. You may find, however, that some clients benefit from direct instruction in open questions focusing on others' thoughts and opinions rather than their own. Those who talk too much about themselves find this skill useful in breaking through their self-absorption. At the same time, please point out the dangers of too many questions, especially that *why* question, which can put others on the spot and make them defensive.

❑ Ability to teach clients in a helping session the social skill of questioning. You may either tell clients about the skill or practice a role-play with them.
❑ Ability to teach small groups the skills of questioning.

▶Determining Your Own Style and Theory: Critical Self-Reflection on Questioning

This chapter has focused on the pluses and minuses of using questions in the session. While we, as authors, obviously feel that questions are an important part of the counseling process, we have tried to point out that there are those who differ from us. Questions clearly can get in the way of effective relationships in counseling and psychotherapy.

Regardless of what any text on counseling and psychotherapy says, the fact remains that it is *you* who will decide whether to implement the ideas, suggestions, and concepts. What single idea stands out for you among all those presented in this chapter, in class, or through informal learning? What stands out for you is likely to be important as a guide toward your next steps. What are your thoughts on multiculturalism and how it relates to your use of questions? How might you use ideas in this chapter to begin the process of establishing your own style and theory?

▶Our Thoughts About Benjamin

We would probably start the session by explaining to Benjamin that we'd like to know what he is thinking about his future after he completes school. We would begin the session with some informal conversation about current school events or something personal we know about him. The first question might be stated something like this: "You'll soon be starting your senior year; what have you been thinking about doing after you graduate?" If this question opens up some tentative ideas, we'd listen to these and ask him for elaboration. If he focuses on indecision between volunteering for the army or entering a local community college or the state university, we'd likely ask him some of the following questions:

"What about each of these appeals to you?"
"Could you tell me about some of your strengths that would help you in the army or college?"
"If you went to college, what might you like to study?"
"How do finances play a role in these decisions?"
"Are there any negatives about any of these possibilities?"
"How do you imagine your ideal life 10 years from now?"

On the other hand, Benjamin just might say to any of these, "I don't know, but I guess I better start thinking about it" and look to you for guidance. You sense a need to review his past likes and dislikes as possible clues to the future.

"What courses have you liked best in high school?"
"What have been some of your activities?"
"Could you tell me about the jobs you've had in the past?"
"Could you tell me about your hobbies and what you do in your spare time?"
"What gets you most excited and involved?"
"What did you do that made you feel most happy in the past year?"

Out of questions such as these, we may see patterns of ability and interest that suggest actions for the future.

If Benjamin is uncomfortable in the counseling office, all of these questions might put him off. He might feel that we are grilling him and perhaps even see us as intruding in his world. Generally speaking, getting this type of important information and organizing it requires the use of questioning. But questions are effective only if you and the client are working together and have a good relationship.

CHAPTER 6

Encouraging, Paraphrasing, and Summarizing
Key Skills of Active Listening

The possibility is only one sentence away. . . . Our goal is to make the eyes shine!

—Andrew Zander

Mission of "Encouraging, Paraphrasing, and Summarizing"

Clients need to know that the counselor or psychotherapist hears what they say, sees their point of view, and feels their world as they retell their experience. Encouraging, paraphrasing, and summarizing are active listening skills that are the heart of the basic listening sequence and help to build empathy. When clients sense that their story is heard, they open up and become more ready for change.

Chapter Goals and Competency Objectives Awareness, knowledge, skills, and actions in active listening skills will enable you to

▲ Help clients clarify what they are trying to say.

▲ Encourage clients to elaborate key points and provide further information, thus clarifying understanding and meaning.

▲ Check on the accuracy of your listening by saying back the key words and ideas in client stories.

▲ Build a base for constructive change, by listening for strengths and demonstrating positive regard.

▲ Help clients organize the key aspects of their issues and concerns through periodic summarization.

Jennifer: (enters the room and starts talking immediately) I really need to talk to you. I don't know where to start. This term things were going pretty well, but I just got my last exam back and it was a disaster, maybe because I haven't studied much lately. I've been sort of going out with a guy for the last month—we were hooking up, but that's over as of last night. . . . (pause) But what really bothers me is that my Mom and Dad called last Monday and they are going to separate. I know that they have fought a lot, but I never thought it would come to this. I'm thinking of going home, but I'm afraid to . . .

Jennifer continues for another 3 minutes in much the same vein, repeating herself somewhat, and seems close to tears. At times the data are coming so fast that it is hard to follow her. Finally she stops and looks at you expectantly.

What might be going through your mind about Jennifer at this moment? What would you say and do to help her feel that you understand her and empathize? Given our commitment to emphasizing strengths and positive assets, do you have any ideas on when and how to talk about something other than problems? Compare your thoughts with ours below.

When working with Jennifer, a useful first step would be to say, "I hear that you're really hurting right now," and then to summarize the essence of Jennifer's several issues and say them back to her. As part of this initial response, use a **checkout** (e.g., "Have I heard you correctly?") to see how accurate your listening was. The checkout (sometimes called a *perception check*) lets you confirm the accuracy of your summary. You could follow this by asking her, "You've talked about many things. Where would you like to start today?"

Choosing to focus first on the precipitating crisis is another possible strategy. We could start with the breakup or, more likely, Jennifer's parents' separation, as the latter seems to be what bothers her the most, but she needs to tell us what is most important to her. Then, we can restate, paraphrase, and summarize some of her key ideas. Doing this is likely to help her focus on one key issue before turning to the others. The other concerns, particularly the breakup, could relate to the parental separation. Once a direction is found, other issues can be discussed later.

We need to keep in mind Jennifer's potential strengths and resources. Obviously, at the moment she doesn't feel that she has any. But, if you listen carefully, fairly soon you will find a list of strengths. There is a clue in her first words, "things were going pretty well." At an appropriate point, paraphrasing back those key words may open the way for learning resources and strengths. Here you may find a number of friends, a previously solid academic record, and before the separation, a good relationship with her parents. All these can be summarized as positive assets that can be used to help her deal with the immediate and longer-term issues that she currently faces.

▶Introduction: Active Listening

Listening is an active process. You do not just sit and listen to a story. Whether using attending skills or the skills presented in this chapter, you are actively involved in the session. If you use encouraging, paraphrasing, or summarizing skills as defined here, you can *anticipate* how clients will respond.

ENCOURAGING (using encouragers and restatements)	ANTICIPATED RESULT
Give short responses that help clients keep talking. They may be verbal restatements (repeating key words and short statements) or nonverbal actions (head nods and smiling).	Clients elaborate on the topic, particularly when encouragers and restatements are used in a questioning tone of voice.
PARAPHRASING (also known as reflection of content)	**ANTICIPATED RESULT**
Shorten, clarify the essence of what has just been said, but be sure to use the client's main words when you paraphrase. Paraphrases are often fed back to the client in a questioning tone of voice.	Clients will feel heard. They tend to give more detail without repeating the exact same story. If a paraphrase is inaccurate, the client has an opportunity to correct the counselor.
SUMMARIZING	**ANTICIPATED RESULT**
Summarize client comments and integrate thoughts, emotions, and behaviors. This technique is similar to paraphrase but used over a longer time span. Important in the summary is that you seek to find strengths and resources that support the client.	Clients will feel heard and often learn how the many parts of their stories are integrated. The summary tends to facilitate a more centered and focused discussion. The summary also provides a more coherent transition from one topic to the next or a way to begin or end a full session.

Active listening demands that you participate fully by helping the client clarify, enlarge, and enrich the story. It requires that you be able to hear small changes in thoughts, feelings, and behaviors. It asks that you walk in the other person's shoes. Active listening demands serious attention to empathy—truly being with and understanding the client as fully as possible.

Encouraging, paraphrasing, and summarizing are basic to empathic understanding and enable you to communicate to clients that they have been heard. In these accurate listening skills, you do not mix your own ideas with what clients have been saying. You say back to the clients what you have heard, using their key words. You help clients by distilling, shortening, and clarifying what has been said.

Encouragers are a variety of verbal and nonverbal means that the counselor or therapist can use to prompt clients to continue talking. They include head nods, open-handed gestures, phrases such as "uh-huh," and the simple repetition of key words the client has uttered. *Restatements* are extended encouragers, the repetition of two or more words exactly as used by the client. In addition, appropriate smiling and interpersonal warmth are major encouragers that help clients feel comfortable and keep talking in the session.

Paraphrasing, sometimes called reflection of content, feeds back to the client the essence of what has just been said. The listener shortens and clarifies the client's comments. Paraphrasing is not parroting; it is using some of your own words plus the exact main words of the client.

Summarizations are similar to paraphrases but are used to clarify and distill what the client has said over a longer time span. Summarizations may be used to begin or end a session, to move to a new topic, or to clarify complex issues. Most important, summarizing helps both the client and you organize thinking about what is happening in the session.

▶Empathy, Unconditional Positive Regard, and Active Listening Skills

Empathy is based on attending and observing, as explained in Chapter 3. But there are additional dimensions that help clarify empathy. Carl Rogers gave extensive attention to **unconditional positive regard**. Another word for this concept is **acceptance**, a term that is perhaps a little easier to understand and work with.

People nurture our growth by being accepting—by offering us what Rogers called unconditional positive regard. This is an attitude of grace, an attitude that values us even knowing our failings. It is a profound relief to drop our pretenses, confess our worst feelings, and discover that we are still accepted. In a good marriage, a close family, or an intimate friendship, we are free to be spontaneous without fearing the loss of others' esteem (Myers, 2013).

Acceptance of the client as he or she *is* can be a challenge. It is easy to offer unconditional positive regard to the child who has been bullied, but far more of an effort to accept the bully. People whose thoughts and behaviors are troublesome to you require a special effort to be empathic. It helps if you think of accepting the person, but not the thoughts or behaviors that confront you. If you are to reach the child or teen diagnosed with conduct disorder, it does no good to become angry or blame. Start by listening to where the client comes from—you likely have more potential for acceptance than you thought.

Empathy can be rated as interchangeable, additive, or subtractive. In a subtractive response, the counselor focuses too much on the negative, forgetting possibility. In an interchangeable response, the counselor notes or reflects accurately what the client has talked about. This is the foundation of effective empathy—hearing what the client has said. This is a good place to start with that difficult client. In an additive response, positives are included. Even the most difficult client has some good qualities and strengths. Listen for them, and make the positives part of your listening skills.

A special type of additive response occurs when the counselor points out that even in the most difficult situation, the client has something to offer. For example, let's assume that Jennifer, introduced above, has spent 10 minutes searching through her thoughts and feelings about her parents' separation. There have been tears, some expression of anger, but also a few moments when she felt that her parents were trying to support her, even though they themselves seemed finished. Search constantly for positives and strengths in your clients. First, summarize the negative story accurately and succinctly, but then, even more important, add the search for strengths. Here is a possibility that includes additive dimensions.

Counselor: (after summarizing Jennifer's story) Jennifer, in the middle of all this, I hear strengths. You are close to both your parents and they both want to be with you, despite their issues. I hear your caring, and I hear them caring for you as well. And, fortunately, the rest of your grades are fine, so you can pick things up there as well. How does that sound to you?

Such responses do not solve difficulties instantly, but they can instill hope and provide a framework from which clients can explore issues and move toward resolution.

To understand the importance of active listening, let us take a moment to consider the opposite—subtractive skills that may look like listening but minimize client concerns, distort what has been said, or fail to connect with anything positive in the client or the situation.

Jennifer, I hear what you are saying. It's tough, but I am sure that you will get over this. (Early reassurance minimizes her issues and is seldom helpful.)

I understand that your parents are considering separating and you want to stop them. (This is a distortion. The parents have decided to separate and Jennifer has said nothing about her interest or ability to stop them.)

I can see that you are hurting and not sure where to turn next as you talk about your parents and failing that exam. But you also broke up with this guy. Tell me more about what happened. (While this issue likely needs to be explored at some point, this lead focuses again on the negative, leaving the client hanging alone without positive supports.)

In all three of these examples, the counselors have indeed "mixed their own things" with those of the client. Listening to oneself is not listening to the client.

▶Example Counseling Session: They Are Teasing Me About My Shoes

The following sample session is an edited version of a videotaped intervention conducted by Mary Bradford Ivey with Damaris, a child actor, role-playing the issue. Damaris is an 11-year-old sixth grader. This session presents a child's problem, but all of us, regardless of age, experience nasty teasing and put-downs, often in our closest relationships. Mary first draws out the child's story about teasing and then Damaris's thoughts and feelings about the teasing. Mary follows with a focus on the child's strengths, an example of the wellness approach.

Mary uses many encouragers and restatements. A review of the entire video transcript reveals nine minimal encouragers ("oh . . . ," "uh, huh," and single word utterances), two positive encouragers ("that's great," "nice"), four additional brief restatements, and numerous smiles and head nods. Children demand constant involvement, and showing your interest and good humor is even more essential. Active listening is especially key with children, as they tend to respond more briefly than adults.

COUNSELOR AND CLIENT CONVERSATION	PROCESS COMMENTS
1. *Mary:* Damaris, how're you doing?	The relationship between Mary and Damaris is already established; they know each other through school activities.
2. *Damaris:* Good.	She smiles and sits down.
3. *Mary:* I'm glad you could come down. You can use these markers if you want to doodle or draw something while we're talking. I know—you sort of indicated that you wanted to talk to me a little bit.	Mary welcomes the child and offers her something to do with her hands. Many children get restless just talking. Damaris starts to draw almost immediately. You may do better with an active male teen by taking him to the basketball court while you discuss issues. It can also help to have things available for adults to do with their hands.

COUNSELOR AND CLIENT CONVERSATION	PROCESS COMMENTS
4. *Damaris*: In school, in my class, there's this group of girls that keep making fun of my shoes, just 'cause I don't have Nikes.	Damaris looks down and appears a bit sad. She stops drawing. Children, particularly the "have-nots," are well aware of their economic circumstances. Some children have used sneakers; Damaris, at least, has newer sneakers.
5. *Mary*: They "keep making fun of your shoes"?	Encourage in the form of a restatement using Damaris's exact key words. (Interchangeable empathic response)
6. *Damaris*: Well, they're not the best; I mean—they're not Nikes, like everyone else has.	Damaris has a slight angry tone mixed with her sadness. She starts to draw again.
7. *Mary*: Yeah, they're nice shoes, though. You know?	It is sometimes tempting to comfort clients rather than just listen. Mary offers reassurance; a simple "uh-huh" would have been more effective. However, reassurance later in the session may be a very effective intervention. (This is slightly subtractive, but unlikely to harm the flow of the session. Do not expect every response to reach level 3.)
8. *Damaris*: Yeah. But my family's not that rich, you know. Those girls are rich.	Clients, especially children, hesitate to contradict the counselor. Notice that Damaris says, "But. . . ." When clients say, "Yes, but . . . ," counselors are off track and need to change their style.
9. *Mary*: I see. And the others can afford Nike shoes, and you have nice shoes, but your shoes are just not like the shoes the others have, and they tease you about it?	Mary backs off her reassurances and paraphrases the essence of what Damaris has been saying using her key words. (Interchangeable)
10. *Damaris*: Yeah. . . . Well, sometimes they make fun of me and call me names, and I feel sad. I try to ignore them, but still, the feeling inside me just hurts.	If you paraphrase or summarize accurately, a client will usually respond with *yeah* or *yes* and continue to elaborate the story.
11. *Mary*: It makes you feel hurt inside that they should tease you about shoes.	Mary reflects Damaris's feelings. The reflection of feeling is close to a paraphrase and is elaborated in the following chapter. (Interchangeable)
12. *Damaris*: Mmm-hmm. (pause) It's not fair.	Damaris thinks about Mary's statement and looks up expectantly as if to see what happens next. She thinks back on the basic unfairness of the whole situation.
13. *Mary*: So far, Damaris, I've heard how the kids tease you about not having Nikes and that it really hurts. It's not fair. You know, I think of you, though, and I think of all the things that you do well. I get . . . you know . . . it makes me sad to hear this part because I think of all the talents you have, and all the things that you like to do and—and the strengths that you have.	Mary's summary covers most of Demaris's comments, and Mary also discloses some of her own feelings. Sparingly used self-disclosure can be helpful. Mary begins the positive asset search by reminding Damaris that she has strengths to draw from. (Interchangeable; the self-disclosure focusing on strengths—if accepted—can possibly be additive.)
14. *Damaris*: Right. Yeah.	Damaris smiles slightly and relaxes a bit. Mary's statement was additive.
15. *Mary*: What comes to mind when you think about all the positive things you are and have to offer?	An open question encourages Damaris to think about her strengths and positives. (Additive)

continued on next page

COUNSELOR AND CLIENT CONVERSATION	PROCESS COMMENTS
16. *Damaris*: Well, in school, the teacher says I'm a good writer, and I want to be a journalist when I grow up. The teacher wants me to put the last story I wrote in the school paper.	Damaris talks a bit more rapidly and smiles. Virtually all clients feel better able to deal with their issues when they are reminded of their abilities and strengths.
17. *Mary*: You want to be a journalist, 'cause you can write well? Wow!	Mary enthusiastically paraphrases positive comments using Damaris's own key words. (Interchangeable, somewhat additive because of vocal tone.)
18. *Damaris*: Mmm-hmm. And I play soccer on our team. I'm one of the people that plays a lot, so I'm like the leader, almost, but . . . (Damaris stops in mid-sentence.)	Damaris has many things to feel good about; she is smiling for the first time in the session.
19. *Mary*: So, you are a scholar, a leader, and an athlete. Other people look up to you. Is that right? So how does it feel when you're a leader in soccer?	Mary has added *scholar* and *athlete* for clarification and elaboration of the positive asset search. She knows from observation on the playground that other children do look up to Damaris. Counselors may add related words to expand the meaning. Mary wisely avoids leading Damaris and uses the checkout, "Is that right?" Mary also asks an open question about feelings. And we note that Damaris used that important word "but." Do you think that Mary should have followed up on that, or should she continue with her search for strengths? (Interchangeable with some additive dimensions)
20. *Damaris*: (small giggle, looking down briefly) Yeah. It feels good.	Looking down is not always sadness! The spontaneous movement of looking down briefly is termed the "recognition response." It most often happens when clients learn something new and true about themselves. Damaris has internalized the good feelings. Her response reveals that Mary's comment was indeed additive.
21. *Mary*: So you're a good student, and you are good at soccer and a leader, and it makes you feel good inside.	Mary summarizes the positive asset search using both facts and feelings. The summary of feeling *good inside* contrasts with the earlier feelings of *hurt inside*. (Additive)
22. *Damaris*: Yeah, it makes me feel good inside. I do my homework and everything [pause and the sad look returns], but then when I come to school, they just have to spoil it for me.	Again, Damaris agrees with the paraphrase. She feels support from Mary and is now prepared to deal from a stronger position with the teasing. Here we see what lies behind the "but" in 18 above. We believe Mary did the right thing in ignoring the "but" the first time. Now it is obvious that the negative feelings need to be addressed. When Damaris's wellness strengths are clear, Mary can better address those negative feelings.
23. *Mary*: They just spoil it. So you've got these good feelings inside, good that you're strong in academics, good that you're, you know, good at soccer and a leader. Now, I'm just wondering how we can use those good feelings that you feel as a student who's going to be a journalist someday and a soccer player who's a leader. Now the big question is how you can take the good, strong feelings and deal with the kids who are teasing. Let's look at ways to solve your problem now.	Mary restates Damaris's last words and again summarizes the many good things that Damaris does well. Mary changes pace and is ready to move to the restorying phase of the session. Here we see clearly that acceptance and identifying strengths support Damaris as Mary and she look toward how to deal with these issues. Whether you are working with children or adults, seek to give as much or more time to strengths as you do to difficulties. Of course, be sure to hear clients tell you in detail about their concerns and do not disregard their less positive emotions and feelings.

Mary had a good relationship and was able to draw out Damaris's story fairly quickly. She moved to story and strengths and found wellness assets that make it easier to address client problems and challenges.

▶Counseling Children

All clients have a need to know they have been heard. Working with children uses all the microskills, but there is more emphasis on encouraging, paraphrasing, and summarizing skills. Many effective elementary teachers constantly say back to students what the students have just said. These skills reinforce the conversation and help children keep talking from their own frame of reference. Telling your story to someone who hears you accurately is clarifying, comforting, and reassuring. Box 6.1 offers additional key ideas for working with children.

Damaris was stressed because of a seemingly "small" event, being teased about her shoes. You may wonder about the importance of small, repeating, negative events in the lives of children, teens, and adults. Enough repetition can lead to more than just sadness. Box 6.2 reviews this issue.

| **BOX 6.1** | Listening Skills and Children |

Attending
Avoid looking down at children; whenever possible, talk to them at their level. This may mean sitting on the floor or in small chairs. Their energy is such that it helps if they have something to do with their hands; perhaps you can allow them to draw or play with clay as they talk to you. Be prepared for more topic jumps than with adults, but use attending skills to bring them back later to critical issues that need to be discussed. Provide an atmosphere that is suitable for children by using small chairs and interesting objects. Warmth and an actual liking for children are essential. Use names rather than pronouns, as children often get confused when under stress (as do many adults). Smiling, humor, and an active style will help.

Questions and Concreteness
Use short sentences, simple words, and a concrete style of language. Avoid abstractions. Children may have difficulty with a broad open question such as "Could you tell me generally what happened?" Break down such abstract questions into concrete and situational language, using a mix of closed and open questions such as "Where were you when the fight occurred?" "What was going on just before the fight?" "Then what happened?"

"How did he feel?" "Was she angry?" "What happened next?" "What happened afterward?" In questioning children on touchy issues, be especially careful of closed, leading questions. Seek to get their perspective, not yours.

Encouraging, Paraphrasing, and Summarizing
Effective elementary teachers use these skills constantly, especially paraphrasing and encouraging. Seek out a competent teacher and observe for yourself. These skills, coupled with good attending and questioning, are needed to help children get out their stories.

Informed Consent and Working With Children
When you work with children, the ethical issues around informed consent are critical. Depending on state laws and practices, obtain written parental permission before interviewing a child. You must also obtain permission before sharing information about the session with others. The child and family should know exactly how the information is to be shared, and session records should be available to them for their comments and evaluation. An important part of informed consent is stating that they have the right to withdraw their permission at any point.

At one level, being teased about the shoes one wears doesn't sound all that serious—children will be children! However, some poor children go through their entire school life wearing clothes that others tease them about and laugh at, either directly or indirectly. At a high school reunion, Allen talked with a classmate who clearly was disturbed emotionally. During the talk, it became clear that teasing and bullying during schools days were still immediate and painful memories for him.

Small slights become big hurts if repeated again and again over the years. Athletes and "popular" students may talk arrogantly and dismissively about the "nerds," "townies," "rurals," or other outgroup. Teachers, coaches, and even counselors sometimes join in the laughter. Over time, these slights mount inside the child or adolescent. Some people internalize their issues in psychological distress; others may act them out in a dramatic fashion—witness the episodes of shootings throughout the country in both high schools and colleges.

Microaggressions

Discrimination and prejudice are other examples of accumulative stress and trauma. They are also known as *microaggressions* (Sue, 2010a, 2010b). One of Mary's interns, a young African American woman, spoke with her of a recent racial insult. She was sitting in a restaurant and overheard two White people talking about affirmative action and how they missed the days when "Blacks knew their place." The comments were not directed at her, but still such comments hurt and offend. The intern told Mary about how common direct and indirect racial insults affected her life and upset her day. When an incident troubled her, she immediately texted her sister for some relief, but also needed to talk to her later in the day

to seek support. It is hard to be fully effective at school or work when such distractions trouble us emotionally. (See Chapters 9 and 10 for a transcript interview example of how Allen dealt with a related issue in a real interview.)

Out of the continuing indignities of microaggressions come feelings of underlying insecurity about one's place in the world (internalized oppression and self-blame) and/or tension and rage about unfairness (externalized awareness of oppression). Either way, the person who is ignored or insulted feels tension in the body, the pulse and heart rate increase, and—over time—hypertension and high blood pressure may result. The psychological becomes physical, and accumulative stress becomes traumatic.

Soldiers at war, women who suffer sexual harassment, those who are overweight or short in height, the physically disfigured through birth or accident, gays and lesbians, and many others are all at risk for accumulative stress building to real trauma. They all suffer the dangers of posttraumatic stress.

As a counselor, you will want to be alert for signs of accumulative stress and microaggressions in your clients. Are they internalizing the stressors by blaming themselves? Are they externalizing and building a pattern of rage and anger inside that may explode? All these people have stories to tell that are especially significant to them. At first, some of these stories may sound routine. Posttraumatic stress responses in later life may be alleviated or prevented by your careful listening and support.

Finally, social work's position on social justice is that the therapist has the responsibility to act and intervene, where possible, to combat oppression and injustice. The counseling and psychology position on social action are becoming clearer. Where do you stand?

▶Concepts Into Action: The Active Listening Skills of Encouraging, Paraphrasing, and Summarizing

To show that you understand exactly, make a sentence or two which gets exactly at the personal meaning the person wanted to put across. This might be in your own words, usually, but use that person's own words for the touchy main things.

—Eugene Gendlin and Marion Hendricks-Gendlin

Necessary in counseling and psychotherapy is a nonjudgmental attitude in which you simply hear and accept what the client is saying. All your behavior can unintentionally convey judgmental and negative attitudes. A real challenge is to listen nonjudgmentally when you have inner feelings discrepant from those of the client. Recall that your client is often able to catch small facial expressions that reveal your judgments. How can you deal with this? Basically, focus your attention fully on the client in the here and now; try to enter that person's world as he or she sees it. Later, you can separate yourself from this world and more ably help your client.

Encouraging

Encouragers include head nods, open gestures, and positive facial expressions that encourage the client to keep talking. Minimal verbal utterances such as "ummm" and "uh-huh" have the same effect. Silence, accompanied by appropriate nonverbal communication, can be another type of encourager. All these encouragers have minimum effect on the direction of client talk; clients are simply encouraged to keep talking. Restatements, repetition of clients' key words, or brief statements are encouragers that more directly influence what clients talk about.

Let's imagine that Jennifer, the client with multiple issues presented at the beginning of this chapter, focuses on her parents' separating as the major immediate issue.

> I feel like my life is falling apart. I've always been close to both my folks and since they told me that they were breaking up, nothing has been right. When I sit down to study, I can't concentrate. And my roommate says that I get angry too easily. I guess everything upsets me. I was doing OK in my classes until this came along. I'm hurting so much for my Mom.

Jennifer still has a lot going on in her life. Eventually, we will want to focus on some of her strengths, but at the moment, she clearly needs to vent and explore her thoughts and feelings around her parents in more depth. There are several key words and ideas in this statement, and the repetition of any of them is likely to lead Jennifer to expand on current issues. As counselors, we tend to recommend repeating the exact key words: "You're hurting." This provides an opening for her to discuss her feelings or thoughts about Mom, herself, and, if she chooses, even other issues.

"You're hurting for your Mom" would focus more narrowly, but likely would be another good choice. "Falling apart," "close to your folks," "can't concentrate," and "you get angry easily" are other possibilities. All of these will help Jennifer continue to talk, but do lead in varying directions.

Key word encouragers contain one, two, or three words; restatements are longer. Both focus on staying very close to the client's language, most typically changing only "I" to "you." (Jennifer: "I'm hurting so much for my Mom." Counselor: "You're hurting.")

It may be helpful if you reread the paragraphs above, saying aloud the suggested encouragers and restatements. Use different vocal tones and note how your verbal style can facilitate others' talking or stop them cold.

All types of encouragers facilitate client talk unless they are overused or used badly. Excessive head nodding or gestures and too much parroting can be annoying and frustrating to the client. From the observation of many counselors, we know that using too many encouragers can seem wooden and unexpressive. However, too few encouragers may suggest to clients that you are not interested or involved. Well-placed encouragers help to maintain flow and continually communicate that the client is being listened to.

Paraphrasing (Reflection of Content)

At first glance, paraphrasing appears to be a simple skill, only slightly more complex than encouraging. However, if you can give an accurate paraphrase to a client, you are likely to be rewarded with a "That's right" or "Yes . . . ," and the client will go on to explore the issue in more depth. The goal of paraphrasing is facilitating client exploration and clarification of issues. The tone of your voice and your body language while paraphrasing will indicate whether you are interested in listening or wish for the client to move on.

Accurate paraphrasing can help clients complete their storytelling. A client who has been through a trauma may need to tell the story several times. Our goal is not to stop this talk, but paraphrasing can help work through the trauma because each time you repeat what the client has said, the client's story has been told again and *heard*. Friends who have been through a difficult hospital operation need to tell their story several times. Rather than becoming bored and saying "I've heard that before," give full attention and say back or paraphrase what you have heard.

How do you paraphrase? Client observation skills are essential in accurate paraphrasing. You need to hear the client's key words and use them in your paraphrase much as the client does. Other aspects of the paraphrase may be in your own words, but the main ideas and concepts should reflect the client's view of the world, not yours!

An accurate paraphrase, then, usually consists of four dimensions:

1. A *sentence stem*, sometimes using the client's name. Names help personalize the session. Examples would be "Damaris, I hear you saying . . . ," "Luciano, it sounds like . . . ," and "Looks like the situation is . . ." A stem is not always necessary and, if overused, can make your comments seem like parroting. Clients have been known to say in frustration, "That's what I just said; why do you ask?"

2. The *key words* used by the client to describe the situation or person. Again, drawing on client observation skills, the effort is to include key words and main ideas that come from the client. This repetition can be confused with the encouraging restatement. A restatement, however, is almost entirely in the client's own words and covers only a limited amount of material.

3. The *essence of what the client has said* in briefer and clearer form. Here the counselor's skill in transforming the client's sometimes confused statements into succinct, meaningful, and clarifying statements is most valuable to smoothing the counseling process. The counselor has the difficult task of keeping true to the client's ideas but not repeating them exactly.

4. A *checkout* for accuracy. The checkout is a brief question at the end of the paraphrase, asking the client for feedback on whether the paraphrase (or summary or other microskill) was relatively correct and useful. Some examples of checkouts are "Am I hearing you correctly?" "Is that close?" "Have I got it right?" It is also possible to paraphrase with an implied checkout by raising your voice at the end of the sentence as if the paraphrase were a question.

Here is a client statement followed by sample key word encouragers, restatements, and a paraphrase:

I'm really concerned about my wife. She has this feeling that she has to get out of the house, see the world, and get a job. I'm the breadwinner, and I think I have a good income. The children view Yolanda as a perfect mother, and I do too. But last night, we really saw the problem differently and had a terrible argument.

▲ Key word encouragers: "Breadwinner?" "Terrible argument?" "Perfect mother?"
▲ Restatement encouragers: "You're really concerned about your wife." "You see yourself as the breadwinner." "You had a terrible argument."
▲ Paraphrase: "You're concerned about your picture-perfect wife who wants to work even though you have a good income, and you've had a terrible argument. Is that how you see it?"

The key word encourager operates like selective attention. Note that the encouragers above lead the client in very different directions for what is appropriate conversation.

"Breadwinner" leads to talk about the job and possibly responsibility. "Terrible argument" may result in the details of the argument, while "perfect mother" leads to his wife's behavior.

As this example shows, the key word encourager, the restatement, and the paraphrase are all different points on a continuum. In each case, the emphasis is on hearing the client and feeding back what has been said. Both short paraphrases and longer key word encouragers will resemble restatements. A long paraphrase is close to a summary. All can be helpful in a session; or they can be overdone.

Summarizing

Summarizing encompasses a longer period of conversation; at times, it may cover an entire session or even issues discussed by the client over several meetings. In summarizing, the counselor attends to verbal and nonverbal comments from the client over a period of time and then selectively attends to key concepts and dimensions, restating them for the client as accurately as possible. Facts, thoughts, and emotions are included in the summary. A check-out at the end for accuracy is an effective part of the summarization. Following are some examples of summarizations.

To begin a session: Let's see, last time we talked about your angry feelings toward your mother-in-law, and we discussed the argument you had with her around the time the new baby arrived. You saw yourself as terribly anxious at the time and perhaps even out of control. Since then, you haven't gotten along too well. We also discussed your homework as developing ideas of what to do next. How did that go?

Midway in the session: So far, I've seen that the ideas you came up with didn't work too well. You felt a bit guilty and worried thinking you were getting too manipulative, and another argument almost started. "Almost" is better than a "blow up." Yet one idea did work. You were able to talk with her about her garden, and it was the first time you had been able to talk about anything without an argument. You visualize the possibility of following up with new ideas next week. Is that about it so far?

At the end of the session: In this session, we've reviewed your feelings toward your mother-in-law in more detail. Some of the following things seem to stand out: First, our plan didn't work completely, but you were able to talk about one thing without yelling. As we talked, we identified some behaviors on your part that could be changed. They include better eye contact, relaxing more, and changing the topic when you start to see yourself getting angry. I also liked your idea at the end of talking with her about the fact that you really want to forgive and be forgiven so that you two can relate better. Does that sum it up? Well, we have some specifics for next week. Let's see how it goes.

▶Diversity and the Listening Skills

Periodic encouraging, paraphrasing, and summarizing are basic skills that seem to have wide cross-cultural acceptance. Virtually all your clients like to be listened to accurately. It may take more time to establish a relationship with a client who is culturally different from you—but again, never generalize or stereotype.

Women tend to use paraphrasing and related listening skills more than men, whereas men tend to use questions more frequently. You may notice in your classes and workshops that men tend to raise their hands faster and interrupt more often. But there are so many exceptions to this "rule" that it should not be relied on. Nonetheless, differences in gender do exist, and it is essential that we stay aware. Gender differences need to be addressed directly in the session by both men and women.

Some Asian (Cambodian, Chinese, Japanese, Indian) clients from traditional backgrounds may be seeking direction and advice. They are likely to be willing to share their stories, but you may need to tell them why you want to hear more about their issues before the two of you come up with answers. To establish credibility, you may have to commit yourself and provide advice earlier than you wish. In such a case, be assured and confident, but let them know you want to learn more with them and that your thoughts may change as you get to know them better. Ultimately, you want to work with them to decide on their own. Boxes 6.3 and 6.4 review evidence that you can use and applications of the listening skills from an international perspective.

BOX 6.3 Research and Related Evidence That You Can Use

Active Listening

An early foundational study found that microskills training enables counselors to respond in a more culturally appropriate fashion (Nwachuku & Ivey, 1991). Research on encouraging, paraphrasing, and summarizing often treats them as part of a larger whole—empathic listening. Some of the most carefully designed work in this area has been done by Bensing (1999a; Bensing & Verheul, 2009, 2010) on physician–patient relationships. She found that physicians who established a solid relationship and listened carefully tended to be rated more highly, and their patients were more likely to follow their suggestions and directives. She also found that talk time of patients markedly increased with physicians who listened to them. Nine studies on microcounseling with nurses found that they were rated more highly on empathy, focused more on the client, and made fewer therapeutic errors. Similar findings have been reported for counselors and therapists (Daniels & Ivey, 2007).

Smiling is one of the most encouraging things you can do. Many researchers have found that smiling "works" and is a primary way to communicate warmth and openness (Restak, 2003). Although not the result of formal research, think about the following: Allen Ivey is a person who does not physically show a lot of emotion—he can smile, but is a bit shy and reserved. On the other hand, Mary Bradford Ivey is known for her smile and sunny disposition. Guess which person other people talk to when Allen and Mary are together? If you are a naturally warm person, this characteristic communicates itself to your client. If you are more like Allen, it may take you a bit longer, but you can still be a very effective listener!

Neuroscience and Active Listening

Active listening, of course, is central to empathic communication. A review of empathy literature has revealed that the neural architecture of the effective, warm, and competent listener matches that of the client. At the same time, we need to know the difference between ourselves and the other person and not violate boundaries. fMRI data suggest that specific brain regions are important in this process: the insula, the anterior cingulate cortex, and the right temporoparietal region. If a part of the brain is disrupted, empathic understanding will suffer (Decety & Jackson, 2006).

Carter (1999, p. 87) reports:

Expressions can . . . transmit emotions to others—the sight of a person showing intense disgust turns on in the observer's brain areas that are associated with the feeling of disgust. Similarly, if you smile, the world does indeed smile with you (up to a point). Experiments in which tiny sensors were attached to the "smile" muscles of people looking at faces show the sight of another person smiling triggers automatic mimicry—albeit so slight that it may not be visible. . . . the brain concludes that something good is happening out there and creates a feeling of pleasure.

This is a variant of movement synchrony, described previously. It is also possible for you to pick up the depressed mood and style of your client and recommunicate the sadness back to the client, thus reinforcing a cycle of negativity. But if you listen with energy and interest, and this is communicated effectively, expect your client to receive that affect as a positive resource in itself. Active listening is a key to developing a relationship and drawing out the client's story and strengths.

Talking about strengths and resources affects the brain in useful ways. For example, the neurotransmitter dopamine is released when situations are pleasant and positive, preparing the brain for new learning and development of new neural networks. New learning will not occur unless the amygdala has enough stimuli to energize the brain to receive new information and ideas. At the same time, we need to recall that the amygdala is a basic seat of negative emotion—and certain types of stimuli will work against learning and change. Extreme external stimulation (war, rape, home break-in) can prompt too much stimulation, in effect "blowing the "fuse," and result in loss of neural connections.

BOX 6.4 National and International Perspectives on Counseling Skills

Developing Skills to Help the Bilingual Client

Azara Santiago-Rivera, Past President, National Latina/o Psychological Association

It wasn't that long ago that counselors considered bilingualism a "disadvantage." We now know that a new perspective is needed. Let's start with two valuable assumptions: The person who speaks two languages is able to work and communicate in two cultures and, actually, is advantaged. The monolingual person is the one at a disadvantage! Research actually shows that bilingual children have more fully developed capacities and a broader intelligence (Power & Lopez, 1985).

If your client was raised in a Spanish-speaking home, for example, he or she is likely to think in Spanish at times, even though having considerable English skills. We tend to experience the world nonverbally before we add words to describe what we see, feel, or hear. For example, Salvadorans who experienced war or other forms of oppression felt that situation in their own language.

You are very likely to work with clients in your community who come from one or more language backgrounds different from yours. Your first task is to understand some of the history and experience of these immigrant groups. Then, we suggest that you learn some key words and phrases in their original language. Why? Experiences that occur in a particular language are typically encoded in memory in that language. So, certain memories containing powerful emotions may not be accessible in a person's second language (English) because they were originally encoded in the first language (for example, Spanish). If the client is talking about something that was experienced in Spanish, Khmer, or Russian, the key words are not English; they are in the original language.

Here is an example of how you might use these ideas in the session:

Social worker: Could you tell me what happened for you when you lost your job?

Maria (Spanish-speaking client): It was hard; I really don't know what to say.

Social worker: It might help us if you would say what happened in Spanish and then you could translate it for me.

Maria: *Es tan injusto! Yo pensé que perdí el trabajo porque no hablo el ingles muy bien. Me da mucho coraje cuando me hacen esto. Me siento herida.*

Social worker: Thanks; I can see that it really affected you. Could you tell me what you said now in English?

Maria: (more emotionally) I said, "It all seemed so unfair. I thought I lost my job because I couldn't speak English well enough for them. It makes me really angry when they do that to me. It hurts."

Social worker: I understand better now. Thanks for sharing that in your own language. I hear you saying that

injusto hurts and you are very angry. Let's continue to work on this and, from time to time, let's have you talk about the really important things in Spanish, OK?

This brief example provides a start. The next step is to develop a vocabulary of key words in the language of your client. This cannot happen all at once, but you can gradually increase your skills. Following are some Spanish key words that might be useful with many clients.

Respeto: Was the client treated with respect? For example, the social worker might say, "Your employer failed to give you *respeto*."

Familismo: Family is very important to many Spanish-speaking people. You might say, "How are things with your *familia*?"

Emotions (see next chapter) are often experienced in the original language. When reflecting feeling, you could learn and use these key words with clients:

Aguantar: endure	*Miedo*: fear	*Amor*: love
Orgullo: proud	*Cariño*: like	*Sentir*: feel
Coraje: anger		

We also recommend learning key sayings, metaphors, and proverbs in the language(s) of your community. *Dichos* are Spanish proverbs, like the following examples:

Al que mucho se le da, mucho se le demanda.
The more people give you, the more they expect of you.

Vale más tarde que nunca.
Better late than never.

No hay peor sordo que el que no quiere oír.
There is no worse deaf person than someone who doesn't want to listen.

En la unión está la fuerza.
In unity there is strength.

Consider developing a list like this, learn to pronounce the words correctly, and you will find them useful in counseling Spanish-speaking clients. Indeed, you are giving them *respeto*. You may wish to learn key words in several languages.

Note: We have produced a Spanish version of the *Basic Attending Skills*, *Las Habilidades Atencionales Básicas: Pilares Fundamentales de la Comunicación Efectiva* (Zalaquett, Ivey, Gluckstern-Packard, & Ivey, 2008), for both monolingual and bilingual helpers. The attending skills are illustrated with examples provided by Latina/o professionals from different Latin American countries. Using the information and exercises included in the book, you can sharpen your counseling and psychotherapeutic tools to provide effective services to clients who speak Spanish.

▶Summary: Practice, Practice, and Practice

We have stressed the importance of three major listening skills in this chapter—encouraging, paraphrasing, and summarizing. These skills are central to effective counseling and psychotherapy, regardless of your theory of choice and your personal integration of these microskills into your own natural style.

Intentional competence in these skills requires practice. Basic competence comes when you use the skills in a session and expect them to be helpful to your clients. Every client needs to be heard; demonstrating that you are listening carefully often makes a real difference. Advanced intentional competence requires deliberate and repeated practice.

At this point in this book, we want to share the following story, as it really drives home the importance of continuing to practice the skills. You can pass exams without practice, but if you are serious about helping, learning these skills to full mastery is critical.

Amanda Russo was a student in a counseling course at Western Kentucky University taught by Dr. Neresa Minatrea. Amanda shared with us how she practiced the skills and gave us her permission to pass this on to you. As you read her comments, ask yourself if you are willing to go as far as she did to ensure expertise.

> For my final project I selected a practice exercise titled "The Positive Asset Search: Building Empathy on Strengths." I chose this exercise early on in the book because I do not have much experience with counseling and I wanted to try a fairly simple exercise to start out. I performed the same exercise on five different people to see if I would get the same results.
>
> The exercise consists of asking the client what some of their areas of strength are, getting them to share a story regarding that strength, and then for the counselor to observe the client's gestures and be aware of any changes. The first person I tried this exercise on was Raphael, a dormitory proctor. Some of his strengths were family, friends, working out, and that he had a good inner circle/support group. As he talked about his support group and how they reminded him of the positives, he started to sit in a less tense position. He seemed very relaxed, yet excited about his topic of discussion, and I noticed a lot of hand gestures. In a matter of seconds I saw him change from tense and unsure to relaxed and enthusiastic about what he was saying.
>
> The next person I practiced this exercise on was my roommate Karol. She was a bit nervous when we started and had a difficult time thinking of strengths. Once I asked her to share a story with me, she became very animated. As she spoke, I could see a sparkle in her eyes. Her voice became stronger and her hands were moving every which way. She feels strongly about doing well at work, giving advice, working out, playing music, and finishing the song she is currently writing.
>
> Once she gave me a couple of strengths, they started her wheels turning and she was coming up with more and more. She felt very good about jazz practice earlier that day. She introduced a new song to the group and they really enjoyed it. She also shared a story with me about a huge accomplishment at work that day. Karol was definitely the person whose mood/persona changed the most in this exercise.

Amanda went on to interview three more people to practice her skills and reported in detail on each one. If you seek to reach intentional competence, the best route toward this is systematic practice. For some of us, one practice session may be enough. For most of us, it will take more time. What commitments are you willing to make?

	Key Points
Active listening	Clients need to know that their story has been heard. Attending, questioning, and other skills help the client open up, but accurate listening through the skills of encouraging, paraphrasing, and summarizing is needed to communicate that you have indeed heard the other person fully. All of these skills involve active listening, encouraging others to talk freely. They communicate your interest and help clarify the world of the client for both you and the client. Active listening skills are some of the most difficult in the microtraining framework.
Active listening skills	Three skills of accurate listening help communicate your ability to listen: 1. *Encouragers* are a variety of verbal and nonverbal means the counselor or psychotherapist can use to encourage others to continue talking. They include head nods, an open palm, "uh-huh," and the repetition of key words the client has uttered. Selective attention to various client key words can have a profound impact on client direction. Restatements are extended encouragers using the exact words of the client and are less likely to determine what the client might say next. 2. *Paraphrases* feed back to the client the essence of what has just been said by shortening and clarifying client comments. Paraphrasing is not parroting; use some of your own words plus the important main words of the client. 3. *Summarizations* are similar to paraphrases except that a longer time and more information are involved. Attention is also given to emotions and feelings as they are expressed by the client. Summarizations may be used to begin a session, for transition to a new topic, to provide clarity in lengthy and complex client stories, and, of course, to end the session. It can be wise to ask clients to summarize the session and the important points that they observed.
Checkout	Use of this "perception check" gives you an opportunity to determine the accuracy of your summary. It also offers clients a chance to think about what they have said.
The "how" of active listening	Paraphrasing and summarizing usually involve four dimensions: 1. *A sentence stem.* Often you will want to use the client's name: "Jamilla, I hear you saying . . . ," "Carlos, sounds like. . . ." 2. *Key words.* The clients' exact key words that they use to describe their situation. 3. *The essence of what the client has said in distilled form.* Here you use the client's key words in a brief clarification of what the client has said. Summaries are longer paraphrases that often include emotional dimensions as well. 4. *A checkout.* Implicitly or explicitly, check with the client to see that what you have fed back to him or her is accurate. "Have I heard you correctly?"
A caution	These skills are useful with virtually any client. However, if you do not do well or you seem mechanical, clients will find repetition tiresome and may ask, "Didn't I just say that?" Consequently, when you use the skill, you should also employ your client observation skills. As you listen to clients, seek to maintain a nonjudgmental, accepting attitude. Even the most accurate paraphrasing or summarization can be negated if you lack supporting nonverbal behaviors.
A commitment	Action is central for mastering these skills. Deliberate, repeated practice is essential to master the use of encouraging, paraphrasing, and summarizing in an effective way. With practice, you will reduce your chances to sound mechanical and increase your capacity to help your client.

▶Competency Practice Exercises and Portfolio of Competence

 Go to CengageBrain.com to access Counseling CourseMate, where you will find an interactive ebook, quizzes, videos, interactive counseling and psychotherapy exercises, the Portfolio of Competence, case studies, a practice test, and more.

The three skills of encouraging, paraphrasing, and summarizing are much less controversial than questions. Virtually all counseling and psychotherapy theories recommend and endorse these key skills of active listening.

Individual Practice

Exercise 1. Identifying Skills   Below is a client statement followed by several alternative counselor responses. Identify encouragers (E), restatements (R), paraphrases (P), and summaries (S).

Client:	The visit went well. I've decided to go back and finish college. But how will I pay for it? I've got a good job now, but I'll have to move to part-time and I'm not sure that they will keep me. Getting all this financed will be difficult. It is kind of scary.

_____ "Uh-huh."

_____ Silence with facilitative body language.

_____ "Scary?"

_____ "You're not sure that they will keep you."

_____ "Sounds like you've made up your mind to finish college, but financing it will be a major challenge."

_____ "In the last session, we talked about your going back to visit the school, and so far it sounds as if it went well and you really want to do it. But at the same time, now, it is a little scary when you think of all the financial issues. Have I heard what's been happening correctly?"

Exercise 2. Generating Written Encouragers, Paraphrases, and Summarizations
a. "Chen and I have separated. I couldn't take his drinking any longer. It was great when he was sober, but it wasn't that often he was. Yet that leaves me alone. I don't know what I'm going to do about money, the kids, or even where to start looking for work."

Write three different types of key word encouragers for this client statement:

_____

Write a restatement/encourager:

_____

Write a paraphrase (include a checkout):

_____

Write a summarization (generate data by imagining previous sessions):

_____

_b._ "And in addition to all that, we worry about having a child. We've been trying for months now, but with no luck. We're thinking about going to a doctor, but we don't have medical insurance."

Write three different types of key word encouragers for this client statement:

_____

Write a restatement:

_____

Write a paraphrase (include a checkout):

_____

Write a summarization (generate data by imagining previous sessions):

_____

Exercise 3. Practice Skills in Other Settings   During conversations with friends or in your own sessions, deliberately use single word encouragers and brief restatements. Note their impact on your friends' participation and interest. You may find that the flow of conversation changes in response to your brief encouragers. Write down your observations.

Group Practice

Exercise 4. Practice Skills With Another Person or Persons   Experience has shown that the skills of this chapter are often difficult to master. It is easy to try to feed back what another person has said, but to do it accurately, so that the client feels truly heard, is another matter.

Step 1: Divide into practice groups. Include triads as a possibility.

Step 2: Select a group leader.

Step 3: Assign roles for the first practice session.

▲   Client
▲   Counselor
▲   Observer 1 (uses Feedback Form, Box 6.5)
▲   Observer 2 (uses Feedback Form)

Step 4: Plan.   Establish and state clear goals for the practice session. For real mastery, try to use only the three skills in this chapter; use questions only as a last resort. The counselor should plan a role-play in which open questions are used to elicit the client's concern. Once this is done, use encouragers to help bring out more details and deeper meanings. Use open and closed questions as appropriate, but give primary attention to the paraphrase and the encourager. End the session with a summary (this is often forgotten). Check the accuracy of your summary with a checkout ("Am I hearing you correctly?").

The suggested topic for this practice session is the story of a past or present stressful experience that may relate to the idea of accumulative trauma. Examples include teasing, bullying, or being made the butt of a joke; an incident in which you were seriously misunderstood or misjudged; an unfair experience with a teacher, coach, or counselor; or a time you experienced prejudice or oppression of some type.

Emotions may appear as the story unfolds. Feel free to paraphrase or summarize these emotions, but this time focus first on the story and the event or situation itself. We suggest that you repeat this story again when you practice the skill of reflection of feeling in the next chapter.

All of the suggested incidents or topics will provide observers with the opportunity to observe nonverbal behaviors and discrepancies, incongruity, and conflict. Does the client internalize or externalize responsibility and blame? Internal attribution occurs when clients see themselves as being at fault. External attribution occurs when "they" or external forces are seen as the cause. Most clients will demonstrate some balance of internal and external attribution—self-blame versus other blame.

Step 5: Conduct a 3-minute practice session.

Step 6: Review the practice session and provide feedback to the counselor for 12 minutes. Be sure to use the Feedback Form (Box 6.5) to ensure that the counselor's statements are available for discussion. This form provides a helpful log of the session, which greatly facilitates discussion. Give special attention to feedback from the client, perhaps using the Client Feedback Form from Chapter 1. If you have made an audio or video recording, start and stop the playback periodically and repeat it to hear and observe the session. Did the counselor achieve his or her goals? What mastery level was demonstrated?

Step 7: Rotate roles.

General reminders.   Encourage clients to talk freely in the role-plays. As you become more confident in the practice session, you may want your clients to become more "difficult" so you can test your skills in more stressful situations. You'll find that difficult clients are often easier to work with after they feel they have been heard.

Portfolio of Competence

Active listening is one of the core competencies of intentional counseling and psychotherapy. Please take a moment to review where you stand and where you plan to go in the future.

Use the following checklist to evaluate your present level of competence. Check those dimensions that you currently feel able to do. Those that remain unchecked can serve as future goals. Do not expect to attain intentional competence on every dimension as you work through this book. You will find, however, that you will improve your competencies with repetition and practice.

Level 1: Identification and classification

- ❑ Identify and classify encouragers, paraphrases, and summaries.
- ❑ Discuss issues in diversity that occur in relation to these skills.
- ❑ Write encouragers, paraphrases, and summaries that might predict what a client will say next.

Level 2: Basic competence.   Aim for this level of competence before moving on to the next skill area.

- ❑ Use encouragers, paraphrases, and summaries in a role-played session.
- ❑ Encourage clients to keep talking through the use of nonverbals and the use of silence, minimal encouragers ("uh-huh"), and the repetition of key words.

Feedback Form: Encouraging, Paraphrasing, and Summarizing

You can download this form from Counseling CourseMate at CengageBrain.com.

_____   _____ (DATE)

_____   _____

(NAME OF COUNSELOR)   (NAME OF PERSON COMPLETING FORM)

Instructions: Write below as much as you can of each counselor statement. Then classify the statement as a question, an encourager, a paraphrase, a summarization, or other. Rate each of the last three skills on a scale of 1 (low) to 5 (high) for its accuracy.

Counselor statement	Open question	Closed question	Encourager	Paraphrase	Summarization	Other	Accuracy rating
1.							
2.							
3.							
4.							
5.							
6.							
7.							
8.							
9.							
10.							
11.							
12.							
13.							
14.							

1. What were the key discrepancies demonstrated by the client?

2. General session observations. Was responsibility for the concern placed internally or externally, or with some balance between the two?

- ❑ Discuss cultural differences with the client early in the session, as appropriate to the individual.

Level 3: Intentional competence

- ❑ Use encouragers, paraphrases, and summaries accurately to facilitate client conversation.
- ❑ Use encouragers, paraphrases, and summaries accurately to keep clients from repeating their stories unnecessarily.
- ❑ Use key word encouragers to direct client conversation toward significant topics and central ideas.
- ❑ Summarize accurately longer periods of client utterances—for example, an entire session or the main themes of several sessions.
- ❑ Communicate with bilingual clients using some of the key words and phrases in their primary language.

Level 4: Teaching competence. Teaching competence in these skills is best planned for a later time, but a client who has particular difficulty in listening to others may benefit from careful training in paraphrasing. Some individuals often fail to hear accurately and distort what others have said to them.

- ❑ Teach clients in a helping session the social skills of encouraging, paraphrasing, and summarizing.
- ❑ Teach these skills to small groups.

▶Determining Your Own Style and Theory: Critical Self-Reflection on the Active Listening Skills

This chapter has focused on the active listening skills of encouraging, paraphrasing, and summarizing, which are critical to obtaining a solid understanding of what clients want and need. Active listening is central, and these three skills are key.

What single idea stands out for you among all those presented in this chapter, in class, or through informal learning? What stands out for you is likely to be important as a guide toward your next steps. What do you think about the use of checkouts? How are you planning to engage in intentional and deliberate practice? What are your thoughts on race/ethnicity? What other points in this chapter struck you as most useful and interesting? How might you use ideas in this chapter to begin the process of establishing your own style and theory? If you are keeping a journal, what trends do you see as you progress this far?

▶Our Thoughts About Jennifer

Active listening requires actions and decisions on our part. What we listen to (selective attention) will have a profound influence on how clients talk about their concerns. When a client comes in full of information and talks rapidly, we often find ourselves confused and, we admit, a bit overwhelmed. It takes a lot of active listening to hear this type of client accurately and fully.

If our work was personal counseling, we would most likely focus on Jennifer's parents' separation and use an encourager by restating some of her key words, thus helping her focus on what may be the most central issue at the moment (e.g., "Your Mom and Dad called earlier this week and are separating"). We'd likely get a more focused story and could learn more about what's happening. As we understand this issue more fully, we could later move to discussing some of the other problems.

Another possibility would be to summarize the main things that Jennifer was saying as succinctly and accurately as possible. We'd do this by catching the essence of her several points and saying them back to her. Most likely, we'd use a checkout to see if we have come reasonably close to what she thinks and feels (e.g., "Have I heard you correctly so far?"). We would then ask her, "You've talked about many things. Where would you like to start today?"

If we were academic counselors, not engaging in personal issues, we'd likely selectively attend to the area of our expertise (study issues) and refer Jennifer to an outside source for personal counseling.

How does this compare to what you would have done?

CHAPTER 7

Reflecting Feelings
A Foundation of Client Experience

Reflection of Feeling

Encouraging, Paraphrasing, and Summarizing

Open and Closed Questions

Client Observation Skills

Attending Behavior and Empathy

Ethics, Multicultural Competence, and Wellness

Reflection of feeling is followed by continued self-exploration and insight, [and] evaluative responses by abandonment of exploration.

—Carl Rogers (paraphrase)

Mission of "Reflecting Feelings"

The mission of reflecting feelings is to clarify emotional life—discovering the "heart of the matter." To many theorists and practitioners, this is the most important microskill for counseling and therapy competence. Underlying clients' words, thoughts, and behaviors are feelings and emotions that motivate and drive action.

Chapter Goals and Competency Objectives Awareness, knowledge, skills, and actions in reflection of feelings will enable you to

▲ Ground the client in basic experience. Sessions can become too verbal and intellectualized, thus moving away from the deeper feelings that guide us.

▲ Bring out the richness of the client's emotional world and increase your empathic understanding.

▲ Help clients explore and sort out mixed or ambivalent feelings toward the issues and decisions they face, toward significant others, and toward the way they understand themselves.

Thomas: My Dad drank a lot when I was growing up, but it didn't bother me so much until now. (pause) But I was just home and it really hurts to see what Dad's starting to do to my Mum—she's awful quiet, you know. (Looks down with brows furrowed and tense) Why she takes so much, I don't figure out. (Looks at you with a puzzled expression) But, like I was saying, Mum and I were sitting there one night drinking tea and he came in, stumbled over the doorstep, and then he got angry. He started to hit my mother, and I moved in and stopped him. I almost hit him myself, I was so angry. (Anger flashes in his eyes.) I worry about Mum. (A slight tinge of fear seems to mix with the anger in the eyes, and you notice that his body is tensing.)

Paraphrasing, as presented in Chapter 6, is concerned with feeding back the key points of what a client has said. Reflection of feeling, in contrast, involves observing emotions, naming them, and repeating them back to the client. The two are closely related and often will be found together in the same statement, but the critical distinction is emphasis on content (paraphrase) and emotion (reflection of feeling).

To clarify the distinction, write a paraphrase of Thomas's comments above with an emphasis on content; then write a reflection of feeling, focusing on emotion. You have not yet been asked to write a reflection of feeling, so use your intuitions and note the main feeling words of the client. Two possible sentence stems are provided for your consideration.

Paraphrase: Thomas, I hear you saying . . .
Reflection of feeling: Thomas, I sense that you are feeling . . .

You may want to compare your response with our thoughts, presented in the following introductory section.

▶Introduction: Reflection of Feeling

Carl Rogers is *the* theorist/practitioner/author who has made us fully aware of the importance of listening. We can thank him for our current use of the term "reflection of feeling." *Becoming a Person* (Rogers, 1961) is the book that brought listening, particularly reflection of feeling, to center stage.

The definition of reflection of feeling is presented in the following table. If you use this listening skill as defined here, you may anticipate specific results.

REFLECTION OF FEELING	ANTICIPATED RESULT
Identify the key emotions of a client and feed them back to clarify affective experience. With some clients, the brief acknowledgment of feeling may be more appropriate. Often combined with paraphrasing and summarizing.	Clients will experience and understand their emotional state more fully and talk in more depth about emotions and feelings. They may correct the counselor's reflection with a more accurate descriptor.

▶Comparing Paraphrasing and Reflection of Feeling

Paraphrasing client statements focuses on the content and clarifies what has been communicated. In the case of Thomas, the content includes the father's drinking history, Mum being quiet and taking it, and of course, the specific situation when the client was last home. The paraphrase will indicate to the client that you have heard what has been said and encourage the client to move further into the discussion.

> *Paraphrase:* Thomas, your father has been drinking a long time and your Mum takes a lot. But now he's started to be violent and you've been tempted to hit him yourself. Have I heard you right?
>
> *Reflection of feelings:* Thomas, you feel real *hurt, anger,* and *worry* because of what happened. The most basic reflections of feeling would be "It really *hurt,*" "You felt *angry,*" and "You are *worried.*"

The first task in eliciting and reflecting feelings is to recognize the key emotional words used by the client. You can know with some certainty that the client has these feelings, as they have been made explicit. At another level, reflections may help the client explore deeper unsaid things. For example, "I hear a lot of *caring* and that you are *anxious* to help resolve the situation."

▶The Techniques of Reflecting Feelings

Somewhat like the paraphrase, reflection of feelings involves a typical set of verbal responses that can be used in a variety of ways. Keep in mind that we will often be using the words *feeling* and *emotion* interchangeably. The classic reflection of feelings consists of the following elements:

Sentence stem. Choose a sentence stem such as "I hear you are feeling . . . ," "Sounds like you feel . . . ," "I sense you are feeling. . . ." Unfortunately, these sentence stems have been used so often they can sound like comical stereotypes. As you practice, you will want to vary sentence stems and sometimes omit them completely. Using the client's name and the pronoun *you* helps soften and personalize the sentence stem.

Feeling label. Add an emotional word or feeling label to the stem ("Jonathan, you seem to feel bad about . . . ," "Looks like you're happy," "Sounds like you're discouraged today; you look like you feel really down"). For mixed feelings, more than one emotional word may be used ("Maya, you appear both glad and sad . . .").

Context or brief paraphrase. You may add a brief paraphrase to broaden the reflection of feelings. The words *about, when,* and *because* are only three of many that add context to a reflection of feelings ("Jonathan, you seem to feel bad about all the things that have happened in the past two weeks," "Maya, you appear both glad and sad because you're leaving home").

Tense and immediacy. Reflections in the present tense ("Right now, you are angry") tend to be more useful than those in the past ("You felt angry then"). Some clients have difficulty with the present tense and talking in the "here and now." Occasionally, a "there and then" review of past feelings can be helpful and feel safer for the client.

Checkout. Check to see whether your reflection of feelings is accurate. This is especially helpful if the feeling is unspoken ("You feel angry today—am I hearing you correctly?"). But there are also many unspoken feelings expressed in client statements—and the client may or may not be fully aware of them. These unspoken or implicit feelings are often, but not always, expressed nonverbally. For example, Thomas looked down with brows furrowed and body tense (an indication of likely tension and confusion); anger and fear flashed in his eyes as he was talking about hitting, and fear in his eyes was mixed with the anger. Note that the client says that his father's drinking didn't bother him until recently. But this seems unlikely, and it may be useful at a later point to explore his family life while he was growing up. Is the client denying underlying long-term deep emotions about family drinking? At this point, however, the main issue is drawing out the story and noting the client's emotions associated with the story. As you move beyond crisis, what does all this mean to the client?

Feelings are also layered, like an onion. Clients may talk about emotional tones such as feeling confused, lost, or frustrated; or they may be direct and forthright with a single clear emotion. However, further listening and reflection often reveals underlying complex and sometimes conflicting emotions. For example, clients may say that they are frustrated in their relationships with a partner. Reflecting that frustration may lead to discussions in which the client talks about anger at lack of attention, fear of being alone if the relationship breaks up, and residual deep caring for the partner. And in the middle of all this, the hurt will likely remain important.

While we believe that focusing first on the potential violence is critical, combining the paraphrase with feelings by repeating the client's stated key feeling words is likely also appropriate. For example, "You're really hurting with it all right now," "You're angry because your Dad hit your Mum," "You're worried that your Dad's drinking is getting worse." Combining the feeling with the paraphrase acknowledges the client's emotions and may encourage a fuller telling of the story.

> Right now, Thomas, you're hurting about the situation. I also see some anger. *Is that right?* (Uses a key word that reflects the underlying emotion.)
>
> Stopping your Dad from hitting your Mum brought out a lot of emotion—I see some anger, perhaps even a little fear about what's going on. *Am I close to what you're feeling?* (The focus here is on unspoken emotions, seen more nonverbally than verbally. The check-out is particularly necessary here to make sure that you are really with the client.)
>
> Thomas, I hear that your Dad has been drinking for many years. (Paraphrase) And I hear many different feelings—anger, sadness, confusion—and I also hear that you care a lot both for your Mum and Dad. *Am I close to what you are feeling?* (This is a broader reflection of feeling that summarizes several emotions and feelings and encourages the client to think more broadly.)

Later in the session, after the story is told more completely through your listening and reflecting, you can help the client sort through the many and often conflicting emotions, set goals, and develop an action plan, all of which will help him move to useful and constructive action for his Mum and the family as a whole.

▶Empathy and Warmth

If you are counseling a client such as Thomas, you can readily see that he needs your warmth and support. Warmth, appropriate smiling, and showing that you respect and care for the client are basic to empathic understanding. Your warmth, expressed through words and the accompanying nonverbals (especially vocal tone and facial support), may be described as the glue that holds empathy and the possibility of positive change together. Change can happen without warmth and empathy, but it will be less effective and is likely to take a longer time.

We suggest that you capture yourself on video in a session. Review the session and obtain feedback from a friend or colleague. Your smile and warmth will be carried by the client beyond the session.

►Example Counseling Session: My Mother Has Cancer, My Brothers Don't Help

The discovery of cancer, AIDS, or other major physical illness brings with it an immense emotional load. Busy physicians and nurses sometimes fail to deal with emotions in their patients. And certainly, they have little time to help family members of those who are ill. Illness can be a frightening experience, and family, friends, and neighbors as well as professionals may have trouble dealing with it.

The following transcript illustrates reflection of feeling in action. This is the second session and Jennifer has just welcomed the client, Stephanie, into the room. They had a brief exchange of personal greetings and it was clear nonverbally that the client was ready to start immediately.

COUNSELOR AND CLIENT CONVERSATION	PROCESS COMMENTS
1. *Jennifer:* So, Stephanie, how are things going with your mother?	Jennifer knows what the main issue is likely to be, so she introduces it with her first open question. Throughout difficult sessions such as this, empathic warmth becomes especially important.
2. *Stephanie:* Well, the tests came back and the last set looks pretty good. But I'm upset. With cancer, you never can tell. It's hard . . . (pause)	Stephanie speaks quietly and as she talks, she speaks in an even softer tone of voice. At the word "cancer," she looks down.
3. *Jennifer:* You're really upset and worried right now.	Jennifer uses the client's emotional word ("upset") but adds the unspoken emotion of worry. With "right now," she brings the feelings to here-and-now immediacy. She did not use a checkout. Was that wise? (Interchangeable empathy with necessary verbal and nonverbal warmth and personal authenticity)
4. *Stephanie:* That's right. Since she had her first bout with cancer . . . (pause), I've been really concerned and worried. She just doesn't look as well as she used to, she needs a lot more rest. Colon cancer is so scary.	Often if you help clients name their unspoken feelings, they will say "That's right" or something similar, or they may nod their head. Naming and acknowledging emotions helps clarify them.
5. *Jennifer:* Scary?	Repeating the key emotional words used by clients may help them elaborate on issues in more depth. (Interchangeable)
6. *Stephanie:* Yes, I'm scared for her and I'm scared for me. They are saying that it can be genetic. She had Stage 2 cancer and we have really got to watch things carefully.	The anticipated result comes true. Stephanie elaborates on the scary feelings and where they come from. She has a frightened look on her face, and she looks physically exhausted.
7. *Jennifer:* So, we've got two things here. You've just gone through your mother's operation and that was scary. You said earlier that they got the entire tumor, but your Mom really had trouble with the anesthesia and that was frightening for a while. You had to do all the caregiving because the rest of the family is far away and you felt pretty lonely. That is scary enough. And the possibility of your inheriting the genes is pretty terrifying. Putting it all together, you feel overwhelmed. Is that the right word to use—overwhelmed?	At this point, Jennifer decides to summarize what has been said. She repeats some key feelings identified in the first session as well as in this session. She uses a new word, *overwhelmed*, which comes from her observations of the total situation and how very tired Stephanie appears. The counselor took a chance with the word *overwhelmed*. It might produce too much emotion in the client at the moment. Wisely, she included a checkout. (Interchangeable with additive aspects and warmth)

COUNSELOR AND CLIENT CONVERSATION	PROCESS COMMENTS
8. *Stephanie:* (immediately) Yes, I'm overwhelmed. I'm so tired, I'm scared, and I'm furious with myself. (pause) But I can't be angry; my mother needs me. It makes me feel guilty that I can't do more. (starts to sob)	This reflection of feeling seems to have brought out more emotion than the counselor expected this early in the session. Stephanie is now talking about her issues, and she is in touch with more basic emotions. This is probably OK as the relationship is solid, Stephanie has not cried before, and she likely needs to allow herself to cry and let the emotions out. Caregivers such as Stephanie often burn out and need care themselves. The primary focus is on the person with illness and often little attention is given to the person who suddenly finds herself with all the responsibilities.
9. *Jennifer:* (sits silently for a moment) Stephanie, you've faced a lot and you've done it alone. Allow yourself to pay attention to you for a moment and experience the hurt. (As Stephanie cries, Jennifer comments.) Let it out . . . that's OK.	Stephanie has held it all in and needs to experience what she is feeling. If you personally are comfortable with emotional experience, this ventilation of feelings can be helpful. At the same time, there is a need at some point to return to discussion of Stephanie's situation from a less emotional frame of reference. Again, note the importance of empathy and warmth.
10. *Stephanie:* (continues to cry, but the sobbing lessens)	See Box 7.3 for ideas in helping clients deal with emotional experience.
11. *Jennifer:* Stephanie, I really sense your hurt and aloneness. I admire your ability to feel—it shows that you care. Could you sit up now and take a breath?	The client sits up, the crying almost stops, and she looks at the counselor a bit cautiously. She wipes her nose with a tissue and takes a deep breath. Jennifer did three things here: (a) she reflected Stephanie's here-and-now emotions; (b) she identified a positive asset and strength in those emotions; and (c) she suggested that Stephanie take a breath. Conscious breathing often helps clients bring themselves together. (Additive)
12. *Stephanie:* I'm OK. (pause)	She wipes her eyes and continues to breathe. She seems more relaxed now that she has let out some of her emotions.
13. *Jennifer:* You've been holding that inside for a long time. That's the first time I've seen you cry. You had to have a lot of strength and power to do what you did. Your caring and strength really show. It's also strong to cry.	Jennifer provides feedback to Stephanie on her observations and outlines some positive strengths that she has seen. She provides a reframe (see Chapter 11), pointing out the positives inherent in Stephanie's caring attitude. (Additive)
14. *Stephanie:* Thanks, but I still feel so guilty.	Stephanie is now back in control of herself.
15. *Jennifer:* You feel guilty?	This is the most basic reflection of feeling and it appears in the form of a restatement. The expectation is that Stephanie will elaborate on the meaning of the feeling. (Interchangeable)
16. *Stephanie:* Who am I to cry? My mother is the one with the pain and Stage 2 cancer. I just wish I had been able to talk my brothers into coming home to see her at least.	The expectation is confirmed and we see Stephanie elaborating on her guilt. We are beginning to see indications of Stephanie being an "overfunctioning" individual who takes on more responsibility than is needed.
17. *Jennifer:* Your mother has gone through cancer, but you also have pain and fear, although in a different way. Do I hear you saying that you feel guilty because your brothers didn't come home?	First, Jennifer reflects Stephanie's feelings of pain and fear. She separates out the guilt, however, with a reflection of feeling in the form of a closed question. In this case, the question serves as a checkout. (Interchangeable)

continued on next page

COUNSELOR AND CLIENT CONVERSATION	PROCESS COMMENTS
18. *Stephanie:* Well, I called them daily and told them what was happening. They were fine on the phone, but they simply wouldn't come.	Stephanie talks a little faster and her fists tighten.
19. *Jennifer:* I sense a little anger at them. Is that close?	Jennifer draws on the nonverbal observations for this reflection of unspoken feelings. She includes a checkout as it is possible that Stephanie will deny the anger. (Additive as Jennifer is observing the more basic emotions)
20. *Stephanie:* I feel so guilty that I couldn't talk them into coming. (thoughtful pause) No, that's right. They should have come. (angrily) I know they have jobs and it's hard to get away, but this is their mother. (pause) And, it's not just this time. They hardly ever call. They just seem to be in their own world. I wonder if they'll even show up this year for the holidays. They didn't last year.	Could it be that Stephanie has taken the caregiver role for the entire family? As she explores her feelings, she is beginning to make new discoveries.
21. *Jennifer:* Stephanie, right now I hear that you're really angry with them because they don't help and aren't involved.	A classic reflection of feeling involves 1. A sentence stem (*I hear you*), usually using the client's name or the pronoun *you*. 2. The naming of the feeling (*angry*) or feelings. 3. The underlying facts or reasons behind the feeling. 4. Bringing the emotion to the here and now of the immediate moment (*right now*). (Interchangeable)
22. *Stephanie:* And you know what else they did? (Continues with another story)	The reflection unleashes Stephanie to share some long held-back stories of frustration and anger with her brothers.
[23–30. Omitted.]	In this exchange, Stephanie explores her feelings toward her brothers.
31. *Jennifer:* So, Stephanie, we've talked about your anger and disappointment with your brothers. You seem very much in touch with something you weren't really aware of before. You seem to be saying, also, that feeling guilty about not getting them to shape up and take their share doesn't make sense anymore. At the same time, I sense that you still have hopes and want to involve them more. Before we go further, I wonder if we can change focus. So far, we've been talking about your concerns and difficulties. At the same time, you've managed to do a lot. You care a lot and you've managed this past month. Could you share with me some of the things and the strengths that have enabled you to manage during this past month?	The intervening discussion is summarized. Stephanie's feelings of guilt and anger are better understood. This should free her for more open discussion in the future. Note how summaries help punctuate the session and aid the transition to other stages and issues. (Interchangeable and potentially additive) Jennifer has listened to problems and challenges for the entire first session and all of this second session thus far. She decides that it is time for the positive asset search. If Jennifer can help the client identify some positive feelings, thoughts, and behaviors associated with this situation, then Stephanie will be better prepared and stronger to deal with the many issues, challenges, and problems that she faces.

You may have noted that the major skill Jennifer uses throughout this session is reflection of feeling, accompanied by a few questions to help bring out emotions. This skill is central in all theories of counseling and therapy, as human change and development are often rooted in emotional experience.

You are most likely beginning your work and starting to discover the importance of reflecting feelings. It will probably take you some time before you are fully comfortable using this skill. This is so because it is less a part of daily communication than the other skills of this book, but reflecting feelings is central to every helping professional.

We would suggest that you start practice by first simply noting emotions and then reflecting them back through short acknowledging reflections. As you gain confidence and skill, you will eventually decide the extent and place of this skill area in your helping repertoire.

Before moving further, please reflect on yourself and your own personal style. How comfortable are you with emotional expression? If discussing feelings was not common in your experience, this skill may be difficult for you. As you work through the ideas presented here, reflect on your own personal history and ability to deal with emotion. If you find this area uncomfortable, you may have difficulty in helping clients explore their issues in depth. The exercises here and throughout the book may help you gain greater access to your own experiential and emotional world. In your own practice with reflection of feeling, attempt to use the skill as frequently as possible. In the early stages of mastery, it is wise to combine the skill with questioning, encouraging, and paraphrasing.

▶Becoming Aware of and Skilled With Emotional Experience

The artistic counselor catches the feelings and emotions of the client. Our emotional side often guides our thoughts and actions, even without our conscious awareness.

—Allen Ivey

People in virtually all cultures and societies do not give much attention to emotions and feelings in day-to-day conversation, yet competence and skill in this area is basic to your becoming an effective helping professional. There is a very real and very large difference between reading this chapter, passing an exam, and actually being able to help clients work with their foundational emotional and feeling experience. If you decide to practice only one skill, this is the one that we'd most recommend, because it is so basic to a real working alliance and a solid relationship.

▶The Language of Emotion

People constantly express emotions and feelings verbally and nonverbally. General social conversation ignores emotions and feelings unless they are especially prominent. Thus, many of us are trained not to focus on the other person's emotional experience, and we may even be unaware of what is happening before our eyes.

We have been using the words *emotion* and *feeling* interchangeably, as in common usage, and we will continue to do so. Counseling historically has identified and taught four basic feelings: sad, mad, glad, and scared. This ancient rubric turns out to be accurate across all cultures throughout the world. However, cross-cultural research has added three more basic feelings: surprise, disgust, and contempt (Ekman, 2009). Disgust is closely related to contempt, and they are often discussed together.

Feelings are physical as well as mental. When Thomas expresses anger at his father's behavior, this is more than a thought. The amygdala energizes the limbic HPA axis, and hormones lead to a faster heartbeat, higher blood pressure, breathing changes, and muscular tension. If you follow discussion of Thomas's difficulties with an emphasis on his strengths and the positive asset search, you will see him start to relax. The positive feelings of the executive frontal cortex (TAP) will be energized.

On this basis, Damasio (2003) has identified the feelings mentioned previously as the most basic and also the **social emotions** (guilt, shame, pride, embarrassment) as cultural blends of the basic feelings. For example, guilt involves limbic fear, but is modified by executive TAP decision making and cognitive experience with culture. For example, the positive emotions related to guilt could be love for one's parents or group. The behavior of something seen as wrong results in the emotion. If you were to reflect feelings of a client experiencing guilt, you would encounter the mixed emotions and seek to help the client sort out what happened and what he or she wants to do about it.

Disgust and surprise have not received as much attention in the counseling process as the traditional sad, mad, glad, and scared. Disgust is closely related to fear and anger. It is believed that disgust originally was protective evolutionarily against rotten food and excrement, as it also relates to the sense of smell. It soon evolved to affect a wide variety of issues. When you engage in marriage counseling, you may discover this feeling; working it through to save a relationship is especially difficult. Disgust tends to be relatively permanent once it appears.

Surprise may be positive, such as discovering something new, fun, and stimulating. It could be the first moments of a new relationship or a present from a friend. On the other hand, surprises can be shocking and startling, even harmful—for example, a dog coming running out of the dark barking loudly, a sound like a bomb nearby, or an earthquake. After the first few moments of surprise, positive or negative feelings as appropriate to the situation will follow.

Surprise is often an opening for change in counseling and therapy. A new insight or a useful confrontation can serve as a surprise to the client. In a solid helping relationship, the surprise becomes a useful tool.

If we are to reflect feelings, we need to deepen our understanding by increasing our emotional vocabulary. A good way is to brainstorm emotional words. Listed below are the four traditional basic feelings (sad, mad, glad, scared) plus disgust/contempt and surprise, with room for you to write related social emotional words. Think particularly of different intensities of the same feeling. For example, *mad* might lead you to think of *annoyed*, *angry*, and *furious*.

Sad	*Mad*	*Glad*	*Scared*	*Disgust/Contempt*	*Surprise*

Now that you have completed your list, you can turn to page 181 and compare your lists with ours. Note that the feeling categories are predominately negative. It is argued that disgust, anger, and fear are in different ways allied to protect us from danger, and they are located primarily in the amygdala and limbic system. It seems possible that sadness is a failure to gain protection and thus may be more closely related to the executive prefrontal area (TAP). Glad and the positive happy emotions are located primarily in the TAP. Surprise seems basically a protective emotion, but may later be evaluated as part of the glad feeling.

We can discover feelings through the language of the client and effectively reflect their understanding—and in that process clarify their emotions and feelings. At times, however, nonverbal behavior may be more important than what the client is saying. Are your client's words in attunement with vocal tone, visual eye contact, and body language? Observing these three may be the best way to discover what is really going on inside your client. Box 7.1 presents some basics of nonverbal emotion that may be useful.

BOX 7.1   The Nonverbal Language of Emotion: Micro and Macro Feelings

Macro nonverbals are those that are relatively easy to see. The client may drop the eyes downward, twist away from you, and talk very quietly, a fairly clear indication of some version of fear. Some specific dimensions of expected presentation of nonverbal emotion are summarized below.

But, micro nonverbals can be equally or more important. Not surprisingly, this is an area to which the Federal Bureau of Investigation has given special attention in its search for clues to deception and subtle behaviors that may be revealing. As we go through the day, these micro nonverbals occur in front of us constantly, but for the most part we don't notice them—and, if we do we don't say anything.

"Microexpressions are fleeting expressions of concealed emotion, sometimes so fast that they happen in the blink of an eye—as fast as one-fifteenth of a second. . . . This results from the individual's attempt to hide them. . . . [They are] a powerful tool for investigators because facial expressions of emotion are the closest thing humans have

to a universal language" (Matsumoto, Hwang, Skinner, & Frank, 2011). With practice we can learn to observe these as they can be as reliable indicators of underlying feelings as macro nonverbals—and they can be even more valuable than more overt, easily observable client behavior.

While observing micro nonverbals is valuable, reflecting them needs to be done carefully, as often the client will not be aware of these underlying feelings. Generally, it is best simply to note them and then watch for a time that these observations may be shared in the session. Micro nonverbals may be examples of the major underlying issues, or they may be minor parts of a larger story.

Following are some examples of what you might expect nonverbally with the seven major feelings. Facial expressions are considered the most important, as the clearest indication of accompanying feelings. And the fleeting, quick micro nonverbals may be most easily noted in the face.

Sad   The mouth curves down and the upper eyelids droop. A raise of the inner brows is considered one of the best indications of sadness. The body may slump or the shoulders drop, while vocal tone may be soft and speech rate slow. The arms may be crossed along with the hunching behavior.

Mad   Anger is typically expressed with an upright body position, frowning, and a louder or forced vocal tone; the mouth and jaws may be tense and lips tightened, fists clenched or the palms down. Other nonverbals may be rapid foot tapping, hands on hips, and in situations of danger, possibly moving toward you. There is also the "anger grin," which may indicate a desire to hide underlying feelings.

Glad   Happiness shows itself in smiling, a general picture of relaxation, open body posture and direct eye contact, typically with pupil dilation. It tends to be a holistic state that usually does not show in micro nonverbals. The smiley face's mouth curves up. The client will often move forward in the chair and use open gestures with the palms up.

Scared   Fear may be indicated by general tension and increased breathing rate, averted eyes or raised eyebrows, furrowed brow, biting the lips, crossed arms, or anxious playing with fingers. The pupils may contract (sometimes this may be the only real clue to fear or the desire to avoid a subject). Vocal tone may waver, with possible stammering or clearing of the throat. There is also the fear grin, which is closely related to the anger grin.

Disgust The nose tends to be wrinkled and the upper lip raised as the lips are pursed. Some believe that this evolved through the smell and awareness of rotten food and served as a protective dimension. Cultures vary in what they find disgusting. Of course, disgust shows in interpersonal communication through "disgusting" behavior or ideas, showing up even in politics. If you work with a client who feels disgust for her or his partner, repair of the relationship may be difficult.

Contempt Though closely related to disgust, the feeling of contempt has slightly different, but identifiable facial features, which reflect an attitude of disdain and disrepect toward another person. The chin is raised, which gives the appearance of looking down one's nose at the other. One lip corner may be tightened and slightly raised. A slight smile is often interpreted as a sneer.

Surprise This emotion typically lasts for only a few seconds, significantly less if the person seeks to hide the surprise. Eyes wide open, the eyebrows raised, and a crinkly forehead are typical. Think of the "jaw-dropping" experiences that have surprised you from time to time. Surprise may show as a fleeting micro nonverbal when you have helped the client discover a new insight or by the effective use of the skill of confrontation (Chapter 10). Surprise can lead to cognitive and emotional change.

▶Helping Clients Increase or Decrease Emotional Expressiveness*

We often want to encourage clients to express their feelings more fully, but there are times when emotion becomes too overwhelming and we want to help the client slow down and regain control. The following list should be helpful.

Observe nonverbals. Breath directly reflects underlying feelings. Rapid or frozen breath signals contact with intense body experience. Also note facial flushing, pupil contraction/dilation, body tension, and changes in vocal tone; note especially speech hesitations. You may also find words about emotions and feelings absent as the client talks about issues at a purely cognitive level. This is a clear clue that the client is avoiding dealing with feelings. Remain aware that the expression of emotion is culturally inappropriate for this client.

Join clients where they are and pace them appropriately. You join clients by listening and clarifying their stories. Once you have that solid relationship and trust, you can pace clients and then lead them to more expression and awareness of affect. Many people get right to the edge of a feeling, and then back away with a joke, change of subject, or intellectual analysis. Following are some ways to pace the client and encourage more or less emotion.

Discuss some positive aspect of the situation. This can free the client up to face the negative. You as counselor also represent a positive asset through your relationship.

Say to the client that she looked as though she was close to something important. "Would you like to go back and try again?" "Would you say that word (or phrase) again?"

Consider asking questions. Used carefully, questions may help some clients explore emotions. Use present tense here-and-now techniques. "What are you feeling right now—at this moment?" "What's occurring now in your body as you talk about this?" "What are you seeing? Hearing? Feeling?" Note that the words *are* and *now* are often best for present tense experiencing.

On the other hand, use the words *do* and *then* if you find your client needs to distance a bit from emotion or you yourself are uncomfortable with emotion. "What do you feel?" or "What did you feel then?" helps move the client away from the here and now.

Use Gestalt exercises. These exercises enable a client to become more aware of the feelings experienced through their bodies.

When clients are directly experiencing emotions such as tears, rage, despair, joy, or exhilaration, your own comfort level will affect how your client faces these issues. If you aren't comfortable with a particular emotion or feeling, this is very likely to show in your nonverbal behavior, and your client may avoid further talk involving this emotion or issue. A balance is needed between, on the one hand, being very present with your own breathing and showing culturally appropriate and supportive eye contact and, on the other, still allowing room to sob, yell, or shake. You can also use phrases such as these:

I'm here. I've been there too.
I'm standing right with you in this. Let it out . . . that's OK.
These feelings are just right. I hear you.
I see you. Breathe with it.

Emotional expression needs to be kept within a fixed time; 2 minutes is a long time when you are crying. Afterward, helping the person reorient to the here and now is important.

---

*This section is based on thoughts of Leslie Brain, a wonderful graduate student who worked with us. We have added new information and thoughts, but we do want to acknowledge her insights.

Tools for reorienting the session include the following:

Slowed, rhythmic breathing
Counselor and client discussion of positive strengths inherent in the client and situation
Discussion of direct, empowering, self-protective steps that the client can take in response to the feelings expressed
Standing and walking, or centering the pelvis and torso in a seated position
Positive reframing of the emotional experience
Commenting that the story needs to be told many times and each time helps

As a caution, when you work with emotion and feeling, there is always the possibility of reawakening issues in a client who has a history of painful trauma. This is an area where the beginning counselor often needs to refer the client to a more experienced professional. Even the 2-minute expression of emotion suggested here may be too long. In such situations, seek supervision and consultation.

Neuroscience research has shown that people have measurable body/feeling reactions to others. When those others are people different from them, in terms of factors such as race or even politics, these reactions can lead to the assigning of certain social emotional terms and cognitive beliefs to others. A person's true feelings may be hidden by language and behavior— for example, claiming to like a person or denying having racist attitudes. But fMRI measures may show something vastly different—areas of negative feeling are activated in the brain, revealing what the real feelings are (Vedantam, 2010; West, 2007; Westen, 2007).

Feelings guide our cognitive decisions, and the prefrontal TAP seeks to modulate and regulate them. At the same time, fixed memories in the hippocampus are often hard to change. One example is the attitudes and beliefs that people hold about race. Box 7.2 presents research examples with cognitive statements that are based on deeply held emotional attitudes.

BOX 7.2 | ## National and International Perspectives on Counseling Skills

The Invisible Whiteness of Being and Feelings Toward Another Race

Derald Wing Sue

Multicultural issues remain a "hot button" topic and often bring out emotions that may surprise us, whether client or counselor. I posed this open question to people on the street in San Francisco: "What does being White mean?" Here are some responses. Note the feelings that this question brought out.

42-Year-Old White Businessman

Q: What does it mean to be White?

A: Frankly, I don't know what you're talking about.

Q: Aren't you White?

A: Yes, but I come from Italian heritage. I'm Italian, not White.

Q: Well then, what does it mean to be Italian?

A: Pasta, good food, love of wine. (obviously agitated) This is getting ridiculous.

Theme: This person denies the color of his skin and speaks superficially about his Italian heritage. He shows feelings of anger around the issue.

26-Year-Old White Female College Student

Q: What does it mean to be White?

A: Is this a trick question? . . . I've never thought about it . . . Well, I know that lots of Black people see us as prejudiced and all that stuff. I wish people would just forget about race differences and see one another as human beings. People are people and we should be proud to be Americans.

Theme: She never thinks of being White and shows feelings of defensiveness and confusion. She focuses on "people are people."

29-Year-Old Latina Administrative Assistant

Q: What does it mean to be White?

A: I'm not White, I'm Latina!

Q: Are you upset with me?

A: No . . . it's just that I'm light, so people always think I'm White. It's only when I speak that they realize I'm Hispanic.

(continued)

BOX 7.2 (continued)

Q: Well, what does it mean to be White?

A: Do you really want to know? . . . OK, it means you're always right. It means that you never have to explain yourself or apologize. . . . You know that movie, Love is never having to say you're sorry? Well, being White is never having to say you're sorry. It means you think you're better than us.

Theme: Anger at being misidentified and anger and frustration with what she sees as dominant Whites who feel superior.

39-Year-Old African American Salesman

Q: What does it mean to be White?

A: If you're White, you're right. If you're Black, step back.

Q: What does that mean?

A: White folks are always thinking they know all the answers. A Black man's word is worth less than a White man's. When White customers come into our dealership and see me standing next to cars, I become invisible to them. Actually, I think they see me as a well-dressed janitor. They seek out White salesmen. I talked to my boss about this, but he says I'm oversensitive. That's what being White means. It's having the authority or power to tell me what's really happening even though I know it's not. Being White means you can fool yourself into thinking that you're not prejudiced, when you are. That's what it means to be White.

Themes: Feelings of frustration and anger around Whites who view minorities as less competent and capable; also, Whites have the power to define the world for others.

These four examples illustrate the deep-seated bodily feelings and social emotions underlying issues of race. The Latina and African American comments illustrate feelings that many people of color have toward White people. Whites, on the other hand, are often oblivious to how color affects their lives. Being White brings with it a privilege that European Americans are often oblivious of and unaware of having.

This box by Derald Wing Sue is partially adapted from *Overcoming Our Racism: The Journey to Liberation*, Jossey-Bass, 2003. Reprinted by permission of John Wiley & Sons Inc.

▶The Place of Positive Emotions in Reflecting Feelings

Contrasting the negative emotions such as sadness with joy can lead to inner peace and stoic equanimity.

—Antony Damasio

Positive emotions, whether joyful or merely contented, are likely to color the ways people respond to others and their environments. Research shows that positive emotions broaden the scope of people's visual attention, expand their repertoires for action, and increase their capacities to cope in a crisis. Research also suggests that positive emotions produce patterns of thought that are flexible, creative, integrative, and open to information (Gergen & Gergen, 2005). *Sad, mad, glad, scared* is one way to organize the language of emotion. But perhaps we need more attention to glad words—*pleased, happy, love, contented, together, excited, delighted, pleasured,* and the like.

When you experience emotion, your brain signals bodily changes. The positive executive TAP axis is activated when the prefrontal cortex recalls and labels variations of the basic feeling

REFLECTION *Here and Now*

Take just a moment now and think of specific situations in which you experienced each of the positive emotions listed immediately above. What changes did you notice in your body and mind? Compare these with the bodily experience of negative feelings and emotions. Very likely, if you smiled, your body tension was reduced and even your blood pressure lowered. Perhaps you even found a little bit of inner peace, as suggested by Anthony Damasio. How was your experience? What do you think of the changes you noticed in your body? In your mind?

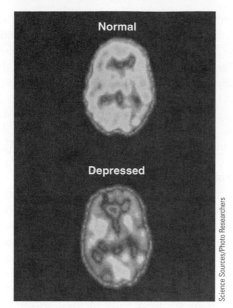

FIGURE 7.1 Brain activation in normal and depressed individuals.

of glad. The amygdala energizes this process, and positive body feelings of relaxation and pleasure result. (See Chapter 1, Figure 1.3, page 21.) When you feel sad or angry, a set of chemicals floods your body, and usually these changes will show nonverbally. Emotions change the way your body functions and thus are a foundation for all your thinking experience (Damasio, 2003). As you help your clients experience more positive emotions, you are also facilitating wellness and a healthier body. Cognitive positive emotional input leads to positive emotional and feeling output. The route toward good health, of course, also often entails confronting negative emotions.

Research on the long-lived Mankato nuns found that women who had expressed the most positive emotions in early life lived longer than those who had experienced a difficult past (Danner, Snowdon, & Friesen, 2001). Research on stress reactions to the 9/11 attacks found that people who had access to the most positive emotions showed fewer signs of depression (Fredrickson, Tugade, Waugh, & Larkin, 2003). A resilient affective lifestyle results in a faster recovery and lower damaging cortisol levels.

These examples of well-being and wellness are located predominantly in the executive TAP prefrontal cortex, with lower levels of activation in the amygdala (Davidson, 2004, p. 1395). Drawing on long-term memory for positive experiences is one route toward well-being and stress reduction.

Figure 7.1 shows the level of brain activity of a depressed and a nondepressed individual. In this picture, we see what happens in the absence of positive feelings and emotions. Working with depression can be challenging, but our goal is to increase positive functioning. If we focus only on negative, problematic stories and reflect these negative feelings, we are reinforcing depression. While it is necessary to hear these stories, ultimately our goal with any client expressing sadness or depression is to find positives. (Needless to say, major depression will often require medication before counseling and therapy can be fully effective.)

▶Strategies for Positive Reflection

Several techniques can be utilized to help clients reflect on positive emotions and feelings. The first is searching for wellness and positive assets, as emphasized in Chapter 2. As part of a wellness assessment, be sure that you reflect the positive feelings associated with aspects of wellness. For example, your client may feel safety and strength in the spiritual self, pride in gender and/or cultural identity, caring and warmth from past and/or present friendships, and the intimacy and caring of a love relationship. It would be possible to anchor these emotions early in the session and draw on these positive emotions during more stressful moments. Out of a wellness inventory can come a "backpack" of positive emotions and experiences that are always there and can be drawn on as needed.

When reflecting feelings, listen carefully and elicit strengths. Make this part of your reflection of feeling strategy. For example, the client may be going through the difficult part of a relationship breakup and crying and wondering what to do. We don't suggest that you should interrupt the emotional flow, but with appropriate timing, reflecting back the positive feelings that you have observed can be helpful.

Couples with relationship difficulties can be helped if they focus more on the areas where things are going well—what remains good about the relationship. Many couples focus on the 5% where they disagree and fail to note the 95% where they have been successful or enjoyed each other. Some couples respond well when asked to focus on the reasons they got together in the first place. These positive strengths can help them deal with very difficult issues.

When providing your clients with homework assignments, ask them to engage daily in activities associated with positive emotions. For example, it is difficult to be sad and depressed when running or walking at a brisk pace. Meditation and yoga are often useful in generating more positive emotions and calmness. Seeing a good movie when one is down can be useful, as can going out with friends for a meal. In short, help clients remember that they have access to joy, even when things are at their most difficult.

Finally, service to others often helps people feel good about themselves. When one is discouraged and feeling inadequate, volunteering for a church group, working on a Habitat for Humanity home, or giving time to work at an animal shelter can all be helpful in developing a more positive sense of self. See Box 7.3 for some research on reflection of feeling.

| BOX 7.3 | Research and Related Evidence That You Can Use |

Reflection of Feeling

Emotional processing—the working through of emotions and the ability to examine feelings and body states—has been found to be fundamental in effective experiential counseling and therapy. People with lower emotional awareness are more likely to make errors than those who are in touch with their emotions (Szczygieł, Buczny, & Bazińska, 2012). Gains in treatment of depressed clients were found to be highly related to emotional processing skills (Pos, Greenberg, Goldman, & Korman, 2003). As you work with all your clients, your skill in reflecting feelings can be a basic factor in helping them take more control of their lives.

Dealing with feeling and emotion is not only a central aspect of counseling, and psychotherapy; it is also key to high-quality interviewing with medical patients (Bensing, 1999b; Bensing & Verheul, 2009, 2010). Working with emotion requires attention to nonverbal dimensions. Head nodding, eye contact, and especially smiling are facilitative. Clearly, warmth, interest, and caring are communicated nonverbally as much as, or more than, verbally. Moreover, Hill (2009) found that using questions oriented toward affect increased client expression of emotion. However,

once a client has expressed emotion, continued use of questions may be too intrusive and the more reflective approach will be more useful.

"Several studies have shown that between 30 and 60 percent of patients in general practice present health problems for which no firm diagnosis can be made" (Bensing, 1999a). Be ready to look to emotions in clients who have medical issues. Older persons tend to manifest more mixed feelings than others (Carstensen, Pasupathi, Mayr, & Nesselroade, 2000). Perhaps this is because life experience has taught them that things are more multi-faceted than they once thought. Helping younger clients become aware of emotional complexity may also be a goal of some counseling sessions.

Tamase, Otsuka, and Otani (1990), through their work in Japan, have provided clear indication that the reflection of feelings is useful cross-culturally. Reporting on a series of studies in this area, Hill (2009) notes the facilitating impact of reflective responses. She reports that clients are usually not aware when helpers are using good restatements and reflections. Effective listening facilitates exploration.

Caution: Please do not use the above paragraphs as a way to tell your clients that "everything will be OK." Some counselors and psychotherapists are so afraid of negative emotions that they never allow their clients to express what they really feel. *Do not minimize difficult emotions by too quickly focusing on the positive.*

▶Summary: Reflection of Feelings Is a Key Skill, but Still Use It With Some Caution

Focusing on emotions and feelings is a basic and necessary feature of the counseling and psychotherapy process and has practical implications beyond the session. Working with emotions and more basic feelings varies according to the client, the situation, and the issue. Helping clients become aware of their emotions may be helpful in its own right. As clients move toward complex issues, the sorting out of mixed emotions is a central ingredient of successful counseling, be it school counseling, vocational counseling, personal decision making, or in-depth individual counseling and therapy.

Many times a short and accurate reflection may be the most helpful. With friends, family, and fellow employees, a quick acknowledgment of feelings ("If I were you, I'd feel angry about that . . ." or "You must be tired today") followed by continued normal conversational flow may be most helpful in developing better relationships. In an interaction with a harried waiter or salesperson, an acknowledgment of feelings may change the whole tone of a meal or business interchange.

Similarly, with many clients a brief reflection of feeling may be more useful. This may be especially true with clients with whom trust and a fully empathic relationship have not yet been established—or those who may be culturally different from you in some way.

Keep in mind that clients are not always ready to address or share with you the way they feel. Clients tend to disclose feelings only after rapport and trust have been developed. Remember that not all clients will appreciate or welcome your commenting on their feelings. Also, an empathic reflection can sometimes have a confrontational quality that causes clients to look at themselves from a different perspective; it may therefore seem intrusive to some. Less verbal clients may find reflections puzzling at times or may say, for instance, "Of course I'm angry; why did you say what I just said?" With some cultural groups, reflection of feeling may be inappropriate and represent cultural insensitivity. Some men, for example, may believe that expression of feelings is "unmanly," yet a brief acknowledgment may be helpful to them. Though noting feelings in the session is essential, acting on your observations may not always be in the best interests of the client. Timing is particularly important with this skill.

▼▼▼▼	Key Points
Feelings and social emotions	Feelings are basic to what we call emotional experience. The research-based basic feelings are sad, mad, glad, and scared, plus surprise and disgust/contempt. Social emotions (for example, guilt, compassion, love) are developed in a cultural context and may be blends of basic feelings. Feelings are not only cognitive; they are found in bodily expressions at multiple levels, from observable body language to subtle changes in heart rate and blood pressure.
Naming	You will want to develop an array of ways to observe and describe feelings and emotions. In naming client feelings, note the following: ▲ Emotional and feeling words used by the client ▲ Implicit emotional and feelings not actually spoken ▲ Observation of nonverbally expressed emotions and feelings. ▲ Mixed verbal and nonverbal emotional cues, which often represent conflict

Reflecting feelings	Emotions may be observed directly, or sometimes you will want to draw them out through questions ("How do you feel about that?" "Do you feel angry?") or feedback ("You look uncomfortable when you say that"; "I sense right now that you are anxious"). As discussion continues, reflect feelings through the following steps:
	▲ Begin with a sentence stem such as "You feel . . . " or "Sounds like you feel . . ." or "Could it be you feel . . . ?" Use the client's name.
	▲ Add a feeling word or words (sad, happy, glad).
	▲ Add context through a paraphrase or repetition of key content ("Looks like you feel happy about the excellent rating").
	▲ In many cases, a reflection in the present tense is more powerful than one in the past or future tense ("You feel happy right now" rather than "You felt" or "You will feel").
	▲ Use a checkout, especially following identification of an unspoken feeling ("Am I hearing you correctly?" "Is that close?"). This lets the client correct you if you are either incorrect or uncomfortably close to a truth that he or she is not yet ready to admit.
Acknowledgment of feelings	Brief acknowledgment of feeling may be helpful. This skill may be particularly useful in working with clients with whom trust has not been fully developed. Deeper reflections may be appropriate in many counseling situations, but they often require a relatively verbal client. The acknowledgment of feeling puts less pressure on clients to examine their emotions and may be especially helpful in the early stages of counseling a client who is culturally different from you. Later, as trust develops, you can explore emotion and feelings in more depth.
Positive emotions in reflecting feelings	Positive emotions affect the ways people respond to others and their environments. They activate our brains, broaden our perceptions, let us think more flexibly, increase our capacity to deal with crises, make us happier, and improve our wellness. Seek out these emotions constantly. Help the client activate the positive areas of the brain in the prefrontal cortex.
	Reflecting feelings helps us reach both negative and positive emotions and feelings. Recognize and reflect the negative, but search for positive strengths and feelings. It can take five or more positives to counteract a negative (Gottman, 2011). A single emotionally laden damaging comment or negative life experience can last a lifetime and change one's self-view. If we wish to build effective emotional and self-regulation, as well as intentionality, the positive approach becomes essential.
Implications of neuroscience	Research on the brain has validated many traditional beliefs and teachings of the counseling field around emotions and feelings. Neuroscience's focus on the limbic HPA as being the first to react, before the TAP executive prefrontal cortex, supports the idea that reflection of feeling the most valuable listening skill after attending behavior (if one isn't attending, feelings will not be observed). Approaches such as cognitive behavioral therapy (CBT) will be more effective if emotion and feeling are recognized.

▶Competency Practice Exercises and Portfolio of Competence

Go to CengageBrain.com to access Counseling CourseMate, where you will find an interactive ebook, quizzes, videos, interactive counseling and psychotherapy exercises, the Portfolio of Competence, case studies, a practice test, and more.

We observe feelings in many daily interactions, but we usually ignore them. In counseling and helping situations, however, they can be central to the process of understanding another person. Further, you will find that increased attention to feelings and emotions may enrich your daily life and bring you to a closer understanding of those with whom you live and work.

Individual Practice

Exercise 1. Increasing Your Emotional and Feeling Vocabulary   Return to the list of affective words you generated earlier in this chapter. Take some more time to add to that list. One way to lengthen the list is to consider two categories of feeling words that would provide additional ideas about how the client feels about the world.

The first category is words that represent mixed or ambivalent emotions and feelings. Your task is to help the client sort out the deeper emotions underlying the surface, expressed word. Write a list of words that represent confused or vague feelings (for instance, *confused, anxious, ambivalent, torn, ripped, mixed*).

A common mistake is to assume that these words represent the root feelings. Most often they cover deeper feelings. The word *anxiety* needs to be considered in this context. Anxiety is often a vague indicator of mixed feelings, perhaps combining fear and sadness, or at another level love or caring and a desire to please. This also can be at the root of what is called an anxiety disorder, as here the feelings of confusion take over the client's being. If you accept client anxiety as a basic feeling, counseling may proceed slowly. An important task of the counselor when noting mixed feeling words is to use questions and reflection of feeling to help the client discover the deeper bodily feelings underlying the surface ambivalence. Underlying anxiety or confusion, for example, you may find anger and fear mixed with love.

Now take two of the words from your mixed emotions list and lay out more basic words that might underlie affectively oriented words such as *confused* or *frustrated*. Again, the basic words *mad, sad, glad,* or *scared* may be helpful in this process. See pages 181–182 for a list of emotion-laden words.

Emotions and feelings are often presented through metaphors and similes, concrete examples, and body language. It is often more descriptive of your emotions to say that you feel like a limp dishrag than to say you are tired and exhausted. Other examples might include "down in the pits," "high as a kite," "crashed worse than a computer," or "proud as a peacock." Because metaphors are often masks for more complex feelings, you may want to search for the underlying feelings through careful listening and perhaps some questions. After you have developed a list of at least five metaphors, you may wish to generate a list of basic feeling words underlying the metaphors.

"Metaphors Make Brains Touchy Feely" is the title of a *Science Now* article (Tellis, 2012). We know that language areas of the brain respond to verbalization in general. Recent research now tells that when we use sensory metaphors, the brain also activates in the area dealing with touch (Lacey, Stilla, & Sathian, 2012). Term a client "soft-hearted" and both the language and touch areas respond. Anticipate that further research on metaphors and key emotional and feeling words will provide further information useful in reflecting feelings.

Generate a list of metaphors and similes that may be useful in the counseling process—for example, "You feel limp as a dishrag"; "The confusion leaves you feeling like you're in a revolving washing machine"; "You look as joyful as a child on a swing."

Exercise 2. Distinguishing a Reflection of Feeling From a Paraphrase   The key feature that distinguishes a reflection of feeling from a paraphrase is the affective word. Many paraphrases contain reflection of feeling; such counselor statements are classified as both. In the example that follows, indicate which of the responses is an encourager (E), which a paraphrase (P), and which a reflection of feeling (RF).

"I am really discouraged. I can't find anywhere to live. I've looked at so many apartments, but they are all so expensive. I'm tired and I don't know where to turn."

Mark the following counselor responses with an E, P, RF, or combination if more than one skill is used.

_____ "Where to turn?"
_____ "Tired . . ."
_____ "You feel very tired and discouraged."
_____ "Searching for an apartment simply hasn't been successful; they're all so expensive."
_____ "You look tired and discouraged; you've looked hard but haven't been able to find an apartment you can afford."

For the next example, write an encourager, a paraphrase, a reflection of feeling, and a combination paraphrase/reflection of feeling in response to the client.

> "Right, I do feel tired and frustrated. In fact, I'm really angry. At one place they treated me like dirt!"

Exercise 3. Acknowledgment of Feeling   We have seen that the brief reflection of feeling (or acknowledgment of feeling) may be useful in your interactions with busy and harried people during the day. At least once a day, deliberately tune in to a server/waitstaff person, teacher, service station attendant, telephone operator, or friend, and give a brief acknowledgment of feeling ("You seem terribly busy and pushed"). Follow this with a brief self-statement ("Can I help?" "Should I come back?" "I've been pushed today myself, as well") and note what happens in your journal.

Assume you are working with one of the preceding clients who avoids really looking at here-and-now emotional experiencing. How would you help this client increase affect and feeling?

Group Practice

Exercise 4. Practicing Reflection of Feelings   One of the most challenging skills is reflection of feeling. Mastering this skill, however, is critical to effective counseling and psychotherapy.

Step 1: Divide into practice groups.

Step 2: Select a group leader.

Step 3: Assign roles for the first practice session.

▲ Client
▲ Counselor
▲ Observer 1, who gives special attention to noting client feelings, using the Feedback Form in Box 7.4. The focus of microsupervision needs to be on the counselor's ability to bring out and deal with emotions and feelings.
▲ Observer 2, who gives special attention to counselor behavior and writes down each specific counselor lead.

Step 4: Plan.   We suggest that you examine a past or present story of a stressful experience (bullying, teasing, being seriously misunderstood, an unfair situation with school personnel, going through a hurricane or flood, or a time when you experienced prejudice or oppression of some type). Please do not just focus on negative emotions. Spend time drawing out positive stories of strength, and reflect these positive feelings.

Many students feel anxious and tense around substantial student loans. Role-play a student facing these issues. Reflect the emotions and feelings, and encourage the client to

Feedback Form: Observing and Reflecting Feelings

You can download this form from Counseling CourseMate at CengageBrain.com.

_____ (DATE)

_____   _____
(NAME OF COUNSELOR)   (NAME OF PERSON COMPLETING FORM)

Instructions: Observer 1 will give special attention to client feelings in notations of verbal and nonverbal behavior below. On a separate sheet, Observer 2 will write down the wording of counselor reflections of feeling as closely as possible and comment on their accuracy and value.

1. Verbal feelings and emotions expressed by the client. List here all related words.

2. Nonverbal indications of feeling states in the client. Facial flush? Body movements? Others? Later check this out with the client. What does the client recall feeling? What did you as counselor feel emotionally through the process of listening and reflecting?

3. Implicit feelings not actually spoken by the client. Check these out with the client later for validity.

4. Reflections of feelings used by the counselor. As closely as possible, use the exact words and record them on a separate sheet of paper.

5. Comments on the reflections of feeling. What were the strengths of the session? Was the counselor's use of the skill accurate and valid? Was the checkout used?

continue. Again, follow this with the positive asset search. What strengths does the student have to deal with the loan?

Establish clear goals for the session. You can use questioning, paraphrasing, and encouraging to help bring out data. Periodically, the counselor reflects feelings. This may be facilitated by one-word encouragers that focus on feeling words and by open questions ("How did you feel when that happened?"). The practice session should end with a summarization of both the feelings and the facts of the situation. Examine the basic and active mastery goals in the Portfolio of Competence to determine your personal objectives for the session.

The observers should use this time to examine the feedback forms and to plan their own sessions.

Step 5: Conduct a 5-minute practice session using this skill.

Step 6: Review the practice session and provide feedback to the counselor for 10 minutes.
How well did the counselor achieve goals and mastery objectives? What feedback does the client provide verbally and through the Client Feedback Form from Chapter 1? As skills and client role-plays become more complex, you'll find that this time is not sufficient for in-depth practice sessions and you'll want to contract for practice time outside the session with your group. Again, what level of mastery did the counselor achieve? Was the counselor able to achieve specific objectives with a specific impact on the client?

Step 7: Rotate roles.

A reminder.　The client may be "difficult" if he or she wishes, but must be talkative. Remember that this is a practice session, and unless affective issues are discussed, the counselor will have no opportunity to practice the skill.

Portfolio of Competence

Skill in reflection of feeling rests in your ability to observe client verbal and nonverbal emotions. Reflections of feeling can vary from brief acknowledgment to exploration of deeper emotions. You may find this a central skill as you determine your own style and theory.

Use the following checklist to evaluate your present level of mastery. As you review the items below, ask yourself, "Can I do this?" Check those dimensions that you currently feel able to do. Those that remain unchecked can serve as future goals. Do not expect to attain intentional competence on every dimension as you work through this book. You will find, however, that you will improve your competencies with repetition and practice.

Level 1: Identification and classification

☐ Ability to generate an extensive list of affective words.
☐ Ability to distinguish a reflection of feeling from a paraphrase.
☐ Ability to identify and classify reflections of feeling.
☐ Ability to discuss, in a preliminary fashion, issues in diversity that occur in relation to this skill.
☐ Ability to write reflections of feeling that might encourage clients to explore their emotions.

Level 2: Basic competence.　Aim for this level of competence before moving on to the next skill area.

☐ Ability to acknowledge feelings briefly in daily interactions with people outside of counseling situations (restaurants, grocery stores, with friends, and the like).
☐ Ability to use reflection of feeling in a role-played session.
☐ Ability to use the skill in a real session.

Level 3: Intentional competence. The following skills are all related to predictability and evaluation of the effectiveness of your abilities in working with emotion. These are skill levels that may take some time to achieve. Be patient with yourself as you gain mastery and understanding.

❑ Ability to facilitate client exploration of emotions. When you observe clients' emotions and reflect them, do clients increase their exploration of feeling states?
❑ Ability to reflect feelings so that clients feel their emotions are clarified. They may often say, "That's right, and . . ." and then continue to explore their emotions.
❑ Ability to help clients move out of overly emotional states to a period of calm.
❑ Ability to facilitate client exploration of multiple emotions one might have toward a close interpersonal relationship (confused, mixed positive and negative feelings).

Level 4: Pychoeducational teaching competence. A client who has particular difficulty in listening to others may indeed benefit from training in observing emotions. Many individuals fail to see the emotions occurring all around them—for example, one partner may fail to understand how deeply the other feels. Empathic understanding is rooted in awareness of the emotions of others. All of us, including clients, can benefit from bringing this skill area into use in our daily lives. There is clear evidence that people diagnosed with antisocial personality disorder have real difficulty in recognizing and being empathic with the feelings of others. You will also find this problem in some conduct disorder children. Here psychoeducation on empathy and recognizing the other person can be a critical treatment. A good place to start is help them observe and name feelings, followed by acknowledgment of feelings.

❑ Ability to teach clients in a helping session how to observe emotions in those around them.
❑ Ability to teach clients how to acknowledge emotions—and, at times, to reflect the feelings of those around them.
❑ Ability to teach small groups the skills of observing and reflecting feelings.

▶Determining Your Own Style and Theory: Critical Self-Reflection on Reflection of Feeling

This chapter has focused on emotion and the importance of establishing a foundation between counselor and client. Special attention has been given to identifying seven basic feelings, along with many more examples of the social emotions, as well as how you might help clients express more or less emotion as appropriate to their situation.

What single idea stands out for you among all those presented in this chapter, in class, or through informal learning? What stands out to you is likely to be useful as a guide toward your next steps. What are your thoughts on diversity? What other points in this chapter strike you as particularly useful in your future practice? How might you use ideas in this chapter to begin the process of establishing your own style and theory?

▶List of Feeling Words

Again, note that our vocabulary is more extensive for the negative feelings than the positive, again illustrating the importance of being with your clients with an optimistic tone.

Sad: unhappy, depressed, tearful, uninterested, blue, bored, cheerless, dismal, dispirited, dull, gloomy, grief, grieving, miserable, anguished, sorrow, regret, sorry, guilty, deplorable, devastated, devalued, pitiful, derided, joyless, melancholy, melancholic, dejected, desolate, heavy-hearted, low, spiritless, rejected, woebegone, falling apart, wistful, wretched, moping.

Mad: angry, annoyed, insulted, irritated, indignant, irate, hostile, offended, ripped, displeased, aggressive, furious, ferocious, rabid, stormy, inflamed, infuriated, hatred, strongly opposed, antagonistic, uncompromising, dislike, animosity, distaste, threatening, dissatisfied, undesirable, unfair, unreasonable, rude, insensitive.

Glad: happy, relaxed, safe, comfortable, calm, at ease, pleased, feeling of "wholeness," valued, accepted, "together," interesting, excited, confident, cheerful, spirited, joy, joyful, heartfelt, appreciative, grateful, cheery, pleasure, bright, contented, satisfied, delight, delighted, enjoy, thankful, relieved, fascinating, lovely, light, cared for, caring, pleasing, eager, compliant, festive, tickled, merry, fortunate, lucky, chipper, lighthearted, esteemed, respected, honored, cherished, welcomed.

Scared: afraid, fretting, fright, frightened, threatened, anxious, anxiety, apprehension, dangerous, concerned, worried, worrisome, agitated, alarmed, dread, horror, panic, terror, trepidation, distressed, troubled, tormented, angst, disquieted, unease, nervous, brooding.

Disgust: revulsion, stinking, repugnance, sicken(ing), loathing, repelling, nauseating. Associated social emotions are sleazy, slimy, dirty, greedy—many things that disagree with our moral or value system.

Contempt: utter disgust, disdain, dislike, disrespect, disapproval, scorn, hatred, derision, condescension.

Surprise: astonishment, amazement, revelation, startled, wonderment, shock, interruption, bombshell.

CHAPTER 8

How to Conduct a Five-Stage Counseling Session Using Only Listening Skills

The Five-Stage Interview Structure

Reflection of Feeling

Encouraging, Paraphrasing, and Summarizing

Open and Closed Questions

Client Observation Skills

Attending Behavior and Empathy

Ethics, Multicultural Competence, and Wellness

A leader is best when people barely know that he [or she] exists,

Of a good leader who talks little,

When the work is done,

The aim is fulfilled

They will say, "We did this ourselves."

—Lao Tse

Mission of "How to Conduct a Five-Stage Counseling Session Using Only Listening Skills"

This chapter provides the basis for you to conduct a full session using the empathic listening skills presented in previous chapters. Basic to this process are the five stages of a well-formed session (*empathic relationship—story and strengths—goals—restory—action*). In addition, the basic listening sequence will be reviewed and decisional counseling as a basic framework for many counseling theories will be introduced.

Chapter Goals and Competency Objectives Awareness, knowledge, skills, and actions in conducting a five-stage counseling session using only listening skills will enable you to

▲ Develop further competence with the basic listening sequence (BLS), the foundation of effective counseling and psychotherapy.

▲ Examine listening skills and empathy as they relate to neuroscience and the brain.

▲ Demonstrate fluid and intentional responsiveness to clients, whether or not your anticipated results from using a skill are successful.

▲ Understand and become competent in the five stages of the well-formed session: empathic relationship—story and strengths—goals—restory—action.

▲ Conduct a complete session using only listening skills.

▲ Become aware of how Benjamin Franklin's original problem-solving model relates to the five stages and how Franklin and the five-stage model clarify what is happening in many theories of counseling and psychotherapy.

> ⬚ Go to CengageBrain.com to access Counseling CourseMate, where you will find an interactive eBook, quizzes, videos, interactive counseling and psychotherapy exercises, the Portfolio of Competence, case studies, a practice test, and more.

▶The Basic Listening Sequence as a Foundation for Effective Communication in Many Areas of Life

To review, the listening skills of the first section of this book contain the building blocks for establishing empathic relationships and effective counseling and psychotherapy. In addition, empathic understanding and careful listening are valuable in all areas of human communication. When you go to see your physician, you want the best diagnostic skills, but you also want the doctor to listen to your story with understanding and empathy. A competent teacher or manager knows the importance of listening to the student or employee. The microskills represent the specifics of effective interpersonal communication and are used in many situations, ranging from helping a couple communicate more effectively to enabling a severely depressed client to make contact with others to training AIDS workers in Africa and interviewing refugees around the world. Furthermore, the teaching of listening skills has become a standard and common part of individual counseling and psychotherapy.

The basic listening sequence, or BLS, helps you and other professionals to understand the client story. These skills have been discussed separately in previous chapters; now their full impact in the session will be realized by using them together. When you use the BLS, you can anticipate how clients are likely to respond.

BASIC LISTENING SEQUENCE (BLS)	ANTICIPATED RESULT
The basic listening sequence (BLS) uses the microskills of using open and closed questions, encouraging, paraphrasing, reflecting feelings, and summarizing. These are supplemented by attending behavior and client observation skills. Select and practice all elements of the basic listening sequence.	Clients will discuss their stories, issues, or concerns, including the key facts, thoughts, feelings, and behaviors. Clients will feel that their stories have been heard.

To review, resting on a foundation of attending and observation, these are the skills of the BLS and the central purpose of each:

▲ Questioning—open questions followed by closed questions to bring out the story and for diagnosis
▲ Encouraging—used throughout the session to help evoke details
▲ Paraphrasing—catches the essence of what the client is saying
▲ Reflecting feelings—examines the complexity of emotions
▲ Summarizing—brings order and makes sense of client conversation

When you add the checkout to the basic listening sequence, you have the opportunity to obtain feedback on the accuracy of your listening. The client will let you know how accurately you have listened.

The skills of the BLS need not be used in any specific sequence, but it is wise to ensure that all are used in listening to client stories. Each person needs to adapt these skills to meet the requirements of the client and the situation. The competent counselor uses client observation skills to note client reactions and intentionally *flexes* to change style, thus providing the support the client needs.

Examples of how the BLS is used in counseling, management, medicine, and general interpersonal communication are shown in Table 8.1.

Counseling and psychotherapy can be difficult experiences for some clients. They have come to discuss their issues and resolve conflicts, so the session can rapidly become a

TABLE 8.1   Four Examples of the Basic Listening Sequence

Skill	Counseling	Management	Medicine	Interpersonal Communication (Student listens to friend.)
Open questions	"Thomas, could you tell me what you'd like to talk to me about . . . ?"	"Assad, can you tell me what happened when the production line went down?"	"Ms. Santiago, could we start with what you think is happening with your headache? Is this OK?	"Kiara, how did the session with the college loan officer go?"
Closed questions	"Did you graduate from high school?" "What specific careers have you looked at?"	"Who was involved with the production line problem?" "Did you check the main belt?"	"Is the headache on the left side or on the right? How long have you had it?"	"Were you able to get a loan that covers what you need? What interest rate are they using?"
Encouragers	Repetition of key words and restatement of longer phrases.			
Paraphrases	"So you're considering returning to college."	"Sounds like you've consulted with almost everyone."	"I hear you saying that the headache may be worse with red wine or too much chocolate."	"Wow, I can understand what you are saying, Kiara, two loans are a lot."
Reflection of feeling	"You feel confident of your ability but worry about getting in."	"I sense you're upset and troubled by the supervisor's reaction."	"You say that you've been feeling very anxious and tense lately."	"I sense you feel somewhat anxious and worried when you think of paying it back."
Summarization	In each case the effective listener summarizes the story from the client's or other person's point of view *before* bringing in the listener's own point of view or perhaps an influencing skill.			

© Cengage Learning

depressing litany of failures and fears. Remember always to use the BLS to help the client identify strengths and resources. To ensure a more optimistic and directed session, use the positive asset search and wellness approach. Rather than just ask about issues and concerns, the intentional counselor or psychotherapist seeks to identify the client's positives and strengths. Even in the most difficult situation, find good things about the client and resources for later problem resolution. Emphasizing positive assets also gives the client a sense of personal worth as he or she talks with you.

Box 8.1 presents a brief discussion on demystifying the helping process.

BOX 8.1 National and International Perspectives on Counseling Skills

Demystifying the Helping Process

Allen E. Ivey and Carlos P. Zalaquett

Demystify: make less mysterious or remove the mystery from

—Webster's Online Dictionary

Demystify: to make something easier to understand

—Cambridge Advanced Learner's Dictionary & Thesaurus

What Makes Counseling and Psychotherapy Successful? For years counseling and psychotherapy remained a mysterious, puzzling, and difficult to decipher process. Carl Rogers was the first to demystify the counseling process. Through the then new technology of the wire recorder, he showed us what is actually done in the session. Before that, therapists only reported what they remembered. Early research after Rogers's contribution found that what counselors said they did in the session was not what they actually did.

Rogers brought us the first conceptions for understanding the therapeutic process and empathy. However, the *behaviors* of a quality session remained vague. What are the *observable* counselor actions that make a session successful? What are the key therapist *behaviors* facilitating client growth? With the number of new therapies increasing over time, the uniqueness of the therapist and client, and the diversity of perspectives, issues, and goals, identifying the central components of the counseling and therapy relationship remained elusive.

Demystifying the Counseling Process Until the microskills approach came along, the counseling and psychotherapy field had not identified the behaviors of the effective helper. With colleagues at Colorado State University, Allen obtained the first video grant to examine more precisely what are the *observable behavioral* skills of the competent counselor and therapist. For the first time, videotaping equipment enabled the recording of sessions for nonverbal as well as verbal behavior.

Attending, the First Behavioral Skill The importance of attending behavior came to our attention almost by chance.

We videotaped our secretary in a demonstration session. She totally failed to attend to the student she was counseling—looking down, nervous movements, and topic jumping. On reviewing the video, we identified the concrete specifics of attending behavior for the first time: appropriate eye contact, comfortable body language, a pleasant and smooth vocal tone, and verbal following—staying with the client's topic. When she went back for another video practice session, she listened effectively, and even looked like a counselor. All that happened in a half-hour!

Taking Microskill Learning Home Via Action The next level of demystification came when our secretary returned after the weekend. "I went home, I attended to my husband, and we had a beautiful weekend!" We had not anticipated that session behavior would generalize to real life. I became aware of the importance of teaching communication microskills to clients and patients. Children, couples, families, management trainees, psychiatric patients, refugees, and many others have now been taught specifics of communication using the skills taught in this book.

Microtraining Goes Viral and Multicultural Soon after the microskills and microtraining became available, reverberations began. More than 450 data-based studies followed, and Allen's work was translated into multiple languages, becoming a regular part of the university curriculum in counseling, social work, and psychology. The adaptability to many different situations and cultures continues even today—ranging from Aboriginal social workers in Australia to AIDS peer counselors in Africa to managers in Japan, Sweden, and Germany. I (Carlos) was trained in the microskills as part of my graduate program and had the privilege of teaching them as part the first course on effective psychotherapy I taught in Chile. My students liked the framework and demonstrated the competencies.

Transtheoretical Character of the Microskills Microskills bridge the gap between what is said is to be done in the session and what actually happens. Following the tradition of Carl Rogers, the session is demystified. At the same time, the microskills provide a foundation for understanding widely different theories of therapy. Virtually all use some variation of the BLS and some form of the five-stage session, but you will find widely varying beliefs and actions when it comes to the influencing and action skills of the second half of this book. The five stages are also a basic decisional system. The discussion of decisional counseling in Chapter 14 emphasizes this point.

Video Your Sessions   This is why we hope that you will video your own sessions as frequently as possible, both during this course and when you move on as a professional. Video is now easily available through inexpensive video cams, miniature cameras, and even cell phones. Demystify your own behavior in the session. Watching the video together with your clients is invaluable. Use the microtraining model to become aware of your natural competencies and learn additional skills to work effectively with more people.

Becoming More Specific About Multicultural Issues   About a year after the identification of the attending skills, paraphrasing, reflecting feelings, and summarizing, Allen was enthusiastically teaching a workshop on microskills. After he provided information on attending behavior and the importance of eye contact, a quiet student came up to him and described her experience in Alaska with Native Inuits. She commented that direct eye contact could be seen as an uncaring or even hostile act by traditional people. We can still attend, but we need to consider the natural communication style of each culture. This led Allen to a personal commitment to discover more and more about how culture affects the person and how it must be considered in all sessions.

And Now, Neuroscience   As a group, our most recent venture has been into this newly relevant field, and we now present workshops and teaching on "brain-based counseling and therapy." Research in neuroscience is further demystifying the helping process. Not too surprising is the fact that almost all of what has been done in our field is validated by neuroscience. But we now can be more precise in understanding what is going on inside the client, thus leading to the most appropriate interventions. Neuroscience research has led us to an increased belief in the need for a positive approach to counseling, with minimal attention to the negative. For practical purposes, expect your clients to come to you with a fair amount of brain knowledge, just from the popular media and TV. Neuroscience is leading toward a paradigm shift in our field, and we need to be ready to use this research in our practice. You will find that including instruction on how the brain can change in the process of counseling helps clients become more committed and hopeful as we work together in the session.

To Be Continued   The learning process of demystification constantly brings something new and exciting. Professional research and practice, along with student feedback, change what we say in this book. Join us in the learning process and share ideas with us. You can make a difference in our exciting field.

▶Concepts Into Action: The Five-Stage Model for Structuring the Session With Many Theories of Counseling and Psychotherapy

The five stages of the well-formed session (empathic relationship—story and strengths—goals—restory—action) provide an organizing framework for using the microskills with multiple theories of counseling and psychotherapy. Regardless of your present or future theoretical orientation, the five-stage session structure can serve as a foundation for using skills that are often quite different from one another. For example, all counselors and therapists need to establish an empathic relationship and draw out the client's story.

The five stages are also a structure for decision making. Eventually, all clients will be making decisions about behavior, thoughts, feelings, and meanings. Each theory gives different attention to these, and they do it using different language and method. But the client makes the decisions, not us.

How are the five stages specifically related to decisions? Many see Benjamin Franklin as the originator of the systematic decision-making model. He suggested three phases of problem solving: (1) identify the problem clearly (draw out the story and strengths, along with goal setting), (2) generate alternative answers (restory), and (3) decide what action to take (action). The ancient Franklin model misses only the importance of empathic relationship (stage 1), the need for clearer goal setting, and ensuring that the client takes action after the session in the real world. Another term for decisional counseling is problem-solving counseling. The essential issue is the same regardless of the terms we use: How can we help clients work through issues and come up with new answers?

Thus, the Franklin problem-solving model still has relevance today, and variations of that framework permeate our counseling and psychotherapy theories. The five stages are a counseling and psychotherapy extension of the Franklin model, making the decision process involved in all theories clearer.

At this point, please review Table 8.2, which summarizes the five stages of the session in detail. Note that listening skills are central at each stage. If you are sufficiently skilled in the microskills and the five stages, you are ready to complete a full session using only attending, observation, and the BLS. Many clients can make decisions without your direct intervention. This was one of Carl Rogers's major goals in his person-centered approach. However, Rogers favored using as few questions as possible.

The second half of this book focuses on influencing skills, which are used primarily in stages 4 and 5. Thus, brief mention of some of the influencing skills is included in the table.

After you have mastered the five stages step by step, consider them a checklist to ensure that you have covered all the bases in any session. However, following the stages in a specific order is not a necessity. Many clients will discuss their issues moving from one stage to another and then back again, and you will frequently want to encourage this type of recycling. New information revealed in later counseling stages might result in the need for more data about the basic story, thus redefining client concerns and goals in a new way. Often, you will want to draw out more strengths and wellness assets.

The circle of the five stages of a counseling session in Figure 8.1 reminds us that helping is a mutual endeavor between client and counselor. We need to be flexible in our use of skills and strategies. A circle has no beginning or end; rather, a circle is a symbol of an egalitarian relationship in which counselor and client work together. The hub of the circle is wellness and the positive asset search, a central part of all stages.

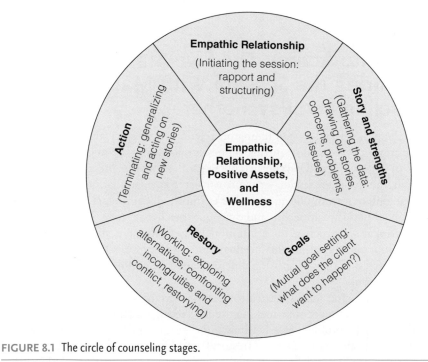

FIGURE 8.1 The circle of counseling stages.

TABLE 8.2   The Five Stages/Dimensions of the Microskills Session

Stage/Dimension	Function and Purpose	Commonly Used Skills	Anticipated Result
1. *Empathic relationship.* Initiate the session. Develop rapport and structuring. "Hello, what would you like to talk about today?"	Build a working alliance and enable the client to feel comfortable with the counseling process. Explain what is likely to happen in the session or series of sessions, including informed consent and ethical issues.	*Attending, observation skills, information giving* to help *structure* the session. If the client asks you questions, you may use *self-disclosure.*	The client feels at ease with an understanding of the key ethical issues and the purpose of the session. The client may also know you more completely as a person and a professional.
2. *Story and strengths.* Gather data. Use the BLS to draw out client stories, concerns, problems, or issues. "What's your concern?" "What are your strengths and resources?"	Discover and clarify why the client has come to the session and listen to the client's stories and issues. Identify strengths and resources as part of a wellness approach.	*Attending* and *observation* skills, especially *the basic listening sequence* and the *positive asset search.*	The client shares thoughts, feelings, and behaviors; tells the story in detail; presents strengths and resources.
3. *Goals.* Set goals mutually. The BLS will help define goals. "What do you want to happen?" "How would you feel emotionally if you achieved this goal?"	*If you don't know where you are going, you may end up somewhere else.* In brief counseling (Chapter 15), goal setting is fundamental, and this stage may be part of the first phase of the session.	*Attending* skills, especially the *basic listening sequence;* certain *influencing skills,* especially *confrontation* (Chapter 10), may be useful.	The client will discuss directions in which he or she might want to go, new ways of thinking, desired feeling states, and behaviors that might be changed. The client might also seek to learn how to live more effectively with situations or events that cannot be changed at this point (rape, death, an accident, an illness). A more ideal story ending might be defined.
4. *Restory.* Explore alternatives via the BLS. Confront client incongruities and conflict. "What are we going to do about it?" "Can we generate new ways of thinking, feeling, and behaving?"	Generate at least *three* alternatives that may resolve the client's issues. Creativity is useful here. Look for at least three alternatives to give the client a choice. One choice may be to do nothing and accept things as they are. *Restorying will vary extensively with different theories and approaches.*	*Summary* of major discrepancies with a supportive *confrontation.* More extensive use of *influencing skills,* depending on theoretical orientation (e.g., *interpretation, reflection of meaning, feedback*). But this is also possible using only *listening skills.* Use *creativity* to solve problems.	The client may reexamine individual goals in new ways, solve problems from at least those alternatives, and start the move toward new stories and actions.
5. *Action.* Plan for generalizing session learning to "real life." "Will you do it?" Use BLS to assess client commitment to taking action after the session.	Generalize new learning and facilitate client changes in thoughts, feelings, and behaviors in daily life. Commit the client to homework and action. As appropriate, plan for termination of sessions.	Influencing skills, such as directives and instruction/psychoeducation, plus attending and observation skills and the basic listening sequence. For example, "Can we set up specific goals or things to do tomorrow or next week?"	The client demonstrates changes in behavior, thoughts, and feelings in daily life outside of the session.

Stage 1. Empathic Relationship—Initiating the Session: Rapport, Trust Building, Structuring, and Preliminary Goals ("Hello")

Introducing the session and building rapport are most critical in the first session, and they will remain central in all subsequent sessions. Most sessions begin with some variation of "Could you tell me how I might be of help?" or "What would you like to talk about today?" "Hello, Lynette." "It's good to meet you, Marcus." A prime rule for establishing rapport is to use the client's name and repeat it periodically throughout the session, thus *personalizing the session.* Once you have completed the necessary structuring of the session (informing the client about legal and ethical issues), many clients are immediately ready to launch into a discussion of their issues. These clients represent instant trust. Our duty is to honor the client and continue to work on relationship issues throughout the session.

Some situations require more extensive time and attention to the rapport stage than others. Rapport building can be quite lengthy and blend into treatment. For example, in reality therapy with a delinquent youth, playing Ping-Pong or basketball and getting to know the client on a personal basis may be part of the treatment. It may take several sessions before clients who are culturally different from you develop real trust.

Structure. Structuring the session includes informed consent and ethical issues, as outlined in Chapter 2. Clients need to know their rights and the limitations of the session. If this is part of an ongoing series of sessions, you can help maintain continuity by summarizing past sessions and integrating them with the current session. At times, even telling clients about the stages of the session may be useful, so that they know what they are about to encounter.

Listen for preliminary goals. Although more definitive setting of short- and long-term goals happens at stage 3, initial goals are helpful during the first session(s). These early goals provide an initial structure for you as you seek to understand and empathize with the client. The early goals are revised and clarified after a fuller story has been brought out. Goal setting at this first stage is particularly important in brief counseling and coaching, described in Chapter 15.

Share yourself as appropriate. Be open, authentic, and congruent. Encourage clients to ask you questions; this is also the time to explore your cultural and gender differences. What about cross-cultural counseling when your race and ethnicity differ significantly from clients? Authorities increasingly agree that cultural, gender, and ethnic differences need to be addressed in a straightforward manner relatively early in counseling, often in the first session (for example, see Sue & Sue, 2013).

Observe and listen. The first session tells you a lot about the client. Note the client's style and, when possible, seek to match her or his language. As the relationship becomes more comfortable, you may note that you and the client have a natural mirroring of body language. This indicates that the client is clearly ready to move on to telling the story and finding strengths (stage 2).

Stage 2. Story and Strengths—Gathering Data: Drawing Out Stories, Concerns, and Strengths ("What Is Your Concern?" "What Are Your Strengths and Resources?")

Draw out the client's story. What are the client's thoughts, feelings, and behaviors related to her or his concern? We draw out stories and concerns using the skills of the basic listening sequence. Open and closed questions will help define the issue as the client views it. Encouragers, paraphrases, and checkouts will provide additional clarity and an opportunity for you to verify whether you have heard correctly. Reflection of feeling will provide understanding of the emotional underpinnings. Finally, summarization provides a good way to put the client's conversation into an orderly format.

Elaborate the story.   Next, explore related thoughts, feelings, and interactions with others. Gather information and data about clients and their perceptions. The basic journalistic outline of *who, what, where, when, how,* and *why* provides an often useful framework to make sure you have covered the most significant items. In your attempts to define the central client concerns, always ask yourself, what is the client's real world and current story?

Draw out strength and resource stories.   Clients grow best when we identify what they can do rather than what they can't do. Don't focus just on the difficulties and challenges. The positive asset search should be part of this stage of the session. This might be the place for a comprehensive wellness search, as described in Chapter 2.

Failure to treat.   Failure to treat can be the cause of a malpractice suit. This most often occurs when counselors fail to draw out stories and clients are unclear about what they really want, often resulting in no clear goals for the session. When the story is clear and strengths are established, you can review and clarify goals. Clients who participate in goal setting and understand the reasons for your helping interventions may be more likely to participate in the process and be more open to change.

Stage 3. Goals—Mutual Goal Setting ("What Do You Want to Happen?")

Mutuality and an egalitarian approach.   Your active involvement in client goal setting is necessary. If you and the client don't know where the session is going, the session may wander with no particular direction. Too often the client and counselor assume they are working toward the same outcome when actually each of them wants something different. A client may be satisfied with sleeping better at night, but the counselor wants complete personality reconstruction. The client may want brief advice about how to find a new job, whereas the counselor wants to give extensive vocational testing and suggest a new career.

Refining goals and making them more precise.   If you searched for broad goals early in the relationship, they can provide a focus and general direction. In stage 3, it helps to review early goals, divide them into subgoals if necessary, and make them truly clear and doable. It has been suggested that if you don't have a goal, you're just complaining. Authorities on brief counseling and coaching favor setting goals in the first part of the session, together with relationship building. With high school discipline problems, less verbal clients, and members of some cultural groups, setting a clear joint goal may be the key factor in relationship building. If you adapt your counseling style to each client, you will have a better chance of succeeding.

Summarizing the differences between the present story and the preferred outcome.   Once the goal has been established, a brief summary of the original presenting concern as contrasted with the defined goal can be very helpful. Consider the model below as a basic beginning to working through client issues. This is a possible opening to stage 4, restorying, using a supportive confrontation.

> "*Kaan*, on the one hand, your concern/issue/challenge is [summarize the situation briefly], but on the other hand, your goal is [summarize the goal]. What occurs to you as possibilities for resolution?"

Needless to say, this will be expressed in more words and differently than in these model sentences. Nonetheless, as part of goal setting, the present situation and the desired situation need to be contrasted. Interesting, even Carl Rogers has been known to ask clients what their session goals might be.

Define both the problem and the desired outcome in the client's language. The summary confrontation should list several alternatives that the client has considered. The client ideally should generate more than one possibility before moving on to stage 4. You may want to use hand movements, as if balancing the scales, to present the real and the ideal. Using such physical movements can add clarity to the summary confrontation of key issues.

Define a goal, make the goal explicit, search for assets to help facilitate goal attainment, and only then return to examine the nature of the concern.

Stage 4. Restory—Working: Exploring Alternatives, Confronting Client Incongruities and Conflict, Restorying ("What Are We Going to Do About It?")

Starting the exploration process.   How does the counselor help the client work through new solutions? Summarize the client conflict as just described in the preceding section, "On the one hand. . . ." Be sure that the summarization of the issue is complete and that the facts of the situation and the client's thoughts and feelings are part of this summary.

Use the BLS to facilitate the client's resolution of the issue(s). Imagine a school counselor talking with a teen who has just had a major showdown with the principal. Establish rapport, but expect the teen to challenge you; he or she likely expects you to support the principal. Do not judge, but gather data from the teen's point of view. If you have developed rapport (stage 1) and listened during data gathering (stage 2), the teen will likely search for solutions in a more positive fashion. Follow by asking what he or she would like to have happen in terms of a positive change. Work with the teen to find a way to "save face" and move on.

Encourage client creativity.   Your first goal in restorying is to encourage your clients to discover their own solutions. To explore and create with the teen above, listen well and use summarization: "You see the situation as . . . and your goal is. . . . The principal tells a different story and his goal is likely to be. . . ." If you have developed rapport and listened well, many teens will be able to generate ideas to help resolve the situation.

The basic listening sequence and skilled questioning are useful in facilitating client exploration of answers and solutions. Here are some useful questions to assist client problem solving. The last two focus on a wellness approach and would be common in brief counseling.

- ▲ "Can you brainstorm ideas—just anything that occurs to you?"
- ▲ "What other alternatives can you think of?"
- ▲ "Tell me about a success that you have had."
- ▲ "What has worked for you before?"
- ▲ "What part of the problem is workable if you can't solve it all right now?"
- ▲ "Which of the ideas that we have generated appeals to you most?"
- ▲ "What are the consequences of taking that alternative?"

Counselors and psychotherapists all try to resolve issues in clients' lives in a similar fashion. The counselor needs to establish rapport, define the issue, and help the client identify desired outcomes.

Relate client issues and concerns to desired outcomes.   The distinction between the problem and the desired outcome is the major incongruity that may be resolved in three basic ways. First, the counselor uses attending skills to clarify the client's frame of reference and then feeds back a summary of client concerns and the goal. Often clients generate their own synthesis and resolve their challenges. Second, counselors can use information, directives, and psychoeducational interventions to help clients generate new answers. Third, if clients do not generate their own answers, the counselor can use interpretation, self-disclosure, and other influencing skills to resolve the conflict. Finally, in systematic problem solving and decision making, counselor and client generate and brainstorm alternatives for action and set priorities among the most promising possibilities.

Aim for a decision and a new story. This exploration/brainstorming/testing of theoretical strategies facilitates client decision making and the generation of a new story. Once a decision has been made or a new workable story developed, see that plans are made to put these ideas into action in the real world. You need to help clients generalize feelings, thoughts, behaviors, and a plan for action beyond the session itself.

The above outline is specific to decisional counseling, but you will find it virtually identical to motivational interviewing and crisis counseling. Different theories will vary in their use of skills and emphasis at this stage. However, the earlier parts of the session tend to be relatively similar among varying theoretical viewpoints.

Stage 5. Action—Concluding: Generalizing and Acting on New Stories ("Will You Do It?")

The complexities of life are such that taking a new behavior back to the home setting may be difficult. How do we generalize thoughts, feelings, and behaviors to daily life? Some counseling theories work on the assumption that behavior and attitude change will come out of new unconscious learning; they "trust" that clients will change spontaneously. This indeed can happen, but there is increasing evidence that planning for change greatly increases the likelihood that it will actually occur in the real world.

Consider the situation of the teen in conflict with the principal. Some good ideas may have been generated, but unless the teen follows up on them, nothing is likely to change in the conflict situation. Find something that "works" and leads to changes in the repeating behavioral problems. As you read through the list of generalization suggestions below, consider what you would do to help this teen and other clients restory and change their thoughts, feelings, and behaviors.

Change does not always come easily, and many clients revert to earlier, less intentional behaviors. Work to help the client plan for change to ensure real-world relevance. Following are three techniques that you can use to facilitate the transfer of learning from the session, increasing the likelihood that the client will take new learning home and try new behaviors.

Contracting. The most basic way to help clients maintain and use new learning is the establishment of an informal or written contract to do something new and different. Ideally, this should be clear and specific enough that the client can actually do it easily. With more complex issues, contract for things that represent part of the solution. If we ask clients to change their behavior totally, they likely will fail and may not return for another session.

Homework and journaling. Assigning homework so that the effect of the session continues after the session ends has become increasingly standard. Some counselors assign "personal experiments" to clients who do not like the idea of doing "homework." Negotiate specific tasks for the client to try during the week following the session. Use very specific and concrete behavioral assignments, such as "To help your shyness, you agree to approach one person after church/synagogue/mosque and introduce yourself." Ask the client to keep a journal of key thoughts and feelings during the week; this can become the basis of the follow-up session. Another possibility is paradoxical intention: "Next week, I want you to deliberately do the same self-defeating behavior that we have talked about. But take special notice of how others react and how you feel." This helps the client become much more aware of what he or she is doing and its results.

Follow-up and support. Ask the client to return for further sessions, each with a specific goal. The counselor can provide social and emotional support through difficult periods.

Follow-up is a sign that you care. Use the telephone for behavior maintenance checks. Using email is possible, but may result in loss of privacy. Many counselors and therapists call clients but do not give them their email. If you work in an agency and give a phone number to a client, use an office number that is always attended.

Chapter 13 offers a wide array of strategies and techniques to ensure that clients can use learning from the session more effectively. There you will find specifics regarding assertiveness training, communication skills psychoeducation, and an array of stress management strategies. Special attention will be paid to research-based therapeutic lifestyle changes (TLC), including how to suggest that clients engage in healthful behaviors such as exercise and meditation and use their spiritual resources. For maximal impact and behavior transfer, we suggest a combination of several techniques and strategies over time.

Behaviors and attitudes learned in the session do not necessarily transfer to daily life without careful planning. Consider asking your client at the close of the interview, "Will you do it?"

▶Example Five-Stage Decisional Counseling Session: I Can't Get Along With My Boss

It requires a verbal, cooperative client to work through a complete session using only listening skills. This session has been edited to show portions that demonstrate skill usage and levels of empathy. Machiko is a graduate student seeing if she can indeed use only listening skills and still get "somewhere." Robert, the client, is 20, a part-time student who is in conflict with his boss at work. He is relatively verbal and willing to work on a real issue. Thus, the first stage is very short in this case.

Based on the five-stage model, decisional counseling is a practical approach that is both a separate theory and also inherent in all theories of counseling and therapy. Why is this so? Virtually all of our clients will be making life decisions, whether about careers, interpersonal conflict, or how to cope with anxiety, depression, or posttraumatic stress. Clients will be deciding whether or not to continue coming to talk with you and whether or not to follow up on new ideas for behavioral, thought, or emotional change.

A brief summary of decisional counseling's practical theory will be presented in Chapter 14. Here we see Machiko illustrating how a client can make decisions when the counselor uses only listening skills.

Stage 1: Empathic Relationship

COUNSELOR AND CLIENT CONVERSATION	PROCESS COMMENTS
1. *Machiko*: Robert, do you mind if we tape this session? It's for a class exercise in counseling. I'll be making a transcript of the session, which our professor will read. Okay? We can turn the recorder off at any time. I'll show you the transcript if you are interested. I won't use the material if you decide later you don't want me to use it. Could you sign this consent form?	Machiko opens with a closed question followed by structuring information. Obtain client permission and offer client control over the material before recording. As a student you cannot legally control confidentiality, but your responsibility is to protect your client. No rating for empathy as this starts the session.
2. *Robert*: Sounds fine; I do have something to talk about. Okay, I'll sign it. [pause as he signs]	Robert seems at ease and relaxed. As the recording was presented casually, he is not concerned about the use of the recorder. Rapport was easily established. As Robert is obviously anxious to start, this part of the session is briefer than usual. Expect to spend more time here.

COUNSELOR AND CLIENT CONVERSATION	PROCESS COMMENTS
3. *Machiko*: What would you like to share?	The open question, almost social in nature, is designed to give maximum personal space to the client.
4. *Robert*: My boss. He's pretty awful.	Robert indicates clearly through his nonverbal behavior that he is ready to go. Machiko observes that he is comfortable and decides to move immediately to gather data (stage 2). With some clients, several sessions may be required to achieve rapport and a working relationship.

Stage 2: Story and Strengths

The session begins, and Machiko draws out the basics of the narrative. Robert's story comes out easily and clearly. Of course, this does not always happen. Many times, you will have to piece together parts of the narrative before a complete story unfolds.

COUNSELOR AND CLIENT CONVERSATION	PROCESS COMMENTS
5. *Machiko*: Could you tell me what's going on with your boss?	This open question is oriented toward obtaining a general outline of the issue that the client brings to the session. (Interchangeable empathy as it builds on previous comments)
6. *Robert*: Well, he's impossible.	Instead of the expected general outline of the concern, Robert gives a brief answer. The anticipated result didn't happen.
7. *Machiko*: Impossible? . . . Go on. . .	Encourager. Intentional competence requires you to be ready with another response. An open physical posture and reflective tone of voice is especially relevant here in communicating to the client. Encouragers that are appropriate are frequently additive, but we can only tell when we see how the client responds to them.
8. *Robert*: Well, he's impossible. Yeah, really impossible. It seems that no matter what I do, he is on me, always looking over my shoulder. I don't think he trusts me. He really bugs me.	We are seeing the story develop. Clients often elaborate on the specific meaning of a concern if you use the encourager. In this case, the prediction from the use of the encourager holds true; 7 above turns out to be somewhat additive.
9. *Machiko*: I hear your frustration and perhaps even anger? So, he's impossible and looking over your shoulder all the time. Could you give me a more specific example of what he is doing to indicate he doesn't trust you?	Robert is a bit vague in his description. Machiko first acknowledges the emotion, even though this is not fully clear. Then she asks an open question seeking concreteness in the story. (Interchangeable empathy)
10. *Robert*: Well, maybe it isn't trust. Like last week, I had this customer lip off to me. He had a complaint about a shirt he bought. I don't like customers yelling at me when it isn't my fault, so I started talking back. No one can do *that* to me! And of course the boss didn't like it and chewed me out. It wasn't fair.	As events become more concrete through specific examples, we understand more fully what is going on in the client's life and mind. We also begin to see the decisions that Robert is going to have to make as he copes with his boss and his own feelings and emotions.

continued on next page

COUNSELOR AND CLIENT CONVERSATION	PROCESS COMMENTS
11. *Machiko:* As I hear it, Robert, it sounds as though this guy gave you a bad time and it made you angry, and then the boss came in.	Machiko's response is relatively similar to what Robert said. Her paraphrase and reflection of feeling again represent basic interchangeable empathy.
12. *Robert:* Exactly! It really made me angry. I have never liked anyone telling me what to do. I left my last job because the boss was doing the same thing.	With that reflection of feeling, Robert's anger becomes much clearer. Accurate listening often results in the client's saying "exactly" or something similar.
13. *Machiko:* So your last boss wasn't fair either?	Machiko's vocal tone and body language communicate nonjudgmental warmth and respect. She brings back Robert's key word *fair* by paraphrasing with a questioning tone of voice, which represents an implied checkout. (Interchangeable empathy)

The session then continues to explore Robert's conflict with customers, his boss, and past supervisors. There appears to be a pattern of conflict with authority figures over the past several years. This is a common pattern among young males in their early careers. Robert needs to decide what to do in this situation. After a detailed discussion of the specific conflict situation and several other examples of the pattern, Machiko decides to conduct a positive asset search to discover strengths. Decisions are made most easily when clients are aware of their strengths and resources.

14. *Robert:* You got it.	
15. *Machiko:* Robert, we've been talking for a while about difficulties at work. I'd like to know some things that have gone well for you there. Could you tell me about something you feel good about at work?	Paraphrase, structuring, open question, and beginning positive asset search. (Additive as new information is being brought to the session)
16. *Robert:* Yeah; I work hard. They always say I'm a good worker. I feel good about that.	Robert's increasingly tense body language starts to relax with the introduction of the positive asset search. He talks more slowly.
17. *Machiko:* Sounds like it makes you feel good about yourself to work hard.	Reflection of feeling, emphasis on positive regard. (Interchangeable empathy)
18. *Robert:* Yeah. For example, . . .	

Robert continues to talk about his accomplishments. In this way, Machiko learns some of the positives Robert has in his past and not just his difficulties. She has used the basic listening sequence to help Robert feel better about himself. Machiko also learns that Robert has real strengths, such as determination and a willingness to work hard. These and his natural intelligence will help him resolve his issues.

Stage 3: Goals

COUNSELOR AND CLIENT CONVERSATION	PROCESS COMMENTS
19. *Machiko*: Robert, given all the things you've talked about, could you describe an ideal solution? How would you like things to be?	Open question. The addition of a new possibility for the client represents the potential beginning of additive empathy. It enables Robert to think of something new. The goals, of course, are the basis of the decisions that Robert will be making.
20. *Robert*: Gee, I guess I'd like things to be smoother, easier, with less conflict. I come home so tired and angry.	Here we see a key result of the five-stage session. Robert is making a specific decision saying where he wants to go in this session.
21. *Machiko*: I hear that. It's taking a lot out of you. Tell me more specifically how things might be better.	Paraphrase, open question oriented toward concreteness.
22. *Robert*: I'd just like less hassle. I know what I'm doing, but somehow that isn't helping. I'd just like to be able to resolve these conflicts without always having to give in.	Robert is not as concrete and specific as anticipated. But he brings in a new aspect of the conflict—giving in. What does that concept mean?
23. *Machiko*: Give in?	Encourager.
24. *Robert*: Yeah. I really like to stand up for myself. Sometimes I feel as if it is a competition between me and the boss. I hate to give him the idea that he can control me.	Here Robert opens up more fully.

Machiko learns another dimension of Robert's conflict with others. Subsequent use of the basic listening sequence brings out this pattern with several customers and employees. As new data emerge in the goal-setting process, you may find it necessary to change the definition of the concern and perhaps even return to stage 2 for more data gathering.

25. *Machiko*: So, Robert, I hear two things in terms of goals. One, that you'd like less hassle, but another, equally important, is that you don't like to give in. And, having personal control of what is happening. Have I heard you correctly?	Machiko uses a summary to help Robert clarify his concerns, even though no resolution is yet in sight. She checks out the accuracy of her hearing. (Interchangeable empathy)
26. *Robert*: You're right on, but what am I going to do about it? I would like to have more control, but I certainly would like less hassle and I need that job. It doesn't do me much good to get angry.	Robert is starting to gain some insight into his issues.
27. *Machiko*: Getting angry doesn't get you where you really want to be. OK, sounds like we have a goal that we can work on today. Let's work together to make that goal more specific, concrete, and doable. OK?	This reflection of feeling also has elements of reframing Robert's issue, thus making concretizing the goal more possible. Her statement about working together reminds the client that this is an egalitarian relationship and that she is involved in seeking an answer. We also sense Machiko's genuineness and authenticity. (Additive empathy)

Stage 4: Restory

COUNSELOR AND CLIENT CONVERSATION	PROCESS COMMENTS
28. *Machiko*: So, Robert, on the one hand, I heard you have a long-term pattern of conflict with supervisors and customers who give you a bad time. On the other hand, I also heard just as loud and clear your desire to have less hassle and gain more personal control. It seems that being in control of things is the real word instead of "giving in." We also know that you are a good worker and like to do a good job. Given all this, what do you think you can do about it?	Machiko remains nonjudgmental and appears to be very congruent with the client in terms of both words and body language. This is a major empathic summary. With her help, Robert now knows that the larger issue is being in personal control and simultaneously avoiding hassles. She ends with an open question asking Robert what comes to his mind as possible solutions. (Additive empathy, as giving in has been reframed as a control issue and the story is summarized in clear form)
29. *Robert*: Well, I'm a good worker, but I've been fighting too much. I let the boss and the customers control me too much. I think the next time a customer complains, I'll keep quiet and fill out the refund certificate. Why should I take on the world?	Robert talks more rapidly. He, too, leans forward. Many clients, if someone listens carefully, will start generating their own solutions. But it is not always this easy or quick. However, Robert's brow is furrowed, indicating some tension. He is "working hard."
30. *Machiko*: I sense your desire to do something different. So one thing you can do is keep quiet. You could maintain control in your own way, and you would not be giving in. (pause)	Machiko begins by acknowledging Robert's nonverbal emotions. She then paraphrases using Robert's key words, including the new word "control" from earlier in the session, to reinforce his present thinking. But she waits for Robert's response. (Interchangeable)
31. *Robert*: (short pause) Yeah, that's what I'll do, keep quiet.	He sits back, his arms folded. This suggests that the "good" response above was in some way actually subtractive. There is more work to do. What looked like a change was, at best, only a beginning.
32. *Machiko*: Sounds like a good beginning, but I'm sure you can think of other things as well, especially when you simply can't be quiet. Can you brainstorm more ideas?	Machiko gives Robert brief feedback. Her open question is adding to the session. She is aware that his closed nonverbals suggest more is needed. (Interchangeable with dimensions of additive empathy)

Clients are often too willing to seize the first idea as a way to resolve their issues and thus avoid further thought. Use a variety of questions and listening skills to draw out the client more fully. Referring to strengths and wellness assets periodically often centers clients and enables them to return to attack difficult and challenging issues.

Later in this part of stage 4, Robert was able to decide on two possibilities for action: (1) talk frankly with his boss and seek his advice and feedback about his performance; and (2) look into an anger management program offered at the campus counseling center. As a result of this session, Robert began to realize that his challenges with his boss were only one example of a continuing problem with authority and anger management. He and Machiko discussed the possibilities of continuing their discussions or for Robert to visit a professional counselor for additional help. Robert decided he'd like to talk with Machiko a bit more. A contract was made: If the situation at work did not improve within two weeks, Robert would visit the counseling center, in addition to the anger management course.

Stage 5: Action—Generalization and Transfer of Learning

If there is no follow-up plan—even a written contract for action—clients often fail to follow up and actually do what was agreed on. The conversation looks as though it has been effective. But too many counselors and psychotherapists stop at this point. Reserve at least 5 minutes at the end of each session to work out an action plan so that the client can take new learning home.

COUNSELOR AND CLIENT CONVERSATION	PROCESS COMMENTS
33. *Machiko*: So you've decided that the most useful steps are to talk with your boss, but also to take a deep breath when you do feel hassled and use your first suggestion, simply keep quiet. But the big question is "Will you do it?"	Paraphrase, open question. (Additive)
34. *Robert*: Sure, I'll do it. The first time the boss seems relaxed.	Robert is moving toward a decision.
35. *Machiko*: As you've described him, Robert, that may be a long wait. Could you set up a specific plan so we can talk about it the next time we meet?	Paraphrase, open question. To generalize from the session, encourage specific and concrete action by your clients to help ensure that they do something about their decisions. (Additive)
36. *Robert*: I suppose you're right. First, I need to lay back and just observe the scene and what the boss is doing with other workers. If he comes to me and I feel hassled, I will really try to keep quiet. Okay, one thing is that we are both football fans, and occasionally he and I drink coffee in the late afternoon at Rooster's on Monday after the Sunday games. That might be the best time to approach him.	A specific decisional plan to take action on the new story is developing.
37. *Machiko*: Wow, that is an amazing plan you have come up with—it sounds like a good decision. Laying back and seeing what is happening should be helpful, and the Rooster's setting is ideal for that chat. What, specifically, are you going to say?	Machiko offers enthusiastic feedback for Robert's plan followed by an open question, again eliciting concreteness. This is another helping lead that is additive, as Robert needs even further clarity in this plan.
38. *Robert*: I could tell him that I like working there, but I'm concerned about how to handle difficult customers. I'll ask his advice and how he does it. In some ways, it worries me a little; I don't want to give in to the boss . . . but maybe he will have a useful idea.	Robert is able to plan something that might work. With other clients, you may want to role-play a practice session in which you play the boss and give Robert a chance to develop a working script for the conversation. You could also give advice or assign homework practice sessions to try before talking with the boss. You will note that Robert is still concerned about "giving in." Control issues remain that can be addressed in a later session.
39. *Machiko*: Would you like to talk more about this the next time we meet? Maybe through your talk with your boss we can figure out how to deal with this in a way that makes you feel more comfortable. Sounds like a good contract. Robert, you'll talk with your boss, and we'll meet later this week or next week.	Open question, structuring. If Robert decides to talk to his boss and listens to his advice—and actually changes his behavior—then this session could be rated holistically as quite additive. If not, then the rating will be lower. Machiko wisely leaves the broader issue of control for a later session.

It would have been more helpful to specify the follow-up contract even more precisely, but this would most likely entail the use of influencing skills, homework, and so forth. You will find that concreteness and specificity are very useful in helping clients make and act on decisions. It was an especially good response when Machiko asked Robert what he was specifically going to do.

The first time, you may find it challenging to work through the systematic five-stage session using only attending, observation, and the basic listening sequence, yet it can be done. This is a useful format to use with individuals who are verbal and anxious to resolve their own issues. You will also find this decisional structure useful with resistant clients who want to make their own decisions. By acting as a mirror and asking questions, you can encourage many of your clients to find their own direction. More information on the five stages and decisional counseling will be presented in Chapter 14, along with a transcript example.

▶Taking Notes in the Session

Beginning helpers typically question whether they should take notes during the session, and it's not hard to find opinions for and against note taking. We are going to share our opinions based on our experience, recognizing that individuals vary in their thoughts. Most important, follow the directions of your agency.

Intentionality in counseling and psychotherapy requires accurate information. Therefore, we recommend that you listen intentionally and take notes. This is our opinion, but some people will disagree with us. You and your client can usually work out an arrangement suitable for both of you. If you personally are relaxed about note taking, it will seldom become an issue. If you are worried about taking notes, it likely will be a problem. When working with a new client, obtain permission early about taking notes. We suggest that any case notes be made available to clients and in practice sessions. Volunteer client feedback on the session can be most helpful in thinking about your own style of helping. Using your own natural style, you might begin:

> "I'd like to take a few notes while you talk. Would that be OK? I'll also write down your exact key words so both of us can refer to them. I'll make a copy of the notes before you leave, if you wish. As you know, all notes in your file are open to you at any time."

In-session note taking is often most helpful in the initial portions of counseling and psychotherapy and less important as you get to know the client better. Audio and video recording the session follow the same guidelines. If you are relaxed and provide a rationale to your client, making this type of record of the session generally goes smoothly. Some clients find it helpful to take audio recordings of the session home and listen to them, thus enhancing their learning from the session. There is nothing wrong with *not* taking notes in the session, but records typically need to be kept, and we recommend writing session summaries shortly after the session finishes. In these days of performance accountability for your actions, a clear record can be helpful to you, the client, and the agency with which you work.

HIPAA (Health Insurance Portability and Accountability Act) legal requirements regarding note taking are detailed and specific, but not always clear. Some rules give certain aspects of counseling more detailed protection than general medical records, but these rules are sometimes written vaguely, and they mention the possibility of maintaining dual records of psychotherapy. The agency you work with in practicum or internship can guide you in this area. You will also find Zur's (2011) *The HIPAA Compliance Kit* helpful.

▶Summary: The Well-Formed Session Using the Basic Attending Skills

The five-stage structure of microskills decisional counseling has been outlined in the preceding discussion and example, showing that it is possible to integrate all the microskills and concepts of the basic listening sequence into a meaningful, well-formed session. It may be challenging to work through the empathic relationship—story and strength—goals—restory—action sequence without using advice and influencing skills, yet it can be done. The five stages are a useful format with individuals who are verbal and who are anxious and able to resolve their own issues—and with resistant clients who want to make their own decisions. By acting as a mirror and asking questions, we can encourage many clients to find their own direction.

Theoretically and philosophically, the five-stage decisional style is similar to Carl Rogers's (1957) person-centered therapy. Rogers developed guidelines for the "necessary and sufficient conditions of therapeutic personality change," and the empathic constructs described in this book are derived from his thinking. Rogers originally was opposed to the use of questions but in later life modified his position. Implicit in your ability to conduct a session without using information, advice, and influencing skills is a respect for the person's ability to find her or his own unique direction. In conducting a session using only attending, observation, and the basic listening sequence, you are using a very person-centered approach to counseling.

Perhaps even more important is your awareness that the five-stage structure can be used with all theories of counseling and therapy. This point is outlined in much more detail in Chapter 15, where you will see a number of theories and strategies presented briefly with accompanying transcripts. For example, the popular cognitive behavioral therapy (CBT) session needs to be built on an empathic relationship. Without a good working alliance between counselor and client, change will be slow, perhaps even impossible. The CBT counselor needs to draw out the client's concerns, and increasingly CBT is learning that a positive approach focusing on strengths improves success rate. Goal setting has always been a central feature of CBT. Whereas the goal of a person-centered session may be to develop a better and more accepting self-concept or to understand emotions, CBT will typically zero in on clearly defining specific behavioral goals and/or goals oriented around thinking more rationally and effectively. In the restory phase, CBT is unique in employing specific and well-researched strategies. CBT has developed real strength in the action phase, and **relapse prevention**, discussed in Chapter 14, has become a standard system also used by many other theories of helping.

▼▼▼▼	Key Points
Basic listening sequence	Draw out the client's story and strengths by questioning, encouraging, paraphrasing, reflecting feelings, and summarizing. This sequence is used in multiple settings, not just in counseling and psychotherapy.
Five stages of the session	Stage 1. *Empathic relationship*: rapport and structuring ("Hello.") Stage 2. *Story and strengths*: gathering information and defining issues ("What's your concern?" "What are your strengths?") Stage 3. *Goals*: determining outcomes ("What do you want to happen?") Stage 4. *Restory*: exploring alternatives and client incongruities ("What are we going to do about it?") Stage 5. *Action*: generalization and transfer of learning ("Will you do it?")
Circle of counseling and decision making	The five stages of counseling need not always follow the five steps in order. Think of the stages as dimensions that need to be considered in each session. Also, give continuous attention to relationship, positive assets, and wellness at the hub of the circle.

▶Competency Practice Exercises and Portfolio of Competence

 Go to CengageBrain.com to access Counseling CourseMate, where you will find an interactive ebook, quizzes, videos, interactive counseling and psychotherapy exercises, the Portfolio of Competence, case studies, a practice test, and more.

Mastery of the skills of this chapter is a complex process that you will want to work on over an extended period of time. Some basic exercises for the individual and systematic group practice in each skill area follow.

Individual Practice

Exercise 1. Illustrating How the Basic Listening Sequence (BLS) Functions in Different Settings   Write counseling leads as they might be used to help a client solve the problem "I don't have a job for the summer." Imagine a full statement of issues and write responses that represent the BLS.

Open question _____

Closed question _____

Encourager_____

Paraphrase _____

Reflection of feeling _____

Summary _____

Now imagine you are talking with a client who has just been told that her or his parents are getting a divorce after more than 25 years of marriage. Your task is to use the BLS to find out how the client is thinking, feeling, and behaving in reaction to this news.

Open question _____

Closed question _____

Encourager_____

Paraphrase _____

Reflection of feeling _____

Summary _____

Finally, how would you use these skills in talking with an elementary school student who has come to you crying because no one will play with her or him?

Open question _____

Closed question _____

Encourager _____

Paraphrase _____

Reflection of feeling _____

Summary_____

Exercise 2. The Positive Asset Search and the Basic Listening Sequence (BLS)   Imagine that you are role-playing a career counseling session. The client says, "Yes, I am really confused about my future. One side of me wants to continue a major in psychology, while the other—thinking about the future—wants to change to business." Use the BLS to draw out this client's positive assets. In some cases, you will have to imagine client responses to your first question.

Open question _____

Closed question _____

Encourager _____

Paraphrase _____

Reflection of feeling _____

Summary _____

You are counseling a couple considering divorce. The husband says, "Somehow the magic seems to be lost. I still care for Chantell, but we argue and argue—even over small things." Use the positive asset search to bring out strengths and resources on which they may draw to find a positive resolution to their problems. In marriage counseling in particular, many counselors err by failing to note the strengths and positives that originally brought the couple together.

Open question _____

Closed question _____

Encourager _____

Paraphrase _____

Reflection of feeling _____

Summary _____

Group Practice

Exercise 3. Practicing the Well-Formed Five-Stage Session and Person-Centered Counseling   Now is your turn to practice with the five-stage structure of the microskills decisional counseling model. Acting on your newly acquired knowledge is what will make the five stages fully relevant. Find someone who is willing to role-play a client who wishes to make a decision or has an opportunity or an issue to discuss. Interview that "client" for at least 15 minutes using the skills you have learned so far.

Read again pages 31–38 and follow the ethical guidelines as you work with a volunteer client. Structure the session by asking the client's permission to record the session. Inform the client that the recorder may be turned off at any time. Common sense demands ethical practice and respect for the client. Discuss issues of confidentiality and ask your client to sign the Chapter 2 permission form (page 36).

Strive to integrate all the microskills and concepts presented into a meaningful, well-formed session. It may be challenging to work through the empathic relationship—story and strengths—goals—restory—action sequence without using advice and influencing skills, yet it can be done. The five stages represent a useful therapeutic format for individuals who are verbal and anxious to resolve their own issues—and with resistant clients who want to make their own decisions.

When you complete the session, ask your client to fill out the client feedback form. In practice sessions, aim to get immediate feedback, which will provide you with information on how the client sees you. Continue this periodically with future clients. Please transcribe this session for later study and analysis. You'll want to compare your first recorded session and this session; you will have another opportunity at the end of the course.

Review an audio or video recording of the session and ask yourself the following questions:

1. Were you able to integrate all the microskills and concepts learned so far into a meaningful, well-formed five-stage session?
2. What did you do that you think was effective and helpful?
3. What stands out for you from the Client Feedback Form and any other comments the client may have said to you about the session?

Use the Feedback Form provided in Box 8.2 to evaluate your session, or ask a classmate to review your recording using that form.

BOX 8.2   Feedback Form: Practice Session Using Only the Basic Listening Sequence

You can download this form from Counseling CourseMate at CengageBrain.com.

_____ (DATE)

_____   _____
(NAME OF COUNSELOR)   (NAME OF PERSON COMPLETING FORM)

Instructions: *Counselor:* Conduct a brief five-stage session using only the skills of the basic listening sequence. We suggest sharing the steps beforehand with your volunteer client. Suggested topics are making a career decision, finding a balance between work and play, or working on a wellness issue from Chapter 2. You and your client can take any current life issue that is not too complex.

Observer: Please provide feedback and commentary to the counselor. Was this counselor able to conduct a session using only listening skills?

1. *Empathic relationship.* Initiating the session. Nature of rapport? Was a relationship established before the session continued to the next stage? Did the counselor provide structuring? Was rapport maintained throughout the session? Observations on BLS?

2. *Story.* Gathering data, defining concerns, and identifying assets. Did the counselor draw out the story from the client using only listening skills? Was at least one positive supportive asset, resource, or strength of the client examined? Observations on BLS?

3. *Goals*. Mutual goal setting. Was a specific outcome or goal outlined for the client through use of listening skills? Was it concrete and doable? Observations on BLS?

4. *Restorying*. Working. Was the counselor able to assist the client in generating new ideas through the use of listening skills only? Did the session move toward achieving the goal, or was it too broad for the client to accomplish? Observations on BLS?

5. *Action*. Terminating and generalizing. Were specific plans made and contracted for with the client for taking ideas home? Is there a systematic plan of action for follow-up and maintenance? Observations on BLS?

6. *Intentional competence*. What did the counselor do right? Did use of the basic listening sequence result in predicted outcomes? When the expected result did not occur, was the counselor able to flex intentionally and use different listening skills?

Portfolio of Competence

A lifetime can be spent increasing one's understanding and competence in the ideas and skills from this chapter. You are asked here to learn and even master the basic ideas of predictability from skill usage in the session, continue empathic concepts, and master the five stages of the well-formed session. We have learned that student mastery of these concepts is possible, but for most of us (including the authors) we find that reaching beginning competence levels makes us aware that we face a lifetime of practice and learning.

You should feel good if you can conduct a session using only listening skills. Focus on that accomplishment, and use it as a building block toward the future. As you do, you are even better prepared for developing your own style and theory.

Use the following checklist to evaluate your present level of mastery. As you review the items below, ask yourself, "Can I do this?" Check those dimensions that you currently feel able to do. Those that remain unchecked can serve as future goals. Do not expect to attain intentional competence on every dimension as you work through this book. You will improve your competencies with repetition and practice.

Level 1: Identification and classification

- ❑ Identify and classify the microskills of listening.
- ❑ Identify and define empathy and its accompanying dimensions.
- ❑ Identify and classify the five stages of the structure of the session.
- ❑ Discuss, in a preliminary fashion, issues in diversity that occur in relation to these ideas.

Level 2: Basic competence. Aim for this level of competence before moving on to the next skill area.

- ❑ Use the microskills of listening in a real or role-played session.
- ❑ Demonstrate the empathic dimensions in a real or role-played session.
- ❑ Demonstrate the five stages of a well-formed session in a real or role-played session.

Level 3: Intentional competence. Ask yourself the following questions, all related to predictability and evaluation of your effectiveness in working with the five stages and the basic listening skills. These are skill levels that may take some time to achieve. Be patient with yourself as you gain mastery and understanding.

- ❑ Anticipate predicted results in clients using the listening microskills.
- ❑ Facilitate client comfort, ease, and emotional expression by being empathic.
- ❑ Enable clients to reach the objectives of the five-stage session process: (a) *empathic relationship*—develop rapport and feel that the session is structured; (b) *story and strengths*—share data about the concern and also positive strengths to facilitate problem resolution; (c) *goals*—identify and even change the goals of the session; (d) *restory*—work toward problem resolution; (e) *action*—generalize ideas from the session to their daily lives.

Level 4: Psychoeducational teaching competence. Teaching competence in these skills is best planned for a later time, but those who run meetings or do systematic planning can profit from learning the five-stage session process. It serves as a checklist to ensure that all important points are covered in a meeting or planning session.

- ❑ Teach clients the five stages of the session, emphasizing listening skills.
- ❑ Teach small groups this skill.

▶Determining Your Own Style and Theory: Critical Self-Reflection on Integrating Listening Skills

You are now at the stage to initiate construction of your own counseling process. You certainly cannot be expected to agree with everything we say. You likely have found that some skills work better for you than others, and your values and history deeply affect the way you conduct a session. Some skills you'd like to keep, and some you might like to change.

We encourage you to look back on these first eight chapters as you consider the following basic questions leading toward your own style and theory.

What single idea stands out for you among all those presented in this text, in class, or through informal learning? Allow yourself time to really think through this one key idea or concept—it may be something you discovered yourself. What stands out for you is likely useful as a guide toward your next steps.

Continue your development of your own style and theory through writing.

Focusing and Empathic Confrontation: Neuroscience, Memory, and the Influencing Skills

Our interaction with clients changes their brain (and ours). In a not too distant future, counseling and psychotherapy will finally be regarded ideal ways for nurturing nature.

—Oscar Gonçalves

We have noted that counseling changes the brain. Through neurogenesis, we develop new neurons and neural connections throughout life and manage to keep the number of neurons in the brain at approximately 100 billion. Although there is evidence that sections such as the prefrontal cortex and olfactory area do as well, research indicates that the memory center of the hippocampus is the area that develops most new neurons (Gould, Beylin, Tanapat, Reeves, & Shors, 1999; Seki, Sawamoto, Parent, & Alvarez-Buylla, 2011).

Thus, it is accurate to state that counseling changes memory. It is through memory change and its many facets that counseling and therapy seek to facilitate growth and change in thoughts, emotions, and behaviors. Changes in memory reverberate throughout the holistic brain. In particular, changes in memory affect the executive prefrontal cortex and the TAP, enabling changes in thought, feelings, and behavior.

The two chapters of Section III provide a transcript of two live sessions from a DVD that illustrates how counseling skills, using neuroscience concepts, can indeed change memory (Ivey, Ivey, Gluckstern-Packard, Butler, & Zalaquett, 2012). You will see that memories are explored through the listening skills and then changed via focusing and empathic confrontation. The client, Nelida Zamora, develops new cognitions leading to change in memory plus change in emotions and behaviors.

Chapter 9. Focusing the Counseling Session: Exploring the Story From Multiple Perspectives Here you will see the client, Nelida, present a troubling and oppressive classroom encounter that, through surprise, was immediately imprinted in her long-term memory. Through the use of listening skills and focusing, the counselor facilitates review of the client's story from multiple perspectives and ensures a comprehensive examination of concerns and issues. Skilled focusing is based on listening; it enables clients to view their stories in new ways without your supplying the answers for them.

Chapter 10. Empathic Confrontation and the Creative New: Identifying and Challenging Client Conflict A second interview with Nelida shows how clarification and empathic confrontation can lead to permanent change in memory. Many theorists and practitioners consider empathic confrontation the key stimulus enabling client change and development. Empathic confrontation builds on your ability to listen empathically and observe client conflict. Skilled supportive confrontation enables resolution of conflict and incongruity, leading toward new behaviors, thoughts, feelings, and meanings.

▶How Memory Changes Are Enacted in the Session

The following two chapters, on focusing and confrontation, illustrate the drawing out of a negative story and the resulting reframing and change in meaning, enabling the client to feel better about herself and think and act more effectively. Nelida Zamora, a superstar graduate student at the University of South Florida, has kindly given permission for us to

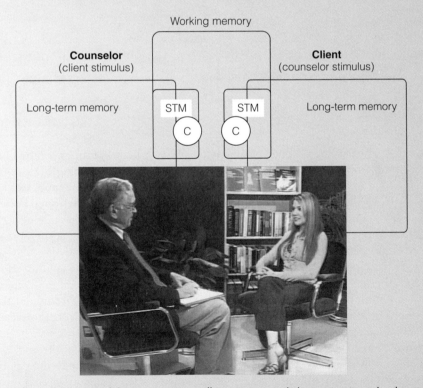

Counselor (Allen) and client (Nelida) in conversation, illustrating nonverbal mirroring (note the almost exact body mirroring) and depicting the interaction of short-term and long-term memories.

share two sessions she videotaped with Allen. In reviewing these sessions, you will see how listening and influencing skills can be combined.

Let us look at this counselor and client from a brain-based skills approach. The client, Nelida, shares a story; Allen listens, reflects, and seeks to help restory the memory and its meaning. Two brains are active in the session, and each person's brain, including short- and long-term memory, may change during the interaction. Two sets of memories in the hippocampus meet in the here and now of conscious conversation. Working memory brings life and the possibility of change to the session. Ultimately, counseling change is an interactive process of influencing client working memory in positive ways. We use working memory as our access to long-term memory, and significant change in long-term memory leads to changes in thoughts, feelings/emotions, and behaviors. Counseling is not a one-way process. We counselors learn and change as we work with clients.

Working memory is the integrated centerpiece of action in counseling conversation. Working memory can be defined as the area where we store high-speed data from here-and-now consciousness, as well as information from short- and long-term memory. Nelida and Allen each likely can store at most 18 items in working memory. However, the amount of information in working memory can change at any moment. For example, a highly emotional experience may leave the client having as few as one or two items in working memory.

In considering the change process that occurs through the interaction of client and counselor, we are also dealing with the relationship of the executive CEO prefrontal cortex to the amygdala and the limbic HPA. Attending behavior (attentional processes heavily controlled by the thalamus and the prefrontal TAP) remains foundational in determining whether or not long-term memories are solidified in the hippocampus. But emotional involvement through the energizing amygdala is necessary for working memory to function and to make this happen.

Authorities vary in how many items one can keep in working memory at one time. The original and often used definition of working memory was that we can hold seven plus or minus two (7±2) items (Miller, 1956). Seven is typically the maximum number of digits one can hold in working memory, but with practice, some people can hold 20 or more. Talking about difficult experiences, as often happens in counseling, may reduce the span of working memory. At times, strong emotions and memories from the long-term memory unconscious may blot out what is occurring in the external here and now. And immediate consciousness in the here-and-now of meditation seems to put working memory at rest.

From a time-based information-processing approach (Ivey, 2000), consciousness (C) represents the psychological present, which ranges in length from 100 to 750 milliseconds and has access to short- and long-term memory. A person who meditates, someone experiencing a real "runner's high," or a ballerina, tennis star, or serious painter is very close to living the here and now of consciousness. Of course, few reach these goals without practice, which involves the executive TAP and the limbic HPA, plus the amygdala's stimulation, memory in the hippocampus, and the holistic brain. The long-term memory has been automated in procedural memory, thus allowing the person to be fully in the here and now. Interestingly, this is also a goal of counseling and therapy—to help clients learn new ways of being that eventually become so much a part of them that they seldom have to think about their actions.

Short-term memory (STM) holds impressions immediately accessible to consciousness for approximately 10 seconds and can hold 100 items. If learning occurs, information moves into long-term memory (LTM). LTM is our storehouse of declarative (episodic, semantic) and nondeclarative (procedural, perceptual, classical conditioning, emotional) information. Deeper in long-term memory is the unconscious, life experience, which is less accessible, but with appropriate stimulation it can be brought to short-term working memory and consciousness. We have potential access to unconscious material, if an appropriate event happens or a counselor provides a key stimulus. Working memory could be called the "action" foundation for psychotherapeutic change, integrating the immediate here-and-now consciousness with STM and LTM.

A key task of counseling is to help the client restory past experience and develop new memories and connections (behaviors, thoughts, feelings, meanings). Successful counseling and psychotherapy change the client and LTM in significant ways and even build new neural networks in the brain (brain plasticity). The attending microskills presented earlier provide the cognitive/emotional "charge" to promote understanding and change. The influencing skills introduced in this section of the book both start and solidify the change process.

CHAPTER 9

Focusing the Counseling Session
Exploring the Story From Multiple Perspectives

Focusing

The Five-Stage Interview Structure

Reflection of Feeling

Encouraging, Paraphrasing, and Summarizing

Open and Closed Questions

Client Observation Skills

Attending Behavior and Empathy

Ethics, Multicultural Competence, and Wellness

Yo soy yo y mi circunstancia.

—José Ortega y Gasset

Mission of "Focusing the Counseling Session"

Focusing is a skill you may not see in other books. Its mission is to help you and your clients think of creative new possibilities for restorying and action. The systematic framework of focusing can help in reframing and reconstructing problems, concerns, issues, and challenges. The skill of focusing is the clearest way to (1) stress the importance of the individual and (2) expand awareness of how individual clients develop in social context—especially community and family.

Chapter Goals and Competency Objectives
Awareness, knowledge, skills, and actions developed through this chapter will enable you to

▲ Help clients tell their stories and describe their issues from multiple frames of reference, a valuable method for creative change.

▲ Increase clients' cognitive and emotional complexity, thus expanding their possibilities for restorying and resolving issues.

▲ Enable clients to see themselves as selves-in-relation and persons-in-community through community and family genograms.

▲ Facilitate clients to take action to address their concerns, issues, and challenges.

▲ Include advocacy, community awareness, and social change as part of your counseling or psychotherapy practice.

Go to CengageBrain.com to access Counseling CourseMate, where you will find an interactive ebook, quizzes, videos, interactive counseling and psychotherapy exercises, the Portfolio of Competence, case studies, a practice test, and more.

The most direct translation of the opening quotation from the Spanish philosopher José Ortega y Gasset is "I am I, and my circumstance." But direct translation from one language to another does not always convey the full meaning. What it means to us is "I am me and my cultural/environmental/social context."

Nelida Zamora, a former graduate student at the University of South Florida, has given us permission to use her name and her community genogram in this chapter.* She has a real concern, that of being made uncomfortable in an introductory counseling class. This is not an unusual situation; many students who come from non-European backgrounds do not always believe that they "fit in." Beyond that, as you know, many "majority" students also may not feel totally welcomed.

Nelida: Here I am, a grad student in counseling. I did well in college in Miami, and thought it was no big deal because I was only four and a half hours away. But my first day of class I raised my hand, made a comment that very first class, and a classmate asked me if I was from America (nervous laugh) or a native (nervous laugh). Yeah, and I said well I'm . . . , I was just four and a half hours away, and he just found it very hard to believe. So, after that comment was made, it kind of made me a little bit more hesitant to participate in discussions. It made me more self-conscious. (Here we see that a single comment affects the amygdala and prefrontal cortex so that cognitive and emotional memories are stored immediately and permanently in the hippocampus.)

Allen: It made you self-conscious. Could we explore that a little bit more? Ah, first of all, in English, what were the feelings that went with that? (As you will see later in the example transcript, those feelings are soon explored in Spanish.)

Nelida: Well, I was surprised because being from Miami a lot of my family members have recently come from Cuba, so there they look at me as the American girl and they make fun.

Allen: . . . and that embarrasses you.

Nelida: Exactly, so when I'm in Miami, my family and friends tease saying that I'm the American who can't speak Spanish a hundred percent correctly 'cause I've forgotten a lot of it because of the English. Then, now, I move here to Tampa, I'm the Cuban girl who can't speak English, so it seems like I'm torn. You know, I don't know where I belong sometimes.

What are the issues that Nelida faces? What internal and external factors affect how she thinks, feels, and acts (behaves)? List as many items as you can before moving on.

---

*This transcript of a real interview held between Allen Ivey and Nelida Zamora is also available on DVD: Ivey, A., Ivey, M., Gluckstern-Packard, N., Butler, K., & Zalaquett, C. (2012). *Basic Influencing Skills* [DVD]. Alexandria, VA: Microtraining/Alexander Street Press. By permission of Microtraining/Alexander Street Press, www.alexanderstreet.com.

►Introduction to Focusing

We should first note that Nelida faces the stress of not belonging and feeling different, plus being caught between the culture of Cuban Miami and a university in Tampa. These stressors affect her emotionally and send damaging cortisol to her brain, imprinting her long-term memory with a negative picture of herself. She is in a high state of incongruence. Focusing will help clearly identify the major areas of conflict and discrepancy, and help determine which ones will be approached first. Listening and using supportive challenges will help her clarify her situation and move more readily to problem solution.

Over this and the next chapter on empathic confrontation we will present an example of how counseling changes memory. This first session introduces the community genogram, a systematic way to review old positive memories and help clients see themselves in social context. The community genogram provides a visual picture that helps us understand the client's personal and cultural background.

A central current issue for Nelida is cultural oppression, which she has internalized; she has come to "blame" herself for being different. Rather than focusing just on Nelida as an individual, if you help her see other perspectives, such as being able to name the oppression of the classroom, she is better prepared to reframe and change the negative memory. In addition, focusing on family and cultural background will facilitate her pride in her Cuban family and culture and provide positive assets, strengths, and resources to deal more effectively with the cutting comments she has experienced.

If you use focusing skills as defined below, you can *anticipate* how clients may respond.

FOCUSING	ANTICIPATED RESULT
Use selective attention to focus the counseling session on the client, theme/concern/issue, significant others (partner/spouse, family, friends), a mutual "we" focus, the counselor, or the cultural/environmental context. You may also focus on what is going on in the here and now of the session.	Clients tend to focus their conversation or story on the areas that the counselor responds to. As the counselor brings in new focuses, the story is elaborated from multiple perspectives. If you selectively attend only to the individual, the broader dimensions of the social context are likely to be missed.

Selective attention (Chapter 3) is basic to focusing but works in different ways. We all tend to focus on or listen to different topics. Clients tend to talk about that which you give your primary attention. Through your attending skills (visuals, vocal tone, verbal following, and body language), you indicate to your client that you are listening and what you are paying attention to. Be aware of both your conscious and unconscious patterns of selective attention; clients may follow your lead rather than talk about what they really want to say.

Counseling is, first and foremost, for the individual. When you focus on individual issues, clients will talk about themselves from their personal frame of reference. Thus, the first focus dimension is on the individual client before you. Using the client's name and the word "you" helps personalize the counseling. While it is essential that you draw out the client's story, don't become so fascinated with the details of that story that you forget about the person talking to you. Some therapists have been known to become voyeurs, prying more deeply into personal affairs than necessary.

A second area of focusing is attending to the *theme*, or central topic(s), of the session. Drawing out strengths from the client's memory through listening to and focusing on the theme, story, or concern is a valuable part of the process.

Here we draw out client stories, issues, or concerns, but also always search for client strengths and positive assets. Traditionally, this was termed drawing out the "problem." We have noted that the word *problem* is itself a problem, tending to put the client in a one-down position. A wellness approach handles these issues quite differently. If a client has gone through a breakup of a significant relationship, has study difficulties, has cancer or another serious illness, we need to hear the details, and often we need to hear a lengthy story. Just telling the story is relieving. We feel better when someone seriously tries to listen and understand. Too many beginners and even professionals become transient voyeurs, so interested in the problematic story that they fail to focus on the unique client before them and their personal strengths to facilitate resolving issues.

Nelida lives in a broad context of multiple systems. The concept of *self-in-relation* may be helpful. The idea of *person-in-community* was developed from an Afrocentric frame by Ogbonnaya (1994), who pointed out that our family and community history and experiences live within each of us. Since that time, the idea that we are persons-in-community has taken hold, and we often hear "It takes a village to raise a child." The client brings to you many community voices that influence the client's view of self and the world. The debriefing of Nelida's community genogram in this chapter shows how this strategy is a useful way to understand your client's history and a good place to identify strengths and resources.

Individual counseling usually focuses on issues of conflict, incongruity, and discrepancies between the individual and family and friends. But, in addition, many client problems are caused by and related to issues and events in the broader context (e.g., poor schools, floods, economic conditions), which can be missed if you focus only on the individual and the first stories that you hear. If you help clients see themselves and their issues as *persons-in-community*, they can learn new ways of thinking about themselves and use existing support systems more effectively. The following list offers potential comments and questions that allow the counselor to focus the session in a specific area.

Significant others (partner, spouse, friends, family)
"Nelida, tell me a bit more about your relationship with your family in Miami."
(Positive memory focus)
"Your grandmother was very helpful to you in the past. What would she say to you about all this?"
"How are your friends helpful to you?"

Mutual focus ("we" statements about client, therapist, or group)
(Early in the session) "Nelida, you have something that's been bothering you for over a year, but *we* will work through this. What would be the best way for us to work together in therapy?"

Counselor focus (sharing one's own experiences and reactions)
"It really bothers me to hear what happened in that first class."
(Later) "I feel good to hear that you are taking charge of your Latina identity and have become aware that it was a form of racism that you experienced in that class."

Cultural/environmental/contextual focus (including broader issues such as the impact of the economy)
"Let's look at your community genogram so that I can learn more about what makes Nelida, Nelida."
"What are some strengths that you gain from your church and community?"
"What are some differences between Miami and Tampa, and how do these differences affect you?"

Immediacy focus (talking about what is going on in that moment in the session)
You may also focus on what is going on in the here and now of the session. Here the focus
    is on the issue, but supplemented by the positive relationship.
"Nelida, right now I can sense you are hurting."
(Later sessions) "Nelida, I sense at this moment that you feel puzzled at what I just said.
    I'm glad that you can openly express your feelings to me."

As a counselor, be aware of how you focus a counseling session and how you can
broaden the session so that clients are aware of themselves more fully in relation to others
and social systems: persons-in-relation, persons-in-community. In a sense, you are like an
orchestra conductor, selecting which instruments (ideas) to focus on, enabling a better
understanding of the whole. Some of us focus exclusively on the client and the issues that
the client faces, neglecting to recognize the total context of client concerns.

Memories take in many things. If we just focus on one aspect, we miss the total
picture. One route to ensure that the client and we have an understanding of the total
picture is the community genogram.

▶The Community Genogram

Clients bring us many stories. Most often we tend to work with only one individual story.
But stories and issues of many others (e.g., friends, family, unique factors of diversity)
deeply affect the client's narrative. There are many other factors we can focus on as well if
we are to help the client deal with complexity in living and personal decisions.

A good way to develop an understanding of the value of focusing and enriching client
stories is the community genogram, which can give us a good picture of a client's cultural
background and history, thus enabling us to view the client in social context. By working
with you on their community genograms, your clients will gain a richer understanding of
themselves as persons in relation to others.

The community genogram is a "free-form" activity in which clients are encouraged to
present their community of origin or their current community, using their own unique style.
Some visual examples of community genograms are shown in Box 9.1. Through the community genogram, we can better grasp the developmental history of our clients and identify
client strengths for later problem solving. Clients may construct a genogram by themselves
or be assisted by you through questioning and listening to the things that they include.

Developing Your Own Community Genogram

Let's start the examination of focusing with you completing your own community genogram, following the steps outlined below. If you take time to develop the genogram, you
will be better prepared to help your clients consider themselves as persons-in-community
and see themselves in social context. The community genogram provides a snapshot of the
culture from which you and your clients come.

▲ Select the community in which you were primarily raised. The community of origin
    is where you tend to learn the most about culture, but any other community, past or
    present, may be used.
▲ Use a large poster board or flipchart paper. Representing yourself or the client with a
    significant symbol, place yourself or the client either at the center or at another appropriate
    place. Encourage clients to be innovative and represent their communities in a format that
    appeals to them. Possibilities include maps, constructions, or star diagrams (see Box 9.1).
▲ Place family or families, nuclear or extended, on the paper, represented by the symbol
    that is most relevant for you or the client. Different cultural groups define family in

BOX 9.1 | The Community Genogram: Three Visual Examples

We encourage clients to generate their own visual representations of their "community of origin" and/or their current community support network. The examples presented here are only three of many possibilities.

1. *Nelida Zamora's community genogram.* Nelida gave considerable thought to this genogram, which she then shared with Allen. She used computer-generated images to describe her community of origin. Note that she presents only a few key dimensions, and one gets the sense of the fairly small Latina/o community in which she was raised. Each of these images contains valuable stories that give us a better understanding of Nelida as a holistic person in her home community.

2. *The map.* The client draws a literal or metaphoric map of the community, in this case a rural setting. Note how this view of the client's background reveals a close extended family and a relatively small experiential world. The absence of friends in the map is interesting. Church is the only outside factor noted.

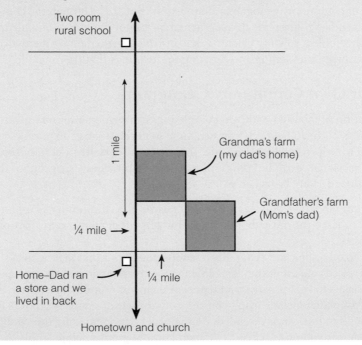

3. *The star.* Janet's world during elementary school tells us a good bit about a difficult time in her life. Nonetheless, pay equal attention to support systems and positive memories.

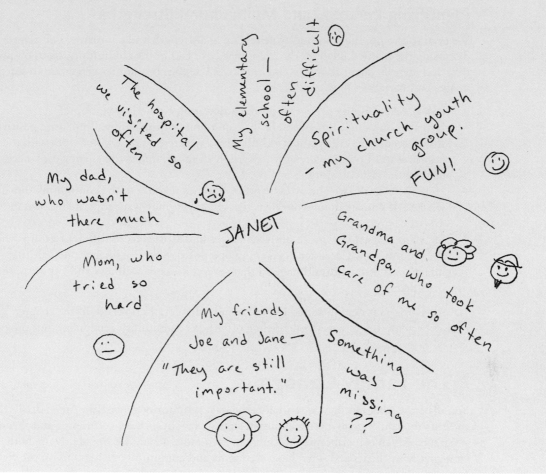

varying ways. This may provide sufficient family information, or you may want to add the family genogram shown in Appendix B.

▲ Place the most influential groups on the community genogram, representing them with distinctive visual symbols. School, family, neighborhood, and spiritual groups are most often selected. For teens, the peer group is often particularly significant. For adults, work groups and other special groups tend to become more central.

▲ You may wish to suggest relevant aspects of the RESPECTFUL model discussed in Chapter 1. In this way, diversity issues can be included in the genogram. Nelida's Latina background is central to her self-concept. All of your clients are deeply affected by their race, ethnicity, social class, and other factors, but they are often unaware of how these factors affect who they are. The community genogram makes it possible to understand where the individual came from.

We can bring broader understanding and multiple perspectives to the session by what we focus on in the client's life and social context. Part of what leads us to focus on certain issues is our own social context. Your developmental past and present issues can affect the counseling. You may consciously or unconsciously avoid talking about certain subjects that make you uncomfortable. You may do the same thing with clients. Becoming aware of your possible biases will free you to understand the uniqueness of each individual more fully.

Identifying Personal and Multicultural Strengths

We urge you to use the community genogram as a strength and a positive asset. Rather than discussing the many difficulties the client may have had in the community, focus on positives and identify client strengths and resources. Use the community genogram to search for images and narratives of strengths:

▲ Post the community genogram on the wall during counseling sessions.
▲ Focus on one single dimension of the community or the family. Emphasize positive stories even if the client wants to start with a negative story. Do not work with the negatives until positive strengths are solidly in mind, unless the client clearly needs to tell you the difficult story.
▲ Help the client share one or more positive stories relating to the community dimension selected. If you are doing your own genogram, you may want to write it down in journal form.
▲ Develop at least two more positive images and stories from different groups within the community. Consider developing one positive family image, one spiritual image, and one cultural image so that several areas of wellness and support are included.

This process is demonstrated in Nelida and Allen's transcript later in the chapter. The transcript analysis contains the debriefing of Nelida's community genogram, thus providing you with ideas on how you can use this strategy.

▶The Family Genogram

You may note that family was central in Nelida's community genogram. Appendix B presents the family genogram, a common strategy taught in most counseling and therapy programs, which can elaborate the family in even more detail. We frequently use both strategies with clients and often hang the family and community genograms on the wall in our office during the session, thus indicating to clients that they are not alone in the counseling session. Many clients find themselves comforted by our awareness of their strengths and social context.

Many of us have memories of family stories that are passed down through the generations. These can be sources of strength (such as a story of a favorite grandparent or ancestor who endured hardship successfully). Family stories are real sources of pride and can be central in the positive asset search. There is a tendency among most counselors and therapists to look for problems in the family history and, of course, this is appropriate. Be sure to search for positive family stories as well as problems.

Children often enjoy the family genogram and a simple adaptation called the "family tree" makes it work for them. The children are encouraged to draw a tree and put their family members on the branches, wherever they wish.

▶Debriefing a Community Genogram

Debriefing the community genogram is your chance to learn about the developmental history and cultural background of your client. It will provide you with considerable data so that you can spotlight and focus on key issues. Start by asking clients to describe the community and things that they consider most significant in their past development. Obtain an overview of the client's community.

Follow this by asking for a story about each element of the genogram. Seek to obtain positive stories of fun and support, strength, courage, and survival. Bring out the facts, feelings, and thoughts within the client's story. Many of us have been raised in communities that have been challenging and sometimes even oppressive, so a positive orientation can focus on the positives and strengths that have helped the client. That platform of positives makes it possible to explore problematic issues with a greater sense of hope. (For more ideas on the community genogram, see Rigazio-DiGilio, Ivey, Grady, & Kunkler-Peck, 2005.)

Following is a portion of the actual debriefing of Nelida's genogram. Here we see some stories from the genogram and how different issues of focus were brought up. How you work with the community genogram is most important. Clients can learn that their issues were developed in a context. The debriefing that emphasizes strengths helps the client learn that they are able and that you respect them. Armed with these positives, Nelida is better able to face some of the challenges in Tampa, and perhaps other difficulties as well.

CLIENT AND COUNSELOR CONVERSATION*	PROCESS COMMENTS
1. *Allen*: Just before the spring break we talked about you doing a community genogram. Were you able to put it together?	Focus on Nelida and the theme and concerns. The theme in this session is the community genogram plus Nelida's individual perceptions of her background. The session began with a brief greeting and check-in as to what was happening with Nelida. She had gone home to Miami. The session starts with an open question focusing on the theme of the session—debriefing the genogram.
2. *Nelida*: I was. Let me show you what I did.	Nelida shows some enthusiasm as she brings her community genogram forward (see Box 9.1).
3. *Allen*: You used the computer to make a gorgeous thing for us. So could you tell me a little bit about what's here?	Focus on individual and theme. Feedback, open question on theme. In using the community genogram, we recommend a focus on positives and strengths, rather than concerns. These positives can be used to ground clients in wellness assets as they move to resolving their challenges.
4. *Nelida*: Sure. I chose to include important parts of my community as I was growing, which shaped me into the person I am today. The solid lines symbolize positive connections, the black lines symbolize both positive and negative connections, and the jagged line is more negative connection.	There are many ways to construct a community genogram; encourage each person to define it in her or his own way. Nelida used a computer and then enlarged it on poster board. Drawing from family therapy genograms, she added the solid and jagged lines to indicate types of relationships.
5. *Allen*: I see the positive there: the university, the church, the local park, and your grandparents. Your mother's home, which is up and down. Well, we've talked in the past about some of the issues with your parents, but today we really wanted to focus on strengths and positives from your background. I'd like to hear a very brief positive story about each one of these, and then we'll select one or two to look at in a little more depth. So let's take the difficult one first, the high school with that real broken line.	Focus on theme and significant others. Paraphrase of what Allen views on the genogram; again, the focus is on the positives of theme with a minor focus on her mother. He first structures the debriefing by mentioning the plan of a positive focus, but he starts the debriefing with a suggestion that we start with one of the broken lines. (Up to this point, there is no specific empathic response, but the relationship seems solid, so likely we could rate the session as interchangeable empathy so far.)

continued on next page

---

*Slightly edited for clarity, this transcript of a real interview held between Allen Ivey and Nelida Zamora is also available on DVD: Ivey, A., Ivey, M., Gluckstern-Packard, N., Butler, K., & Zalaquett, C. (2012). *Basic Influencing Skills,* 4th ed. [DVD]. Alexandria, VA: Microtraining/Alexander Street Press. By permission of Microtraining/Alexander Street Press. (http://alexanderstreet.com/products/microtraining)

CLIENT AND COUNSELOR CONVERSATION	PROCESS COMMENTS
6. *Nelida*: Well, when I was in high school, I wasn't sure if I was going to pursue a college education, just because my family is very traditional. They're Latino, so it's more accepted for a woman to stay at home and be a wife and not pursue an education. So, I really didn't have my parents' and my grandparents' support to pursue that at first. So that's why my senior year was a little more tumultuous.	Here we see Nelida's past conflict with her family before she went to college. Multicultural and gender issues are quite clear here.
7. *Allen*: A little more tumultuous and the Latino tradition was not supportive of women's education. But were there any positive things that happened? Something good must have happened or you wouldn't even be here now.	Allen briefly focuses on the cultural/environmental context, but then moves to asking for an emphasis on positive memories, so critical in the community genogram process. We all have a history of some difficulties in our home communities, and these can be explored later, if appropriate. The goal here is a positive asset search that will bring out strengths that Nelida can use in the future. In later chapters, you will see the cultural/environmental/contextual focus taking central importance.
8. *Nelida*: Well, I was lucky enough to have a very good counselor in high school, who pretty much guided me in the direction that I thought I needed to go. She just put things into perspective for me, so that was very helpful.	The importance of a supportive counselor is mentioned. This is a good illustration of how other people in our historical community have helped us reach where we are. This counselor was a positive resource and still might be helpful now.
9. *Allen*: It's good you found a counselor who helped. Small wonder you end up in the counseling field. So even though you had that jagged line for the high school, there are some real strengths that are there in your community that helped you keep going.	Focus on Nelida and the theme. Allen supports the positive by focusing on a key person in Nelida's past. He adds the word "strengths" as part of his paraphrase. (This is slightly additive empathy.)
10. *Nelida*: Luckily my relationship with my mom right now is much better, so it's always good to visit her whenever I get a chance.	Nelida is talking more positively as she considers memories that support her in the past and present. Too many helpers might search for or underline problematic memories, thus giving a negative tone to the session. We can best work through our issues with the strengths of what we can do and our positive memories.
11. *Allen*: And one thing I hear as you talk about your mom is that even though you put that as sort of semi-conflictual, I see your eyes almost dancing.	Focus on Nelida. Feedback of nonverbal communication. (Additive empathy)
12. *Nelida*: Do you? It honestly feels good for me too because at high school we had a very difficult relationship, but when I decided to pursue college and later graduate school, my mother was one of my main support systems.	Despite the jagged relationship in the genogram, Nelida has reframed her view of her mother as more positive. If Allen had focused on the negative, likely we would have had a series of sad, perhaps even depressing, stories that likely would be of little help to the client. At this point, you can see that the negative story from the classroom is being compared to positive life experiences, a useful way to restory or rewrite negative memories in the brain.
13. *Allen*: So, now things are much better with your mother. Okay, and ah, your grandparents, they were something very special in growing up.	Focus on significant others. Paraphrase, emphasis on positives. (Interchangeable empathy)

CLIENT AND COUNSELOR CONVERSATION	PROCESS COMMENTS
14. *Nelida*: Um-hum. Yeah. You know they raised me, so even though they are very traditional and conservative, not too open-minded about certain things such as a woman pursuing a higher level of education. But, regardless, they've always supported me in my decisions and been great strength and support.	Nelida again restates the support she now gets from her family. It is good for clients to repeat positive strengths and assets, thus reinforcing resource development. What is occurring here is building emotional awareness of strengths, thus increasing the possibility of change from a base of strength.
15. *Allen*: They've really been important to you and provided critical support, even though they were so conservative. I imagine that word "conservative" also means you can count on them when you're facing difficulty. Is that right?	Focus on significant others and Nelida. A brief paraphrase followed by a reframing of the word "conservative" in a more positive light. This is an example of respect not only for the client, but also for her family, and for Latino culture. (Additive empathy)
16. *Nelida*: Yes, I can really count on them.	
17. *Allen*: The next thing I see is the local park and church.	Focus on theme—the community genogram. Allen topic jumps and notes some more positive connections in the community genogram.
18. *Nelida*: These bring back good memories of my childhood. You know, it was right across the street from my grandparents' house. It's not something that you really think about too much, but as I was putting this together, I always kept going back to the park, good memories that I had growing up. I remember when my grandfather would take me to the park and we'd go on bike rides together and things like that. Um, so that definitely made me feel good, thinking about it. I decided to include it.	Respect can be shown by drawing out positive stories, and memories of relationships and good experiences give the client a chance to show the good things in their lives. Our clients are not just a long litany of problems.
19. *Allen*: The park is important with your grandfather. Do you have any particular visual image of the park when you think about it?	Focus on significant others and Nelida. Restatement/paraphrase, open question. Here we see Allen moving to the use of imagery in connection with the community genogram. (See Chapter 13 for more specifics on this influencing strategy.) Visual memories often encapsulate life's events.
20. *Nelida*: Just how tranquil it was.	Images are not always visual. In this example, Nelida speaks to a feeling that brings her peace.
21. *Allen*: Tranquil.	Focus on Nelida. Encourager. *Tranquil* is obviously what we call a key word, which is often representative of positive here-and-now memories from the past. In our discussion of reflection of feeling, we noted that the technical definitions of emotion and feeling differ. Emotions are partially cognitive constructions, while feelings are more associated with the body. A goal of the imagery exercise is to put people more in touch with their feelings. (Interchangeable empathy)

continued on next page

CLIENT AND COUNSELOR CONVERSATION	PROCESS COMMENTS
22. *Nelida*: Um-hum. There was always discord between my grandparents and my mom, which of course consequently affected me, but the park was like a getaway, you know, so it was a just calm, tranquil atmosphere. My grandfather would take me on bike rides, and it was just a chance to kind of leave the issues at home, you know, and just kind of go on a mini-vacation and get away someplace.	We hear the cognitions that go with the feelings.
23. *Allen*: A place where you could really feel tranquil and at peace. I'd like to stay with that feeling of tranquility. Can you kind of get a visual image of a time when you're in that park and you have a specific time with your grandfather that you really felt peaceful and tranquil?	Focus on Nelida. Reflection of feeling/paraphrase, followed by a directive associated with imagery. This is additive as it encourages Nelida to go into her experience in more concrete depth. Seeking concreteness is often associated with additive helping. Note that Allen ignores the discord issue, which has already been explored. (Interchangeable empathy; the request for the visual image is potentially additive.)
24. *Nelida*: It must have been, you know, when I would just take my bike down there and just go on the hills and ride back and forth. Um, I just felt very at peace and free to be able to do that safely with my grandfather there.	Nelida seems to be almost totally "into" the recollection, and it clearly is in her working memory. (As Nelida was able to use the visual image, Allen's lead at 23 was additive.)
25. *Allen*: Could you say those words "at peace and free"?	Focus on Nelida. Encourager in the form of a question seeking more immediacy and in-the-moment experiencing.
26. *Nelida*: I just felt at peace and free.	All the next exchanges between Nelida and Allen are very brief, indicating that the session is in the here-and-now moment.
27. *Allen*: How does it feel when you say those words?	Focus on Nelida. Open question directed toward basic feelings.
28. *Nelida*: Soothing.	This is a clear example of a more basic feeling as compared to emotion.
29. *Allen*: Soothing. Where do you feel that soothing physically in your body?	Focus on Nelida. Encourager. Question directs Nelida to the here-and-now feelings in her body. (Another possibly additive comment)
30. *Nelida*: Here on my chest.	Nelida also shows a real feeling of peace and tranquility, likely very similar to the memory of feelings she had with her grandfather in the park. Feeling and emotions are more than cognitive; they are also felt physically at some level.
31. *Allen*: You feel that tranquility in your chest. One of the purposes of the community genogram is to find strengths that we get from past events or present events and then try to locate them in our body. When we are stressed, we can draw on past stories of support and strength, which can help us deal with difficult issues as they come up. Does that make sense?	Focus on Nelida and her concerns. Reflection of feeling followed by explanation of the value of positive events located in the body. It might have been wise for Allen to have encouraged more time with those positive feelings to anchor them more fully. (Interchangeable empathy)

CLIENT AND COUNSELOR CONVERSATION	PROCESS COMMENTS
32. *Nelida*: It does. Um-hum, that feeling in my body just helped me put it into perspective because I didn't realize how much I thought about it till I actually did this, so yeah, I do find myself going back to those memories and those visuals and that calming and soothing feeling that I had when I was there.	This is not a new memory for Nelida, but the exercise has brought its importance to her more fully.
33. *Allen*: Yeah. A couple things as we move on. First of all, when you find yourself feeling stress and tension, you always have that tranquil feeling in your chest of you and your grandfather. If you take a breath (breathing), let it go, and visualize that park, that's what we call our resource.	Focus on Nelida, theme, and concerns. Information giving and suggestion. The emphasis here is to help Nelida generalize this experience for action in the real world. If our clients have several positive physical feeling resources in their bodies, they can draw on them to help them move through times of stress. (Additive empathy)
34. *Nelida*: It feels right. Thanks.	

▶Using Focusing to Examine Your Own Beliefs and Approach

Let us turn to a more general consideration of how focusing can be used to broaden client thinking. Before you continue with your reading, take some time to think through your own thoughts on a difficult and challenging issue—abortion. It will help if you take time to write your responses to the *italicized* questions below.

As a counselor or psychotherapist, you will encounter controversial cases and work with clients who have made different decisions than perhaps you would. Abortion is part of what is sometimes called the "culture wars." There are deeply felt beliefs and emotions around this issue. Even the language of "pro-choice" and "pro-life" can be upsetting to some. *What is your personal position around this challenging issue?*

What do your family, your friends, and others close to you think about abortion? What do your community and church, both past and present, say and think? And how does your understanding of state laws and the extensive national media coverage affect your thinking? *From a more complex, contextual point of view, spend a little time thinking about what has influenced your thinking on this issue; record what you discover. Who decided? You or your family/community context?*

As a counselor, it is vital that you understand the situations, thoughts, and feelings of those who take varying positions around abortion or any other controversial issue, whether you agree with them or not. *Can you identify some of the thoughts and feelings of those who have a different position from your own? How do they think and feel?*

Counseling is not teaching clients how to live or what to believe. Rather, we seek to enable clients make their own decisions. Regardless of your personal position, you may find yourself using the counseling session to further that position. Most would agree that counselors should avoid bias in counseling. You may need to help your clients understand more than one position on abortion or recognize and deal with their conscious or unconscious sexism, racism, anti-Semitism, anti-Islamism, or other forms of intolerance. The art and mastery of effective counseling merge awareness of and respect for beliefs with unbiased probing in the interest of client self-discovery, autonomy, and growth.

▶Applying Focusing With a Challenging Issue

Some school systems and agencies have written policies forbidding any discussion of abortion. Further, if you are working within certain agencies (e.g., a faith-based agency or a pro- or antiabortion counseling clinic), the agency may have specific policies regarding counseling around abortion. Ethically, clients should be made aware of specific agency beliefs before counseling begins. Again, write your answers to the questions below, but there are no necessarily "right" answers to these difficult issues.

Imagine that a client comes to you who just terminated a pregnancy. How would you help this client, who clearly needs to tell her story? Below are several issues; we ask you to think through how you would respond using different focus dimensions.

Focus on the Individual and on Significant Others

> *Teresa*: I just had an abortion and I feel pretty awful. The medical staff was great and the operation went smoothly. But Cordell won't have anything to do with me, and I can't talk with my parents.

> What would you say to focus on Teresa as an individual?
> What could you say to focus on Cordell, the significant other?
> How might you focus on the attitudes and possible supports from her friends?

Choosing an appropriate focus can be most challenging. Too many beginners focus only on the problem. The prompt "Tell me more about the abortion" may result in drawing out details of the abortion but may reveal little about the client's distinctive personal experience. "I'd like to hear *your* story" or "What do *you* want to tell me?" There are no final rules on where to focus, but generally, we want to hear the client's unique experience. Focusing on the individual is usually where to start—note the emphasis on the words *you* and *your*.

Other key figures (Cordell, family, friends) are part of the larger picture. What are their stories? How do they relate to Teresa as a person-in-relation? You can more fully understand her situation when you draw out other stories and viewpoints. Keep all significant others in mind in the process of problem examination and resolution. For a full understanding of the client's experience, all pertinent relationships eventually need to be explored.

Focus on Family

> *Teresa*: My family is quite religious and they have always talked strongly against abortion; it makes me feel all the more guilty. I could never tell them.

> How might you focus on the family in response to her statement?
> How would you search for others in the family who might be helpful or supportive?

The family is where personal values and ethics are first learned. How does Teresa define "family"? There are many styles of family beyond the nuclear. African American and Hispanic clients may think of the extended family; a lesbian may see her supportive family as the gay community. Issues of single parenthood and alternative family styles continue to make the picture of the family more complex. Developing a community or family genogram may help Teresa locate resources and models that might help her. If her parents are not emotionally available, perhaps an aunt or grandmother might help.

Mutuality Focus

> *Teresa:* I feel everyone is just judging me. They all seem to be condemning me. I even feel a little frightened of you.

> How would you appropriately focus on the relationship between yourself and the client? What might you say to Teresa that focuses on here-and-now feelings?

A mutual immediate focus often emphasizes the *we* in a here-and-now relationship. Working together in an egalitarian relationship can empower the client. Also, helping clients recognize the depth of their feelings in the here and now can be immensely valuable and powerful. "Right now at this moment, *we* have an issue." "Can *we* work together to help you?" "What are some of your thoughts and feelings about how *we* are doing?" The emphasis is on the relationship between counselor and client. Two people are working on an issue, and the counselor accepts partial ownership of the problem.

In feminist counseling, the "we" focus may be especially appropriate: "*We* are going to solve this problem." The "we" focus provides a sharing of responsibility, which is often reassuring to the client regardless of his or her background. Many feminist counselors emphasize "we." In some counseling theories and Western cultures, emphasizing the distinction between "you" (client focus) and "me" (counselor focus) is more common and "we" would be considered inappropriate.

The mutual focus often includes a here-and-now dimension and brings immediacy to the session. To focus on the here and now, you can choose several different types of responses. "Teresa, right now you are really hurting and sad about the abortion." "I sense a lot of unsaid anger right now." There is also the classic "What are you feeling right now, at this moment?"

Counselor Focus

> *Teresa:* What do you think about what I did? What should I do?

What would you say? A counselor focus could involve self-disclosure of feelings and thoughts or personal advice about the client or situation: "*I* feel concerned and sad over what happened"; "Right now, *I* really hurt for you, but I know that you have what it takes to get through this"; "*I* want to help"; or "*I*, too, had an abortion . . . *my* experience was . . ." Opinions vary on the appropriateness of counselor or psychotherapist involvement, but the value and power of such statements are increasingly being recognized. They must not be overused; keep self-disclosures brief.

> How might you share your own thoughts and feelings appropriately?
> Would you give advice from your frame of reference? What would it be?

Cultural/Environmental/Contextual Focus

Given Teresa's discussion so far, what would you say to bring in broader cultural/environmental/contextual issues?

Perhaps the most complex focus dimension is the cultural/environmental context. Some topics within these broad areas are listed here, along with possible responses to the client. A key cultural/contextual issue in discussing abortion will often be religion and spiritual orientation. Whether the client is a conservative or liberal Christian, a Jew, Hindu, Muslim, or nonbeliever, discussing the values issue from a spiritual perspective may be central to her thinking and being.

▲ *Moral/religious issues*: "What can you draw from your spiritual background to help you?"
▲ *Legal issues*: "The topic of abortion brings up some legal issues in this state. How have you dealt with them?"

- ▲ *Women's issues*: "A support group for women is just starting. Would you like to attend?"
- ▲ *Economic issues*: "You were saying that you didn't know how to pay for the operation."
- ▲ *Health issues*: "How have you been eating and sleeping lately? Do you feel aftereffects?"
- ▲ *Educational/career issues*: "How long were you out of school/work?"
- ▲ *Ethnic/cultural issues*: "What is the meaning of abortion among people in your family/church/neighborhood?"

Any one of these issues, as well as many others, could be important to a client. With some clients, all of these areas might need to be explored for satisfactory problem resolution. The counselor or psychotherapist who is able to conceptualize client issues broadly can introduce

---

BOX 9.2 National and International Perspectives on Counseling Skills

Where to Focus: Individual, Family, or Culture?

Weijun Zhang

Case study: Carlos Reyes, a Latino student majoring in computer science, was referred to counseling by his adviser because of his recent academic difficulties and psychosomatic symptoms. The counselor was able to discern that Carlos's major concern was his increasing dislike of computer science and growing interest in literature. While he was intrigued about changing his major, he felt overwhelmed by the potential consequences for his family, in which he is the oldest of four siblings. He is also the first person in his family to attend college. Carlos has received some limited financial support from his parents and one of his younger siblings, and the family income is barely above the poverty line. The counseling was at an impasse, for Carlos was reluctant to take any action and instead kept saying, "I don't know how to tell this to my folks. I'm sure they'll be mad at me."

During class discussion of this case, almost everyone argued that Carlos's problem is that he does not give priority to his personal career interests, that he should learn to think about what is good for his own mental health, and that he needs assertiveness training. I did not quite agree with my fellow students, who are all European Americans. I thought they were failing to see a decisive factor in the case: Carlos is Latino!

In traditional Hispanic culture, the extended family, rather than the individual, is the psychosocial unit of cooperation. The family is valued over the individual, and subordination of individual wants to family needs is assumed. Also, traditional Hispanic families are hierarchical in form; parents are authority figures and children are supposed to be obedient. Given this cultural background, to encourage Carlos to make a major career decision totally by himself was impossible. Any counseling effort that does not focus on the whole family is doomed to fail.

Because financial support from the family made his college education possible, Carlos may be expected to contribute to the family when he graduates. This reciprocal relationship is a lifelong expectation in Hispanic culture,

and the oldest son is especially responsible in this regard. Changing his major in his junior year not only means postponing the date when he will be able to help his family financially, but it also means he may not be able to do so at all, for we all understand how hard it is to find a job that pays well in the field of literature. When interdependence is the norm among Hispanic Americans, how can we expect Carlos to focus entirely on his personal interests without giving more weight to his family's pressing economic needs?

If I were Carlos's counselor, rather than focusing immediately on his needs, I would first support him with his family loyalty and then help him understand that there are not just two solutions: either/or. Together, we might brainstorm to generate some alternatives, such as having literature as his minor now and as his pastime after he graduates, changing his career when his younger siblings are off on their own, or exploring possibilities that might combine the two. He could, for example, design computer programs to help schoolchildren learn literature. Each of these takes into account family needs as well as his own.

The professor praised me highly for my "different and sensitive perspective," but I shrugged it off; this is just common sense to most Third World people and probably many Italian and Jewish Americans as well. (I remember years ago, when I was trying to make major career decisions with my parents, at least ten of my relatives were involved. And these days, I am still obligated to help anyone in my extended family who is in financial need.)

If the meaning of family in Hispanic culture is confusing to many counselors, the traditional extended family clan system of Native American Indians, Canadian Dene, or New Zealand Maori can be even more difficult for them to grasp. This family extension at times can include several households and even a whole village. Unless majority group counselors are aware of these differences in family structure, they may cause serious harm through their own ignorance.

BOX 9.3 Research and Related Evidence That You Can Use

Focusing

Training students to focus on cultural/environmental/contextual issues resulted in greater awareness and willingness to discuss racial and gender differences early in the session and to make these issues a consistent part of the counseling or therapy (Zalaquett, Foley, Tillotson, Hof, & Dinsmore, 2008).

In a classic review of the contextual focus, Moos (2001) has reviewed much of the contextual literature and points out that the way we appraise a situation can be self-centered or environmental/contextual. Clients often come to the counseling session with a focus that may work against their own best interest. Too much of an "I" focus may result in self-blame and lack of awareness of context. On the other hand, too much of a "they" focus may mean that clients are avoiding responsibility or their part in the conflict. As you know, there are two or more stories as people look at the same event. Moos noted that teaching clients the context of their issues helps them understand themselves in new ways and "makes possible a transformative experience."

Educating and training students in multicultural counseling provides an excellent model for the future (Sue & Sue, 2013). A study that tested a set of multicultural skill training videos found that working with these videotapes of culture-specific counseling increased students' multicultural effectiveness and understanding (Torres-Rivera, Pyhan, Maddux, Wilbur, & Garrett, 2001).

---

many valuable aspects of the problem or situation. Note that much of cultural/environmental/contextual focusing requires sensitive leading and influencing from the counselor.

Box 9.2 illustrates some issues in working with diverse clients. Box 9.3 provides a summary of research related to the skill of focusing.

▶The Cultural/Environmental Context, Advocacy, and Social Justice

What is the role of the counselor or psychotherapist in advocacy and social justice?

You are going to face situations in which your best counseling efforts are insufficient to help your clients resolve their issues and move on with their lives. The social context of homelessness, poverty, racism, sexism, and other contextual issues may leave clients in an impossible situation. The problem may be bullying on the playground, an unfair teacher, or an employer who refuses to follow fair employment practices. Helping clients resolve issues is much more challenging when we examine the societal stressors that they may face.

Advocacy is speaking out for your clients; working in the school, community, or larger setting to help clients; and also working for social change. What are you going to do on a daily basis to help improve the systems within which your clients live? Following are some examples showing that simply talking with clients about their issues may not be enough.

▲ As an elementary school counselor, you counsel a child who is being bullied on the playground.

▲ You are a high school counselor working with a 10th grader who is teased and harassed about being gay while the classroom teacher quietly watches and says nothing.

▲ As a personnel officer, you discover systematic bias against promotion for women and minorities.

▲ Working in a community agency, you are counseling a client who speaks of abuse in the home but fears leaving because she sees no future financial support.

▲ You are working with an African American client who has dangerous hypertension. You know that there is solid evidence that racism influences blood pressure.

The elementary school counselor can work with school officials to set up policies concerning bullying and harassment, actively changing the environment that allows bullying to occur. The high school counselor faces an especially challenging issue as session confidentiality may preclude immediate classroom action. If this is not possible, then the counselor can initiate school policies and awareness programs against oppression in the classroom. The passive teacher may be made more aware through training you offer to all the teachers. You can help the African American client understand that hypertension is not just "his problem," but rather that his blood pressure is partially related to racism in his environment, and you can work to eliminate oppression in your community.

"Whistle-blowers" who name problems that others like to avoid can face real difficulty. The company or agency may not want to have their systematic bias exposed. On the other hand, through careful consultation and data gathering, the human relations staff may be able to help managers develop a more fair, honest, and equitable style. Again, the issue of policy becomes important. Counselors can advocate policy changes in work settings and equal pay for equal work. You can help the client who suffers racial, gender, and sexual orientation harassment. You can speak to employers about how they can employ more people with disabilities.

The counselor in the community agency knows that advocacy is the only possibility when a client is being abused. For clients in such situations, advocacy in terms of support in getting out of the home, finding new housing, and learning how to obtain a restraining order against the abusing person may be far more important than self-examination and understanding.

Counselors who care about their clients also act as advocates for them when necessary. They are willing to move out of the counseling office and seek social change. You may work with others on a specific cause or issue to facilitate general human development and wellness (e.g., pregnancy care, child care, fair housing, shelter for the homeless, athletic fields for low-income areas). These efforts require you to speak out, to develop skills with the media, and to learn about legal issues. *Ethical witnessing* moves beyond working with victims of injustice to the deepest level of advocacy (Ishiyama, 2006). Counseling, social work, and human relations are inherently social justice professions. Speaking out for social concerns needs our time and attention.

▶Summary: Being-in-Relation, Becoming a Person-in-Community

Focusing helps the client see issues and concerns in a broader setting. While the "I" focus remains central, we are also *beings-in-relation*. We must start, of course, by focusing on the special individual before us. Counseling and psychotherapy are for the person. However, by focusing on various dimensions, we can help clients expand their horizons. Connection and interdependence are as necessary for mental health as are independence and autonomy.

The community genogram places the client in connection with the family in a cultural context. Rather than focusing on the many possible negatives in our communities of origin, the genogram reveals that these are places where we also learned strengths. The imaging of stories about community and cultural strengths can be highly useful in counseling and psychotherapy.

We recommend that you consider developing family and community genograms with your clients and placing them on the wall throughout the counseling series. In this way, both you and the client are reminded of the self-in-relation and the need to take multiple perspectives on any issue.

It is not necessary to generate the genograms with every client. What is most important is to be aware that for any client issue, there may be multiple explanations and multiple new stories.

▼▼▼▼	**Key Points**
Selective attention	The way you listen can and does influence clients' choice of topics and responses. Listening exclusively to "I" statements affects the way clients talk about their issues. Listening to culture, gender, and context also affects the way they respond.
The skill of focusing	Focusing is a form of selective attention that enables multiple views of client stories. This skill will help you and clients think of creative new possibilities for restorying and action. It emphasizes the importance of both the individual/issue and the social/cultural context. Focusing enables both the client and the counselor to explore the context of past memories more fully.
Draw out stories with multiple focusing	Client stories and issues have many dimensions. It is tempting to accept problems as presented and to oversimplify the complexity of life. Focusing helps counselor and client to develop an awareness of the many factors related to an issue as well as to organize thinking. Focusing can help a confused client zero in on important dimensions. Thus, focusing can be used to either open or tighten discussion.

Use selective attention to focus the session on the client, issue/concern, significant others (partner/spouse, family, friends), a mutual "we" focus, the counselor, or the cultural/environmental/contextual issues. You may also focus on what is going on in the here and now of the session. |
| **Seven focus dimensions** | There are seven types of focuses. The one you select determines what the client is likely to talk about next, but each offers considerable room for further examination of client issues. As a counselor or psychotherapist, you could say many things, including the following:

▲ Focus on client: "Tari, you were saying last time that you are concerned about your future."
▲ Focus on the main theme or problem: "Tell me more about your getting fired. What happened specifically?"
▲ Focus on others: "So you didn't get along with the sales manager. I'd like to know a little more about him." "How supportive has your family been?"
▲ Focus on mutual issues or group: "We will work on this. How can you and I (our group) work together most effectively?"
▲ Focus on counselor: "My experience with difficult supervisors was . . ."
▲ Focus on cultural/environmental/contextual issues: "It's a time of high unemployment. Given that, what issues will be important to you as a woman seeking a job?"
▲ Focus on the here and now (immediacy): "You seem disappointed right now. Can you share with me what came to your mind right now?" |
Community and family genograms	Genograms are visual maps to help clients gain new perspectives on themselves and their relationships to their families and their communities. They can bring to life the "internalized voices" affecting the client. A community genogram will help you and your clients understand their relation to their environment and show both stressors and assets in their lives. A family genogram will help in understanding a client's family history and current relationships. Both represent useful ways to understand the client's history and identify strengths and resources.
Apply focusing to examine your own beliefs	As a counselor, explore your own beliefs and compare these with the views of others. Use the focus dimensions to explore other people's views. What do your family, your friends, and others close to you think? Awareness of yours and others' views will help your work with your clients.
Focusing and other skills	Focusing can be consciously added to the basic microskills of attending, questioning, paraphrasing, and so on. Careful observation of clients will lead to the most appropriate focus. In assessment and problem definition, consciously and deliberately assist the client to explore issues by focusing on all dimensions, one at a time. Advocacy and social action may be necessary when you discover that the client's issues cannot be resolved through the session alone. Counseling could be described as a social justice profession.

Multicultural issues	Focusing will be useful with all clients. With most clients the goal is often to help them focus on themselves (client focus), but for many people, particularly those of a Southern European or African American background, the family and community focuses may sometimes be more appropriate. The goal of much North American counseling and therapy is individual self-actualization, whereas among other cultures it may be the development of harmony with others—self-in-relation. Deliberate focusing is especially helpful in problem definition and assessment, where the full complexity of the problem is brought to light. Moving from focus to focus can help increase your clients' cognitive complexity and their awareness of the many interconnecting issues in making decisions. With some clients who may be scattered in their thinking, a single focus may be wise.
Social justice and advocacy	Sometimes working only with the client may not be enough. Helping clients navigate an unfair situation, working with the school to provide needed accommodations, and working for social change may be appropriate to help improve the systems within which your clients live.
The importance of the individualistic "I" focus	Recall that counseling is for the client. Though expanding awareness of context and self-in-relation and understanding alternative stories of a situation can be very useful, ultimately the unique client before you will be making decisions and acting. The bottom line is to assist that client in writing her or his own new story and plan of action.

▶Competency Practice Exercises and Portfolio of Competence

Go to CengageBrain.com to access Counseling CourseMate, where you will find an interactive ebook, quizzes, videos, interactive counseling and psychotherapy exercises, the Portfolio of Competence, case studies, a practice test, and more.

Awareness, knowledge, and skills are central, but action is essential. Mastering the skills of counseling and psychotherapy is achieved through intentional practice and experience. Reading and understanding are at best a beginning. Some find the ideas here relatively easy and think that they can perform the skills, but what makes one competent in basic skills is practice, practice, practice.

Individual Practice

Exercise 1. Writing Alternative Focus Statements   A 35-year-old client comes to you to talk about an impending divorce hearing. He says the following:

> I'm really lost right now. I can't get along with Elle, and I miss the kids terribly. My lawyer is demanding an arm and a leg for his fee, and I don't feel I can trust him. I resent what has happened over the years, and my work with a men's group at the church has helped, but only a bit. How can I get through the next 2 weeks?

Fill in the client's main issue as you see it; then write several alternative focus statements in the spaces below. Be sure to brainstorm a number of cultural/environmental/contextual possibilities.

Main issue as presented _____

Client focus _____

Theme, concern, story focus _____

Others focus _____

Family focus _____

Mutual, group, "we" focus _____

Counselor focus _____

Cultural/environmental/contextual focus _____

Immediacy focus _____

Exercise 2. Developing a Community Genogram This chapter presented specific step-by-step instructions for developing a community genogram (see Exercise 9.1). Most of your classmates will have completed a genogram by now. With one of them, debrief the genogram using the ideas from pages 218–223. Also consider the Nelida/Allen session; you may want to try the imagery exercise at the end of that session. Present the completed genogram and briefly summarize what you learned.

Exercise 3. Developing a Family Genogram Using information from the chapter and the illustrations in Box 9.1, develop a family genogram with a volunteer client or classmate. After you have created the genogram, ask the client the following questions and note the impact of each. Change the wording and the sequence to fit the needs and interests of the volunteer.

> What does this genogram mean to you? (individual focus)
> As you view your family genogram, what main theme, problem, or set of issues stands out? (main theme, problem focus)
> Who are some significant others, such as friends, neighbors, teachers, or even enemies who may have affected your own development and your family's? (others' focus)
> How would other members of your family interpret this genogram? (family, others' focus)
> What impact does your ethnicity, race, religion, and other cultural/environmental/contextual factors have on your own development and your family's? (C/E/C focus)
> What I have learned as counselor working with you on this genogram is . . . (state your own observations). How do you react to my observations? (counselor focus)

Summarize in journal form some of what you've learned from this exercise. What questions did you find most helpful?

Group Practice

Exercise 4. Practicing Focusing Skills

Step 1: Divide into groups.

Step 2: Select a group leader.

Step 3: Assign roles for the first practice session.

▲ Client, who has completed a family or community genogram.
▲ Counselor, who will use focusing to bring out past memories using a positive wellness orientation.
▲ Observer 1, who will give special attention to focus of the client, using the Feedback Form, Box 9.4. The key microsupervision issue is to help the counselor continue a central focus on the client while simultaneously developing a comprehensive picture of the client's contextual world.
▲ Observer 2, who will give special attention to focus of the counselor, using the Feedback Form, Box 9.4.

BOX 9.4 Focus

Feedback Form: Focus

You can download this form from Counseling CourseMate at CengageBrain.com.

_____ (DATE)

_____ (NAME OF COUNSELOR)   _____ (NAME OF PERSON COMPLETING FORM)

Instructions: Observer 1 will give special attention to the client and Observer 2 to the counselor. Note the correspondence between counselor and client statements. In the space provided, record the main words used. Classify each statement by checking a box.

Main words	Client							Counselor						
	Client (self)	Concern/problem	Significant others	Family	Mutual "we"	Counselor	Cultural/environmental/contextual	Client	Concern/problem	Significant others	Family	Mutual "we"	Counselor (self)	Cultural/environmental/contextual
1.														
2.														
3.														
4.														
5.														
6.														
7.														
8.														
9.														
10.														
11.														
12.														
13.														
14.														

Observations about client verbal and nonverbal behavior:

_____

_____

Observations about counselor verbal and nonverbal behavior:

_____

_____

Step 4: Plan. Establish clear goals for the session. The task of the counselor in this case is to go through all seven types of focus, systematically outlining the client's issue. If the task is completed successfully, a broader outline of memories related to the client's concern should be available.

A useful topic for this role-play is a story from your family or community. Your goal here is to help the client see the issues in broader perspective.

Observers should take this time to examine the feedback form and plan their own sessions. The client may fill out the Client Feedback Form from Chapter 1.

Step 5: Conduct a 5-minute practice session using the focusing skill.

Step 6: Review the practice session and provide feedback for 10 minutes. Give special attention to the counselor's achievement of goals and determine the mastery competencies demonstrated.

Step 7: Rotate roles.

General reminders: Be sure to cover all types of focus; many practice sessions explore only the first three. In some practice sessions, three members of the group all talk with the same client, and each counselor uses a different focus.

Portfolio of Competence

The history of counseling and therapy has provided the field with a primary "I" focus in which the client is considered and treated within a totally individualistic framework. The microskill of focusing is key to the future of culturally competent counseling and psychotherapy, as it broadens the way both counselors and clients think about the world and review memories. This does not deny the importance of the "I" focus. Rather, the multiple narratives made possible by the use of microskills actually strengthen the individual, for we all live as selves-in-relation. We are not alone. The collective strengthens the individual.

At the same time, the above paragraph represents a critical theoretical point. Some might disagree with the emphasis of this chapter and argue that only the individual and problem focus are appropriate. What do you think? As you work through this list of competencies, think ahead to how you would include or adapt these ideas in your own Portfolio of Competence.

Use the following checklist to evaluate your present level of mastery of the competencies presented here. As you review the items below, ask yourself, "Can I do this?" Check those dimensions that you currently feel able to do. Those that remain unchecked can serve as future goals. Do not expect to attain intentional competence on every dimension as you work through this book. You will find, however, that you will improve your competencies with repetition and practice.

Level 1: Identification and classification. You will be able to identify seven types of focus as counselors and clients demonstrate them. You will note their impact on the conversational flow of the session.

❑ Identify focus statements of the counselor.
❑ Note the impact of focus statements in terms of client conversational flow.
❑ Write alternative focus responses to a single client statement.

Level 2: Basic competence. You will be able to use the seven focus types in a role-play session and in your daily life.

❑ Demonstrate use of focus types in a role-play session and draw out multiple stories.
❑ Use focusing in daily life situations.

Level 3: Intentional competence. Use the seven types of focus in the session, and clients will change the direction of their conversation as you change focus. Maintain the same focus as your client if you choose (that is, do not jump from topic to topic). Combine this skill with earlier skills (such as reflection of feeling and questioning) and use each skill with alternative focuses. Check those skills you have mastered, and provide evidence via actual session documentation (transcripts, recordings).

❑ My clients tell multiple stories about their issues.
❑ I maintain the same focus as my clients.
❑ During the session, I observe focus changes in the client's conversation and change the focus back to the original one if it is beneficial to the client.
❑ I combine this skill with skills learned earlier. Particularly, I can use focusing together with confrontation to expand client development.
❑ I use multiple focus strategies for complex issues facing a client.

▶Determining Your Own Style and Theory: Critical Self-Reflection on Focusing

What single idea stands out for you among all those presented in this chapter, in class, or through informal learning? What stands out for you is likely to be a guide toward your next steps. What do you think of the concept of selective attention and its role in focusing? Focusing places attention on individual memories as well as their relations, situation, and context. What are your thoughts and feelings on this approach? What are your thoughts on multicultural issues and the use of the focusing skill? What other points in this chapter struck you as most memorable? What are your thoughts and experiences with regard to the community and family genograms? How might you use ideas in this chapter to begin the process of establishing your own style and theory?

CHAPTER 10

Empathic Confrontation and the Creative *New*

Identifying and Challenging Client Conflict

Empathic Confrontation

Focusing

The Five-Stage Interview Structure

Reflection of Feeling

Encouraging, Paraphrasing, and Summarizing

Open and Closed Questions

Client Observation Skills

Attending Behavior and Empathy

Ethics, Multicultural Competence, and Wellness

Conditions for creativity are to be puzzled, to concentrate, to accept conflict and tension, to be born every day, to feel a sense of self.

—Erich Fromm

Mission of "Empathic Confrontation and the Creative *New*"

Empathically confronting client issues directly through clarification and supportive challenge promotes creative change when said in a nonjudgmental manner. Skills in empathic confrontation will help minimize resistance and facilitate creativity, growth, and development in new areas. Facing challenges is basic to client restorying.

Chapter Goals and Competency Objectives Awareness, knowledge, skills, and actions in confrontation will enable you to

▲ Identify conflict, incongruity, discrepancies, ambivalence, and mixed messages in behavior, thought, and feelings/emotions.

▲ Encourage and facilitate exploration and creative resolution of conflict and discrepancies.

▲ Evaluate client creative change processes occurring during the session and throughout treatment sessions, using the Client Change Scale.

▲ Consider multicultural and individual differences when using confrontation.

 Go to CengageBrain.com to access Counseling CourseMate, where you will find an interactive ebook, quizzes, videos, interactive counseling and psychotherapy exercises, the Portfolio of Competence, case studies, a practice test, and more.

▶Defining Empathic Confrontation

Many clients come to counseling "stuck"—having limited alternatives for resolving their issues. Our task is to assist in freeing the client from stuckness and facilitate the development of creative thinking and expansion of choices. **Stuckness** is an inelegant but highly descriptive term coined by the Gestalt theorist Fritz Perls to describe the opposite of intentionality, or a lack of creativity. Other words and phrases that represent stuckness include *immobility and ambivalence, blocks, repetition compulsion, inability to achieve goals, lack of understanding, limited behavioral repertoire, limited life script, impasse,* and *lack of motivation.* Stuckness may also be defined as an inability to resolve conflict, reconcile discrepancies, and deal with incongruity. In short, clients often come to counseling because they are stuck for a variety of reasons and seek the ability to move, expand alternatives for action, and become motivated to do something to rewrite their life stories.

Empathic confrontation is an influencing skill that invites clients to examine their stories for discrepancies between verbal and nonverbal communication, between expressed attitudes and behaviors, or conflict with others. Effective confrontation leads clients to creative new ways of thinking and increased intentionality.

Empathic confrontation is not a direct, harsh challenge. It is a gentle skill that involves first listening to client stories carefully and respectfully and then encouraging the client to examine self and/or situation more fully. Confrontation is not "going against" the client; rather, it represents "going with" the client, seeking clarification and the possibility of a creative *New*, which enables resolution of difficulties. However, with some clients, you will find that rather direct and clear confrontation will be required before they can hear you.

Empathic confrontation is based on careful listening. Paraphrasing is particularly useful when the conflict or incongruity involves a decision and the pluses and minuses of the decision need to be outlined. Reflection of feeling is important for emotional issues, particularly when clients have ambivalent or mixed feelings ("on one hand, you feel . . . , but on the other, you also feel . . ."). A summary is a good choice for bringing together many conflicting strands of thoughts, feelings, and behaviors. At times your observations of verbal and nonverbal behavior and personal awareness of what might be happening with the client may lead you to *carefully* add your thoughts to the confrontation. Your own words can be additive and enrich the client's world. Or, if not in tune, very subtractive.

When you use confrontation with intentionality and effectiveness, you can anticipate the following:

CONFRONTATION	ANTICIPATED RESULT
Supportively challenge the client to address observed discrepancies and conflicts. 1. Listen, observe, and note client conflict, mixed messages, and discrepancies in verbal and nonverbal behavior. 2. Summarize and clarify internal and external discrepancies by feeding them back to the client, usually through summarization. 3. Evaluate how the client responds and whether the confrontation leads to client movement or change. If the client does not change, flex intentionally and try another skill.	Clients will respond to the confrontation of discrepancies and conflict by creating new ideas, thoughts, feelings, and behaviors, and these will be measurable on the five-point Client Change Scale. Again, if no change occurs, *listen.* Then try an alternative style of confrontation.

▶Multicultural and Individual Issues in Confrontation

Cultural intentionality has been stated as a central goal for counselors and psychotherapists. The ability to respond to unique clients and their changing needs and issues requires us to be able to flex and have creative multiple responses in the session. Intentionality and creativity are central to the change process. Clients need to be able to flex and meet the multiple challenges of life, constantly create the *New.* When client discrepancies, mixed messages, and conflicts are confronted skillfully and nonjudgmentally, clients are encouraged to talk in more detail and to resolve their problems and issues.

Confrontation is relevant to all clients, but it must be worded to meet individual and cultural needs if real creativity is to occur. You will find that not all clients respond to this skill, and you should be ready to use other listening skills. A narcissistic or self-centered client may resist confrontation; with this client, the microskills of interpretation and feedback may be more helpful. Other clients will need and even prefer a more direct challenge. For example, if you are working with an acting-out or antisocial client, a firm and more solid confrontation may be necessary. The client may sneer at and manipulate "nice" helpers, but may be more likely to respect and work with a counselor who listens and offers respect but takes no "garbage" from the client. Clients from more direct and outspoken cultures, such as European Americans and African Americans, may respond well to appropriate confrontations. Cultures that place more emphasis on subtlety and an indirect approach, such as Asian groups, may prefer gentler, more polite confrontations. Do not expect individuals in any cultural group always to follow one pattern; avoid stereotyping. Modification in style to accommodate individual and cultural differences will be necessary.

▶Empathy and Nonjudgmental Confrontation

If you don't maintain the relationship, you likely will lose the client.

—Allen Ivey

If you are to challenge the client with a confrontation, an empathic relationship is essential, along with good listening skills. As you draw out client stories and strengths, you will be looking for verbal and nonverbal conflict and discrepancy. An essential part of confrontation is paraphrasing the conflict, reflecting the feelings of mixed messages, and providing an accurate summary of the situation. Keep in mind that the majority of clients need to be comfortable with the challenge. However, many clients need the prod and even the anxiety that a confrontation can produce. Even so, in all of these situations, a nonjudgmental approach is key.

Even with empathic listening, a client who is challenged or confronted may feel put on the spot, perhaps even that you are attacking them. This is where Carl Rogers's **nonjudgmental** empathy will be most helpful. Closely related to positive regard and respect, a nonjudgmental attitude requires that you suspend your own opinions and attitudes and assume value neutrality in relation to your client. Many clients have attitudes toward their issues and concerns that may be counter to your own cherished beliefs and values. But people who are working through serious difficulties do not need to be judged or evaluated, and your neutrality is necessary if you want to maintain the relationship.

A nonjudgmental attitude is expressed through vocal qualities and body language and by statements that indicate neither approval nor disapproval. However, as with all qualities and skills, there are times when your judgment may facilitate client exploration. There are no absolutes in counseling and therapy.

Stop and think for a moment of a person whose behavior troubles or angers you personally. It may be someone whom you regard as dishonest, one who perpetrates violence, or one who clearly demonstrates sexism or racism. These are challenging moments for the nonjudgmental attitude. You do not have to give up your personal beliefs to maintain a nonjudgmental attitude; rather, you need to suspend your private thoughts and feelings. You do not have to agree with or approve of the thoughts and behaviors of the client, but if you are to help these people change and become more intentional, presenting yourself as nonjudgmental is critical. You may have an obligation to educate the client and help her or him move to new understandings and new stories. Nonetheless, you need to be nonjudgmental in expressing yourself, as change requires a basis of trust and honesty.

At times, however, judgment may indeed be called for. For example, the Nelida session in the preceding chapter is basically nonjudgmental and supportive, but Allen clearly is judging those who have not respected her Latina heritage. This type of judgment may be appropriate here. But joining clients too soon and agreeing with their views could distort the facts of issues and concerns. It also could be a violation of counseling boundaries.

Also, at times, psychoeducation and information giving (Chapter 13) may be wise when the client expresses oppressive racist, sexist, or other discriminatory comments. Needless to say, this must be done in a nonjudgmental and supportive fashion, or the client will not return.

▶The Skills of Empathic Confrontation: An Integrated Three-Step Process

Empathic confrontation is best described as an integrative, combination skill involving both listening and influencing. You can identify the confrontation skill most easily when the counselor paraphrases or summarizes observed ambivalence or conflict in some form of the classic "On one hand . . . , but on the other hand . . . ; how do you put that together?" In this form, the conflict, discrepancy, or mixed message is said back to the client clearly.

The story is brought out, and conflict in that story is identified. Also, drawing out positive stories is very helpful in the confrontation process. For example, "On one hand, you speak about your inability to defend yourself from your partner, but on the other hand, I also heard you talking earlier about how you were able to handle bullying in high school so successfully. What does that high school story say to the present situation?"

Virtually all counseling and psychotherapeutic leads, both listening and influencing, have the goal of enabling clients to explore their ambivalence and conflict, rather than just complaining. Moreover, they also seek to facilitate clients in finding their own resolution—the creative *New*. At times you may have to use reframing or self-disclosure to help the client "put it together," but ultimately resolving discrepancies and conflict almost always is the client's issue.

However, when you face situations of abuse or danger to the client, or the client faces a severe crisis and cannot act, or the client is oppressed by racism, sexism, classism, and the like, then it may be necessary that you take action and work both inside and outside the session in the community to help find a satisfactory resolution. For example, a school bullying situation may require you to intervene in the school and community.

Confrontation, a supportive challenge, has three stages.

1. *Listen.* Identify conflict by observing conflict, incongruities, discrepancies, ambivalence, and mixed messages.
2. *Summarize.* Clarify issues of internal and external conflict and work toward resolution.
3. *Evaluate.* Evaluate the change process using the Client Change Scale.

Step 1: Listen

Identify conflict by observing incongruities, discrepancies, ambivalence, and mixed messages.

The previous chapter on focusing presented Nelida, who gave permission for us to use her session with Allen, enabling us to present counseling skills as they occur immediately in a real-life session. Here you see Nelida again; we are repeating the early part of her session following the community genogram, presented in the previous chapter. The transcript here is an edited and markedly shortened version of the original.*

What types of conflict and challenges do you see Nelida facing? List as many as you can.

CLIENT AND COUNSELOR CONVERSATION	PROCESS COMMENTS
1. *Nelida*: Here I am, a grad student in counseling. I did well in college in Miami, and thought it was no big deal because I was only four and a half hours away. But my first day of class I raised my hand, made a comment . . . , and a classmate asked me if I was from America (nervous laugh) or a native (nervous laugh). Yeah, and I said well I'm . . . , I was just four and a half hours away, and he just found it very hard to believe. So, after that comment was made, it kind of made me a little bit more hesitant to participate in discussions. It made me more self-conscious.	There are several dimensions of conflict in Nelida's words. How many can you identify? Note how powerful one negative comment can be. The emotional impact of this comment brought it to immediate memory in the hippocampus and prefrontal cortex (PFC). The message in the executive PFC was to keep quiet and not talk in class. There is already an implicit goal—to facilitate Nelida's being proud of her cultural heritage and her skill in English, to help her build self-esteem and self-confidence and speak up for herself.
2. *Allen*: It made you self-conscious. Could we explore that a little bit more? Ah, first of all, in English, what were the feelings that went with that?	Encouraging in the form of restatement around emotions and open questions. As you will see later, those feelings will soon be explored in Spanish. (Interchangeable empathy with an attempt to add emotional dimensions)
3. *Nelida*: Well, I was surprised because being from Miami a lot of my family members have recently come from Cuba, so there they look at me as the American girl and they make fun.	Can you identify cultural/environmental/contextual issues that add to the conflict?
4. *Allen*: . . . and that embarrasses you.	Drawing on nonverbals, Allen supplies an emotion word that acknowledges feeling, but was that the right word? We shall see in Nelida's next statement. (Interchangeable)
5. *Nelida*: Exactly, so when I'm in Miami, my family and friends tease me saying that I'm the American who can't speak Spanish a hundred percent correctly 'cause I've forgotten a lot of it because of the English. Then, now, I move here to Tampa, I'm the Cuban girl who can't speak English, so it seems like I'm torn. You know, I don't know where I belong sometimes.	What is the cultural conflict here?

---

*Slightly edited for clarity, this transcript of a real interview held between Allen Ivey and Nelida Zamora is also available on DVD: Ivey, A., Ivey, M., Gluckstern-Packard, N., Butler, K., & Zalaquett, C. (2012). *Basic Influencing Skills,* 4th ed. [DVD]. Alexandria, VA: Microtraining/Alexander Street Press. By permission of Microtraining/Alexander Street Press. http://alexanderstreet.com/products/microtraining,

Before moving on, identify as many issues of conflict as you can. Perhaps even list them here. What ideal goals or results from the session might you hope that Nelida achieves?

_____

_____

The first task in effective confrontation is to *listen and identify conflict in clients' mixed messages, discrepancies, and incongruity.* This is where metacognition comes in, as you wish to be thinking about what is happening in the here and now of the session, as well as how to respond to the client. While you listen, silently search in your mind for what is "going on now" with the client. Listen and *think* before you help clients clarify their issues.

Use the following questions to practice your metacognitive skills. What conflicts came to your mind while reading Nelida's transcript? What did you think?

_____

_____

What emotions or body reactions do you recall?

_____

_____

Did Nelida's conversation remind you of something from your own life?

_____

_____

How did what Nelida said affect your thoughts, feelings, and behaviors?

_____

_____

Now let's examine the internal and external conflicts affecting Nelida. Focusing on aspects of both can eventually help Nelida resolve her feelings and thoughts about these issues—and perhaps even lead to changes in behavior. In step 2, you will see how issues of conflict play themselves out in the Nelida session.

Step 2. Summarize

Summarize and clarify issues of internal and external conflict and work toward resolution through further observation and listening skills.

Internal conflicts are those that reside primarily within the client's thoughts and feelings. What internal conflicts does Nelida face? "I'm torn" represents a central issue that needs to be addressed. There are mixed and conflicting thoughts of embarrassment, being different, self-consciousness, and not being fully capable. Out of these internal conflicts has come a decision not to speak up in class—another internal conflict, as she would rather say what she thinks.

The key **external conflicts** are those between the client and the surrounding world. These, of course, provide some internal struggles for the client as well. Sources of external conflicts include the class members who have made her feel either singled out or excluded because of her accent, and her family and friends in Miami who call her "the American girl." Implicit in this, and explored later, are issues in the cultural/environmental context: What does it mean to be Cuban American? How does my background relate to me where I am now? How do I relate to others in this new context? And, internally, "How do I keep it all together and still feel OK about myself?" because the external conflicts almost always becomes internal as well.

Focusing can be very helpful in identifying and working with conflict. While our central focus always is on the client, Nelida's conflicts and internal incongruity relate to cultural/environmental/contextual issues (Cuban American culture and "American" Tampa classroom culture) and her family. Thus, part of the session needs to give these focus areas central attention.

We now return to Nelida's early session comments, this time attending to issues of conflict and discrepancy, which are starting to move to confrontation. This continues the part of the session presented earlier in this chapter. (The conversational exchange in items 4 and 5 are repeated here to provide context for this segment of the session.) Note that Nelida's community genogram, presented in the previous chapter, served as a background for the session and provided several positive stories on which change could be built. The strengths and resources identified in the community genogram were reviewed again in the following session, but these have been edited out to save space and focus on confrontation skills.

Note that Allen seeks throughout this interchange to draw out further aspects of the conflict, but typically focuses on strengths that Nelida brings for creative resolution. We solve our difficulties best from our strengths, resources, and positive assets.

CLIENT AND COUNSELOR CONVERSATION	PROCESS COMMENTS
4. *Allen*: . . . and that embarrassed you.	Repeat of Allen's comment from prior transcript.
5. *Nelida*: Exactly, so when I'm in Miami, my family and friends tease me saying that I'm the American who can't speak Spanish a hundred percent correctly 'cause I've forgotten a lot of it because of the English. Then, now, I move here to Tampa, I'm the Cuban girl who can't speak English, so it seems like I'm torn. You know, I don't know where I belong sometimes.	Nelida identifies her central conflict clearly, but she still needs to tell her story in more detail.
6. *Allen*: So, on one hand, you feel challenged about your English here and, on the other hand, when you go home, you get challenged on your Spanish. You feel torn. . . . How would you describe that feeling of tornness in Spanish?	Confrontation in the form of paraphrase and reflection of feeling emphasizes that key word "torn." Conflict inevitably has an emotional dimension. Working with just the cognitive decision-making issues will be less effective. The open question asks Nelida to describe her feelings in Spanish. Clients who use English as a second language often will feel more comfortable describing the larger issues in their first language and then translating them for you. Allen actually used movement from hand to hand as he made that comment.
	Encouraging clients to use their home language is a sign of empathic respect and shows an authentic openness on the part of the counselor.
7. *Nelida*: *Muy conflictiva.* Just very conflicting.	Nelida shows more tension and frustration in her body language. The confrontation did not enable her to resolve these issues at all, but the tornness and the conflict are now clearer. Here we see that feelings and emotions are physical, as well as cognitive.

continued on next page

CLIENT AND COUNSELOR CONVERSATION	PROCESS COMMENTS
8. *Allen*: Did you notice any difference between English and Spanish when you said it?	Closed question to check out the importance of Spanish with this client. Not all clients will feel comfortable with talking about their issues in this way, but most will. This in itself is a confrontation that points out the conflict between the two languages and seeks to facilitate Nelida's learning the power of her own language. She had felt strange and that Spanish was somehow a disability in graduate school.
9. *Nelida*: Again, it almost felt more real when I said it in Spanish . . . like truer. I'm comfortable with both languages, but like I said, my primary language is Spanish, so I guess to an extent it does feel more real, truly to myself, when I do say it in Spanish.	A new story is being created here as Nelida creates a new meaning for Spanish. Whereas Spanish was described earlier as kind of a handicap, we see Nelida moving to realizing that her English is good (despite the student comments), and also that her native Spanish is respected and valuable. Bilingualism is a strength and actually builds wider neural networks in the brain than monolingualism.
10. *Allen*: It feels more real when you say it in Spanish. (pause) I'd like to hear more about your story, what it meant for you to come from Miami to Tampa and how it went for you.	Restatement of feelings followed by a statement that really is an open question about *meaning* of the situation. Specifically, how does Nelida frame or interpret what happened? Chapter 11 will discuss issues of reframing and interpreting meaning. (Interchangeable empathy)
11. *Nelida*: Well, I was surprised because being from Miami a lot of my family members have recently come from Cuba, so they look at me as the American girl and they make fun. When I'm over there, I'm the American who can't speak Spanish a hundred percent correctly 'cause I've forgotten a lot of it because of the English. Then, now, I move here and I'm the Cuban girl who can't speak English, so it seems like I'm torn. You know, I don't know where I belong sometimes.	Nelida now is clearer about her major conflict and will gain more understanding through these brief segments of the session.
12. *Allen*: You seem to face conflict with your family at home, with your classmates here in Tampa—and then the two conflicts actually seem to conflict against each other as well and really add to the tornness.	Summary of the confrontation issues—on one hand, at home . . . on the other hand, here. (Interchangeable)
13. *Nelida*: Absolutely. But then I have my grandparents tell me that I'm forgetting my Spanish, you know, also. Kind of confusing and I feel torn. Um-hum.	Expressions like "absolutely," "exactly," or "yes" confirms accuracy of the summary.
14. *Allen*: Um . . . hummm.	Allen leans forward with a minimal encourager anticipating that Nelida will continue processing the issue.
15. *Nelida*: It was pretty bad my first semester in graduate school, maybe even throughout the first year. I think the accent's kind of gone away a little, living on campus here a while, but I still get it every now and then. Not as often, though. And, then, my family in Tampa has gradually become more supportive as they've seen my successes.	Here we see Nelida starting to synthesize discrepancies and resolve part of the issue around language. She demonstrates that clients can find their own way to resolve contradiction.

CLIENT AND COUNSELOR CONVERSATION	PROCESS COMMENTS
16. *Allen*: From your community genogram and the way you talked about the last visit home, it sounds like your grandparents and your mother have become even more supportive, even though they may tease you occasionally about your "American accent."	Summary from previous session. There is a mild confrontation here as we see both sides of Nelida's relationship with her grandparents. Allen follows up with information from the last meeting and the community genogram. (Additive because of linking a past session with the present)
17. *Nelida*: Yes, perhaps I should not feel so torn. Things really are getting better, but I don't know how to deal with those comments about accent. They don't come as often, but . . .	Having worked through a contradiction with her family and remembering the supports and resources in her family genogram, Nelida is prepared to explore the more immediate issues.

Step 3: Evaluate

Evaluate the change process using the Client Change Scale.

The effectiveness of a confrontation is measured by how the client responds. If you observe closely in the here and now of the session, you can rate how effective your interventions have been. You will discover if your attempt at confrontation is subtractive, interchangeable, or additive. With a facilitative confrontation, you will see the client change (or not change) language and behavior in the session. When you don't see the change you anticipate or think is needed, it is time for creative intentionality, flexing, and having another response, skill, or strategy available.

▶The Client Change Scale (CCS)*

Imagine that you have provided a confrontation by summarizing a client conflict ("On one hand, you feel this . . . , but on the other hand, you think. . . . How do you put that together?"). The Client Change Scale gives you a framework for evaluating how the client responds to your confrontation. Did the client deny that a conflict, discrepancy, or mixed messages exist; show minor movement toward synthesis; or actually use the confrontation in a way that leads to significant change in thoughts and feelings, so that later these new discoveries can lead to behavioral change?

A summary of the Client Change Scale follows. Figure 10.1 illustrates how clients can move through the various levels of change.

CLIENT CHANGE SCALE (CCS)	ANTICIPATED RESULT
The CCS helps you evaluate where the client is in the change process. Level 1. Denial. Level 2. Partial examination. Level 3. Acceptance and recognition, but no change. Level 4. Creation of a new solution. Level 5. Transcendence.	The CCS can help you determine the impact of your use of skills. This assessment may suggest other skills and strategies that you can use to clarify and support the change process. You will find it invaluable to have a system that enables you to (1) assess the value and impact of what you just said; (2) observe whether the client is changing in response to a single intervention; or (3) use the CCS as a method for examining behavior change over a series of sessions.

*A paper-and-pencil measure of the Client Change Scale was developed by Heesacker and Pritchard and was later replicated by Rigazio-Digilio (cited in Ivey, Ivey, Myers, & Sweeney, 2005). A factor analytic study of more than 500 students and a second study of 1,200 revealed that the five CCS levels are identifiable and measurable.

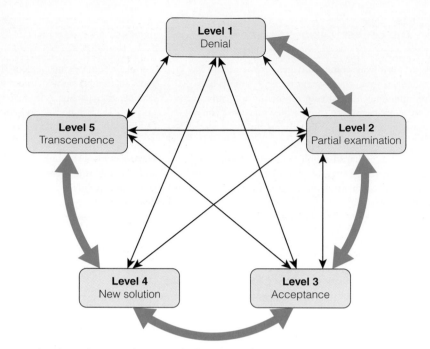

FIGURE 10.1 The Client Change Scale (CCS). The five stages of creative change may occur in order. However, as the arrows indicate, there can be movement back and forth among the stages. In fact, the entire process can go back to the beginning as clients discover new thoughts, feelings, behaviors, and meanings.

Although the progression from denial through acceptance through significant change can be linear and step by step, this is not always the case, as suggested in Figure 10.1. Think of a client working through the expected death of a loved one or a contested divorce. One possibility is for the client to move through the CCS stages one at a time. But often a client who seems to be moving forward will suddenly drop back a level or two. At one session the client may seem acceptant of what is to come, but in the next session move back to partial examination or even denial. Then we might see a temporary jump to transcendence, followed by a return to acceptance.

The Client Change Scale is also useful as a broad measure of success in several types of counseling and psychotherapy. The CCS provides a useful framework for accountability and measuring client growth. The clearest example is that of clients with substance abuse issues. Such clients may come voluntarily or be referred by the court for cocaine or alcohol abuse (often both). They may also be depressed and use the drugs to alleviate pain. Often these clients start by denying that they really have a problem (we call this Level 1 on the Client Change Scale). If we are successful in challenging, supporting, and confronting, we will see the client move to Level 2, admitting that there may be a problem. Some call this a "bargaining" stage, in which the client moves back and forth between denial and recognizing that something needs to be done.

Acceptance and recognition of the issue occur at Level 3, where the alcoholic admits that he or she is indeed an alcoholic and the cocaine abuser acknowledges addiction. But acceptance of the problem and recognition are not resolution. The client may be less depressed but still continue to drink and use drugs.

Change occurs when the client reaches Level 4 and actually stops the substance use, after which the depression usually lifts. Although substance abuse and depression are often co-occurring mental health issues, there is some risk that depression will remain and continued treatment will be necessary. Even so, this is real success in counseling and therapy; it is not easy to achieve, but clearly observable and measurable. Level 5, transcendence and the development of new ways of being and thinking, may occur, but not all clients will achieve this level. Level 5 is represented by the user who becomes fully active in support groups, helps others move away from addiction, and continues to work on feelings, behaviors, actions, and relationships that led to the alcohol abuse and addiction. This person achieves changes in life's meaning and a much more positive view of self and the world—far more than just "getting by."

The following table presents the Client Change Scale with examples of each of the five levels (a) as a client responds to issues around a divorce and (b) as Nelida thinks about the hurtful student comments. Anytime clients are working through change, they talk about their issues with varying levels of awareness. The client may be in denial one moment, the next minute talk as if he or she accepts the problem, and then return to bargaining to avoid change.

CLIENT CHANGE SCALE STAGES	CLIENT EXAMPLES (A) CLIENT'S STATED REACTIONS TO DIVORCE (B) NELIDA'S INTERNAL (UNSAID) THOUGHTS AND FEELINGS
Level 1. Denial. The individual denies or fails to hear that an incongruity or mixed message exists.	(a) I'm not angry about the divorce. These things happen. I do feel sad and hurt, but definitely not angry. (b) There is something wrong with me because I don't speak English like everyone else here.
Level 2. Partial examination. The client may work on a part of the discrepancy but fail to consider the other dimensions of the mixed message.	(a) Yes, I hurt and perhaps I should be angry, but I can't really feel it. Perhaps we should look at it again. (b) I don't think they meant it to hurt me, but it sure did. (Nelida's emotional attitudes toward herself are negative.)
Level 3. Acceptance and recognition, but no change. The client may engage the confrontation but makes no resolution. Until the client can examine incongruity, stuckness, and mixed messages accurately, real change in thoughts, feelings, and behavior is difficult. Coming to terms with anger or some other denied emotion is a sufficient breakthrough and often is a sufficient solution for the client.	(a) I guess I do have mixed feelings about it. I certainly do feel hurt about the marriage. I hurt, but now I realize how really angry I am. (b) OK, I see that they were insensitive and should not have said that. On the other hand, it is OK that I speak Spanish. (Thoughts about the students, whom she allowed to cause her to feel uncomfortable, are now changing and Nelida emotionally feels better about Spanish, but only "OK.")
Level 4. Creation of a new solution. The client moves beyond recognition of the incongruity or conflict and puts things together in a new and productive way. Needless to say, this usually does not happen immediately. It can take several sessions over a period of weeks and sometimes months.	(a) Yes, I've been avoiding my anger, and I think it's getting in my way. If I'm going to move on, I will accept the anger and deal with it. (b) Actually, I am starting to feel proud of my language and Cuban cultural background. The students here really don't understand and perhaps are a bit racist. (Move to emotion of pride and awareness of anger toward students.)

continued on next page

CLIENT CHANGE SCALE STAGES	CLIENT EXAMPLES (A) CLIENT'S STATED REACTIONS TO DIVORCE (B) NELIDA'S INTERNAL (UNSAID) THOUGHTS AND FEELINGS
Level 5. Transcendence—development of new, larger, and more inclusive constructs, patterns, or behaviors. Many clients will never reach this stage. A confrontation is most successful when the client recognizes the discrepancy and generates new thought patterns or behaviors to cope with and resolve the incongruity.	(a) You helped me see that mixed feelings and thoughts are part of every relationship. I've been expecting too much. If I expressed both my hurt and anger more effectively, perhaps I wouldn't be facing a divorce. But now I know that I am more in touch with myself, and this new relationship looks like it may work. (b) I think I am now going to have to educate them, but being sharp and angry will do no good. I'm not going to take it anymore. Perhaps I may even find other Latino students and start a group discussion of what's going on here. (A cognitive move to action beyond just thought, based on awareness of oppression and positive emotions toward self and Cuban culture.)

When you confront clients, ask them a key question, or provide any intervention, they may have a variety of responses. Ideally, they will actively generate new ideas and move forward, but much more likely they will move back and forth with varying levels of response. The idea is to note how clients respond (at what level they answer) and then intentionally provide another lead or comment that may help them grow. Clients do not work through the five levels in a linear, straightforward pattern; they will jump from place to place and often change topic on you. Be empathic and patient and keep on task to help them move forward to creating change.

Depending on the issue, change may be slow. For some clients, movement to partial acceptance (Level 2) or acceptance but no change (Level 3) is a real triumph. For a variety of issues, acceptance represents highly successful counseling and therapy. For example, the client may be in a situation in which change is impossible or really difficult. Thus, accepting the situation "as it is" is a good result. A client facing death is perhaps the best example and does not always have to reach new solutions and transcendence. Simply accepting the present situation may be enough. There are some things that cannot be changed and need to be lived with. "Easy does it." "Life is not fair." "There is a need to accept the inevitable." For someone whose partner or parent is an alcoholic, a major step is to realize that the situation cannot be changed; acceptance is the major breakthrough that will lead later to new solutions. The newly found solutions would facilitate the mediation process and help resolve the conflict.

Now, let us return to Nelida and focus on Step 3: Evaluate the change process using the Client Change Scale. Here you will see her move to Level 5 on the Client Change Scale.

COUNSELOR AND CLIENT CONVERSATION	PROCESS COMMENTS
1. *Allen:* The grandparents who raised you came from Cuba—could you tell me about their story?	Open question, focused both on family and on the cultural/environmental context.

COUNSELOR AND CLIENT CONVERSATION	PROCESS COMMENTS
2. *Nelida*: They were 27 years old and had good jobs when Castro came into power. They weren't in agreement with Communism, so that's why they decided to move to New York. My grandmother was a seamstress. My grandfather was cleaning and mopping floors because that's the only job he was able to get when he moved here. He didn't really have time to learn English properly, so even still today they struggle and they don't speak it fluently at all. And I guess that's something that goes back to my reason for wanting to communicate with them and speak Spanish well . . . (pause)	Client Change Scale (CCS) Level 3, recognition of what is really going on. Client stories may originally start with Level 1 denial of the problem, as Nelida did when she accepted the negative microaggressions such as "Where are you from?" But with a review of strengths provided by the family genogram, Nelida is able to challenge her old thoughts, and her story line now recognizes things "as they are."
3. *Allen*: Talking to them in good Spanish is important.	Restatement. (Interchangeable empathy)
4. *Nelida*: Um-hum, I don't want them to think that I'm forgetting or not valuing the culture as much as I used to, because I still do.	CCS Level 3, recognition. The restatement reinforced Nelida's awareness that she needed to be aware of and value her culture. Because of this, the restatement above has some additive aspects. Here we see that her attitudes and emotions have changed in her long-term memory.
5. *Allen*: Okay. I'd like to go back to strengths for a minute. What do you see are some of the strengths in the Cuban culture, as you've lived with them and it?	This is a longer-term confrontation as Allen is asking Nelida to contrast her former negative beliefs with Cuban strengths. Topic jump, open question, focus on cultural/environmental context. (Potentially additive as it emphasizes respect, but we need to see how Nelida responds. How additive a statement is depends more on the client's reaction than what the counselor says.)
6. *Nelida*: They're very persistent. I know my grandparents have always been extremely persistent in terms of moving ahead, you know. They lived in New York and then they moved to Miami, and they're much better off now than what they used to be, but it took them a lot of hard work to get to where they are today.	CCS Level 3. Here we see a very concrete example of the strength of her resources. Allen is also learning more about the cultural context and family in Nelida's background.
7. *Allen*: Okay. They are persistent. Who else is persistent?	A positive confrontation in which Nelida is compared to her grandparents. Encourager picking up the key word "persistent" followed by an open question. (Potentially additive, also indicates a real sense of respect for Nelida and her culture.)
8. *Nelida*: According to the genogram, I am persistent too (laughs). When I get up here in Tampa, I find the culture being denigrated. And when I think of their story, it makes me proud.	CCS Level 4, creation of new thoughts and feelings. Nelida is reframing her thoughts about the relevance of her cultural and family background.

continued on next page

COUNSELOR AND CLIENT CONVERSATION	PROCESS COMMENTS
9. *Allen*: Proud and persistent. Pretty impressive. You've got a lot to be proud of, given what they went through, and now you seem able to do the same.	Encourager, self-disclosure, feedback with additive empathy. Nelida had previously seen her Miami Cuban background as a problem, but now she is confronted with the idea that she has much to be proud of. The major conflict could be summarized as "On one hand, Nelida, you've been put down for your Cuban American background, but on the other hand, you now are aware that you have much to be proud of and that other people may be the problem, not you." Allen will not lay out the confrontation that clearly, but encourage Nelida to make these discoveries herself.
10. *Nelida*: (pause) Now that I think about it, perhaps I have done better than I thought. They made me feel devalued and, over time, I continued to lose confidence in myself. I did OK in classes, but never felt really good about myself.	CCS Level 3, recognition that will lead to further change. Nelida herself is stating the confrontation that Allen above hoped for. She is naming the "on one hand . . . , but on the other . . ." herself. Facilitate clients' confronting their own conflict in a positive, strength-based fashion. Allen's comment at 9 is now apparent as additive, and respect is a helpful part of the process.
11. *Allen*: Let's explore that a little bit more. You lost confidence, but now you are seeing yourself a little differently. I'll ask you a question. How do you feel about people that treat others unfairly, particularly if they are talking about one's ethnicity, religion, or sexual orientation?	Confrontation. Allen summarizes the old and new views that Nelida has of herself. Building on that positive self-view, he confronts her with how she feels about unfairness. (Additive)
12. *Nelida*: It's not right.	CCS Level 4, a new view of the situation. Nelida is succinct and clear, and she is resolving the conflict in her own mind.
13. *Allen*: It's not right. How about their treatment of you?	Encourager and further confrontation through the restatement and question.
14. *Nelida*: It's not right.	CCS Level 4. Affirmation of the change by repetition. Once a new idea is accepted and reinforced, it becomes part of one's self-concept, but needs further reinforcement by both herself and Allen.

The conversation between 15 and 21 has been deleted to save space, but is available on the CD-ROM. In those exchanges, Nelida reaffirms her idea that what happened to her was not right, and through Allen's reflecting feeling, she becomes aware of her anger over what she now sees as harassment. She has moved from an internal contradiction ("I'm the one who is inadequate") to awareness that the conflict is external and begins to discover how this situation is related to unconscious (or possibly conscious) verbal oppression by the classmates. Nelida also becomes aware that she is paying a price for trying to be empathic, rather than facing up to the harassment.

At this point, what are your thoughts about Nelida and this session? Do you think she is overreacting? How might you handle this session differently from Allen? If you are a woman or aware of women's issues, we suspect that you might add some further items to this session. For example, would the students have treated a male Latino the same way?

The session continues and moves to Level 5 and a later plan for action.

COUNSELOR AND CLIENT CONVERSATION	PROCESS COMMENTS
22. *Nelida*: Maybe how to deal with that feeling instead of always being so empathic, you know, what else can I do? (pause) I should probably address it. I don't want to be rude. You know, because like I said, it might be unintentional, but I should address it.	CCS Level 3, recognition and willingness to explore further.
23. *Allen*: Okay. You should address it. So one possibility, then, is to address it when it happens. How would you name it when you get somebody who talks to you denigrating your culture? Ah, what name could we give it?	Encourager/restatement/paraphrase. Allen then supplies a directive oriented to brainstorming plus an open question. (Most likely this is additive empathy, plus acceptance of Nelida "where she is" as well as respect for her and her cultural background.)
24. *Nelida*: Well, it's victimizing . . . and I've allowed myself to be victimized. Reminds me of my multicultural class—I've internalized a negative view of myself. And that actually is a form of racism that I've taken inside myself without thinking.	CCS Level 5, the development of a new and larger view of herself and her situation. Note the use of the transformational word "victimizing."
25. *Allen*: Victimizing. Okay. So when you're victimized, and historically we find that you've allowed yourself to be victimized, how do you feel about that?	Confrontation in the form of a summary, followed by a checkout/perception check. (Interchangeable empathy)
26. *Nelida*: That's exactly right. I've gone about it the wrong way. (laugh)	CCS Level 4, further integration of the new constructs Nelida created for herself.
27. *Allen*: So you are allowing insensitivity, victimization, and racism oppression to sit inside you and make you feel bad about yourself. 28. *Nelida*: I never thought of it that way. 29. *Allen*: Okay. So seems to me like your thoughts are changing. 30. *Nelida*: Yes. Definitely.	This set of exchanges is grouped together. Allen first summarizes the essence of the conflict with a confrontation. Nelida responds with her awareness of the new way of being. Allen then paraphrases what she just said. Most meaningful, we hear the word "Definitely" said strongly for the first time. This consolidates Level 5 thinking and feeling (but is still not behavioral change).
31. *Allen*: Now, what's the Spanish word for strength and force and the persistence that you and your grandparents have? What's the Spanish word for that? 32. *Nelida*: *Fuerza*. 33. *Allen*: Say it again. 34. *Nelida*: *Fuerza*. 35. *Allen*: Say it loud. 36. *Nelida*: **Fuerza**!	CCS Levels 4 and 5. These exchanges represent a further consolidation of Nelida's resources that will enable her to deal behaviorally with the world around her. The strength within the word *fuerza* is becoming a central part of Nelida's being. Allen used the Gestalt directive of repetition to help Nelida reinforce the creation of the *New*, which in turn leads to a more confident self concept. Again, saying key words, particularly emotional words, in one's native language is highly recommended. Allen's efforts here are additive.

continued on next page

COUNSELOR AND CLIENT CONVERSATION	PROCESS COMMENTS
37. *Allen*: Can you think of that word and how it represents you and your grandparents' pride and force the next time you are put down, like you have been?	
38. *Nelida*: I definitely will.	CCS Levels 4 and 5.
39. *Allen*: *Fuerza*. Okay, *fuerza* is going to protect you from inside. And now how are you going to deal with those who have harassed you? Can we take what you have discovered and use it to make things better for you—and perhaps for others as well?	Summary and a move to go to the next stage of the session, followed by an open question. There is clearly an implied confrontation here: "On one hand, you are feeling better about yourself and your culture, but on the other hand, what are you going to do about it?" All these issues need to be explored in a follow-up third meeting.

Looking back at the first session with Nelida, we have seen her move from CCS Levels 1 and 2 to a significant change in her view of herself and a willingness to take action. We see that Nelida has become much more aware of her personal strengths and resources in her external family and friends. And, clearly, she has a heightened respect for and pride in her cultural heritage and cultural identity. Clearly, she has made considerable progress in her internal feelings about herself and has a better understanding of external factors that have troubled her.

At the end of the session (not shown), Nelida moves to Level 5 on the Client Change Scale. She addresses the need to speak up when others demean her Cuban heritage and educate them when possible. She is aware that bilingualism is an advantage. Finally, she discusses the possibility of bringing other Spanish-speaking students together to provide support for one another.

Some research speaking to the "how" of empathic confrontation is reported in Box 10.1.

BOX 10.1 Research and Related Evidence That You Can Use

Confront, but Also Support

Relatively little of value in research on confrontation existed until 2011, when Mikecz's qualitative research addressed how to work in difficult counseling situations. He identified the following issues as central:

▲ Don't confront unless you have trust and relationship.
▲ Pay attention to and understand the client's point of view and way of thinking about the issue. Summarize the client's interpretation of the situation.
▲ Share with the client only if he or she can listen to and hear you.
▲ The client needs to be in charge of what happens and how things are interpreted.
▲ Knowledge of the client's cultural background and general personal style is essential. If you know even a few words of the client's first language (if different from yours), this will help.
▲ Attending skills such as eye contact are critical in the relationship.

▲ Maintain neutrality; avoid judgments.
▲ Follow up, both in the session to examine how the conflict was resolved or not resolved and after the session to see if new knowledge has generalized into action in the real world.

Attending and listening skills are used frequently in the session, but you'll find that confrontation strategies are used only occasionally. Confrontations account for only 1% to 5% of counselor statements (Hill & O'Brien, 1999). The reviewers noted that confrontations are useful, but they also often make clients uncomfortable. Empathic listening is required; otherwise, clients frequently become defensive and may not deal fully with issues following a confrontation.

Counselor eye contact affects client perception of rapport in the session. Specifically, less direct eye contact early in the session when discussing sensitive matters is helpful; at this point, clients appreciate a nonconfrontational approach. As the session progresses, more eye contact and more confrontations are acceptable (Sharpley & Sagris, 1995).

Confrontation, Creativity, and Neuroscience

Researching creativity is like nailing jelly to a wall.

—Oshin Vartainian

Defining creativity in terms of neural science and the fMRI has turned out to be challenging. In a review of 72 studies, Dietrich and Kanso (2010) leave us with the general impression that "diffuse activity" in the frontal cortex is essential. What we call "creativity" may be located in the connections between the holistic right brain and the linear left brain as well as the participation of the mainly unconscious limbic system (Carter, 1999, 2010; Gazzaniga, 2000). There is also some limited evidence that inhibiting the left hemisphere enables the right to push through with creative ideas. This idea, while not fully substantiated, is attractive, as confrontation is a way to unbalance thinking and "mix things up," thus resulting in diffuse activity. By challenging preexisting thoughts (existing in the hippocampus), confrontations may enable the prefrontal cortex to "loosen up" and create the *New*.

Counseling and psychotherapy are very much concerned with helping clients create the *New* and discover pathways to growth. As part of this, fresh neural networks are created. To facilitate significant change, seek an appropriate balance of stress while supporting the client. Too much stress is damaging, but too little stress likely won't lead to change. Maintain awareness that "released adrenaline (*resulting from stress*) influences almost all regions of the brain—the entire cortex, the hypothalamus, the hind brain, and the brain stem" (Grawe, 2007, p. 220).

Too much stress can flood the brain with damaging cortisol and fix negative memories in the mind (posttraumatic stress). There are a few unethical and charismatic "therapists" who encourage clients to reach strong emotions. They use this here-and-now base to reach back to so-called "long forgotten and repressed" memories of trauma. Unfortunately, this can result in permanently imprinting false memories that do not exist (Loftus, 2003). Clients come to believe that things happened that never did. This type of "therapy" introduces new damaging neural networks in the brain.

▶Cultural Identity Development and the Confrontation Process

We have seen Nelida move from Level 1 to Levels 4 and 5 on the Client Change Scale. Key to this process was her becoming more aware of her identity as a cultural person— a Spanish-speaking Latina, a minority person in a predominantly White environment. Clearly, cultural background is a major part of personal identity, even though she, and many of your own clients, are not aware of this.

Cultural identity developmental theory has useful parallels to the Client Change Scale. Five levels of identity were first identified by William Cross (1971, 1991), who outlined specific and measurable stages of Black identity development. Since that time, several other theorists have explored the Cross five-stage model and applied it not only to racial/ethnic issues, but also to the development of gender awareness, gay/lesbian identity, the disabled, and many other groups. Included in this is White awareness, which focuses on Whiteness and the White experience as a culture.

Moving from one cultural identity stage to another requires confrontation of the discrepancies within life at that stage. For example, the conformity stage is illustrated by the African American who denies racial issues, the woman who accepts male values as "the truth," or the gay male who hides in the closet and denies his sexual orientation. All of these are constantly confronted with the contradictions and discrepancies they see daily as they interact with others. When enough data and emotional impact have come from these encounters, energy to confront the discrepancy mounts and the individual can move to another stage of cultural identity development.

The Racial/Cultural Identity Development (R/CID) model (Sue & Sue, 2013) is one model used to understand where the client is in his or her cultural identity development and to assess progress. The R/CID outlines five stages of development: conformity, dissonance, resistance and immersion, introspection, and integrative awareness. With a focus on racial/cultural issues, like the Cross model, it is adaptable and useful with all groups within the RESPECTFUL framework.

Cultural identity theory is summarized in Box 10.2, where we examine counseling and therapy once again from a cultural perspective.

Maintain awareness that each client you meet, whether a Person of Color or White, has some level of cultural identity. Many White people deny that they have a culture or a cultural identity, and this may be an issue for counseling itself where creating the *New* may be challenging. A good place to start identity development with White individuals is ethnicity or region of the nation. Clients are often willing to explore Irish, Polish, or German backgrounds as identity, but often have more difficulty with that word "Whiteness." Reviewing the history of ethnic and religious prejudice is one way to facilitate awareness of societal oppression and lack of tolerance.

BOX 10.2 | Cultural Identity Development

Conformity Stage Clients at this stage frequently demonstrate (a) self-depreciating attitudes and beliefs, (b) group-depreciating attitudes and beliefs toward members of the same minority group, (c) discrimination toward members of different minorities, and (d) group-appreciating attitudes and beliefs toward members of the dominant group. They prefer dominant cultural values. They may seek to identify with the majority without rejecting their own cultural group. They may develop negative views of their own heritage and cultural group.

Dissonance Stage Clients who prefer dominant cultural values frequently encounter information or experiences that are inconsistent with such values and begin to question them. Clients at this stage usually demonstrate (a) conflict between self-depreciating and self-appreciating attitudes and beliefs, (b) conflict between group-depreciating and group-appreciating attitudes and beliefs toward members of the same minority, (c) conflict between dominant-held views of minority hierarchy and feelings of shared experience, and (d) conflict between group-appreciating and group-depreciating attitudes toward members of the dominant group.

Resistance and Immersion Stage Clients at this stage tend to endorse minority-held views and to reject the dominant societal values. These clients may demonstrate (a) self-appreciating attitudes and beliefs, (b) group-appreciating attitudes and beliefs toward members of the same minority group, (c) conflict between feelings of empathy for other minority group experiences and feelings of culturocentrism, and (d) group-depreciating attitudes and beliefs toward members of the dominant group. Clients may feel guilt, shame, and anger. Motivation to eliminate oppression and racism and a focus on changing external factors are observed.

Introspection Stage Clients at this stage feel uncomfortable with the group views held in the previous stage, which may be perceived as rigid and global. Clients demonstrate concerns about (a) the basis of self-appreciating attitudes and beliefs, (b) the unequivocal nature of group appreciation toward members of the same minority, (c) ethnocentric bias for judging others, and (d) the basis of group depreciation toward members of the dominant group. Clients begin to focus on understanding themselves and their own cultural group.

Integrative Awareness Stage Minority clients at this stage achieve a personal sense of security and an appreciation for both their culture and the dominant culture. Clients usually demonstrate (a) self-appreciating attitudes and beliefs, (b) group-appreciating attitudes and beliefs toward members of the same minority group, (c) group-appreciating attitudes toward members of a different minority, and (d) attitudes and beliefs of selective appreciation toward members of the dominant culture. Clients express a commitment to eliminate all forms of oppression and discrimination.

Understanding cultural identity development enables counselors to assess the role of cultural/contextual factors to better understand culturally diverse clients. With this knowledge, counselors can empower clients to positively embrace their cultural identity and improve their lives.

▶Cultural Identity Development and the Nelida/Allen Counseling Sessions

Nelida and Allen's two sessions highlight the importance of helping clients address conflicts related to racial/cultural identity. Clients may report low self-worth and self-esteem, and blame themselves. These feelings and thoughts may be products of oppression and racism. Nelida reported conflictive negative feelings about herself—she "never felt really good about myself" and was unable to stick up for herself. She made up excuses for the other student's question because it must have been "unintentional."

With the help of Allen, she begins to appreciate her culture and feels "proud" of her grandparents' persistence and hard work when they came to the United States. She begins to appreciate herself, seeing these positive attributes in herself as well. Nelida gains awareness of her resulting negative feelings and sense of being "devalued" as an issue that should have been addressed during class because it was "victimizing" and self-denigrating. With further progress in her identity development, she will be able to eliminate internalized negative views of self and replace them with positive views and a commitment to eliminate racism and oppression. Box 10.3 applies the cultural identity model to Nelida's experience.

To fully achieve complete awareness at the introspection and integrative awareness stages, Nelida will need to take action in the real world. As the session continued from 39 above, Nelida first made it clear that she no longer wanted to continue passive acceptance of things as they were and she wanted to speak up for her grandparents and her culture. Her first action thought was to seek to educate those who might unintentionally* put down her accent and implicitly her culture as well. She commented, "Maybe I should educate that person a little bit about where I come from and what my culture is. And hopefully that person will realize what an ignorant question that is to ask somebody. Maybe tell them a little bit about my culture. I see now that I'm speaking not only for me, but also for my grandparents and for my culture. I need to stand up."

Possibilities for educating those around her were explored, and it was clear that this was her major goal. She was not interested in expressing anger. At times, the effort to educate all those who express an oppressive or racist comment can be very tiring, so Nelida realized that sometimes it would be wise for her just to ignore it—and, while ignoring it, think of the good feelings in her body that represent positive family and cultural experiences.

Beyond educating and ignoring, Nelida realized that she could talk about the incidents with her friends and family. Her grandparents experienced similar racist incidents, particularly in New York. She also considered the possibility of joining a Latina/o action group on her campus so that a larger educational effort could be undertaken.

---

*Showing respect for cultural difference is critical for an empathic relationship. Nelida's fellow students "unintentionally" showed a lack of respect for who she was and where she was from. Those who find themselves in a "minority" culture or group often experience unintentional racism when basically "good" people say things that hurt, which really represent forms of racism or other oppression. For example: "You speak English well" (to a third-generation Asian American, whose English is better than the commenter's). "African Americans have a lot of talent, especially in music and athletics." "I bet you have great Mexican food at home."

BOX 10.3 **National and International Perspectives on Counseling Skills**

A Practical Application of the Racial/Cultural Identity Development Model

With three doctoral students from Carlos's diversity course, Jenna Zucchi, Arianna Witgestein, and Gina Galiano, we applied the cultural identity model to describe Nelida's experience. Our observations of her positive movement while working with Allen are presented below. Reference to specific sections of their conversation are indicated in parentheses.

R/CID	Nelida
Conformity Stage	Nelida's comments at the beginning of the session suggest she may have been in the Conformity Stage before meeting Allen. She seems to have preferred and adopted dominant group attitudes and beliefs. Her "surprise" over being asked where she was from (line 1) and her family looking at her as the "American girl" (3) further suggest her identification with the dominant culture. Also, she does not reject her cultural heritage (4).
Dissonance Stage	Comments from her classmate (1) and family members (5) have conflicted Nelida. She begins to question her beliefs and attitudes and expresses concerns about herself, her identity, and her self-esteem. She feels conflicted (7, 13) because she doesn't know where she belongs (5); she is experiencing some embarrassment over her accent and a lack of clear identity or integration into either culture. She is experiencing a mixture of shame and pride in both aspects of her cultural identity. With the help of Allen, she focuses on her heritage and strengths (12–15 and in the omitted portion of the session). Information from her community genogram provides a positive foundation for this exploration.
Resistance and Immersion Stage	As the sessions progress, Nelida begins to enter the Resistance and Immersion stage. She expresses her appreciation for her grandparents and her willingness to conserve her native language to facilitate their communication (2). She also recognizes persistence and hard work as her family's and her own values (6) and feels pride (9). Nelida's progression is observed in statements such as "I feel more real, truly to myself, when I do say it in Spanish," and when she talks about her grandparents and the struggles they have had in life. Nelida begins to appreciate her culture, and it makes her "proud" to think of her grandparents' persistence and hard work when they began in the United States. In addition, she begins to appreciate herself, seeing these positive attributes in herself as well.

However, she also has conflicted negative feelings about herself, as indicated by her statement that she "never felt really good about myself" and her inability to stick up for herself. Nelida makes up excuses for the other student's question because it must have been "unintentional." Throughout the course of the sessions, Nelida seems to come to grips with her feelings of embarrassment and confusion, and seems to channel those emotions into anger.

She starts to challenge the prejudice and shame that she is feeling as a result of other people's (dominant society's) judgments. Nelida gains awareness of her resulting negative feelings and sense of being "devalued" as an issue that should have been addressed during class because it was "victimizing" and self-denigrating. She begins to focus on the pride that she has in herself and her family (10–27). She realizes that she may be handling the adjustment between integrating the two different cultures better than she thought and indicates that she will address derogatory comments when they occur. She embraces her *fuerza* (31–39). |
| **Introspection Stage** | Although it is impossible to predict the future, Nelida may draw from her experiences in the dominant and Hispanic/Latino worlds, as well as her bilingual capacity, to develop an appreciation for both cultures. Further work with Allen may focus on self-exploration and developing a new sense of identity that values and honors the unique contributions of each culture. |
| **Integrative Awareness Stage** | Further progress in counseling will help Nelida be even more proud of her Cuban and U.S. heritage. She will develop an inner sense of security and an appreciation of the unique aspects of her family's culture as well as those of U.S. culture. Empowered by her renewed cultural identity, she will have a stronger sense of self-worth and confidence, and will seek to eliminate oppression and racism. |

►The CCS as a System for Assessing Change Over Several Sessions

We have seen Nelida move, through the community genogram and the session, from CCS Levels 1 and 2 to Levels 3 and 4 with clear beginnings at Level 5. Now, let's review the CCS as it might appear in a counseling session with virtually any topic. If the client is in the denial stage, the story may be distorted, others blamed unfairly, and the client's part in the story denied. In effect, the client in *denial* (Level 1) does not deal with reality. When the client is confronted effectively, the story becomes a discussion of inconsistencies and incongruity, and we see Level 2 *bargaining and partial acceptance*—the story is changing. At *acceptance* (Level 3), the reality of the story is recognized and acknowledged, and thus storytelling is more accurate and complete. Moreover, it is possible to create *new solutions* and *transcendence* (Levels 4 and 5). When changes in thoughts, feelings, and behaviors are integrated into a new story, we see the client move into major new ways of thinking accompanied by action after the session is completed.

Virtually any problem a client presents may be assessed at one of the five levels. If your client starts with you at *denial* or *partial acceptance* (Level 1 or 2) and then moves with your help to *acceptance* and *generating new solutions* (Level 3 or 4), you have clear evidence of the effectiveness of your therapy process. The five levels may be seen as a general way to view the change process in counseling and therapy. Each confrontation or other counseling intervention in the here and now may lead to identifiable changes in client awareness.

Small changes in the session will result in larger client change over a session or series of sessions. Not only can you measure these changes over time, but you can also contract with the client in a partnership that seeks to resolve conflict, integrate discrepancies, and work through issues and problems. Specifying concrete goals often helps the client deal more effectively with confrontation.

The CCS provides you with a systematic way to evaluate the effectiveness of each intervention and to track how clients change in the here and now of the session. If you practice assessing client responses with the CCS model, eventually you will be able to make decisions automatically "on the spot" as you see how the client is responding to you. For example, if the client appears to be in denial of an issue despite your confrontation, you can intentionally shift to another microskill or approach that may be more successful.

►Conflict Resolution and Mediation: A Psychoeducational Strategy for Creating the *New*

Empathic confrontation serves as a solid base for mediation. In conflict resolution and mediation, whether between children, adolescents, or adults, the following microskills and the five-stage counseling and psychotherapy model provide a useful framework.

Empathic Relationship

Develop rapport and outline the structure of the session.   Needless to say, ensure that each person has a full opportunity to tell his or her point of view, but also seek to balance talk time. Often one person dominates. It is important to keep clients away from anger and nasty/negative comments about the other. Make this an agreement as part of structuring—and when it occurs, name it and seek to calm the discussion. Ask for honesty and real thoughts and feelings. And, drawing from attending behavior, ask each person to listen fully to the other. We have also found that it really helps if you ask each participant to paraphrase what the other has said. This may take time, but it ensures accurate listening, even when the listener does not like what has just been said.

Story and Strengths

Define the problem (concern).   Use the basic listening sequence to clearly and *concretely* draw out the point of view of each person involved in the dispute. To avoid emotional outbursts, acknowledgment of feeling rather than reflection of feeling is recommended. Summarize each person's frame of reference, and carefully check out your accuracy with each one of them. You may ask each disputant to state the opponent's point of view. Outline and summarize the points of agreement and disagreement, perhaps in written form, if the conflict is complex. In this process, include an attitude of respect and spend time on the strengths of each participant.

Goals

Use the basic listening sequence to draw out each person's wants and desires for satisfactory problem solution.   Focus primarily on concrete facts rather than emotions and abstract intangibles. This is the beginning of the negotiation process, in which problems and concerns may be redefined and clarified. Summarize the goals for each person, with attention to possible joint goals and points of agreement.

Restory

Begin negotiation in earnest.   Rely on your listening skills to see whether the parties can generate their own satisfactory solutions. When a level of concreteness and clarity has been achieved, the parties involved may come close to agreement. If the parties are very conflicted, meet each one separately as you brainstorm alternative solutions. With touchy issues, summarize them in writing. Many of the influencing skills discussed later in this book will be useful in the process of negotiation.

Action

Contract and generalize.   Use the basic listening sequence; summarize the agreed-upon solution (or parts of the solution if negotiations are still in progress). Make the solution as concrete as possible and write down touchy main issues to make sure each party understands the agreement. Obtain agreement about subsequent steps. With children, congratulate them on their hard work and ask each child to tell a friend about the resolution.

The Martin Luther King Jr. Center (1989) summarizes six steps for nonviolent change that are closely related to this mediation model: (1) information gathering, (2) education, (3) personal commitment, (4) negotiations, (5) direct action, and (6) reconciliation. When you work on complex issues of institutional or community change, a review of Dr. King's model may help you in thinking through your approach to major challenges.

▶Summary: Supportive Challenge for the Creation of the *New*

We have defined empathic confrontation as a supportive challenge used to help clients note discrepancies and incongruities in their stories. This challenge includes the following three steps:

1. Identify conflict by observing incongruities, discrepancies, and mixed messages.
2. Point out issues of incongruity and work toward resolution.
3. Evaluate the change process using the Client Change Scale.

Empathic confrontation is a set of skills that may be used in different ways. The most common confrontation uses the paraphrase, reflection of feeling, and summarization of discrepancies observed in the client, or between the client and her or his situation. However, questions and influencing skills and strategies can also lead to client change.

Confrontations will typically occur only now and then in session conversation. For example, Nelida could speak of the classroom situation where she felt alone and needed to say no more. Then, perhaps 10 or 20 conversational sequences later, she might talk about how valuable her culture and language are to her. The counselor could then confront by summarizing the two situations and adding "How do you put that together, Nelida?" Similarly, issues to be confronted may be separated by one or more sessions. Through summarization of the two, the client will see new connections and be more able to find a creative synthesis.

The CCS can be used to determine the creative effect of your confrontation. Also, it can be used with all skills and as an informal assessment of the success of your interventions over a series of sessions. Some clients will move rapidly through all five levels in one session. Most clients will move more slowly. If you work with a major grief reaction around divorce or a highly significant change such as stopping drinking, do not expect clients to respond to your confrontations rapidly, but look for client change over the time you are working with them. Action based on a new story may take time and patience.

▼▼▼▼ Key Points	
Confrontation	Clients come to us stuck and immobilized in their developmental processes. Through the use of microskills such as confrontation, we facilitate change, movement, and transformation—restorying and action. Confrontation has been defined as a supportive challenge in which you note incongruities and discrepancies and then feed back or paraphrase those discrepancies to the client. Our task is then to work through the resolution of the discrepancy.
Confrontation and change strategies	An explicit confrontation can be recognized by the model sentence, "On one hand . . . , but on the other hand. . . . How do you put those two together?" In addition, many counseling statements contain implicit confrontations that can be helpful in promoting client growth and developmental movement. For example, you may summarize client conversation, pointing out discrepancies, or use an influencing skill such as the interpretation/reframe or feedback to produce change.
The Client Change Scale	The Client Change Scale is a tool to examine the effect that microskills and confrontation have on client verbalizations immediately in the session. At the lowest level, clients may deny their incongruities; at middle levels, they may acknowledge them; at higher levels, they may transform or integrate incongruity into new stories and action.
Multicultural and individual issues	Confrontation is believed to be relevant to all clients, but it must be worded to meet individual and cultural needs. Do not expect individuals in any cultural group always to follow one pattern; avoid stereotyping; and adapt confrontation to individual and cultural differences.
The Racial/ Cultural Identity Development (R/CID) model	The R/CID model, developed by Sue and Sue (2013), includes five stages of development: conformity, dissonance, resistance and immersion, introspection, and integrative awareness. Use of the R/CID model helps the counselor understand where the client is in his or her cultural identity development and determine the role of cultural/contextual factors. Based on this knowledge, counselors can offer interventions to empower clients to embrace their cultural identity, promote life improvements, and assess progress.

▶Competency Practice Exercises and Portfolio of Competence

 Go to CengageBrain.com to access Counseling CourseMate, where you will find an interactive ebook, quizzes, videos, interactive counseling and psychotherapy exercises, the Portfolio of Competence, case studies, a practice test, and more.

This chapter is designed to help you construct a view of helping oriented toward change. If you master the cognitive concepts of the reading material and engage in deliberate practice of the exercises that follow, you will be able to promote client change and assess the effectiveness of your interventions. Again, this is an area that takes practice and experience. Practice, practice, practice, and apply the ideas here throughout the rest of your work with this book.

Individual Practice

Exercise 1. Identifying Discrepancies, Incongruity, and Mixed Messages, and Strengths Leading Toward Resolution   Please review your first session (Chapter 1) and other practice exercises completed so far. Viewing video recordings of sessions, especially your own, is an effective way to learn about conflict, discrepancies, contradictions, and confrontation. Unless you can identify incongruity in yourself, seeing it in others may be difficult or even inappropriate. The following exercise will advance your learning of this microskill.

Discrepancies internal to the self.   Can you identify specific times when your nonverbal behavior contradicted your verbal statements and gave you away? Are there times when you say two things at once and your verbal statements are incongruous? Have you done one thing while saying another?

Discrepancies between you and the external world.   Part of life is living with contradictions. Many of these are unresolvable, but they can give considerable pain. What are some of the discrepancies between you and other individuals? What are some of the mixed messages, contradictions, and incongruities you face in your world of schooling or work?

Discrepancies between you and the client.   You may have already experienced this and can easily summarize times when you felt out of tune with and discrepant from a client. Or if you have not engaged in counseling extensively, it may be helpful to think of situations in which you had major differences with someone else. Often we have typical situations that "push our buttons" and move us toward actions that are too quick. Self-awareness in this area can be most helpful.

Specific strengths.   Resolution of conflict and discrepancy is often made from a positive frame of reference. Can you identify personal strengths and wellness assets that can help you resolve internal and external differences? What strengths do you admire in others that you might like to add to your repertoire?

Exercise 2. Thinking About Your Thinking   Observing your own thoughts, feelings, and behaviors in the session promotes self-examination. Looking at how you think about your work in counseling and therapy will help you understand your decision-making process and become more effective.

First, review one of your video- or audio-recorded sessions. Ask yourself what you were thinking during the process. Think about what was going on in your mind during this time.

Second, stop the recording when a contradiction or discrepancy is noted. Write down what was happening in your mind at that moment. Use the following questions to facilitate your recollections:

What were you thinking?
What emotions or body experience do you recall?
What do you think of your behavior regarding that discrepancy or contradiction?
What did you like?
What might you want to do differently?
How did what you said affect the client's contradictory thoughts, feelings, and behaviors?

Exercise 3. Practicing Confrontation of Incongruity and Conflict

Write confrontation statements for the following situations. The model sentence "On the one hand . . . , but on the other hand . . ." provides a useful standard format for the actual confrontation, but you may also use variations such as "You say . . . but you do. . . ." Remember to follow up the confrontation with a checkout.

A client breaks eye contact, speaks slowly, and slumps in the chair while saying, "Yes, I really like the idea of getting to the library and getting the career information you suggest. Ah . . . I know it would be helpful for me."

"Yes, my family is really important to me. I like to spend a lot of time with them. When I get this big project done, I'll stop working so much and start doing what I should. Not to worry."

"My partner is good to me most of the time—this is only the second time he's hit me. I don't think we should make a big thing out of it."

"My daughter and I don't get along well. I feel that I am really trying, but she doesn't respond. Only last week I bought her a present, but she just ignored it."

Exercise 4. Practicing With the Client Change Scale

Here are some statements made by clients. Identify which of the five levels each client statement represents.

1. Denial
2. Partial examination
3. Acceptance and recognition
4. Creation of a new solution
5. Transcendence

Health issues. Look for movement from denial to new ways of taking care of one's body.

_____ I can't have a heart attack. It will never happen to me. I need to eat real food.

_____ Oh, I suppose I am overweight, but if I cut down a bit on butter and perhaps no more milk shakes, I'll be okay.

_____ I guess I can see that I need to balance my diet, but the busy life I lead won't really allow that to happen.

_____ I'm now able to cut out fats. At least that's taken care of.

_____ I've completely changed my way of doing things. I eat right—no fat at all—I exercise, and I'm even getting to like relaxation and stress management.

Career planning. Look for a movement from inaction or randomness to action.

_____ Okay, I guess I see your point. I've been released from two work-study programs because I didn't show up on time. But those were the bosses' fault. They should have made what they wanted clearer.

_____ The teacher referred me to you. Everyone has to have a job plan, but I see no need to worry about it so much. I'll be OK.

_____ Yes, I need a job plan. I can see now that is necessary. I'll write one and bring it to you tomorrow.

_____ I've got a job! The plan worked and I interviewed well and now I'm on my way.

_____ The plan has been helpful. I think I see now how to interview more effectively and present myself better.

Awareness of racism, sexism, heterosexism. Look for movement from denial that these issues exist to awareness and action.

_____ I feel committed. I've started action at home and at work, and I'm really going to concentrate on a more active approach to work against discrimination.

_____ Well, some people do discriminate, but I think that many people are just exaggerating.

_____ I don't really believe there is such a thing as racism or sexism. It's just people complaining.

_____ I've started working with my family and children on being more tolerant, fair, and understanding of people different from us.

_____ There is a fair amount of prejudice, racism, and sexism everywhere.

Exercise 5. Writing Model Confrontation Statements Review the Client Change Scale. Then read the following confrontations.

Dominic: How can she expect me to work around the house? That's women's work!

Ryan: On one hand, I hear you wanting that second income she brings in. On the other hand, you seem to want her to keep up her housework as she did in the past without any help. How do you put that together?

Dominic could respond to that confrontation by using denial, or he could work toward new ways of thinking. Can you write below example statements for Dominic representing the five levels of the Client Change Scale?

Level 1 (Denial): _____

_____

Level 2 (Partial examination): _____

_____

Level 3 (Acceptance and recognition): _____

_____

Level 4 (Creation of a new solution): _____

_____

Level 5 (Transcendence: Development of new, larger, and more inclusive constructs, patterns, or behaviors): _____

_____

Client:	I'm getting tired of talking with you. You always seem to think I'm taking the easy way out.
Counselor:	Sounds as if you're telling me that on the one hand you want to change—that's why you started counseling—but on the other, now that change is getting close, you want to leave. That seems similar to the way you handle your relationships with the opposite sex: When someone gets close, you leave. How do you respond to that?

Write statements that would represent each level of the Client Change Scale.

Exercise 6. Thinking About Your Racial/Cultural Identity Development Stage   Review Box 10.2: Cultural Identity Development. Then answer the following questions.

▲ Thinking about the R/CID model, in what stage do you place yourself?
▲ What does this means to you? In which ways might advancing your cultural development awareness help you become a better counselor or psychotherapist?
▲ What are some actions you might take to advance your cultural awareness or move higher in the cultural development model?

Group Practice

Exercise 7. Evaluation Confrontation Leads and Client Responses Using the CCS

Step 1: Divide into groups.

Step 2: Select a group leader.

Step 3: Assign roles for the first practice session.

▲ Client.
▲ Counselor.
▲ Observer 1, who will rate each client statement using the Feedback Form (Box 10.4) and, during the replay of an audio or video recording, will stop the recording after each client statement and rate it carefully.
▲ Observer 2, who will record the key words of each counselor statement on a separate sheet of paper, thus making it possible to construct a picture of the session. This observer will pay special attention to the microskill leads used by the counselor.

Step 4: Plan.   State the goal of the session. The counselor's task is to use the basic listening sequence to draw out a conflict in the client and then to confront this conflict or incongruity. The counselor will observe and note discrepancies on the spot during the session and feed them back to the client.

A useful topic for this practice session is any issue on which the volunteer client feels conflicted within or without. Internal conflict often arises around a difficult decision, past or present. External conflict most often appears when one has difficulty in dealing with a family member, a friend, or someone at work. Usually you will find both internal and external conflict in the client. Potentially useful topics include these:

▲ An important purchase
▲ Taking out a loan versus working part-time
▲ Choosing between two equally attractive majors in college
▲ A career decision involving a choice between a larger income and work that would be enjoyed more fully

BOX 10.4 Feedback Form: Confrontation Using the Client Change Scale

You can download this form from Counseling CourseMate at CengageBrain.com.

_____ (DATE)

_____   _____
(NAME OF COUNSELOR)   (NAME OF PERSON COMPLETING FORM)

Instructions: Video and/or audio recording will be necessary for the best type of feedback. Otherwise, it will be best for the observer(s) to stop the session shortly after a confrontation has occurred and then discuss what was observed. In this practice, we are seeking a review of all leads used by the counselor, but looking for confrontations. Rate how the client responded to the confrontation on the five-point Client Change Scale. 1 = Denial, 2 = Partial examination, 3 = Acceptance and recognition, 4 = Creation of a new solution, 5 = Transcendence and the creation of the _New_.

Counselor Statement (Write key words to help recollection and discussion.)	Client Comment (Write key words to help recollection and discussion.)	CCS Rating
1.		
2.		
3.		
4.		
5.		
6.		
7.		
8.		
9.		
10.		

- ▲ Moral decisions, ranging from telling the truth when one has held it back, to differences of opinion on abortion, divorce, or making a commitment, to issues of diversity or the role of spirituality in one's family
- ▲ Debating whether to go to the best party of the year or study for next day's test
- ▲ Deciding to tell your roommate that he or she needs to clean up his or her mess
- ▲ Deciding how to tell your fiancé(e) that you don't love him (her) anymore
- ▲ Confronting a friend you saw stealing a book from the library
- ▲ Virtually any other type of interpersonal conflict

Step 5: Conduct a 5-minute practice session using confrontation skills as part of your listening and observation demonstration.

Step 6: Review the practice session using confrontation skills.

Step 7: Rotate roles.

Some general reminders: The volunteer client may be asked to complete the Client Feedback Form from Chapter 1. This exercise is an attempt to integrate many of the skills and concepts used so far in this book. Allow sufficient time for thinking through and planning this practice session. Recall the potential value of the positive asset search, coupled with full awareness of wellness potential.

Portfolio of Competence

Skill in confrontation depends on your ability first to listen and then to take an active role in the helping process. This needs to be done in a nonjudgmental fashion, with respect for differences. As you work through this list of competencies, think ahead to how you would include confrontation skills in your own Portfolio of Competence.

Use the following checklist to evaluate your present level of mastery. Check those dimensions that you currently feel able to do. Those that remain unchecked can serve as future goals. Do not expect to attain intentional competence on every dimension as you work through this book. You will find, however, that you will improve your competencies with repetition and practice.

Level 1: Identification and classification

☐ Ability to identify discrepancies and incongruities manifested by a client in the session.
☐ Ability to identify client stage of R/CID development during the session.
☐ Ability to classify and write counselor statements indicating the presence or absence of elements of confrontation.
☐ Ability to identify client change processes on the Client Change Scale through observation.

Level 2: Basic competence

☐ Ability to demonstrate confrontation skills in a real or role-played session.
☐ Ability, in the here and now of the session, to observe and identify client responses on the five levels of the Client Change Scale.
☐ Ability to utilize wellness and the positive asset search to help clients find strengths that might help them move forward toward positive change when confronted.

Level 3: Intentional competence.
You will be able to use confrontational skills in such a manner that clients improve their thinking and behaving as reflected on the CCS.

- Ability to help clients change their manner of talking about a problem as a result of confrontation. This may be measured formally by the CCS or by others' observations.
- Ability to move clients from initial discussion of issues at the lower levels of the CCS to discussion at higher developmental levels at the end of the session, or when the topic has been fully explored.
- Ability to identify client responses inferred from the CCS on the spot in the session and change counseling interventions to meet those responses.

Level 4: Psychoeducational teaching competence.   Are you able to teach change and confrontation concepts to clients and to others?

The basic dimensions of confrontation are really designed more for counselors and psychotherapists than for clients. But there are some very specific ways that psychoeducation will be an important part of your practice. First, those going through the stages of grief associated with death may find it helpful to have the change stages identified for them, thus enabling them to understand their feelings and thoughts more fully. These stages of change will also be helpful in understanding reactions to serious illness, accidents, alcoholism, and traumatic incidents. Second, you can set up change goals with your clients and work with them to discover how far they have progressed in meeting goals and making life changes.

▶Determining Your Own Style and Theory: Critical Self-Reflection on Confrontation

Confrontation is based primarily on listening skills, but it does require you to move more actively in the session by highlighting discrepancies and conflict. The Client Change Scale (CCS) was presented to show that you can assess the influence of your interventions in the here and now of the session. The creative *New* provides a more philosophical dimension to confrontation and change.

What single idea stands out for you among all those presented in this chapter, in class, or through informal learning? What stands out that is likely to be a guide toward your next steps. How might confrontation relate to diversity issues? How would you use mediation as a psychoeducational treatment program? What other points in this chapter strike you as important? How might you use ideas in this chapter to begin the process of establishing your own style and theory?

Interpersonal Influencing Skills for Creative Change

The key to successful leadership is influence, not authority.

—Kenneth Blanchard

All *Intentional Interviewing and Counseling* skills and strategies are based on attending, observation, and listening. The influencing skills presented here are most useful as supplements to the first half of this text. Couple these three chapters with the preceding two on focusing and confrontation, and you have a rich array of ways to work creatively with your clients, increasing their intentionality and ability to take action on their *New* discoveries.

Through attending and listening, we are influencing the client indirectly. Influencing skills must be based on listening, but they take a more direct approach. Do not assume, however, that you, the counselor, are in charge. Clients are the "deciders"; our task is to help provide options, plus supporting and encouraging change. The influencing skills need to be used judiciously and sparingly.

Given that caveat, you will find that an egalitarian, empathic approach to influencing skills will be welcomed by most of your clients. At issue is seeing that their intentionality and creativity are fostered, rather than yours. Interestingly, to accomplish this effectively will require intentionality and creativity on your part as well.

Look for the following skills and strategies in the three chapters of this section.

Chapter 11. Reflection of Meaning and Interpretation/Reframe: Helping Clients Restory Their Lives For many of your clients, this may be the most helpful influencing skill for finding meaning and vision in life as it provides goals that can support them through many difficulties. Here you will examine the relationship between and among behaviors, thoughts, feelings, and their underlying meaning structure. These skills help you

265

gain a deeper understanding of each client's issues and history. Clients will gain valuable new perspectives on their problems and stories.

Chapter 12. Self-Disclosure and Feedback: Immediacy and Genuineness in Counseling and Therapy

These highly useful interviewing skills help ensure that the session is authentic and immediately relevant. Self-disclosure reveals your immediate experiencing of the client's world or a relevant short story about your own life. In feedback, you share your perception of the client's behaviors, thoughts, and feelings. Learning about your experience and perceptions may help clients create new stories and generalize and act on what is discovered in the session.

Chapter 13. Concrete Action Strategies for Client Change: Logical Consequences, Instruction/Psychoeducation, Stress Management, and Therapeutic Lifestyle Changes

The action influencing skills are explored with specific suggestions for facilitating client restorying and action. Here you will see a wide array of alternatives that actively involve the client in thinking and acting differently. Special attention is paid to stress management strategies and therapeutic lifestyle changes. With all of these strategies, empathic, egalitarian relationships with the client are essential.

As you develop competence in influencing skills, you may expect to develop the ability to

1. Help clients move to deeper levels of self-exploration and self-understanding using the skills of reflection of meaning and interpretation/reframe.
2. Facilitate client restorying by using self-disclosure and feedback, in the process creating a more open egalitarian relationship.
3. Use influencing skills and strategies to assist client developmental progress, particularly when the more reflective listening skills fail to produce change and understanding.

Competence in the influencing skills will further advance your intentional competence. The effective interviewer is always in process—growing and changing in response to new learning.

CHAPTER 11

Reflection of Meaning and Interpretation/ Reframe

Helping Clients Restory Their Lives

Reflection of Meaning and Interpretation/Reframe

Empathic Confrontation

Focusing

The Five-Stage Interview Structure

Reflection of Feeling

Encouraging, Paraphrasing, and Summarizing

Open and Closed Questions

Client Observation Skills

Attending Behavior and Empathy

Ethics, Multicultural Competence, and Wellness

Today we seek movies, novels, and "news stories" that put the events of the day in a form that our brains evolved to find compelling and memorable. Children crave bedtime stories; the holy books of the great religions are written in parables; and research in the cognitive sciences has shown that lawyers whose closing arguments tell a story win jury trials against their legal adversaries who just lay out "the facts of the case."

—Drew Westen

Mission of "Reflection of Meaning and Interpretation/Reframe"

Although changes in thoughts, feelings, and behaviors are usually thought of as the major goals of counseling and psychotherapy, you will find clients whose concerns are focused on a vision for life direction and seeking deeper understanding of the "meaning of it all." The mission of interpretation/reframing is to provide a new way of restorying and understanding thoughts, feelings, and behaviors, which often results in new ways of making meaning. These skills are related in that both seek to facilitate new perspectives and new ways of thinking about multiple issues.

This chapter is dedicated to Viktor Frankl. The initial stimulus for the skill of reflection of meaning came from a 2-hour meeting with him in Vienna shortly after we visited the German concentration camp, Auschwitz, where he had been imprisoned in World War II. He impressed on us the central value of meaning in counseling and therapy—a topic to which most theories give insufficient attention. It was his unusual ability to find positive meaning in the face of impossible trauma that impressed us most. His thoughts also affected our wellness and positive strengths orientation. His theoretical and practical approach to counseling and therapy deserves far more attention than it receives. We have often recommended his powerful book *Man's Search for Meaning* (1959) to our clients who face serious life crises.

Chapter Goals and Competency Objectives Awareness, knowledge, skills, and actions in reflection of meaning and interpretation/reframing will enable you to

▲ Understand reflection of meaning and interpretation/reframing and their similarities and differences.

▲ Assist clients, through reflection of meaning, to explore their deeper meanings and values and to discern their own vision, goals, or life purpose.

▲ Realize the power of perceptions. The way you perceive and think about things affects how you think, feel, and behave.

▲ Help clients, through interpretation/reframing, to find an alternative frame of reference or way of thinking that facilitates personal development.

▲ Understand how these skills bring about change that can be measured on the Client Change Scale.

 Go to CengageBrain.com to access Counseling CourseMate, where you will find an interactive ebook, quizzes, videos, interactive counseling and psychotherapy exercises, the Portfolio of Competence, case studies, a practice test, and more.

▶Introduction: Defining the Skills of Reflection of Meaning and Interpretation/Reframe

The two closely related microskills of reflecting meaning and interpretation/reframing seek to enable clients to think differently about themselves, their feelings, and their stories. It is more than just facts; it is the nature of the story and its meaning. You can be similar to the skilled lawyer, but you do not speak for them—your task is to facilitate clients' own reframing of old stories and development of new and positive stories. Interpretation/reframing and reflection of meaning are important routes toward new perceptions of the "facts."

Both interpretation/reframing and reflection of meaning seek implicit issues and meanings below the surface of client conversation. Reflecting meaning is the art of encouraging clients to find new ways of examining their lives through your in-depth listening for deeper issues and visions of the present, past, and future. Interpretation/reframing is the art of supplying clients with new perspectives, words, and ideas that they can use to create new ways of thinking and feeling and ultimately to behave differently.

From those definitions, you can see the logic of presenting the two closely related skills together. Figure 11.1 illustrates the centrality of meaning. Both interpretation/reframe and reflection of meaning may help clients find their "center of being in the world."

Eliciting and reflecting meaning are both skills and strategies. As skills, they are quite straightforward. To elicit meaning, ask the client some variation of the basic question, "What does . . . *mean* to you, your past, or your future life?" At the same time, effective exploration of meaning becomes a major strategy in which you bring out client stories, *past*, *present*, and *future*. You use all the listening, focusing, and confrontation skills to facilitate this self-examination, yet the focus remains on the client's meaning and finding purpose in his or her life.

Interpretations and reframes vary with theoretical orientation. The joint term *interpretation/ reframe* is used because both skills focus on providing a new way of thinking or a new frame of reference for the client, but the word *reframe* is a gentler construct that usually comes from your here-and-now observations. Keep in mind when you use influencing skills that interpretive statements are more directive than reflection of meaning. When we use interpretation/

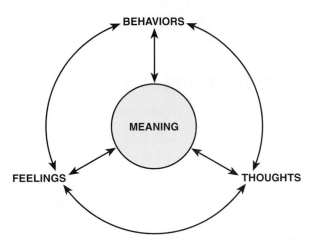

FIGURE 11.1 Pictorial representation of the relationships among behaviors, thoughts, feelings, and meanings.

© Cengage Learning

reframing, we are working primarily from the counselor's frame of reference. This is neither good nor bad; rather, it is something we need to be aware of when we use influencing skills.

If you use reflection of meaning and interpretation/reframing skills as defined here, you can *anticipate* how clients will respond.

REFLECTION OF MEANING	ANTICIPATED RESULT
Meanings are close to core client experiencing. Encourage clients to explore their own meanings and values in more depth from their own perspective. Questions to elicit meaning are often a vital first step. A reflection of meaning looks very much like a paraphrase but focuses on going beyond what the client says. Often the words *meaning, values, vision,* and *goals* appear in the discussion.	The client will discuss stories, issues, and concerns in more depth with a special emphasis on deeper meanings, values, and understandings. Clients may be enabled to discern their life goals and vision for the future.

INTERPRETATION/REFRAME	ANTICIPATED RESULT
Interpretation and reframing can provide the client with a new meaning or perspective, frame of reference, or way of thinking about issues. Interpretations/reframes may come from your observations; they may be based on varying theoretical orientations to the helping field; or they may link critical ideas together.	The client will find another perspective or way of thinking about a story, issue, or problem. The new perspective could be generated by a theory used by the counselor, by linking ideas or information, or by simply looking at the situation afresh.

The case of Charlis will serve as a way to illustrate similarities and differences between reflection of meaning and interpretation.

Charlis, a workaholic 45-year-old middle manager, has had a heart attack. After several days of intensive care, she has been moved to the floor where you, as the hospital social worker, work with the heart attack aftercare team. Charlis is motivated; she is following physician directives and progressing as rapidly as possible. She listens carefully to diet and exercise suggestions and seems to be the ideal patient with an excellent prognosis.

However, she wants to return to her high-pressure job and continue moving up through the company; you observe that she feels some fear and puzzlement about what has happened.

▶Reflection of Meaning

You recognize that Charlis is reevaluating the meaning of her life. She asks questions that are hard to answer: "Why me? What is the meaning of my life? What is God saying to me? Am I on the wrong track? What should I *really* be doing?" You sense that she feels something is missing in her life, and she also wants to reevaluate where she is going and what she is doing. How might you help Charlis? What thoughts occur to you? What do you see as the key issues that relate to the meaning and purpose of her life?

To elicit meaning, we may ask Charlis some variation of a basic meaning question, "What does the heart attack *mean* to you, your past and future life?" We may also ask her if she would like to examine the meaning of her life through the process of *discernment*, a more systematic approach to meaning and purpose defined in some detail later in this chapter. If she wishes, we can share the specific questions of discernment presented there and ask her which areas she'd like to explore. In addition, we'd ask her to think of questions and issues that are particularly important to her as we work to help her discern the meaning of her life, her work, her goals, and her mission. These questions often bring out emotions, and they certainly bring out meaning in the client's thoughts and cognitions. When clients explore meaning issues, the session becomes less precise as the client struggles with defining the almost indefinable. As appropriate to the situation, questions such as the following can address the general issue of meaning in more detail:

> "What has given you the most satisfaction in your job?"
> "What's *been missing* for you in your present life?"
> "What do you *value* in your life?"
> "What *sense* do you make of this heart attack and the future?"
> "What things in the future will be most *meaningful* to you?"
> "What is the *purpose* of your working so hard?"
> "You've said that you wonder what God is saying to you with this trial. Could you share some of your thoughts?"
> "What gift would you like to leave the world?"

Eliciting meaning often precedes reflection. Reflection of meaning as a skill looks very much like a reflection of feeling or paraphrase, but the key words *meaning, sense, deeper understanding, purpose, vision,* or some related concept will be present explicitly or implicitly. "Charlis, I sense that the heart attack has led you to question some basic understandings in your life. Is that close? If so, tell me more." Eliciting and reflecting meaning is an *opening* for the client to explore issues for which there is not a final answer but rather a deeper awareness of the possibilities of life. Both reflecting meaning and interpretation/reframing are designed to help clients look deeper, first by careful listening and then by helping clients examine themselves from a new perspective.

Reflecting meaning involves *client* direction; the interpretation/reframe implies *therapist* direction. The client provides the new and more comprehensive perspective in reflection of meaning, whereas an interpretation/reframe supplies a new way of being as suggested by the counselor.

▶Comparing Reflection of Meaning and Interpretation/Reframe

Here are brief examples of how reflection of meaning and interpretation may work for Charlis as she attempts to understand some underlying issues around her heart attack.

COUNSELOR AND CLIENT CONVERSATION	PROCESS COMMENTS
Charlis: My job has been so challenging and I really feel that pressure all the time, but I just ignored it. I'm wondering why I didn't figure out what was going on until I got this heart attack. But I just kept going on, no matter what.	Stress is something you will encounter in the majority of your clients. Part of managing stressors is learning to reframe, or think differently about the issue, concern, or larger story.

Eliciting and Reflecting Meaning

COUNSELOR AND CLIENT CONVERSATION	PROCESS COMMENTS
Counselor: I hear you—you just kept going. Could you share what it feels like *to keep going on* and what it *means* to you?	Encourager focusing on the key words "keep going on"; open question oriented to meaning. (Interchangeable empathy)
Charlis: I was raised to keep going. My mother always prided herself on doing a good job, even in the worst of times. Grandma did the same thing.	Family history is an important part of the story.
Counselor: Charlis, I hear that keeping going and persistence have been a key family value that remains very important to you. "Hanging in" is what you are good at. Could we focus now on how that value around persistence and *keeping going on* relates to your rehabilitation?	Reflection of meaning emphasizing strength. This is followed by an open question that seeks to use family strengths to help her plan for the future. (Additive)
Charlis: Hummmm. I guess we could. Now, how would I start? Can we talk about that?	The meaning of persistence has been applied to the likely need for a change of lifestyle.

Interpreting/Reframing

COUNSELOR AND CLIENT CONVERSATION	PROCESS COMMENTS
Charlis: My job has been so challenging and I really feel that pressure all the time, but I just ignored it. I'm wondering why I didn't figure out what was going on until I got this heart attack. But I just kept going on, no matter what.	Now, how would interpretations and reframes be different?

continued on next page

COUNSELOR AND CLIENT CONVERSATION	PROCESS COMMENTS
Counselor: I hear you—you just kept going. The heart attack is very scary, but it also provides you with an opportunity to look at what you've been doing and what you want to do with your life.	Interpretation. This meaning and topic come from the counselor. It is a reasonable way to reframe the heart attack in the sense that it provides an opportunity for a life review. If Charlis responds to this, then it is additive.
Charlis: Yes, I've even been thinking about that. Am I really cut out for a pressurized life? But then I have the bills to pay, and I think I really need to get back to work as soon as possible.	Charlis does respond briefly, but seems to be heading back to thinking that more work is the answer. As such, we would change the empathy rating above to interchangeable, as it allowed her to continue talking.
Counselor: So you *have* been thinking about your lifestyle, but realize how hard it would be to change. Tell me more about your thinking about lifestyle and pressure.	First, a paraphrase of the first part of Charlis's comment, while simultaneously ignoring the repetition of the drive and almost workaholic aspects. "Tell me more" represents an open question and return to the topic the counselor has selected for interpretation/reframing. (Potentially additive or subtractive, this is an example of the risk that goes with many interpretations/reframes.)
Charlis: Well, it makes no sense to continue as I have been. I guess that I really need to slow down. I come home so tired and worn out. Maybe there is some other way.	Charlis responds with more thought, and the combination of listening plus the interpretation seem to be taking hold.
Counselor: So I hear you saying that you want to change. Your family value of persistence is great, but how would it be to apply this to slowing down?	Paraphrase with an interpretative flavor that carries it a bit further than Charlis really said. Then this is followed by a reframe of how the family style of persistence could be used in a different direction. (Potentially additive)
Charlis: (pause) Well, I hear what you are saying. Let's give it a try.	The interpretation/reframes seem to have helped her move on and consider more useful life alternatives.

Both reflection of meaning and the interpretation/reframing of the situation ended up in nearly the same place, but Charlis is more in control of the process with reflection of meaning. Whichever approach is used, we are closer to helping Charlis work on the difficult questions of the meaning and direction of her future life. If the client does not respond to reflective strategies, move to the more active reframing or a theoretical interpretation. We need to give clients power and control of the session whenever possible. They can often generate new interpretations/reframes and new ways of thinking about their issues.

Linking is an important part of interpretation, although it often appears in an effective reflection of meaning as well. In linking, two or more ideas are brought together, providing the client with a new insight. The insight comes primarily from the client in reflection of meaning, but almost all from the counselor in interpretation/reframing. Consider the following four examples:

Interpretation/reframe 1.   Charlis, what stands out to me at this moment is how able you are, and we can use your "smarts" and ability to understand situations to find new, more comfortable directions. (Interpretation/reframe with a very positive perspective. Positive reframes in the here and now are often the most useful.)

Interpretation/reframe 2.   Charlis, you seem to have a pattern of thinking that goes back a long way—we could call it an "automatic thought." You seem to have a bit of perfectionism there, and you keep saying to yourself (self-talk), "Keep going no matter what." (Cognitive behavioral theory; links the past to the present perfectionism.)

Interpretation/reframe 3.   It sounds as if you are using hard work as a way to avoid looking at yourself. The avoidance is similar to the way you avoid dealing with what you think you need to change in the future to keep yourself healthier. (Combines confrontation with linking to what is occurring in the interview series, but with a less positive view.)

Interpretation/reframe 4.   The heart attack almost sounds like unconscious self-punishment, as if you wanted it to happen to give you time off from the job and a chance to reassess your life. (Interpretation from a psychodynamic perspective focusing first on negative issues [self-punishment], then on new opportunities that are now open [reassess your life]).

In summary, a reflection of meaning looks very much like a paraphrase but focuses beyond what the client says. Often the words *meaning*, *values*, and *goals* will appear in the discussion. Clients are encouraged to explore their own meanings in more depth from their own perspective. Questioning and eliciting meaning are often vital first steps.

Interpretations/reframes provide the client with a new perspective, frame of reference, or way of thinking about issues. They may come from observations of the counselor, they may be based on varying theoretical orientations to the helping field, or they may link critical ideas together.

The two skills are similar in helping clients generate a new and potentially more helpful way of looking at things. Reflection of meaning focuses on the client's worldview and seeks to understand what motivates the client; it provides more clarity on values and deeper life meanings. An interpretation/reframe results from counselor observation and seeks new and more useful ways of thinking.

▶Example Counseling Session: Travis Explores the Meaning of a Recent Divorce

In the following session, Travis is reflecting on his recent divorce. When relationships end, the thoughts, feelings, and underlying meaning of the other person and the time together often remain an unsolved mystery. Moreover, some clients are likely to repeat the same mistakes in their relationships when they meet a new person.

However, both the interpretation/reframe and reflection of meaning are central skills in helping clients take a new perspective on themselves and their world. Terrell, the counselor, seeks to help Travis think about the word *relationship* and its meaning. Note that Travis stresses the importance of connectedness with intimacy and caring. The issue of self-in-relation to others will play itself out very differently among individuals in varying cultural contexts. Many clients will focus on their need for independence.

COUNSELOR AND CLIENT CONVERSATION	PROCESS COMMENTS
1. *Terrell*: So, Travis, you're thinking about the divorce again . . .	Encourager/restatement. (Interchangeable empathy)
2. *Travis*: Yeah, that divorce has really thrown me for a loop. I really cared a lot about Ashley and . . . ah . . . we got along well together. But there was something missing.	Travis goes into more depth on the issue as anticipated when you use an encourager/restatement.
3. *Terrell*: Uh-huh . . . something missing?	Encouragers appear to be closely related to meaning. Clients often supply the meaning of their key words if you repeat them back exactly. (Interchangeable)
4. *Travis*: Uh-huh, we just never really shared something very basic. The relationship didn't have enough depth to go anywhere. We liked each other, we amused one another, but beyond that . . . I don't know . . .	Travis elaborates on the meaning of a closer, more significant relationship than he had with Ashley. (Travis appears to start this session primarily at Levels 2 and 3 on the Client Change Scale. He is aware of what is going on, but is still very puzzled.)
5. *Terrell*: You amused each other, but you wanted more depth. What sense do you make of it?	Paraphrase using Travis's key words followed by a question to elicit meaning. (Interchangeable empathy)
6. *Travis*: Well, in a way, it seems like the relationship was shallow. When we got married, there just wasn't enough depth for a meaningful relationship. The sex was good, but after a while; I even got bored with that. We just didn't talk much. I needed more . . .	Note that Travis's personal constructs for discussing his past relationship center on the word *shallow* and the contrast *meaningful*. This polarity is probably one of Travis's significant meanings around which he organizes much of his experience.
7. *Terrell*: Mm-hmmm . . . you seem to be talking in terms of shallow versus meaningful relationships. What does a meaningful relationship feel like to you?	Reflection of meaning followed by a question designed to elicit further exploration of meaning. (Potentially additive)
8. *Travis*: Well, I guess . . . ah . . . that's a good question. I guess for me, there has to be some real, you know, some real caring beyond just on a daily basis. It has to be something that goes right to the soul. You know, you're really connected to your partner in a very powerful way.	Connection appears to be a central dimension of meaning. Travis is exploring his meanings in more depth.
9. *Terrell*: So, connections, soul, deeper aspects strike you as really important.	Reflection of meaning. Note that this reflection is also very close to a paraphrase, and Terrell uses Travis's main words. The distinction centers on issues of meaning. A reflection of meaning could be described as a special type of paraphrase. (Interchangeable)

COUNSELOR AND CLIENT CONVERSATION	PROCESS COMMENTS
10. *Travis*: That's right. There has to be some reason for me to really want to stay married, and I think with her . . . ah . . . those connections and that depth were missing. We liked each other, you know, but when one of us was gone, it just didn't seem to matter whether we were here or there.	Here we are beginning to see movement on the Client Change Scale. Travis appears to be thinking more completely at Level 3. He is understanding his situation fairly well, but we see no real change.
11. *Terrell*: So there are some really good feelings about a meaningful relationship even when the other person is not there. You didn't value each other that much.	Reflection of meaning plus some reflection of feeling. Note that Terrell has added the word *values* to the discussion. In reflection of meaning it is likely that the counselor or therapist will add words such as *meaning*, *understanding*, *sense*, and *value*. Such words lead the client to make sense of experience from the client's own frame of reference. (Potentially additive)
12. *Travis*: Uh-huh.	
13. *Terrell*: Ah . . . could you fantasize how you might play out those thoughts, feelings, and meanings in another relationship?	Open question oriented to meaning. (Potentially additive. We seldom really know how effective a question is until we see the client response.)
14. *Travis*: Well, I guess it's important for me to have some independence from a person, but when we were apart, we'd still be thinking of one another. Depth and a soul mate is what I want.	Travis's meaning and desire for a relationship are now being more fully explored. Travis moves a bit further as he defines his life goals more precisely. This could be called a Level 3+ response.
15. *Terrell*: Um-hum.	Encourager.
16. *Travis*: In other words, I don't want a relationship where we always tag along together. The opposite of that is where you don't care enough whether you are together or not. That isn't intimate enough. I really want intimacy in a marriage. My fantasy is to have a very independent partner I care about and who cares about me. We can both be individuals but still have bonding and connectedness.	Connectedness is an important meaning issue for Travis. With other clients, independence and autonomy may be the issue. With still others, the meaning in a relationship may be a balance of the two. We see further progress on the Client Change Scale, but Level 4 (generation of a new solution) clearly has not been achieved yet.
17. *Terrell*: Let's see if I can put together what you're saying. The key words seem to be independence with intimacy and caring. It's these concepts that can produce bonding and connectedness, as you say, whether you are together or not.	This reflection of meaning becomes almost a summarization of meaning. Note that the key words and constructs have come from the client in response to questions about meaning and value. Level 4 on the Client Change Scale will occur when Travis truly incorporates new meanings and acts on them. (Additive)

Further counseling would aim to bring behavior or action into accord with thoughts. Other past or current relationships could be explored further to see how well the client's behaviors or actions illustrate or do not illustrate expressed meaning.

As we look at this session, we see that the counselor used listening skills and key word encouragers to focus on meaning issues. The counselor's open questions oriented to values and to meaning are often effective in eliciting client talk about meaning issues. You may also note that a reflection of meaning looks very much like a paraphrase except that the focus is on implicit deeper issues, often not expressed fully in the surface language or behavior of the client.

▶Concepts Into Action: The Specific Skills of Eliciting and Reflecting Meaning

Meaning issues often become prominent after a person has experienced a serious illness (AIDS, cancer, heart attack, loss of sight), encountered a life-changing experience (death of a significant other, divorce, loss of a job), or gone through serious trauma (war, rape, abuse, suicide of a child). Issues of meaning are also prominent among older clients who face major changes in their lives. These situations cannot be changed; they are a permanent part of the life experience. Box 11.1 presents national and international perspectives on finding meaning in a serious illness.

| BOX 11-1 | National and International Perspectives on Counseling Skills |

What Can You Gain From Counseling Persons With AIDS and Serious Health Issues?

Weijun Zhang

A good friend of mine had just started working with persons with AIDS. When I asked him, "What does this mean to you?" he started to grumble: "It's being around people with serious illness who could die at any time. It also means that no one gets cured, despite my best efforts." What a bleak picture he painted. No wonder some counselors are reluctant to work with AIDS clients or those facing truly serious health issues. Can something as miserable and difficult as AIDS be meaningful? How can one work as a counselor in a kidney dialysis unit? What about working in a hospice where all are expected to die within a reasonably short time? Certainly, there is much to learn and profit from in this type of work. What my friend was missing is that there are precious rewards available.

Years ago, I happened on an article outlining the positive aspects of working with clients who face extremely difficult futures or death.

1. We can help clients appreciate each moment. Working with people who have no choice but to live in the present can change our perspective. It can help us find a deeper meaning in life and learn to watch

the world in wonder and appreciation as if this were our last day.

2. It is possible to learn something from each patient or client if we are willing to listen and be with his or her experience. Having direct contact with clients who face unthinkable pain and suffering but who still strive to live fully can help us understand the great strength that is the human spirit. This courage in the face of adversity can be very contagious.

3. Doing AIDS work or counseling the seriously ill will enable us to witness a lot of truly unconditional love and help us to become bigger-hearted, more loving, and caring persons.

I have always liked doing work with the seriously ill because of the selfless giving needed on the part of the counselor. But I have learned that it can also mean a tremendous taking, learning, and satisfaction for those who help others most in need. Though this taking should never be our primary motive in doing AIDS or related work, it is certainly rewarding to see the light when we deal with the darkness associated with major life challenges. This positive insight I wish to share with my good friends.

Reflecting meaning can also help clients work through issues of daily life. We saw in the sample session how Travis gained understanding of himself as he reflected on the meaning of divorce. Everyday issues of life and many of our typical concerns can be resolved if we turn to serious examination of meaning, values, and life purpose. Religion and spiritual life provide a value base and can be a continuing source of strength and clarity.

Eliciting Client Meaning

Understanding the client is the essential first step. Consider storytelling as a useful way to discover the background of a client's meaning-making. If a major life event is critical, illustrative stories can form the basis for exploration of meaning. Clients do not often volunteer meaning issues, even though these may be central to the clients' concerns. Critical life events such as illness, loss of a parent or loved one, accident, or divorce often force people to encounter deeper meaning issues. If spiritual issues come to the fore, draw out one or two concrete example stories of the client's religious heritage. Through the basic listening sequence and careful attending, you may observe the behaviors, thoughts, and feelings that express client meaning.

Here are some useful questions for eliciting stories and client meaning systems, adapted from the original work of Mary Fukuyama (1990, p. 9):

"When in your life did you have existential or meaning questions? How have you resolved these issues thus far?"

"What significant life events have shaped your beliefs about life?"

"What are your earliest childhood memories as you first identified your ethnic-cultural background? Your spirituality?"

"What are your earliest memories of church, synagogue, mosque, a higher power, or lack of religion?"

"Where are you now in your life journey? Your spiritual journey?"

Reflecting Client Meanings

Say back to clients their exact key meaning and value words. Reflect their own meaning system, not yours. Implicit meanings will become clear through your careful listening and questions designed to elicit meaning issues from the client. Using the client's key words is preferable, but occasionally you may supply the needed meaning word yourself. When you do so, carefully check that the word(s) you use feel right to the client. Simply change "You feel . . ." to "You mean. . . ." A reflection of meaning is structured similarly to a paraphrase or reflection of feeling. "You value . . . ," "You care . . . ," "Your reasons are . . . ," or "Your intention was. . . ." Distinguishing among a reflection of meaning, a paraphrase, and a reflection of feeling can be difficult. Often the skilled counselor will blend the three skills together. For practice, however, it is useful to separate out meaning responses and develop an understanding of their import and power in the session. Noting the key words that relate to meaning (*meaning, value, reasons, intent, cause,* and the like) will help distinguish reflection of meaning from other skills.

Reflection of meaning becomes more complicated when meanings or values conflict. Here concepts of confrontation (Chapter 10) may be useful. Conflicting values, either explicit or implicit, may underlie mixed and confused feelings expressed by the client. For instance, a client may feel forced to choose between loyalty to family and loyalty to spouse. Underlying love for both may be complicated by a value of dependence fostered by the family and the independence represented by the spouse. When clients make important decisions, helping them sort out key meaning issues may be even more important than the many other issues that affect the decision.

For example, a young person may be experiencing a value conflict over career choice. Spiritual meanings may conflict with the work setting. The facts may be paraphrased accurately and the feelings about each choice duly noted, yet the underlying *meaning* of the choice may be most important. The counselor can ask, "What does each choice mean for you? What sense do you make of each?" The client's answers provide the opportunity for the counselor to reflect back the meaning, eventually leading to a decision that involves not only facts and feelings but also values and meaning. And, as in confrontation, you can evaluate client change in meaning systems using the Client Change Scale in Chapter 10.

Discernment: Identifying Life Mission and Goals

Listen. Listen, with intention, with love, with the "ear of the heart." Listen not only cerebrally with the intellect, but with the whole of feelings, our emotions, imaginations, and ourselves.

—Esther de Waal

Discernment is "sifting through our interior and exterior experiences to determine their origin" (Farnham, Gill, McLean, & Ward, 1991, p. 23). The word *discernment* comes from the Latin *discernere*, which means "to separate," "to determine," "to sort out." In a spiritual or religious sense, discernment means identifying when the spirit is at work in a situation— the spirit of God or some other spirit. The discernment process is important for all clients, regardless of their spiritual or religious orientation or lack thereof. Discernment has broad applications to counseling and psychotherapy; it describes what we do when we work with clients at deeper levels of meaning. Discernment is also a process whereby clients can focus on visioning their future as a journey into meaning.

"There is but one truly serious problem, and that is . . . judging whether or not life is worth living" (Camus, 1955, p. 3). Viktor Frankl (1978) wrote about *The Unheard Cry for Meaning*. Frankl claimed that 85% of people who successfully committed suicide saw life as meaningless, and he blamed this perception on an excessive focus on self. He said that people have a need for transcendence and living beyond themselves.

The vision quest, often associated with the Native American Indian, Dene, and Australian traditions, is oriented to helping youth and others find purpose and meaning in their lives. These individuals often undertake a serious outdoor experience to find or envision their central life goals. Meditation is used in some cultures to help members find meaning and direction.

You will not go through the formal discernment process with many clients, only those who are truly interested in understanding themselves and their potential to do good or obtaining a deeper spiritual sense. Relatively healthy clients who may be troubled or depressed over their lives or the world situation or the "meaninglessness of life" can be helped through discerning what is important to them.

Specific discernment questions leading to examination of goals, values, and meaning can be found in Box 11.2. Share the list of questions and encourage the client to participate with you in deciding which questions and issues are most important. You can draw on these questions in any client session, as they will typically bring new and helpful ways of thinking about issues.

Multicultural Issues and Reflection of Meaning

For practical multicultural counseling and psychotherapy, recall the concept of focus. When helping clients make meaning, focus exploration of meaning not just on the individual but also on the broader life context. In much of Western society, we tend to assume that the individual is the person who makes meaning. With most of your clients, you will be helping them find meanings and determine their individual life goals, but cultural context remains important.

In many other cultures—for example, the traditional Muslim world—the individual will make meaning in accord with the extended family, the neighborhood, and religion. If individual meaning is not in accord with cultural beliefs, making that meaning work in daily life will present major challenges to the client. An African American or Latina/o client

You may find it helpful to share this list with the client before you begin the discernment process and identify together the most helpful questions to explore. In addition, we encourage adding topics and questions that occur to you and the client. Discernment is a very personal exploration of meaning, and the more the client participates, the more useful it is likely to be. Questions that focus on the here and now and intuition may facilitate deeper discovery.

Following is a systematic approach to discernment. You or your client may wish to begin by first thinking quietly about what might give life purpose, meaning, and vision. Here-and-now body experience and imaging can serve as a physical foundation for intuition and discernment.

▲ Relax, explore your body, and find a positive feeling of strength that might serve as an anchor for your search. Allow yourself to build on that feeling and see where it goes.

▲ Sit quietly and allow an image (visual, auditory, kinesthetic) to build.

▲ What is your gut feeling? What are your instincts? Get in touch with your body.

▲ Discerning one's mission cannot be found solely through the intellect. What feelings occur to you at this moment?

▲ Can you recall feelings and thoughts from your childhood that might lead to a sense of direction now?

▲ What is your felt body sense of spirituality, mission, and life goal?

Concrete questions leading to telling stories can be helpful.

▲ Tell me a story about that image above. Or a story about any of the *here-and-now* experiences listed there.

▲ Can you tell me a story that relates to your goals/ vision/mission?

▲ Can you name the feelings you have in relation to your desires?

▲ What have you done in the past or are you doing presently that feels especially satisfying and close to your mission?

▲ What are some blocks and impediments to your mission? What holds you back?

▲ Can you tell about spiritual stories that have influenced you?

For self-reflective exploration, the following are often useful.

▲ Let's go back to that original image and/or the story that goes with it. As you reflect on that experience or story, what occurs for you?

▲ Looking back on your life, what have been some of the major satisfactions? Dissatisfactions?

▲ What have you done right?

▲ What have been the peak moments and experiences of your life?

▲ What might you change if you were to face that situation again?

▲ Do you have a sense of obligation that impels you toward this vision?

▲ Most of us have multiple emotions as we face major challenges such as this. What are some of these feelings, and what impact are they having on you?

▲ Are you motivated by love/zeal/a sense of morality?

▲ What are your life goals?

▲ Can you see some specific examples of these goals?

▲ What do you see as your mission in life?

▲ What does spirituality mean to you?

The following questions place the client in larger systems and relationships—the self-in-relation. They may also bring multicultural issues into the discussion of meaning.

▲ Place your previously presented experiences and images in broader context. How have various systems (family, friends, community, culture, spirituality, and significant others) related to these experiences? Think of yourself as a self-in-relation, a person-in-community.

▲ *Family.* What do you learn from your parents, grandparents, and siblings that might be helpful in your discernment process? Are they models for you that you might want to follow, or even oppose? If you now have your own family, what do you learn from them, and what is the implication of your discernment for them?

▲ *Friends.* What do you learn from friends? How important are relationships to you? Recall important developmental experiences you have had with peer groups. What do you learn from them?

▲ *Community.* What people have influenced you and perhaps serve as role models? What group activities in your community may have influenced you? What would you like to do to improve your community? What important school experiences do you recall?

▲ *Cultural groupings.* What is the place of your ethnicity/race in discernment? Gender? Sexual orientation? Physical ability? Language? Socioeconomic background? Age? Life experience with trauma?

(continued)

BOX 11.2 (continued)

▲ *Significant other(s).* Who is your significant other? What does he or she mean to you? How does this person (or persons) relate to the discernment process? What occurs to you as the gifts of relationship? The challenges?

▲ *Spiritual.* How might you want to serve? How committed are you? What is your relationship to

spirituality and religion? What does your holy book say to you about this process?

Discernment questions are from Ivey, A., Ivey, M., Myers, J., & Sweeney, T. (2005). *Developmental Counseling and Therapy: Promoting Wellness Over the Lifespan.* Boston: Lahaska/Houghton Mifflin. Reprinted by permission.

will often feel more comfortable if meanings are made in a broader context. In contrast, many European Americans focus on individual meaning with minimal attention to broader issues—the individualistic culture of the United States.

Cultural, ethnic, religious, and gender groups all have systems of meaning that give an individual a sense of coherence and connection with others. Muslims draw on the teachings of the Qur'an. Similarly, Jewish, Buddhist, Christian, and other religious groups draw on their writings, scriptures, and traditions. African Americans may draw on the strengths of Malcolm X or Martin Luther King Jr., or on support they receive from Black churches as they deal with difficult situations. Women, who are often more relational than men, may make meaning out of relationships, whereas men may focus more on issues of personal autonomy.

You may counsel clients who have experienced some form of religious bias or persecution. As religion plays such an important part in many people's lives, members of dominant religions in a region or a nation may have different experiences from those who follow minority religions. For example, Christianity is privileged in North America, where people of Jewish and other faiths may feel uncomfortable, even unwelcome, during Christian holidays (Blumenfeld, Joshi, & Fairchild, 2009). Anti-Semitism, anti-Islamism, anti–liberal Christianity, anti–evangelical Christianity are all possible results when clients experience spiritual and/or religious intolerance. Christians and other religious groups in countries where they are a minority can suffer serious religious persecution—to the point of death.

▶ Frankl's Logotherapy: Making Meaning Under Extreme Stress

During the Holocaust, Viktor Frankl helped many Jews survive and find meaning while imprisoned in the German concentration camp at Auschwitz:

> Then I spoke of the many opportunities of giving life a meaning. I told my comrades . . . that human life, under any circumstances has meaning. . . . I said that someone looks down on each of us in difficult hours—a friend, a wife, somebody alive or dead, or a God—and He would not expect us to disappoint him. . . . I saw the miserable figures of my friends limping toward me to thank me with tears in their eyes.

If one person were to be identified with meaning and the therapeutic process, that individual would have to be Viktor Frankl, the originator of logotherapy. Frankl (1959) has pointed out the importance of a life philosophy that enables us to transcend suffering and find meaning in our existence. He argues that our greatest human need is for a core of meaning and purpose in life.

Frankl, a survivor of the German concentration camp at Auschwitz, could not change his life situation, but he was able draw on important strengths of his Jewish tradition to change the meaning he made of it. The Jewish tradition of serving others facilitated his survival. When times were particularly bad and prisoners had been whipped and starved, Frankl (1959, pp. 131–133) counseled his entire barracks, helping them reframe their terrors and difficulties, pointing out that they were developing strengths for the future.

Shortly after his liberation, Frankl wrote his famous book *Man's Search for Meaning* (1959) within a 3-week period. This short, emotionally powerful book has remained a constant best seller since that time. Frankl believed that finding positive meanings in the depth of despair was vital to keeping him alive. During the darkest moments, he would focus his attention on his wife and the good things they enjoyed together; or in the middle of extreme hunger, he would meditate on a beautiful sunset.

Mary and Allen spent 2 hours with Dr. Frankl after their lecture tour to Poland, which included a visit to Auschwitz. There we saw the gas chambers and the ovens designed specifically for the complete extermination of Jews, Gypsies, and the handicapped. Unnumbered communists, gays and lesbians, and Polish men, women, and children were also targeted. Frankl shared again the importance of positive meaning for survival. He quoted the German philosopher Nietzsche: "He who has a *why* will find a *how*."

If your clients can find a meaningful vision and life direction (the *why*), they often will bear many difficult things as they seek ways to resolve their issues and continue life. Also memorable is Frankl's comment "The best of us did not survive." It was an incredible experience to be in the presence of the man who was the real forerunner of the cognitive behavioral movement (Mahoney & Freeman, 1985). Frankl was fully aware that meaning in itself is not enough—we also must *act* on our meaning and value system.

We have given *Man's Search for Meaning* to many clients facing a real crisis. We recommend that you read it while studying this chapter. It will make a difference to you and to those clients who face real-life challenges. As you can see, Frankl was influential in our focus on wellness and the positive asset search. The recent trends toward a positive psychology and wellness are other examples of how an emphasis on strengths can aid the client.

Logotherapists search for positive meanings that underlie behavior, thought, and action. Dereflection and modification of underlying attitudes are specific techniques that logotherapy uses to uncover meaning and facilitate new actions. Many clients "hyperreflect" (think about something too much) on the negative meaning of events in their lives and may overeat, drink to excess, or wallow in depression. They are constantly attributing a negative meaning to life. When clients focus solely on negatives, dereflection helps to uncover deeper meanings and enables clients to become more positive in outlook.

The direct reflection of meaning may encourage such clients to continue their negative thoughts and behavior patterns. Dereflection, by contrast, seeks to help them discover the values that lie deeper in themselves. This strategy is similar to positive reframing/interpretation, but the client, rather than the counselor, does much of the positive thinking. The goal is to enable clients to think of things other than the negative issue and to find alternative positive meanings in the same event. The questions listed on page 279 represent first steps in helping clients dereflect and change their attitudes. The following abbreviated example illustrates this approach.

Client: I really feel at a loss. Nothing in my life makes sense right now.

Counselor: I understand that—we've talked about the issues with your partner and how sad you are. Let's shift just a bit. Could you tell me about what has been meaningful and important to you in the past? (The client shares some key supportive religious experiences from the past. The counselor draws out the stories and listens carefully.)

Counselor: (reflecting meaning) So, you found considerable meaning and value in worship and time spent quietly. You also found worth in service in the church. You drifted away because of your partner's lack of interest. And now you feel you betrayed some of your basic values. Where does this lead you in terms of a meaningful way to handle some of your present concerns?

As you may note, the process of dereflection is a special form of the positive asset search. But rather than focusing just on the concretes (spirituality, service to others, walking in the outdoors, enjoying one's friends), the counselor explores the positive meaning of these specifics. "What does spirituality mean to you?" "What sense do you make of a person who finds such joy in walking outdoors and enjoying sunsets?" "What values do you find in service to others?" Out of the exploration of meaning may come data for restorying one's problems and even life-transforming actions.

But Frankl was interested in more than just meaning. He would also discuss specific actions that the client could take in the here and now of daily life. Meaning without implementation and action is not enough. His emphasis on action beyond thinking new thoughts was pathbreaking and innovative.

Box 11.3 presents relevant research regarding reflection of meaning.

| BOX 11.3 | Research and Related Evidence That You Can Use |

Reflection of Meaning and Reframing

Ratey (2008a, p. 41), a leader in neuroscience applications and research, has commented:

> You have to find the right mission, you have to find something that's organic, that's growing, that keeps you focused on and continues to provide meaning and growth and development for yourself.
>
> I see meaning as a big part of neuroscience. We start with neuroscience and now we're talking about transcendence. Spirituality even lights up key centers in the brain. Meaning drives the lower centers and is connected to emotions and motivation areas. It's a huge, huge, human construct that means so much to our race and our species. Obviously it involves memory and learning and remembering the good stuff, remembering what your goals are, remembering what you want to do, and so you need all those things working well to keep you on the right meaning path. If you can get people into a situation where they have the meaning direction provided by their mission or their job or their goal, they don't need medicine.

Carl Rogers brought meaning issues to center stage as part of his work on reflecting feelings. Viktor Frankl provided both philosophical and practical applications of meaning in counseling. A solid relationship with your client helps give meaning to your encounter.

Classic research by Fiedler (1950) and Barrett-Lennard (1962) set the stage for the present when they found that relationship variables (closely related to the listening skills) were vital to the success of all forms of all counseling and therapy, regardless of theory. Now the relationship issues are termed "common factors," and

the idea of relationship as central has become almost universally accepted. Those therapists and researchers working in the "Heart and Soul of Change Project" cite data suggesting that 30% or more of successful therapy is based on relationship (Miller, Duncan, & Hubble, 2005).

Research has demonstrated that "families that seek support and try to accept what happened after a traumatic injury may experience less injury-related stress and family dysfunction over time" (Wade et al., 2001, p. 412). Turning to religion was the second most used strategy among parents whose children suffered traumatic injury. Connectedness with others and the comfort of spirituality can be a most important positive asset and wellness strength for many clients. A classic and often cited study by Probst (1996) found that religiously oriented clients do better in cognitive behavioral therapy when their spirituality becomes part of the process. Recovery from heart surgery has been found to be more rapid among those with religious involvement, particularly among women (Contrada et al., 2004, p. 227).

Lucas (2007/2008) examined experiences of 19 caregivers and teachers working with traumatized children. She found that learning coping strategies of reframing and realistic goal setting helped them reduce emotional exhaustion and increased their personal sense of accomplishment. Li and Lambert (2008) found positive reframing to be one of the best predictors of job satisfaction among 102 intensive care nurses from the People's Republic of China.

Neuroscience and Meaning

At the surface, the broad idea of meaning would appear to be beyond measurement in a physical sense. Our sense of meaning brings our thoughts, feelings, and behavior into a whole, enabling us to make sense of our experience. A useful exploration of the brain and its relation to meaning is provided by Carter (1999, p. 197):

> Meaningfulness is inextricably bound up with emotion. Depression is marked by wide-ranging symptoms, but the cardinal feature of it is the draining of meaning from life. . . . By contrast, those in a state of mania see life as a gloriously ordered, integrated whole. Everything seems to be connected and the smallest events are bathed in meaning.

Creation of new ideas also means that new neural networks are formed in the brain and long-term memory.

Ratey (2008) indicates that there is a key moral and spiritual dimension in the brain that we are close to identifying. Stimulation of a portion of the brain appears to bring out spiritual images in many people. Morality may be partially hardwired. *The Political Brain* (Westen, 2007) follows this logic. Westen speaks of how candidates directly affect the mirror neurons of the public, creating empathy and changing neural connections. Morality as described by neuroscientists is awareness of the Other. An interesting challenge in brain science is explaining the individualist mind versus the collectivist mind. Gene expression is clearly part of this, but gene expressions often require environmental events before they are triggered. Some genes may lie dormant throughout a lifetime.

▶The Skills of Interpretation/Reframing

You can't connect the dots looking forward; you can only connect them looking backward. So you have to trust that the dots will connect in your future. You have to trust in something—your gut, destiny, life, karma, whatever. This approach will never let you down and it has made all the difference in my life.

—Steve Jobs

When you use the microskill of interpretation/reframing, you are helping the client to restory or look at the problem or concern from a new, more useful perspective. This new way of thinking is central to the restorying and action process. In the microskills hierarchy, the words *interpretation* and *reframe* are used interchangeably. Interpretation reveals new perspectives and new ways of thinking beneath what a client says or does. The reframe provides another frame of reference for considering problems or issues. And eventually the client's story may be reconsidered and rewritten as well.

The basic skill of interpretation/reframing may be defined as follows:

▲ The counselor listens to the client story, issue, or problem and learns how the client makes sense of, thinks about, or interprets the story or issue.

▲ The counselor may draw from personal experience and/or observation of the client (reframe) or may use a theoretical perspective, thus providing an alternative meaning or interpretation of the narrative. This may include *linking* together information or ideas discussed earlier that relate to each other. Linking is particularly important as it integrates ideas and feelings for clients and frees them to develop new approaches to their issues.

▲ (Positive reframe from personal experience) "You feel that coming out as gay led you to lose your job, and you blame yourself for not keeping quiet. Maybe you just really needed to become who you are. You seem more confident and sure of yourself. It will take time, but I see you growing through this difficult situation." Here self-blame has been reinterpreted or reframed as a positive step in the long run.

▲ (Psychoanalytic interpretation with multicultural awareness) "It sounds like the guy who fired you is insecure about anyone who is different from him. He sounds as if he is projecting his own unconscious insecurities on you, rather than looking at his own heterosexism or homophobia."

Consider an example of interpretation/reframe developed from the logic of the counselor. Allen, the client, was going through a divorce and was very angry—a common reaction for those engaged in a major breakup, particularly when finances are involved. He was telling his attorney, at some length, about what he wanted and why. Attorneys use a form of interviewing involving many questions, and it sometimes involves informal counseling. After listening carefully to Allen's issues and acknowledging his strong feeling (acknowledging, not reflecting), the attorney got out from behind the desk and stood over Allen saying: "Allen, that's your story. But I can tell you that you won't get what you want. Your wife has a story as well, and what will happen is something between what you both feel you need and deserve. For your own and your children's sake, think about that." This was a rather rough and confrontative reframing of Allen's story. It also changed the focus from Allen and his problems to his wife and children. Fortunately, he heard this powerful reframe, and resolution of differences in the divorce finally began.

This story has several implications. First, even with the most effective listening, clients may still hold onto unworkable stories, ineffective thinking, and self-defeating behaviors. Clearly, they need a new perspective. Respect clients' frame of reference before interpreting or reframing their words and life in new ways. In effect, *listen before you provide your interpretation or reframe.* There will always be some clients who will need the strong, confrontative interpretation that Allen got, but recall that the attorney first listened attentively to Allen.

We may also consider the interpretation/reframe as the creation of the New because we and the client are building another way to think about issues—and ultimately create a more effective and happy self. The value of an interpretation or reframe depends on the client's reaction to it and how he or she changes thoughts, feelings, or behaviors. Think of the Client Change Scale (CCS)—how does the client react to each interpretation? If the client denies or ignores the interpretation, you obviously are working with denial (Level 1 on the CCS). If the client explores the interpretation/reframe and makes some gain, you have moved that client to bargaining and partial understanding (Level 2 on the CCS). Interchangeable responses and acceptance of the interpretation (Level 3) will often be an important part of the gradual growth toward a new understanding of self and situation. If the client develops useful new ways of thinking and behaving (Level 4 on CCS), movement is clearly occurring. Transcendence, perhaps the ultimate creation of the New (Level 5), will appear only with major breakthroughs that change the direction of counseling and psychotherapy. But let us recall that movement from denial (Level 1) to partial consideration of issues (Level 2) may be a major breakthrough, beginning client improvement.

The potential power of the effective interpretation/reframe can be seen in the divorce example above. Allen was in denial about what he could "win" in the divorce and refused even to bargain. But confronted by the attorney towering over him, he moved almost immediately from denial (Level 1) to a new understanding (Level 3) by accepting the attorney's reframe. The real test of change would be whether *he changes his behavior as a result of his new insights.* New solutions (Level 4) are never reached without behavior change. Transcendence (Level 5) is rarely found in complex cases of divorce!

▶Interpretation/Reframing and Other Microskills

Focusing, like reflection of meaning and interpretation/reframing, is another influencing skill that greatly facilitates the creation of new client perspectives. In the story of Allen and his attorney, the focus on the wife and her needs was key to the successful reframe. As another example, you may work with a client who feels that he or she has been subjected to gender discrimination or sexual harassment. If you just focus on the individual, the client may blame himself or herself for the problem. By focusing on gender or other multicultural issues, you are expanding client perspectives, and these clients may generate a new perspective, meaning, or way of solving the problem on their own—again, the creation of the New.

Interpretation may be contrasted with the paraphrase, reflection of feeling, focusing, and reflection of meaning. In those skills, the counselor remains in the client's frame of reference, and effective listening often enables creation of the New. In interpretation/reframing, the frame of reference comes from the counselor's personal and/or theoretical constructs.

The following are examples of interpretation/reframing paired with other skills.

Annaliese: (with a low self-concept)	I just feel so bad about myself. I don't feel that I'm performing at work. I think the boss is going to be down on me pretty soon.
Counselor:	(*Paraphrase/restatement*) You've not doing as well as you'd like, and you know your boss doesn't like it.
Counselor:	(*Eliciting meaning*) Could we move in a different direction for a moment? Annaliese, what does this job really mean to you? Does it fit with your life goals?
Annaliese:	No! I'm bored and frustrated. The job just doesn't make sense to me. I thought it would, but it doesn't. I need something that I care about—that is meaningful to me, so I can go home feeling that I've done something worthwhile. I'd like to care for others instead of working with numbers all day long.
Counselor:	(*Reflection of feeling*) Annaliese, you're really troubled and worried, perhaps even scared.
Counselor:	(*Reflecting meaning*) I hear you, Annaliese. You care. So what we have is a job that has little meaning for you, but it pays well. (*Next the counselor reflects deeper meanings.*) You seem to feel that you would have more value for yourself if you could help others more. (*Next the counselor elicits meaning for more depth.*) Let's explore that vision of caring for others a bit more and how it might lead to more meaningful work. You may be interested in discernment as a way to explore life goals more deeply.
Counselor:	(*Positive reframe/interpretation*) Let's look at another way, Annaliese. The fact that you're bored is a sign that you've accomplished what you need to in that job. You've shown great skills, and you are ready to move on to something new where you can use your strengths more effectively.
Counselor:	(*Interpretation/reframe—linkage*) Annaliese, this seems to tie in with what you said last week about enjoying your volunteer work with children in the inner city several years ago. You got joy in that, but your job eventually took you away. It seems that you may be ready to be taken away to something that is more satisfying at a deeper level . . . something that brings you more fun and joy.

Interpretation has traditionally been viewed as a mystical activity in which the counselor reaches into the depths of the client's personality to provide new insights. However, we can demystify interpretation if we consider it to be merely a new frame of reference. Interpretation reframes the situation. Viewed in this light, the depth of a given interpretation refers to the magnitude of the discrepancy between the frame of reference from which the client is operating and the frame of reference supplied by the therapist. Gradually, *reframing* has become a more prominent term as it provides a more understandable view of interpretation.

To ensure mutuality and not influence the client too much, follow most interpretations/reframes with a checkout—"How does that sound to you?" "What meaning do you take from what I just said?"

▶Theories of Counseling and Interpretation/Reframing

Theoretically based interpretations can be extremely valuable as they provide us with a tested conceptual framework for thinking about the client. Each theory is itself a story—a story told about what is happening—and each theory makes meaning and uses language in its own way. Integrative theories find that each theoretical story has some value. Most likely, as you generate your own natural style, you will develop your own integrative theory, drawing from those approaches that make most sense to you.

Table 11.1 provides several examples of how different orientations to theory might interpret the same information—in this case, Charlis's description of a dream that she had.

TABLE 11.1 Theoretically Based Interpretations

Theory	Counselor Response	Process Notes
Decisional theory. A major issue in counseling for all clients is making appropriate decisions and understanding alternatives for action. Decisions need to be made with awareness of the cultural/environmental context. Interpretation/reframing helps clients find new ways of thinking about their decisions. Linking ideas together is particularly important.	Charlis, you're facing new challenges since the heart attack and have many key decisions to make, including what you want to do with the rest of your life. You feel almost as if you might fall off the cliff if things don't straighten out soon. The whole situation is frightening, and making decisions can make it worse. On the other hand, we have already identified several strengths that will enable you to make the important decisions you have to make.	Interpretation focuses on the parallels to the heart attack and then draws on strengths and wellness.
Person-centered. Clients are ultimately self-actualizing. Our goal is to help them find the story that builds on their strengths and helps them find deeper meanings and purpose. Reflection of meaning helps clients find alternative ways of viewing the situation; interpretation/reframing are not used. Linking can occur through effective summarization.	That dream seems to mean something important to you, Charlis. I hear the terrible fright, and I notice the rage of the sea. You've had the dream many nights, and now you wonder what it means.	Reflection of meaning and reflection of feelings. Interpretations would be very rare in this theory.
Brief solution-focused counseling. Brief methods seek to help clients find quick ways to reach their central goals. The session is conceived first as a goal-setting process and then methods are found to reach goals through time-efficient methods. Interpretation/reframing will be rare except for linking of key ideas.	You're facing new challenges since the heart attack and have some important decisions to make. Charlis, this goal is important; which path do you want to take to get well?	Mild interpretation with a move to goal setting.

Theory	Counselor Response	Process Notes
Cognitive behavioral theory. The emphasis is on sequences of behavior and thinking and what happens to the client, internally and externally, as a result. Often interpretation/reframing is useful in understanding what is going on in the client's mind and/or linking the client to how the environment affects cognition and behavior.	Charlis, the dream seems very close to what you face now. You have told me that you feel rage toward what happened to you and now you are wondering which direction to take. Our next task is to work on some stress management strategies to help you find behaviors to cope with these challenges. Later, let's look at how this might relate to what's going on with your parents.	Here we see the counselor active in linking the dream with present issues.
Psychodynamic theory. Individuals are dependent on unconscious forces. Interpretation/reframing are used to help link ideas and enable the client to understand how the unconscious past and long-term, deeply seated thoughts, feelings, and behaviors frame the here and now of daily experiences. Freudian, Adlerian, Gestalt, Jungian, and several other psychodynamic theories each tell different stories.	You feel rage at your parents, Charlis, and you can't tell them how you really feel. It frightens you. And now you find the people around you force you to keep quiet about your feelings, but the prospect of challenging them is terrifying. This links back to earlier stories you mentioned about not being able to depend on your parents.	Emphasis on how the past affects the present.
Multicultural counseling and therapy (MCT). The person is situated in a cultural/environmental context, and we need to help clients interpret and reframe their issues, concerns, and problems in relation to their multicultural background. (See the RESPECTFUL model, Chapter 1.) MCT is an integrative theory and uses all of the methods above, as appropriate, to facilitate clients' understanding of themselves and how the cultural/environmental context affects them personally.	You felt frightened—I hear that. From what you've told me, sexual harassment was one of the stressors you faced before the heart attack. The cliff could be the hassles you had at the office, and returning to the job clearly is frightening at this point. I also hear a woman who has the courage to get out on those cliffs and face the challenges. Charlis, we will work together to help you find some support here to cope with the challenges.	Feminist frame of reference; issues are interpreted in a multicultural context.

Before the actual interpretation, you will see a brief theoretical paragraph that provides some background for the theory-oriented interpretation that follows.

Imagine that you have worked with Charlis over a longer period, and she has come to you upset over a troubling dream. This dream recurred frequently in childhood; after the heart attack, it returned with a vengeance and would wake Charlis, sweating, in the middle of the night. Charlis tells you her dream story.

> *Charlis*: I dreamed that I was walking along the cliffs with the sea raging below. I felt terribly frightened. There was a path that I could have taken away from the cliffs, but I just felt so undecided about what to do. The dream just went on and on. I woke up in a cold sweat— and I've had that dream almost every night since I last saw you.

Different counseling theories would interpret and work with the dream differently, but each would provide a new frame of reference, a new perspective for the client.

All of the theoretical orientations provide the client with a new, alternative way to consider the situation. In short, interpretation renames or redefines "reality" from a new point of view. Sometimes just a new way of looking at an issue is enough to produce change. Which is the correct interpretation? Depending on the situation and context, any of these interpretations could be helpful or harmful. The first two responses deal with here-and-now reality, whereas psychodynamic interpretation deals with the past. The feminist interpretation links the heart attack with sexual harassment on the job.

▶A Cautionary Comment About Interpretation/Reframing

Interpretation/reframing offers clients redefinitions of challenging or negative feelings, behaviors, or events. Many people use this microskill with the specific goal of making the negative seem more positive and manageable. In fact, doing it for the purpose of altering immediate reactions to a situation or to provide rationalizations for concerning behaviors may resemble what politicians, customer service representatives, and others do to redirect criticisms or challenges. Redefining a suicide attempt as a demonstration of bravery may offend your client or family. Telling your anxious client that the psychiatrist's prescription of medication demonstrates the biological basis of the client's disorder is inappropriate. Intentional interpretation/reframing considers the client and the situation, is supportive and nonjudgmental, and has a clear client-focused purpose.

▶Summary: Facilitating Clients in Finding Their Meaning Core and Developing New Perspectives

Eliciting and reflecting meaning is a complex skill that requires you to enter the sense-making system of the client. Full exploration of life meaning requires a self-directed, verbal client willing to talk. The skill complex is most often associated with an abstract, formal operational counseling style. However, all of us are engaged in the process of meaning-making and trying to make sense of a confusing world. With clients who are more concrete, you will still find that eliciting and reflecting meaning is useful. But these clients may not be able to see patterns in their thinking or be as self-directed and reflective as those who think at a more complex level. You may find that the more directive approach to meaning taken by the cognitive behavioral therapists is more useful with clients who have difficulty reflecting on themselves.

With highly verbal or resistant clients, you may find that they like to spend all their time thinking, reflecting on meaning, and thus end up intellectualizing with little or no action to change their behaviors, thoughts, or feelings. Viktor Frankl was well aware of this possible problem and encouraged his clients to take action on their meanings. Meaning that does not move into the "real world" may become a problem in itself.

Meanings are organizing constructs that are at the core of our being. You will find that exercises with reflection of meaning, if completed in depth, will result in your having a more comprehensive understanding of your client than is possible with most other skills. Mastering the art of understanding meaning will take more time than other skills. The exercises in this chapter are designed to assist you along the path toward this goal.

When you interpret or reframe, first be sure that you have heard the client's story or concerns, and then draw from personal experience or a theoretical perspective to provide the client a new way of thinking and talking about issues. Focusing and multicultural counseling and therapy are the most certain way to discuss multicultural issues. A woman, a gay or lesbian, or a Person of Color may be depressed over what is considered a personal failure. Helping the client see the cultural/environmental context of the issue can reveal a new perspective, providing a totally new and more workable meaning.

The effectiveness of an interpretation/reframe can be measured on the Client Change Scale. The new perspective is useful if the client moves in a positive direction. Each counseling theory provides us with a new and different story about client issues and concerns. Drawing from theory for interpretation/reframing provides a more systematic frame for considering the client. However, logic and your personal experience and observations may be as effective as a theoretically oriented reframe.

▼▼▼▼	Key Points
Meaning	Meaning is not observable behavior, although it could be described as a special form of cognition that reaches the core of our being. Helping clients discern the meaning and purpose of their lives can serve as a motivator for change and provide a compass as to the direction of that change. Meaning organizes life experience and often serves as a metaphor from which clients generate thoughts, feelings, and behaviors. A person with a sense of meaning and a vision for the future can often work through and live with the most difficult issues and problems. Reflections of meaning are generally for more verbal clients and may be found more in their sessions.
The how of meaning	A well-timed reflection of meaning may help many clients facing extreme difficulty. It can help clarify cultural and individual differences, as the same words often have varying underlying meaning for each client. As meaning is often implicit, it is helpful to ask questions that lead clients to explore and clarify meaning.
	Eliciting meaning. "What does 'XYZ' mean to you?" Insert the key important words of the client that will lead to meanings and important thoughts underlying key words. "What sense do you make of it?" "What values underlie your actions?" "Why is that important to you?" "Why?" (by itself, used carefully)
	Reflecting meaning. Essentially, this looks like a reflection of feeling except that the words *meaning*, *values*, or *intentions* substitute for feeling words. For example, "You mean . . . ," "Could it mean that you . . . ," "Sounds like you value . . . ," or "One of the underlying reasons/intentions of your actions was. . . ." Then use the client's own words to describe his or her meaning system. You may add a paraphrase of the context and close with a checkout.
	For example, you could reflect an immediate meaning this way: "Anish, you value service to others and you've enjoyed working in the hospital as a volunteer." However, if there is conflict of values and meanings, the following could be added to confront the discrepancy between individual and family values. "Your family values medical practice as it offers more money, but you want to work in research on cancer as you see that as the best way to help more people in the long run and the financial rewards are not as important to you. Have I summarized the value conflict clearly?"

Interpretation/ reframe	The counselor helps clients obtain new perspectives, new frames of reference, and sometimes new meanings, all of which can facilitate clients' changing their view and way of thinking about their issues. This skill comes primarily from the counselor's observations and occasionally from the client.
	Theoretical interpretations. These come from a specific counseling theory such as psychodynamic, interpersonal, family therapy, or even Frankl's logotherapy. Clients tell their story or speak about their problems and issues. The counselor then makes sense of what they are saying from a particular theoretical perspective. "That dream suggests that you have an unconscious wish to run away from your husband." "Sounds like an issue of what we call boundaries—your husband/wife is not respecting your space." "I hear you saying that you don't know where you are going; it sounds like you lack meaning in your life."
	Reframes. These tend to come from here-and-now experience in the session, or they might be larger reframes of major client stories. The reframes are based on your experience in providing the client with another interpretation of what has happened or how the story is viewed. Effective reframes can change the meaning of key narratives in clients' lives. The positive reframe is particularly important. "Charlis, what stands out to me at this moment is how able you are and we can use your 'smarts' and ability to understand situations to find new, more comfortable directions." Positive reframes in the here and now are often the most useful.
Interpretation/ reframing at the deepest level	Meaning affects interpretation. Viktor Frankl constantly reframed his experience in the concentration camp, integrating here-and-now positive reframing with meaning. In the midst of the terror, he was able to enjoy the beauty of a sunset; he remembered his times with his wife; he was able to enjoy and focus on tasting and eating a small bit of bread. The major reframe of such traumatic experience, of course, is "I survived" or "You survived." Despite the traumatic experience (war, rape, accident), you are still here with the possibility of changing a part of the world.

▶Competency Practice Exercises and Portfolio of Competence

 Go to CengageBrain.com to access Counseling CourseMate, where you will find an interactive ebook, quizzes, videos, interactive counseling and psychotherapy exercises, the Portfolio of Competence, case studies, a practice test, and more.

The concepts of this chapter build on previous work. If you have solid attending and client observation skills, can use questions effectively, and can demonstrate effective use of the encourager, paraphrase, and reflection of feeling, you are well prepared for the exercises that follow.

Individual Practice

Exercise 1. Identification of Skills   Read the following client statement. Which of the following counselor responses are paraphrases (P), reflections of feeling (RF), reflections of meaning (RM), or interpretations/reframes (I/R)?

> I feel very sad and lonely. I thought Jose was the one for me. He's gone now. After our breakup, I saw a lot of people but no one special. Jose seemed to care for me and make it easy for me. Before that I had fun, particularly with Carlos. But it seemed at the end to be just sex. It appears Jose was it; we seemed so close.

_____ "You're really hurting and feeling sad right now."

_____ "Since the breakup you've seen a lot of people, but Jose provided the most of what you wanted."

_____ "Sounds like you are searching for someone to act as the father you never had and Jose was part of that."

_____ "Another way to look at it is that you unconsciously don't really want to get close; and when you get really close, the relationship ends."

_____ "Looks like the sense of peace, caring, ease, and closeness meant an awful lot to you."

_____ "You felt really close to Jose and now are sad and lonely."

_____ "Peace, caring, and having someone special mean a lot to you. Jose represented that to you. Carlos seemed to mean mainly fun, and you found no real meaning with him. Is that close?"

List possible single-word encouragers for the same client statement. You will find that the use of single-word encouragers, perhaps more than any other skill, leads your client to talk more deeply about the unique meanings underlying behavior and thought. A good general rule is to search carefully for key words, repeat them, and then reflect meaning.

Exercise 2. Identifying Client Issues of Meaning

Affective words in the preceding client statement include "sad" and "lonely." Some other words and brief phrases in the client statement contain elements that suggest more may be found under the surface. The following are some key words that you may have listed under possible encouragers: "the one for me," "care for me," "easy for me," "I had fun," "just sex," and "we seemed close." The feeling words represent the client's emotions about the current situation; the other words represent the meanings she uses to represent the world. Specifically, the client has given us a map of how she constructs the world of her relationships with men.

To identify underlying meanings for yourself, talk with a client, or someone posing as a client, observing his or her key words—especially those that tend to be repeated in different situations. Use those key words as the basis of encouragers, paraphrasing, and questioning to elicit meaning. Needless to say, this should be done with considerable sensitivity to the client and her or his needs. Record the results of your experience with this important exercise. You will want to record patterns of meaning-making that seem to be basic and that may motivate many more surface behaviors, thoughts, and feelings.

Exercise 3. Questioning to Elicit Meanings

Assume a client comes to you and talks about an important issue in her or his life (for instance, divorce, death, retirement, a pregnant daughter). Write five questions that might be useful in bringing out the meaning of the event.

Exercise 4. Practice of Skills in Other Settings

During conversations with friends or in your own interviews, practice eliciting meaning through a combination of questioning and single-word encouragers, and then reflect the meaning back. You will often find that single-word encouragers lead people to talk about meaningful issues. Record your observations of the value of this practice. What one thing stands out from your experience?

Exercise 5. Discernment: Examining One's Purpose and Mission

Using the suggestions of Box 11.2, work through each of the four sets of questions. You may do this by yourself, using a meditative approach and journaling, or you may want to do it with a classmate or close acquaintance. Allow yourself time to think carefully about each area. Add questions and topics that occur to you—make this exercise fully personal.

What do you learn from this exercise about your own life and wishes?

Exercise 6. Individual Practice in Interpretation/Reframing   Interpretations provide alternative frames of reference or perspectives for events in a client's life. In the following examples, provide an attending response (question, reflection of feeling, or the like) and then write an interpretation. Include a checkout in your interpretation.

Exercise 6a

"I was passed over for promotion for the third time. Our company is under fire for sex discrimination, and each time a woman gets the job over me. I know it's not my fault at all, but somehow I feel inadequate."

Listening response _____

_____

_____

Interpretation/reframe from a psychodynamic frame of reference (i.e., an interpretation that relates present behavior to something from the past): _____

_____

_____

Interpretation/reframe from a gender frame of reference _____

_____

_____

Interpretation/reframe from your own frame of reference in ways that are appropriate for varying clients _____

_____

_____

Exercise 6b

"I'm thinking of trying some pot. Yeah, I'm only 13, but I've been around a lot. My parents really object to it. I can't see why they do. My friends are all into it and seem to be doing fine."

Listening response _____

_____

_____

Reframe from a conservative frame of reference (one that opposes the use of drugs) _____

_____

_____

Reframe from an occasional user's frame of reference _____

_____

_____

Interpretation from your own frame of reference on this issue _____

_____

_____

The preceding examples of interpretations and reframes are representations of meaning and value issues that you will encounter in counseling and therapy. What are the value issues involved in these examples, and what is your personal position on these issues? Finally, how do you reconcile the importance of a client's responsibility for her or his own behavior with your position? What would you actually do in these situations?

Group Practice

Three group exercises are suggested here. The first focuses on the skill of eliciting and reflecting meaning, the second on the dereflection process as it might be used in logotherapy, and the third on interpretation/reframing.

Exercise 7. Systematic Group Practice in Eliciting and Reflecting Meaning

Step 1: Divide into practice groups.

Step 2: Select a group leader.

Step 3: Assign roles for the first practice session.

▲ Client
▲ Counselor
▲ Observer 1, who observes the client's descriptive words and key repeated words, using Feedback Form: Reflecting Meaning (Box 11.4)
▲ Observer 2, who notes the counselor's behavior, using the same feedback form

Step 4: Plan. For practice with this skill, it will be most helpful if the session starts with the client's completing one of the following model sentences. The session will then follow along, exploring the attitudes, values, and meanings to the client underlying the sentence.

> "My thoughts about spirituality are . . ."
> "My thoughts about moving from this area to another are . . ."
> "The most important event of my life was . . ."
> "I would like to leave to my family . . ."
> "The center of my life is . . ."
> "My thoughts about divorce/abortion/gay marriage are . . ."

A few alternative topics are "My closest friend," "Someone who made me feel very angry (or happy)," and "A place where I feel very comfortable and happy." Again, a decision conflict or a conflict with another person may be a good topic.

Establish the goals for the practice session. The task of the counselor in this case is to elicit meaning from the model sentence and help the client find underlying meanings and values. The counselor should search for key words in the client response and use those key words in questioning, encouraging, and reflecting. A useful sequence of microskills for eliciting meaning from the model sentence is

1. An open question, such as "Could you tell me more about that?" "What does that mean to you?" or "How do you make sense of that?"
2. Encouragers and paraphrases focusing on key words to help the client continue
3. Reflections of feeling to ensure that you are in touch with the client's emotions
4. Questions that relate specifically to meaning (see Box 11.2)
5. Reflecting the meaning of the event back to the client, using the framework outlined in this chapter

It is quite acceptable to have key questions and this sequence in your lap to refer to during the practice session.

BOX 11.4 Feedback Form: Reflecting Meaning

You can download this form from Counseling CourseMate at CengageBrain.com.

_____ (DATE)

_____   _____
(NAME OF COUNSELOR)   (NAME OF PERSON COMPLETING FORM)

Instructions: Observer 1 completes the first part of this form, giving special attention to recording descriptive words the client associates with meaning and to key repeated words. In the second part, Observer 2 notes the counselor's use of the reflection-of-meaning skill, giving special attention to questions that appeared to elicit meaning issues.

Part One: Client Observation

Key words/phrases:

_____

_____

_____

What are the main meaning issues of the session?

Part Two: Counselor Observation

List questions and reflections of meaning used by the counselor, continuing on a separate sheet as needed.

1. _____
2. _____
3. _____
4. _____
5. _____
6. _____

Comment on the effectiveness of the reflection-of-meaning skill.

Examine the basic and active mastery competencies in the Portfolio of Competence and plan your session to achieve specific goals.

Observers should study the feedback form especially carefully.

Step 5: Conduct a 5-minute practice session using the skill.

Step 6: Review the practice session and provide feedback for 10 minutes. This may be a time that the microsupervision process includes a group discussion of the place of values in therapy. The feedback forms are useful here. It is often tempting to just talk, but you might forget to give the counselor helpful and needed specific feedback. Take time to complete the forms before talking about the session. As always, give special attention to the mastery level achieved. The client can complete the Client Feedback Form from Chapter 1.

Step 7: Rotate roles. Remember to share time equally.

Some general reminders. This skill can be used from a variety of theoretical perspectives. It may be useful to see if an explicit or implicit theory is observable in the counselor's behavior.

Exercise 8. Discernment Practice Take another person through the discernment procedure, working carefully with each step. Recall de Waal's statement on how to listen to the other person:

> Listen. Listen, with intention, with love, with the "ear of the heart." Listen not only cerebrally with the intellect, but with the whole of feelings, our emotions, imaginations, and ourselves. (de Waal, 1997, Preface)

Share the list of questions and ideas with your client. Ask your volunteer to suggest additional questions and issues that may be missing from this list. Have the client define which questions he or she may wish to discuss. Use all your listening skills as you help the client find personal direction and meaning. Take your time!

What do you and the client learn?

Exercise 9. Interpretation/Reframing Practice

Steps 1–3. See Exercise 7. Use Feedback Form: Interpretation/Reframe (Box 11.5).

Step 4: Plan. To practice with this skill, ask the client to think about and describe something that is frustrating at the moment. A few alternative topics are "having to move to a different residence hall," "having roommates with religious beliefs different from yours," "trying to adopt healthier eating habits," "taking a challenging course," and "finding that you tend to procrastinate."

Establish the goals for the practice session. The first task of the counselor is to listen to the client story and learn how he or she thinks about or interprets the frustrating issue. The second task is to provide an alternative meaning or interpretation of the narrative; draw from personal experience or a theoretical perspective. Also, you may link critical ideas together (see linking examples in this chapter). Examine the basic and intentional mastery competencies in the Portfolio of Competence and plan your session to achieve specific goals.

The value of an interpretation or reframe depends on the client's reaction to it. Use the Client Change Scale (CCS) in Chapter 10 to assess the client's reaction to your reframe. Observers should study the feedback form before the practice exercise.

Step 5: Conduct a 5-minute practice session using the skill.

Step 6: Review the practice session and provide feedback for 10 minutes. Include a group discussion of the role of reframing in the helping process. Don't forget to use feedback forms here and to give special attention to the mastery level achieved. The client can complete the Client Feedback Form from Chapter 1.

Step 7: Rotate roles. Remember to share time equally.

Some general reminders. When we use interpretation/reframing, we are working primarily from the counselor's frame of reference. Your goal is to help the client to restory or look at the frustrating problem or concern from a new perspective. To accomplish this goal, you need to listen before you provide your interpretation or reframe. Respect clients' frame of reference before interpreting or reframing their words and frustrating situations in new ways. Provides clients with a new perspective or way of thinking about issues.

Portfolio of Competence

As you work through this list of competencies, think about how you would include the ideas related to reflection of meaning in your own Portfolio of Competence.

Use the following checklist to evaluate your present level of mastery. As you review the items below, ask yourself, "Can I do this?" Check those dimensions that you currently feel able to do. Those that remain unchecked can serve as future goals. Do not expect to attain intentional competence on every dimension as you work through this book. You will find, however, that you will improve your competencies with repetition and practice.

Level 1: Identification and classification.   You will be able to differentiate reflection of meaning and interpretation/reframing from the related skills of paraphrasing and reflection of feeling. You will be able to identify questioning sequences that facilitate client talk about meaning. You will be able to provide new ways for clients to think about their issues through interpretation/reframing.

❑ Identify and classify the skills.
❑ Identify and write questions that elicit meaning from clients.
❑ Note and record key client words indicative of meaning.

Level 2: Basic competence.   You will be able to demonstrate the skills of eliciting and reflecting meaning and interpretation/reframing. You will be able to demonstrate an elementary skill in dereflection.

❑ Elicit and reflect meaning in a role-play session.
❑ Examine yourself and discern more fully your life direction.
❑ Use dereflection and attitude change in a role-play interview.
❑ Use interpretation/reframing.

Level 3: Intentional competence.   You will be able to use questioning skill sequences and encouragers to bring out meaning issues and then reflect meaning accurately. You will be able to use the client's main words and constructs to define meaning rather than reframing in your own words (interpretation). You will not interpret but rather will facilitate the client's interpretation of experience.

With interpretation/reframing, you will be able to provide clients with new and fresh perspectives on their issues.

❑ Use questions and encouragers to bring out meaning issues.
❑ When you reflect meaning, use the client's main words and constructs rather than your own.
❑ Reflect meaning in such a fashion that the client starts exploring meaning and value issues in more depth.
❑ In the session, switch the focus of the conversation as necessary from meaning to feeling (via reflection of feeling or questions oriented toward feeling) or to content (via paraphrase or questions oriented toward content).
❑ Help others discern their purpose and mission in life.
❑ When a person is hyperreflecting on the negative meaning of an event or person, find something positive in that person or event and enable the client to dereflect by focusing on the positive.
❑ Provide clients with appropriate new ways to think about their issues, helping them generate new perspectives on their behavior, thoughts, and feelings.

- ❏ Provide a new perspective via interpretation/reframing, using your own knowledge, and help your clients use these ideas to enlarge their thinking on their issues.
- ❏ Use various theoretical perspectives to organize your reframing.

Level 4: Psychoeducational teaching competence

- ❏ Teach clients how to examine their own meaning systems.
- ❏ Facilitate others' understanding and use of discernment questioning strategies.
- ❏ Teach reflection of meaning to others.
- ❏ Teach clients how to interpret their own experience from new frames of reference and to think about their experiences from multiple perspectives.
- ❏ Teach interpretation/reframing to others.

▶Determining Your Own Style and Theory: Critical Self-Reflection on Reflecting Meaning and Interpretation/Reframing

Meaning has been presented as a central issue in counseling and psychotherapy. Interpretation has been presented as an alternative method for achieving much the same objective but with more counselor involvement. What single idea stands out for you among all those presented in this chapter, in class, or through informal learning? What stands out for you is likely to be important as a guide toward your next steps. What are your thoughts on multicultural issues and the use of this skill? What other points in this chapter strike you as important? How might you use ideas in this chapter to begin the process of establishing your own style and theory? Are you able to find new meanings and reinterpret/reframe your own life experience? In particular, what have you learned about discernment and its relation to your own life?

▶Our Thoughts About Charlis

Eliciting and reflecting meaning are both skills and strategies. As skills, they are fairly straightforward. To elicit meaning, we'd want to ask Charlis some variation of the basic question, "What does the heart attack mean to you, your past and future life?" As appropriate to the situation, questions such as the following can address the general issue of meaning in more detail:

"What has given you the most satisfaction in your job?"
"What's been missing for you in your present life?"
"What do you find of value in your life?"
"What sense do you make of this heart attack and the future?"
"What things in the future will be most meaningful to you?"
"What is the purpose of your working so hard?"
"You've said that you have been wondering what God is saying to you with this trial. Could you share some of your thoughts here?"
"What would you like to leave the world as a gift?"

Questions such as these do not usually lead to concrete behavioral descriptions. They may often bring out emotions, and they certainly bring out certain types of thoughts and cognitions. Typically, these thoughts are deeper in that they search for meanings and understandings. When clients explore meaning issues, the therapy session, almost by necessity, becomes less precise. Perhaps this is because we are struggling with defining the almost indefinable.

As part of our work with Charlis, we'd ask if she wants to examine the meaning of her life in more detail through the process of discernment. This is a more systematic approach to meaning and purpose defined in some detail in this chapter. If she wishes, we'd share the specific questions of discernment presented here and ask her which ones she'd like to explore. In addition, we'd ask her to think of questions and issues that are particularly important to her, and we would give these special attention as we work to help her discern the meaning of her life, her work, her goals, and her mission.

Reflection of meaning as a skill looks very much like reflection of feeling or paraphrasing, but the key words *meaning, sense, deeper understanding, purpose, vision,* or some related concept will be present explicitly or implicitly. "Charlis, I sense that the heart attack has led you to question some basic understandings in your life. Is that close? If so, tell me more."

It can be seen that we regard eliciting and reflecting meaning as an opening for the client to explore issues where there often is not a final answer but rather a deeper awareness of the possibilities of life. At the same time, effective exploration of meaning becomes a major strategy in which you bring out client stories, past, present, and future. You will use all the listening, focusing, and confrontation skills to facilitate this self-examination. Yet the focus remains on the client's finding meaning and purpose in his or her life.

CHAPTER 12

Self-Disclosure and Feedback
Immediacy and Genuineness in Counseling and Therapy

Self-Disclosure and Feedback

Reflection of Meaning and Interpretation/Reframe

Empathic Confrontation

Focusing

The Five-Stage Interview Structure

Reflection of Feeling

Encouraging, Paraphrasing, and Summarizing

Open and Closed Questions

Client Observation Skills

Attending Behavior and Empathy

Ethics, Multicultural Competence, and Wellness

Before I open up to you (self-disclose), I want to know where you are coming from. . . . In other words, a culturally different client may not open up (self-disclose) until you, the helping professional, self-disclose first. Thus, to many minority clients, a therapist who expresses his/her thoughts and feelings may be better received in a counseling situation.

—Derald Wing Sue and Stanley Sue

Mission of "Self-Disclosure and Feedback"

If used carefully and sensitively, these two skills can bring here-and-now immediacy to the session. When counselors share their experience of the client, or briefly comment on something from their own lives, the client will listen. Counselor feedback from observations of client behaviors, thoughts, or feelings can be highly beneficial. However, the use of these two skills remains controversial as the power of the therapist may be used inappropriately.

Chapter Goals and Competency Objectives Awareness, knowledge, skills, and actions in self-disclosure and feedback will enable you to:

▲ Use appropriate self-disclosure, which builds a sense of equality and encourages client trust and openness. Disclosure may be about immediate here-and-now feelings, history of the counselor, or thoughts and feelings that the therapist may have.

▲ Offer accurate feedback on how the client is experienced by the counselor in the here and now of the session; how the client is progressing on issues; how others view the client; and thoughts, feelings, and behaviors observed in the session.

▲ Apply these skills empathically with a sense of personal genuineness and authenticity with awareness of the counselor's power in the session.

▲ Work effectively with the empathic qualities of immediacy.

 Go to CengageBrain.com to access Counseling CourseMate, where you will find an interactive ebook, quizzes, videos, interactive counseling and psychotherapy exercises, the Portfolio of Competence, case studies, a practice test, and more.

▶Defining Self-Disclosure

We are all cultural beings; thus, the quote from the Sue brothers above applies to all of our clients. Because trust and openness are often essential if you are to establish a relationship, a working alliance, many effective counselors and therapists allow clients to inquire about them as persons and encourage asking questions early in the process. We are talking here about some immediate specifics about how you, the counselor, respond to personal questions and offer information about yourself in the session. As a counselor, you are in a position of power and influence, and the client can overreact to what you say. Sensitivity and awareness of client needs are central.

Should you share your own personal observations, experiences, and ideas with the client? Self-disclosure by the counselor has been a controversial topic. Some theorists and clinicians argue against counselors' sharing themselves openly, preferring a more distant, objective persona. However, humanistically oriented and feminist counselors have demonstrated the value of appropriate self-disclosure. Multicultural theory considers early self-disclosure as key to trust building in the long run. This seems particularly so if your background is substantially different from that of your client. For example, a young person counseling an older person needs to discuss this issue early in the session. Moreover, research reveals that clients of counselors who self-disclose report lower levels of symptom distress and like the counselor more (Barrett & Berman, 2001).

Following is a brief definition of self-disclosure and the anticipated result when you use it appropriately.

SELF-DISCLOSURE	ANTICIPATED RESULT
As the counselor, share your own related past personal life experience, here-and-now observations or feelings toward the client, or opinions about the future. Self-disclosure often starts with an "I" statement. Here-and-now feelings toward the client can be powerful and should be used carefully.	When used appropriately and empathically, the client is encouraged to self-disclose in more depth and may develop a more egalitarian relationship in the session. The client may feel more comfortable in the relationship and find a new solution relating to the counselor's self-disclosure.

The concept of self-disclosure brings up the important issue of whether you can effectively counsel someone if you have not experienced the client's issues in some way. Many alcoholics are dubious about the ability of nonalcoholics to understand what is occurring for them. A client with a serious fertility problem often feels that no one can really understand her without experiencing her issues.

Imagine that you are counseling an older person coping with a forthcoming divorce after 35 years of marriage. Your background is very different, and truly empathizing with the client may be a challenge. Moreover, the older client may have difficulty trusting you and your expertise. Open discussion, some self-disclosure on your part, and exploration of differences may be essential. Self-disclosure of who you really are can be helpful in those situations in which you have not "been there."

Similar issues will occur if you are counseling an alcoholic or a drug abuser and have not been in their shoes. Women or men may not feel comfortable with those of the opposite gender. If you are significantly different from the client on any dimension of the RESPECTFUL model (Chapter 1), the differences may need to be discussed.

As the session progresses, self-disclosure can encourage client talk, create additional trust between counselor and client, and establish a more equal relationship. Nonetheless, not everyone agrees that this is a wise skill to include among counselors' techniques. Some express valid concerns about counselors' monopolizing conversation or abusing the client's rights by encouraging openness too early; they also point out that counseling and therapy can operate successfully without any self-disclosure at all.

Dangers of Self-Disclosure

Along with the benefits of self-disclosure, there are some real dangers that need to be considered. First among these is **countertransference**, which can be defined as unwise conscious or unconscious entanglement with the clients' issues, causing the boundaries between you and the client to be violated. Your own subdued or even unconscious feelings may resurface through identification with the emotions, experiences, or issues of the person you are working with. The client's problems and issues may remind you of your own, and suddenly you are taking the focus from the client and talking too much about your own issues. Keep self-disclosures short and succinct, and then return the focus to the client. Empathic understanding requires us to separate ourselves from clients' emotional and cognitive issues, while simultaneously being fully conscious of them as we seek to "walk in their shoes" and see things as they see them.

Another type of entanglement and countertransference occurs when your self-disclosure is too distant from the client's concern. For example, the client may be talking about anxiety over a required classroom oral presentation. You may start talking about a recent bout of speech anxiety you experienced when you had to speak on a panel to a large audience. Make your self-disclosures relevant and timely.

Inappropriate and too extensive self-disclosure takes the focus away from the client. The boundaries separating you from the client may be violated, resulting in harm to the client and even ethical violations.

The Skill of Self-Disclosure

Here are five key aspects of self-disclosure:

1. *Listen.* As with all influencing skills, first attend to the client's story carefully, and then assess the appropriateness of any self-disclosure you might make. Be sure that it is relevant to the client's issue.
2. *Self-disclose and share briefly.* Immediately return focus to the client (for example, "How does that relate to you?"), while noting how the self-disclosure is received.
3. *Use "I" statements.* Self-disclosure almost always involves "I" statements or self-reference using the pronouns *I, me,* and *my*—or the self-reference may be implied.
4. *Share and describe briefly your thoughts, feelings, or behaviors.* "I can imagine how much pain you feel." "I feel happy to hear you talk of that wonderful experience—it was a real change!" "My experience of divorce was hurtful too." "I also grew up in an alcoholic family and understand some of the confusion you feel."

5. *Be empathically genuine and use appropriate immediacy and tense.* Are your self-disclosures real or faked? Don't make up things. The most beneficial self-disclosures are usually made in the here and now, the present tense. ("Right now I feel." "I am hurting for you at this moment—I care.") However, variations in tense are used to strengthen or soften the power of a self-disclosure.

▶Empathic Self-Disclosure

Can you be immediate and close to your client? Are you? Can you self-disclose in a way that is meaningful to the client, but not intrusive? Making self-disclosures relevant to the client is a complex task involving the following issues, among others.

Genuineness

To demonstrate genuineness, the counselor must truly and honestly have the feelings, thoughts, or experiences that are shared. Second, self-disclosure must be genuine and appropriate in relation to the client. For example, if you are working with a client who grew up in an alcoholic family and you have had experience in your own family with alcohol, a brief sharing of your own story can be helpful. The danger of storytelling, of course, is that you can end up spending too much time on your own issues and neglect the client. Here the "1-2-3" influencing pattern outlined in Box 12.1 is particularly important.

Immediacy

The following examples show how the use of *here and now* brings immediacy to the session. This may be compared to the *there and then* of the past and future tenses. But recall that all three approaches will be useful in the session—we need to know some past and we also need to anticipate the future.

Madison: I am feeling really angry about the way the way I'm treated by men in power positions.

Counselor: (present tense) You're coming across as really angry right now. I like that you finally are in touch with your feelings.

BOX 12.1 | **The "1-2-3" Pattern of Listening, Influencing, and Observing Client Reaction**

1. Listen
Use attending, observation, and listening skills to discover the client's view of the world. How does the client see, hear, feel, and represent the world through "I" statements and key descriptors for content (paraphrasing), feelings (reflection of feeling), and meaning (encouragers and reflection of meaning)?

2. Assess and Influence
An influencing skill is best used *after* you hear and understand the client's story. Timing is central—when is the client ready to hear or learn a new way to think about

what has been said? An influencing skill such as feedback, logical consequences, or a directive can be an abrupt change from the listening style unless the client is ready.

3. Check Out and Observe Client Response
Use a checkout as you offer an influencing skill ("How does that seem to you?"), then listen and observe carefully. If client verbal or nonverbal behavior seems incongruent or conflicted, return to the use of listening skills. Influencing skills can remove you from being totally "with" clients.

Counselor:	(past tense) I've had the same difficulty expressing feelings in the past. I recall when I would just sit there and take it.
Counselor:	(future tense) This awareness of emotion can help us all be more in touch in the future. I know it will continue to aid me.

Be careful when clients say, "What would you do if you were in my place?" Clients will sometimes ask you directly for opinions and advice on what you think they should do. "What do you think I should major in?" "If you were me, would you leave this relationship?" "Should I indeed have and keep the child?" Effective self-disclosure and advice can potentially be helpful, but it is *not* the first thing you need to do. Your task is to help clients make their own decisions. The right solution for you may not be the right solution for the client, and involving yourself too early can foster dependency and lead the client in the wrong way. Note the following exchange.

Madison:	How do you think I ought to tell my boss what I think of him hitting on me?
Counselor:	I'm not in your position, and I haven't heard too much about your boss yet nor enough about the job. First, could we explore your relationship in more detail? (past tense)
Counselor:	(If you feel forced to share your thoughts when you prefer not to, keep your comments brief and ask the client for her or his reflections.) I sense your anger right now, Madison (here and now, present tense). My own thought would be to share with the boss what he is doing. He may not be fully aware of how he's affecting you. But I'm not you. He may not hear that. What might be the outcome if you told him? (future tense)
Counselor:	(after drawing out more information on the relationship) From what I've heard, it sounds wise to bring up what's going on at a safe time. I think it is important to bring it up directly, and I admire the way you are thinking ahead. How does that sound to you? (future tense moving to the here and now as the issue is explored further)

Timeliness

If a client is talking smoothly about something, counselor self-disclosure is not necessary. However, if the client seems to want to talk about a topic but is having trouble, a slight leading self-disclosure by the counselor may be helpful. Too deep and involved a self-disclosure may frighten or distance the client.

Cultural Implications

You cannot expect to have experienced all that your clients bring to you. After an appropriate time for introducing the session, when a man works with a woman, for example, it may be useful to say, "Men don't always understand women's issues. The things you are talking about clearly relate to gender experience. I'll do my best, but if I miss something, let me know. Do you have any questions for me?" If you are White or African American working with a person of the other race, frank disclosure that you recognize the differences in cultures at the initiation of the session can be helpful in developing trust (see Box 12.2).

BOX 12.2 **National and International Perspectives on Counseling Skills**

When Is Self-Disclosure Appropriate?

Weijun Zhang

My good friend Carol, a European American, has had lots of experience counseling minority clients. She once told me that one of the first questions she asks her minority clients is "Do you have any questions to ask me?" which often results in a lot of self-disclosure on her part. She would answer questions not only about her attitudes toward racism, sexism, religion, and so forth, but also about her physical health and family problems. During the initial interview, as much as half of the time available could be spent on her self-disclosure.

"But is so much self-disclosure appropriate?" asked a fellow student in class after I mentioned Carol's experience.

"Absolutely," I replied. We know that many minority clients come to counseling with suspicion. They tend to regard the counselor as a secret agent of society and doubt whether the counselor can really help them. Some even fear that the information they disclose might be used against them. Some questions they often have in mind about counselors are "Where are you coming from?" "What makes you different from those racists I have encountered?" and "Do you really understand what it means to be a minority person in this society?" If you think about how widespread racism is, you might consider these questions legitimate and healthy. And unless these questions are properly answered, which requires a considerable amount of counselor self-disclosure, it is hard to expect most minority clients to trust and open up willingly.

Some cultural values held by minority clients necessitate self-disclosure from the counselor, too. Asians,

for example, have a long tradition of not telling personal and family matters to "strangers" or "outsiders" in order to avoid "losing face." Thus, relative to European Americans, we tend to reveal much less of ourselves in public, especially our inner experience. A mainstream counselor may well regard openness in disclosing as a criterion for judging a person's mental health, treating those who do not display this quality as "guarded," "passive," or "paranoid." However, traditional Asians believe that the more self-disclosure you make to a stranger, the less mature and wise you are. I have learned that many Hispanics and Native Americans feel the same way.

Because counseling cannot proceed without some revelation of intimate details of a client's life, what can we do about these clients who are not accustomed to self-disclosure? I have found that the most effective way is not to preach or to ask, but to model. Self-disclosure begets self-disclosure. We can't expect our minority clients to do well what we are not doing ourselves in the counseling relationship, can we?

According to the guidance found in most counseling textbooks here, much of self-disclosure by the counselor is considered unprofessional; but if we are truly aware of the different cultural styles, it seems that a different approach may be needed.

The classmate who first questioned the practice asked with a smile, "Why are you so eloquent on this topic?" I said, "Perhaps it is because I have learned this not just from textbooks, but mainly from my own experience as both a counselor and a minority person."

When you are culturally different from your client, self-disclosure requires more forethought, and it is vital that you are personally comfortable with difference. Open discussion and some self-disclosure on your part may be essential. The frank disclosure of differences can be helpful in establishing rapport. Many alcoholics are dubious about the ability of nonalcoholics to understand what is occurring for them. A client with cancer or heart disease may feel that no one can really understand if that person has not experienced this particular illness. On the other hand, if you have had a difficult life experience with alcohol, sharing your story briefly can be very helpful. It often helps to say to the client, "Please feel free to ask questions about me if you wish." Just remember to keep your responses brief! Be sure that what you do or do not do is authentic to your own knowledge and comfort level, but also be sure that the self-disclosure is reasonable and comfortable for the client.

Self-disclosure research presented in Box 12.3 provides a useful summary of how to use self-disclosure wisely.

BOX 12.3 | Research and Related Evidence That You Can Use

Self-Disclosure

A comprehensive review of more than 200 articles by Henretty and Levitt (2010) found that more than 90% of counselors and therapists self-disclose to their clients, and those with the most experience tend to use the most self-disclosure. Some of their other key findings regarding what they term "therapeutic self-disclosure" (TSD) are listed below. Please bear in mind that these studies are talking in general terms and that counselors and clients will always vary uniquely in their interest in and use of TSDs.

Theoretical Orientation
- Different theories vary in their emphasis, but not as much as might be anticipated.
- Psychodynamic and psychoanalytic therapists tend to use the fewest TSDs.

Diversity
- African Americans tend to like self-disclosure, particularly when the client is of a different race. They also appear to be the group that most favors self-disclosure.
- Latinos may prefer nondisclosure, while White clients fall in between in interest and acceptance.
- Male clients endorsed more TSD of sexual material than women.

Expertness, Trust, and Attractiveness
- TSD counselors tend to be seen as more expert, trustworthy, and attractive (key aspects of interpersonal influence).

- If the counselor engaged in TSD in group work, men were more likely to disclose; counselor TSD did not seem to make a difference in women's groups.
- Self-disclosing feelings was rated as more effective than sharing facts.

Empathic Dimensions
- Surprisingly, TSDs did not result in higher ratings of empathy, positive regard, congruence, unconditionally, and warmth.

Impact on Counseling's Success
- Clients were more likely to self-disclose themselves, but different theoretical orientations varied in their ability to predict the effectiveness of TSD.
- Clients were slightly more likely to return when TSDs were used.
- Clients liked the TSD counselor more.
- TSDs resulted in more positive client response.

In summary, self-disclosure is a potentially valuable skill, but only if used appropriately, at the right time, with real sensitivity to both the immediate here-and-now and long-term needs of the client.

▶Defining Feedback

To see ourselves as others see us,
To hear how others hear us,
And to be touched as we touch others . . .
These are the goals of effective feedback.

—Allen Ivey

When you use feedback effectively as described below, you can anticipate the client's response.

FEEDBACK	ANTICIPATED RESULT
Present clients with clear, nonjudgmental information on how the counselor believes they are thinking, feeling, or behaving and how significant others may view them or their performance.	Clients may improve or change their thoughts, feelings, and behaviors based on the therapist's feedback.

Knowing how others see us is a powerful dimension in human change, and it is most helpful if the client solicits feedback. Feedback is an important influencing strategy if you have developed good rapport and enough experience with the client to know that he or she trusts you. The following feedback guidelines are critical:

1. *The client receiving feedback should be in charge.* Listen first to determine whether the client is ready for feedback. Feedback is more successful if the client solicits it. (Listen—provide feedback—use checkout.)

2. *Feedback is best received when you focus on strengths and/or something the client can do something about.* It is more effective to give feedback on positive qualities and build on strengths. Corrective feedback focuses on areas in which the client can improve thinking, feeling, and behaving. Corrective feedback needs to be about something the client can change, or a situation the client needs to recognize and accept as something that can't be changed.

3. *Feedback needs to be concrete and specific.* "You had two recent arguments with Chris that upset both of you. In each case, I hear you giving in almost immediately. You seem to have a pattern of giving up, even before you have a chance to give your own thoughts. How do you react to that?" Avoid abstractions and generalizations, such as "That seems typical of you." "It would be better if you stopped that way of thinking/behaving."

4. *Feedback is best when it is empathic, relatively nonjudgmental, and interactive.* A central dimension of empathy is to be with the client in a nonjudgmental fashion. Stick to the facts and specifics. Facts are friendly; judgments may or may not be. Demonstrate your nonjudgmental attitude through your vocal qualities and body language. "I do see you trying very hard. You have a real desire to accept the way Chris is and learn to live with what you can't change. How does that sound?"—not "You give in too easily, I wish you'd try harder" or the all-too-common "That was a *good* job."

5. *Feedback should be lean and precise.* Don't overwhelm the client; keep corrective feedback brief. Most of us can hear only so much and can change only one thing at a time. Select one or two things for feedback, and save the rest for later.

6. *Check out how your feedback was received.* Involve clients in feedback through the checkout. Their response indicates whether you were heard and how useful your feedback was. "How do you react to that?" "Does that sound close?" "What does that feedback mean to you?"

Positive feedback has been described as the "the breakfast of champions." Your positive, concrete feedback helps clients restory their problems and concerns. Whenever possible, find things right about your client. Even when you have to provide challenging feedback, try to include positive assets of the client. Help clients discover their wellness strengths, positive assets, and useful resources.

Corrective feedback is a delicate balance between negative feedback and positive suggestions for the future. When clients need to seriously examine themselves, corrective feedback may need to focus on things they are doing wrong or behavior that may hurt them in the future. Management settings, correctional institutions, and schools and universities often require the counselor to provide corrective feedback in the form of reprimands and certain types of punishments. When you must give negative corrective feedback, keep your vocal tone and body language nonjudgmental and stick to the facts, even though the issues may be painful. **Praise and supportive statements** ("You can do it, and I'll be there to help") convey your positive thoughts about the client, even when you have to give troubling feedback.

Negative feedback is necessary when the client has not been willing to hear corrective feedback. For example, in cases of abuse, planned behavior that hurts self or others, and criminal behavior, negative feedback—including the negative consequences the client's actions can bring—is necessary and can be beneficial. It is our responsibility to act in these situations, but listening to the client's point of view, even if the client is a perpetrator, remains important.

Vague, judgmental, negative feedback	Madison, I really don't think the way you are dealing with your boss is effective. You come across as just letting things go and accepting what's happening.
Concrete, nonjudgmental, positive feedback	Madison, you can do it; I've seen what you can do in other settings. Your ability to be assertive will be important. May I suggest some specific things that might be helpful the next time you face that situation?
Corrective feedback	Your effort was in the right direction. You can do even more if we set up a role-play to practice new ways of behaving. As we do this, I can offer more suggestions to help you change behavior. (See Chapter 13 for ideas on using role-plays in the session.)

Remember your own positive and negative experiences with feedback. These may have involved feedback from friends, family, teachers, or a work supervisor. What do you notice personally about effective and ineffective feedback? Now that you have read this section of the chapter, what would you do better, or what would you avoid, when giving feedback?

▶Example Counseling Session: How Do I Deal With a Difficult Situation at Work?

The following is an excerpt from the later stages of another session with Madison in which the counselor, Onawumi, uses self-disclosure and feedback. Madison is speaking of difficulties with her supervisor, who has started "hitting" on her.

COUNSELOR AND CLIENT CONVERSATION	PROCESS COMMENTS
1. *Onawumi:* Madison, as we've talked these last few sessions, you come across as very able and sure of yourself. In this situation, you are not so sure. Now, your boss, Jackson, is hitting on you and you're scared for your job, but still angry at what he is doing. The anger makes a lot of sense.	Onawumi starts here with brief feedback on Madison's strengths, followed by contrasting her abilities with her current fears and anger. Two major discrepancies are confronted here: (1) internal strengths shown in the past, as contrasted with current fears and indecision; (2) the conflict with her boss, including her present silence and the need to keep her job. Onawumi also provides feedback on the validity of Madison's anger, giving less attention to the fear. (Additive)
2. *Madison:* That's right, what am I doing? What can I do?	As Madison talked earlier about this issue, it was clear that she was moving between Levels 1 and 2 on the Client Change Scale. At times she denied that anything was happening, but then was able to examine the issues partially. Finally, she moved to Level 3, showing clear awareness of what was happening.

COUNSELOR AND CLIENT CONVERSATION	PROCESS COMMENTS
3. *Onawumi:* You've got considerable strength and women go through this too often, but something can be done. What occurs to you?	Feedback again on strength, but also provides feedback on other women. She then tosses the decision back to Madison, thus providing both respect and potential empowerment. (Additive)
4. *Madison:* Well, talking with you is helping. I think I see what's going on more clearly. I think I can pull myself together after all. I know that Jackson does like the work I am doing. I think I got that raise because my work is good, not because he had his eye on me.	Madison is drawing on the positive feedback and showing more certainty about herself and her abilities. She has moved to Level 3 on the CCS.
5. *Onawumi:* So you know he respects your work; and I gather from what you say, you are holding an important project together for him and the company. Clearly you have some leverage in that. Go on, I think you are on the right track.	Paraphrase, feedback. (Additive)
6. *Madison:* I wonder what would happen if I sat down with him and reviewed the present status on the project and how well things are going. And he has never directly really forced or embarrassed me, but the hints are really there. Perhaps, after we review the project, I could simply say that I like working with him, but feel that we must keep our relationship on a professional level. I've not been comfortable with a few things he has said, but I still respect him. Would he mind if we kept our focus on the project?	Recall that this did not happen that fast; only in the last 15 minutes of the session did Madison show real movement. Here she is at CCS Level 3, but moving toward Level 4.
7. *Onawumi:* Well, if he is a reasonable person, that likely could work. You are giving him respect, but still standing up for yourself. Let's try your idea in a role-play. I'll be Jackson, and you go through and practice what you might say.	Feedback to Madison on her ideas, followed by a directive for testing them out in a role-play (see the next chapter).

The role-play then follows, and Madison clearly summarizes the project and her feelings about the sometimes tense relationship. As she presents the situation, she firmly maintains her "cool," but also shows respect for the boss. She feels good about the way she handled the role-play.

| 8. *Onawumi:* That was great, Madison. I think that is about what you can try. Let's hope that it works. On the other hand, I can say that I've been in a similar situation, and I tried to do what we just role-played. The boss there just ignored what I said, but he did start leaving me alone. Nonetheless, I wonder if I lost some power. You've got some ideas that might well work, but—what if they don't? | Feedback followed by a self-disclosure. The self-disclosure of a parallel situation in Onawumi's life is accurate, but it can also give Madison some additional fear. Nonetheless, it is also important that Madison explore the possible logical negative consequences of speaking up. (Potentially additive) |

continued on next page

COUNSELOR AND CLIENT CONVERSATION	PROCESS COMMENTS
9. *Madison:* Oh, I know that he might become angry, but it's good to know that you had to deal with the same thing. Onawumi, women run into this all too often and we have to start standing up for our rights. I think this boss can take it and may have to. We just had corporate training on sexual harassment and company policy. He may even appreciate my feedback. I suppose it is possible that he doesn't know what he is doing.	More indication of Level 4 responding. Note that the feedback and self-disclosures have still left Madison in control of decisions and future actions.
10. *Onawumi:* So, what if it doesn't go well?	Open question confronting what might happen. (Potentially additive)
11. *Madison:* Actually, I'm not as scared for my job as I was. I think I'll get through this. I have to stand up. He needs me for this project. And if it doesn't work out, I'm ready to leave as I know that I can find something else.	Level 4 on CCS.
12. *Onawumi:* That makes me feel good that you are able to take that risk. And my sense is that you have the power and wisdom to make it work, no matter what the result. I'll be waiting here to learn what happened. You are doing the right thing for you.	Self-disclosure and feedback.

▶Feedback and Neuroscience

Relaxation training, biofeedback/neurofeedback, and stress management via feedback from both electronic skin response and brain activity via fMRI have become extensive. Neurofeedback has been found useful with headaches, depression, and even attention deficit disorder. This is a very different type of feedback from what has been discussed in this chapter, but using electronic feedback of varying types will be an important part of your practice in the future. Here are just a few examples.

Biofeedback instruments measure physiological functions such as brain waves, muscle tone, skin conductance, heart rate, and pain perception. The client is given feedback instruments such as electromyography (EMG), the electroencephalograph (EEG), and the electrocardiograph (ECG). Sensors are placed appropriately on the body, and the client learns to control physiological responses. These strategies are often useful in working with stress and anxiety, headaches, depression, and many other issues.

Neurofeedback, also known as biofeedback for the brain, is one modality based in neuroscience that empowers individuals to recognize, monitor, and self-regulate brain wave activity to create greater wellness, achieve self-regulation, and cope with mental disorders. Neurofeedback can be used to enhance normal human responses (e.g., concentration and work-related performance) and promote well-being (e.g., achieve relaxation). It has the capacity to address, via brain plasticity, underlying functional dysregulation or deficits that may manifest as mental health disorders (e.g., attention deficit disorder and headaches). In many cases, neurofeedback can reduce or eliminate the need for extensive and costly therapies based on medications that can have harmful side effects. Certification in neurofeedback is now available.

Some people predict that within 10 years, counselors will be using advanced technology for diagnosis and suggestions for treatment. Functional resonance magnetic imaging (fMRI) and positron emission tomography (PET), related systems, are used mainly for diagnosis and research. Several such studies have been cited throughout this book, and they have provided valuable information supporting much of what we do in counseling and psychotherapy. Knowledge of this research also helps us refine our practice.

▶Summary: Rewards and Risks of Self-Disclosure and Feedback

Some central potential values of self-disclosure and feedback are that they

- ▲ Provide an opportunity that might otherwise be missed for the client to gain from your experience and knowledge.
- ▲ Can enhance the relationship and encourage the client to be more open and self-disclosing.
- ▲ Bring more here-and-now immediacy to the session.

Following are some of the main cautions in using the sometimes controversial skills of self-disclosure and feedback:

- ▲ Beware of the possibility of counselors' misusing the power they have in the session. Clients tend to trust what counselors have to say, sometimes too readily.
- ▲ Keep both self-disclosures and feedback statements brief and to the point. Return the focus to the client quickly.
- ▲ Tune in what you say to where the client "is." By this we mean make your comments relevant to the life experience of the client. What seems like a parallel experience or useful feedback may be far off from the client's worldview.
- ▲ Consider whether your use of these two skills is appropriate to the multicultural background of your client; be aware of the RESPECTFUL model introduced in Chapter 1.

Where do you personally stand on the use of these skills? If you are particularly interested in humanistic forms of counseling and therapy, you likely will use these skills regularly. If you prefer a cognitive behavioral approach, this is less likely, but you still may decide to use the skills selectively, particularly self-disclosure in the early parts of the session.

▼▼▼▼	Key Points
Put clients first	Ideally, feedback and self-disclosure should be solicited by clients. When clients ask you what you would do in their place, your task is to help them make their own decisions. Involving yourself too early can foster dependency and potentially mislead clients. It is important to discuss multicultural differences openly and with respect.
"1-2-3" pattern	In any interaction with a client, first attend to and determine the client's frame of reference, then assess her or his reaction before using your influencing skills. Finally, check out the client's reaction to your use of the skill.

Self-disclosure	Indicating your thoughts and feelings to a client constitutes self-disclosure, which necessitates the following: 1. Use personal pronouns ("I" statements). 2. Use a verb for content or feeling ("I feel . . ." "I think . . ."). 3. Use an object coupled with adverb and adjective descriptors ("I feel happy about your being able to assert yourself . . ."). 4. Express your feelings appropriately. Self-disclosure tends to be most effective if it is genuine, timely, and phrased in the present tense. Keep your self-disclosure brief. At times, consider sharing short stories from your own life.
Feedback	Feed back accurate data on how you or others view the client. Remember the following: 1. The client should be in charge. 2. Focus on strengths. 3. Be concrete and specific. 4. Be nonjudgmental. 5. As appropriate, provide here-and-now feedback. 6. Keep feedback lean and precise. 7. Check out how your feedback was received. These guidelines are useful for all influencing skills.
Positive feedback	Positive feedback is "the breakfast of champions." Use this skill relatively frequently; it will balance more challenging necessary corrective feedback.
Corrective feedback	Corrective feedback will be necessary at times. It needs to focus on things that the client can actually change. Seek to avoid criticism. Be specific and clear and, where possible, supplement with client strengths that support the change.
Negative feedback	In some cases (e.g., abusive or criminal behavior), negative feedback—including the negative consequences the client's actions can bring—is necessary and can be beneficial. It is our responsibility to act in these situations, but listening to the client's point of view, even if the client is a perpetrator, remains important. Use this skill sparingly.

▶Competency Practice Exercises and Portfolio of Competence

Go to CengageBrain.com to access Counseling CourseMate, where you will find an interactive ebook, quizzes, videos, interactive counseling and psychotherapy exercises, the Portfolio of Competence, case studies, a practice test, and more.

Individual Practice

We strongly suggest that you practice both of the skills presented in this chapter.

Exercise 1. Self-Disclosure   Find a classmate or friend, get the person's permission, and the two of you try the strategy on each other. What happens? What occurs for you? What did you learn? Would you like to continue and practice this skill further?

Exercise 2. Feedback   Again, find a classmate or friend, get the person's permission, and the two of you try the strategy on each other. What happens? What occurs for you? What did you learn? Would you like to continue and practice this skill further?

Group Practice

Exercise 3. Group Practice With Self-Disclosure and Feedback   Small-group work with the influencing skills requires practice with each skill. The general model of small-group work is suggested. Remember to include the Client Feedback Form from Chapter 1 as part of the practice session. This is a particularly important place to practice group supervision, sharing, and feedback.

Step 1: Divide into practice groups.

Step 2: Select a leader for the group.

Step 3: Assign roles for each practice session.

▲ Client
▲ Counselor, who will use listening skills to draw out the client story or issue and then attempt one or more influencing skills such as feedback and self-disclosure
▲ Observers, who will complete the Feedback Form in Box 12.4

Step 4: Plan.   In using influencing skills, the acid test of mastery is whether the client actually does what is expected (for example, does the client follow the directive given?) or responds to the feedback or self-disclosure in a positive way.

A member of the group may present any personal issue that he or she is currently working with or role-play a situation experienced or observed in the past. The task of the counselor is to share something personal that relates to the client's concern and then to provide feedback. Again, the checkout is important in all cases.

Step 5: Conduct a 5- to 15-minute practice session using the skills.   You will find it difficult to use particular influencing skills frequently, as they must be interspersed with attending skills to keep the session going. However, attempt to use the targeted skill at least twice during the practice session.

Step 6: Review the practice session and provide feedback for 10 to 12 minutes.   Remember to stop the recording to provide adequate feedback for the counselor.

Step 7: Rotate roles.

Portfolio of Competence

This chapter is about two key interpersonal influencing skills, self-disclosure and feedback, and it covers considerable material. You cannot be expected to master all these skills until you have a fair amount of practice and experience. At this point, however, it will be helpful if you think about the major ideas presented in this chapter and where you stand currently. Also, where would you like to go in terms of next steps?

Use the following chart to evaluate your present level of mastery. As you review the items below, ask yourself, "Can I do this?" Check those dimensions that you currently feel able to do. Those that remain unchecked can serve as future goals. Do not expect to attain

Feedback Form: Disclosure and Feedback

You can download this form from Counseling CourseMate at CengageBrain.com.

_____ (DATE)

_____   _____
(NAME OF COUNSELOR)   (NAME OF PERSON COMPLETING FORM)

Instructions: The two raters will complete the form and then discuss their observations with the practicing counselor and the volunteer client.

1. Did the counselor use the basic listening sequence to draw out and clarify the client's story or concern? How effectively?

2. Provide nonjudgmental, factual, and specific feedback for the counselor on the use of the specific influencing skill or directive strategy. How empathic was the feedback?

3. As you view the totality of the session, where was the client at the beginning on the Client Change Scale? Where was he or she at the conclusion? What aspects of the skill or strategy impressed you as most useful and effective?

4. Evaluate the effectiveness and empathic level of the use of self-disclosure.

intentional competence on every dimension as you work through this book. You will find, however, that you will improve your competencies with repetition and practice.

Skill/Strategy	Level 1: Can you identify/classify the skill and write example statements?	Level 2: Can you demonstrate the skill in a role-played interview?	Level 3: Can you demonstrate your ability to use this skill in interviews with specific impact on your clients?	Level 4: Can you teach this skill to others?
Self-disclosure				
Feedback				

▶Determining Your Own Style and Theory: Critical Self-Reflection on Self-Disclosure and Feedback Skills

You have encountered and practiced two skills: self-disclosure and feedback. What did you learn about these skills? How are you planning to use them? With which one do you feel more comfortable? How do you feel about the idea of consciously revealing personal aspects of yourself? How do you feel about providing direct feedback to your clients?

What ideas stand out for you among all those presented in this chapter, in class, or through informal learning? What stands out for you is likely to be important as a guide toward your next steps. What are your thoughts on multicultural issues and the use of these skills? What other points in this chapter strike you as important? How might you use ideas in this chapter to begin the process of establishing your own style and theory?

CHAPTER 13

Concrete Action Strategies for Client Change

Logical Consequences, Instruction/Psychoeducation, Stress Management, and Therapeutic Lifestyle Changes

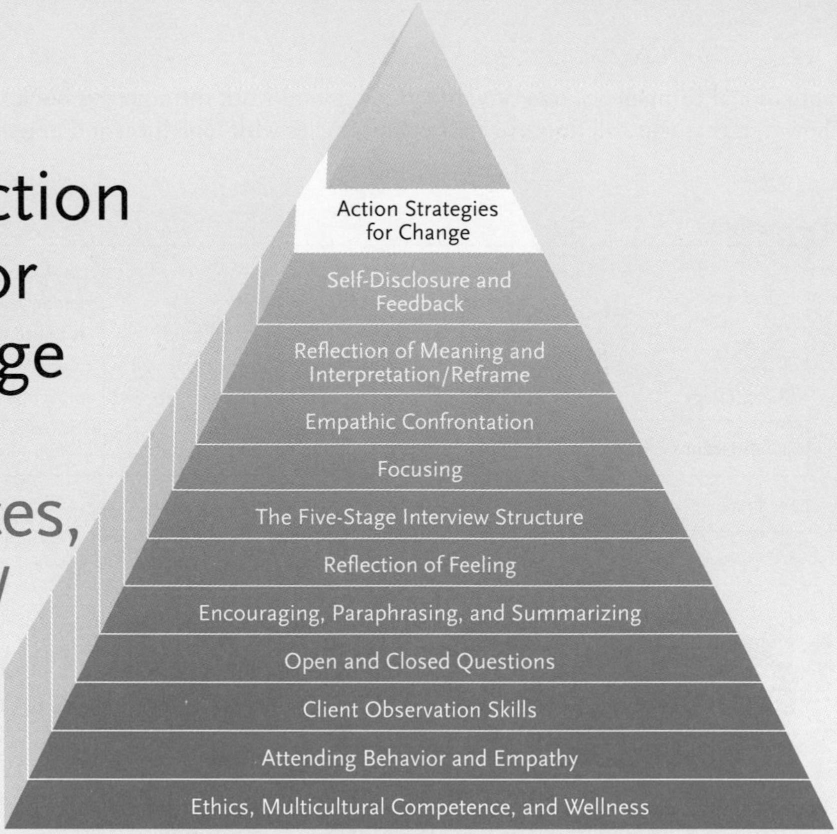
Action Strategies for Change
Self-Disclosure and Feedback
Reflection of Meaning and Interpretation/Reframe
Empathic Confrontation
Focusing
The Five-Stage Interview Structure
Reflection of Feeling
Encouraging, Paraphrasing, and Summarizing
Open and Closed Questions
Client Observation Skills
Attending Behavior and Empathy
Ethics, Multicultural Competence, and Wellness

Blessed is the influence of one true, loving human soul on another.

—George Eliot

Mission of "Concrete Action Strategies for Client Change"

The action skills and strategies of logical consequences, instruction/psychoeducation, stress management, and therapeutic lifestyle changes are designed to help clients examine new possibilities for their behaviors, thoughts, and feelings and then take action in the real world. With competence in listening and other influencing skills, you are well prepared to add these more active skills and strategies to facilitate clients in restorying, dealing with their issues, and making a difference in their lives.

Chapter Goals and Competency Objectives Awareness, knowledge, skills, and actions with the influencing skills and strategies will enable you to

▲ Help clients examine the logical consequences of alternative choices and actions.

▲ Use instructional processes and psychoeducation interactively in the helping process, rather than just telling.

▲ Understand the significance of stress and stress psychobiology underlying most issues in counseling and therapy and how stress management has become an essential part of practice.

▲ Become aware of the strategies of therapeutic lifestyle changes (TLC), which have been demonstrated to improve both mental and physical health, and explore their relationship to stress management.

▲ Connect these skills and strategies to other counseling and therapy theories, particularly decisional counseling and cognitive behavioral therapy.

 Go to CengageBrain.com to access Counseling CourseMate, where you will find an interactive ebook, quizzes, videos, interactive counseling and psychotherapy exercises, the Portfolio of Competence, case studies, a practice test, and more.

▶Defining Logical Consequences

Consider every action in terms of its consequences for seven generations to come.

—Yakima nation proverb

Decisions and actions have logical consequences that reverberate throughout a client's here-and-now life. In addition, critical decisions change one's future life. If important decisions are made poorly, with little attention to the logical consequences of the decisions, lifetime difficulties can result for both clients and those around them. Logical consequences is a strategy appropriate for most theories of counseling and psychotherapy, particularly decisional counseling, cognitive behavioral therapy, motivational interviewing, and brief counseling. Generating clear, logical decisions is critical in crisis counseling.

The strategy of logical consequences is typically a gentle strategy with the goal of clarifying what happens when behaviors don't change, decisions are made without adequate thought, or the client is unaware of dangers that lie ahead. On the other hand, some clients will welcome and respond to a more dynamic and active approach. There are also positive consequences that can result from counseling that increases thoughtful, emotionally aware decisions and behaviors.

The task is to help clients foresee the results and consequences of each possibility for action: *"If you do _____, then _____ will/may result."* When you use this strategy, you can anticipate how clients are likely to respond and how they can anticipate the potential consequences of alternative decisions.

LOGICAL CONSEQUENCES	ANTICIPATED RESULT
Explore specific alternatives with the client and the concrete positive and negative consequences that would logically follow from each one. "If you do _____, then _____ may result."	Clients will change thoughts, feelings, and behaviors through better anticipation of the consequences of their actions. When you explore the positives and negatives of each possibility, clients will be more involved in the process of decision making.

As an example of how the strategy of logical consequences relates to decisional counseling, suppose that a client comes to the session excited about a new job opportunity that offers advancement and higher pay, but fearful of the effects of moving the family to a new city. Through further questioning and discussion, the counselor can help the client clarify the factors and the consequences of the decision.

For many decisions, the various alternatives are apt to have both negative and positive consequences. Potential **negative consequences** of changing jobs could include leaving a smoothly functioning and friendly workgroup, disrupting long-term friendships, and moving teenage children to a new school. **Positive consequences** might be the pay raise, a better school system, and money for a new home, plus the opportunity for further advancement.

The counselor can help clients become aware of the potential negative consequences of actions, including negative results and even punishment. Consideration of possible negative consequences is necessary for a client who is thinking of dropping out of school, a pregnant client who continues smoking, or a client who wants to "tell off" a boss, co-worker, or friend.

We also need to enable clients to anticipate the positive consequences, results, and rewards of decisions and behaviors. Finishing school will lead to a better job, the nonsmoking pregnant woman's baby is likely to be healthier, and a better alternative for handling difficult people may be simply to keep one's mouth shut for the moment. It is best when your clients can generate the likely consequences of any given action through your listening skills, questioning, and confrontation.

Choices and decisions have a central emotional and feeling component. The decision may be rational and seemingly correct, but the client needs to feel satisfaction or a sense of peace before moving forward. There are many examples of tough but wise choices, such as accepting a new job in a new city that is a good move professionally but means leaving behind friends and family. Clients need to balance the advantages and disadvantages. Other difficult decisions center around leaving an abusive partner and facing financial hardship, or saying no to drugs or to cheating at school and then being teased by friends who do. Many ultimately wise and positive choices have some short-term emotional consequences, and we need to prepare our clients to face those issues as well.

How do *strategies* relate to *skills*? A skill is something that you can do, and typically a skill can be identified through counselor verbal and nonverbal behavior (e.g., attending behavior, paraphrasing, feedback). A strategy is a plan for action that may involve the use of several microskills, but may also require unobservable cognitive thought. Many influencing strategies require a combination of skills. For example, logical consequences and providing instruction or psychoeducation typically require us first to listen with empathic understanding, then to make sure that the client is interested and involved in what we are sharing via strategy, and only then should we move to more direct action. Skills are implemented in counseling via thought, which is then turned into action for the client's benefit.

The following brief vignette illustrates the logical consequences strategy in action. Note that the counselor (who has previously listened to the story) uses a combination of questioning and paraphrasing so that the client brings out the answer rather than being told what to do by the counselor.

Counselor: What is likely to happen if you continue smoking while pregnant?

Brightstar: I know that it isn't good, but I can't stop and I really don't want to. Smoking relaxes me and I've done it for years with no harm. (The client partially answers the question and avoids the consequence for the child.)

Counselor: You really feel more relaxed when you smoke. I wonder what you think are the possible negative consequences of continuing to smoke? (Reflection of feeling, followed by a somewhat confrontive open question.)

Brightstar: (pause) I've been told that the baby could be harmed.

Counselor:	Right, is that something you want? What is the benefit of stopping smoking for the baby? (The use of logical consequences can involve too much leading. Be careful or you will soon lose the client.)
Brightstar:	No, I don't want to do harm. I'd be so guilty. But how can I stop smoking?
Counselor:	You would really miss smoking, but you feel a bit guilty about the possible harm. Let's explore those feelings a bit more. Tell me more about what you are thinking . . . (Here the discussion becomes open to a more mutual exploration of the issues, short- and long-term consequences.)

For disciplinary issues or when the client has been required to come to the session, remember that considerable power rests with the counselor, often resulting in a lack of trust. Trust issues and limits of confidentiality need to be explored thoroughly. The school, agency, or court may ask the counselor for recommendations that they may follow. Warnings are a form of logical consequences that may center on *anticipation of punishment*. If used effectively, and coupled with clients' rapport and listening, warnings may reduce dangerous risk taking and produce desired behavior. But warnings are only truly effective if the client is motivated to change and willing to listen.

In disciplinary situations, clients need to think about the fact that they will have happier lives in the long run if they select more positive alternatives for their lives. A school disciplinary official, an attorney, or a correctional officer often needs to help the client see clearly what might be ahead. But people who hold power over others need to *follow through with the consequences* they warned about earlier, or their power to influence will be lost. The "if, then" language pattern is especially useful to summarize the situation: "If we don't make the counseling sessions work, then you know what the judge will do."

Finally, when counseling a child diagnosed with conduct disorder or a teen or adult termed antisocial, your first consideration is safety and preventing harm to self or others. While gentleness remains the basic style, authentic and real firmness and strength are often needed. Counselors and therapists are often seen as easy "cons" by these clients.

Here are some specifics for using the logical consequences skill. Note that you will be using listening skills as part of the process.

1. *Draw out story and strengths.* Through listening skills, make sure you have drawn out the client's strengths and resources that may be useful in resolving issues.
2. *Generate alternatives.* Use questions and brainstorming to help the client generate alternative possibilities for restorying and resolving issues. Where necessary, carefully provide your own additional ideas.
3. *Identify positive and negative consequences.* Work with the client to outline both the positive and negative consequences of any potential decision or action. Ask the client to generate a possible future story of what might happen if a particular choice is made. For example, "Imagine yourself two years from now. What will your life be like if you choose the alternative we just discussed?" In anticipating the future, special attention needs to be paid to likely emotional results of decisions and continued or changed behavior. Emotions are often the ultimate "decider."
4. *Provide a summary.* As appropriate to the situation, provide clients with a summary of positive and negative consequences in a nonjudgmental manner, or ask them to make the summary. Pay special attention to emotional issues. With many people, this step is not needed; they will already have made their own judgments and decisions.
5. *Encourage client decision and action.*

▶Defining Instruction and Psychoeducation Strategies

We don't usually think of counseling and therapy as providing instruction to clients, teaching them new skills, or educating them. However, instruction and psychoeducation are common strategies in career counseling and in providing health and lifestyle information. Community and family genograms require instructional procedures. Stress management, lifestyle changes, crisis counseling, and many counseling and psychotherapy theories will use instruction and psychoeducation.

All instructional and psychoeducational strategies, like logical consequences, require us to work first from the client's frame of reference. Clients who are not interested or involved are not likely to listen to the information being shared long enough to be useful in their daily life.

Closely allied to instruction but somewhat different is psychoeducation, in which the counselor takes on a rather specific teaching role and provides systematic ways for clients to increase their life skills. These life skills intertwine with therapeutic lifestyle changes and stress management, which are discussed later in this chapter. Psychoeducation has become prominent in recent years in cognitive behavioral therapy (CBT) and is now a preferred treatment that prepares clients to change behavior and enables them to meet a wide variety of challenging situations.

Predicting what clients will do when you provide information or suggestions or teach them skills via psychoeducation is somewhat different than with other skills because it takes time to see whether the information "took" and was useful to the client. If you use these skills as described below, you can anticipate how clients are likely to respond, but always be ready with alternatives when they don't show interest or act on what has been presented.

INSTRUCTION AND PSYCHOEDUCATION	ANTICIPATED RESULT
Share specific information with the client, such as career information, choice of major, or where to go for community assistance and services. Offer advice or suggestions on possible ways a client can resolve issues, and support their decisions. Teach clients specifics that may be useful, such as helping them develop a wellness plan, teaching them how to use microskills in interpersonal relationships, and educating them on multicultural issues and discrimination.	If information and ideas are given sparingly and effectively, the client will use them to act in new, more positive ways. Psychoeducation that is provided in a timely way and involves the client in the process can be a powerful motivator for change.

Providing effective instruction and psychoeducation requires a similar approach to that presented for logical consequences. As these are action strategies, it is essential that clients participate fully.

1. *Involve your clients as co-participants in the instructional or psychoeducational strategy or program.* Draw out their story and their strengths. Work with them to find appropriate and doable goals. Out of this will often come areas that could benefit from instruction and psychoeducation. Share what you think (or know) can be helpful, always with awareness that the client has to want to hear what you have to say and believe that it will make a difference in her or his life. Again, we urge *working with*, rather than *working on*, your client.

2. *Be clear and concrete in your verbal expression, and time the instruction to meet client needs.* Directives need to be authoritative and clear but also stated in such a way that they are in tune with the needs of the client. Compare the following:

   I think you can manage some of this by starting to relax. I'd like you to try that next week. (*vague instruction*) Relaxation training and breathing are good ways to deal with stress. We'll focus on some others later, but right now, your body seems tense and your breathing is shallow. Have you done relaxation training before? (*concrete preparation for instruction*)

   I've got a handout that you can take home for practice, but now I'll share some specifics of relaxation training. Are you ready? (*concrete instruction*)

   Sit quietly . . . feel the back of the chair on your shoulders . . . close your eyes . . . notice your breathing and take a deep breath . . . now we will explore the tension in your muscles and work on relaxing them . . . now, tighten your right hand . . . hold it tight . . . now let it relax slowly . . . (the counselor then moves through the body step by step). (*concrete instruction*)

   Relaxation training and meditation will be explored in more detail later in this chapter. Both have been found to be useful in helping clients get started on the way to health. These examples illustrate the importance of indicating clearly to your client what to expect. Know what you are going to say, and say it clearly and explicitly.

3. *Watch for client resistance to instructions and psychoeducation, and adjust your approach accordingly.* Research has found that if the client is angry or resistant, a more nondirective listening approach may be more effective. If the client is interested and cooperative, clear explicit instructions and suggestions are more likely to be accepted (Karno, 2005).

4. *Check out whether your instruction was heard and understood.* Just because you think you are clear doesn't mean the client understands what you said. Explicitly or implicitly

check to make sure you were heard. This is essential when using a more complex strategy. For example, "Could you repeat back to me what I just asked you to do?" or "I suggested three things for you to do for homework this coming week. Would you summarize them to me to make sure I've been clear?" The Client Change Scale (CCS) can be used to determine whether the client actually changed thoughts, feelings, or behaviors as a result of your directive.

Before we move on to a detailed examination of influencing skills, it may be useful to think about cultural issues that continually modify the nature and use of both listening and influencing skills. Box 13.1 speaks to these issues.

BOX 13.1 National and International Perspectives on Counseling Skills

Explorations Around the World: Learning New Approaches to Working With Cultural Strengths

Mary Bradford Ivey

Allen and I have been lucky to be invited frequently to present internationally on microskills, developmental counseling and therapy, and multicultural issues. The microskills approach has always focused on positives. In this chapter, we give special attention to stress management and therapeutic lifestyle changes (TLCs), which are cross-culturally important.

Each time we encounter a new setting, we learn something new that can be added to the strength inventory. At the same time, we inevitably discover that the basic listening sequence (BLS) is central, regardless of the country or culture in which we present. Whether one counsels in Aboriginal Australia, the Canadian Arctic, Tokyo, or Tel Aviv, or consults with those in Africa working with refugees or AIDS issues, listening comes first. Always, the BLS and other microskills need to be adapted to meet the verbal and nonverbal behavior of unique individuals who exist in a cultural context.

Adelaide, Australia This experience was pivotal in our thinking and the development of our view of multicultural issues. We worked with Aboriginal Australian social workers developing microskill videos appropriate for their culture. Among many other things, we learned that counseling must be collaborative and egalitarian. It often takes a half hour or more to establish an empathic relationship—more if you are White or from another culture. Aboriginal people wanted to know who we were and where we came from. It was important to understand their family relationships and connections in some detail. We found that self-actualization in the Aboriginal world is thought of as "selfish-actualization." In many cultures, one can only be self-actualized if one is fully connected to others.

Jerusalem, Israel How can we work with children who may have to wear gas masks when bombs drop on random parts of the country? This was the challenging topic that greeted us as we met with the country's counseling supervisors. Yes, we must listen to the children's fears and concerns, and they need help in anticipating and dealing with emotions. They also need clear and specific directions on how to put on a gas mask and what they should do during a bombing. Children (and adults) can draw on the spiritual and cultural strengths of the Jewish tradition. We do not think of such issues often in counseling training, but we need to be prepared for surprises in both cultures and new situations.

Tokyo, Japan Should we translate the English version of microskills directly into Japanese, or should we make culturally appropriate language modifications for the culture? The answer, of course, is that we must adapt the microskills framework for each setting (and each person). An interesting aspect of Japanese culture is the importance of hierarchy and respect for elders. The older person is the one who is most appropriate to advise and direct younger people and staff members. What does this mean for you? If you are younger and working with an elder in any culture, respect age difference if you are to be seen as a respected and trustworthy counselor.

Izmir, Turkey Turkish counselors are a very select group, as a very low percentage of students are admitted to master's and doctoral programs. The Turkish people are kind to strangers and are great at partying. Allen and I found ourselves most accepted when we joined them in joyful group traditional Turkish dances. A real highlight of learning new cultures sometimes is letting go and joining their way of being.

One of Allen's former students, Nancy Baron, has become an international leader in work with refugees and in training AIDS counselors. Refugees need crisis counseling and advice, plus timely stress management information on how to make it through today and tomorrow. Nancy brings refugees together in small groups as they share their stories of trauma; in addition, she teaches them culturally appropriate listening microskills so that they can talk together and listen more effectively. Nancy has set up a worldwide training center in Cairo for individual and group cultural communication and crisis counseling. Those who face AIDS need clear, understandable instruction, psychoeducation, information, and advice. With cultural adaptations, the microskills are useful throughout the world.

Each client comes to us with a unique and special multicultural background. The RESPECTFUL model tells us that we constantly need to be ready to change and adapt. But, best of all, what a wonderful opportunity for us to learn from and with our clients!

▶Stress and Stressors

No cognition without emotion, no emotion without cognition.

—Jean Piaget

Stress is a psychological and physical response to change, whether that change is actually happening now or anticipated in the future. We all have positive and negative stress in our lives. Positive stressors tend to make us happy and joyful in many ways. Examples are planning for a big date or marriage, being deeply involved at an opera or a baseball game, rock climbing, running, or driving fast on a racetrack. These are sometimes called *eustress*, from the Greek *eu*, meaning "well" or "good," plus *stress*. Eustress can provide fulfillment, such as the satisfaction of a job well done, being able to help another person, learning and completing a new and difficult task, graduating from a master's degree program, or helping build a Habitat for Humanity house.

Each of these good things affects the body as well as the mind. Along with activating the TAP executive functions, these positive stressors raise blood pressure and heart rate, change the pattern of hormones throughout the body, and lead to experiencing positive social emotions and the underlying feeling "glad." Cortisol, necessary for learning, is produced in small bursts, improving immune functioning and balancing metabolism. In times of emergency, cortisol improves functioning and our ability to respond quickly to dangerous and frightening situations.

It is negative stressors that bring most clients to counseling and psychotherapy. Many issues and concerns involve stress in various ways, ranging from the ending of a relationship to failing an exam to depression to thoughts of suicide. Sad, angry, and scared feelings and emotions, associated with the limbic HPA, are usually paired with negative stress. The TAP cognitive functions may be impaired. As in positive stress, blood pressure and heart rate increase, but in damaging ways that can eventually lead to general distress, including anxiety disorders, depression, ADHD, and general ineffective cognitive/emotional functioning. The adrenal glands (the "A" of HPA) produce an overabundance of cortisol, which can injure immune functioning, cause inflammation, strokes, and heart attacks, and increase the possibility of Alzheimer's disease in later life. Out of this understanding has come awareness of the central importance of stress management in counseling and psychotherapy.

Stress can be good or bad. It makes life interesting and exciting, or overwhelming and challenging. Appropriate stress encourages neurogenesis, new synapse connections, and is required for learning and even change in counseling and psychotherapy. Stress management and therapeutic lifestyle changes (TLC) are key to handling stress effectively, and also to preventing harmful stress.

Key to stress management is body awareness. Note when your body and mind are tense, where and how it feels. How does your body feel during positive moments, including relaxation and peaceful feelings? The nature of stress and its body impact will become clearer if you try the following reflective exercise.

REFLECTION   *The Impact of a Simple Stressor*

With a partner: Try this exercise with a friend, taking turns as leader and follower. Have your partner close his/her eyes. Then say in a slow, even voice "NO NO NO NO NO NO." The partner then opens his/her eyes and you debrief the experience. How was the word "NO" felt in the body? What thoughts and feelings occurred for your partner? You will find that repetition of the negative word "NO" has an almost immediate impact.

Now ask your partner to close his/her eyes again. In a slow, even voice, say "YES YES YES YES YES YES." Open the eyes and debrief this experience, comparing it with "NO." Typically, it takes several "YES's" to relax a person from the multiple "NO's."

On your own: Close your eyes and visualize some error or difficult experience from the past, along with the word "NO." Take a moment to know both body and mind. Then follow this with a positive joyful memory. Take some time to debrief.

This exercise provides some understanding of what even the simplest stressor can do to the body.

A child or adult who receives a negative comment or goes through a difficult personal experience experiences stress. These negative events have an impact on the amygdala and imprint memory in the hippocampus at a deeper level than positives. Recall Nelida, the client in transcripts of Chapters 9 and 10—one negative comment in the classroom kept her quiet. Her prefrontal cortex made the decision to stop talking, and the comment was held in memory in the hippocampus, Again, think of the idea that it takes 10 positives to counteract a negative. If there is a trauma, even 10 will not be enough and the negative memory may take over one's life. It has been suggested that clients who have a negative view of life start a journal and note both positive and negative happenings. Usually, they are surprised at how many positives have occurred without their giving full attention. Similarly, when we anticipate a negative experience, we can protect ourselves partially by deliberately planning at least three positives during the day before it happens. If the negative experience is anticipated, rather than an unpleasant surprise, the impact will be less powerful and damaging.

Hans Seyle first brought issues of stress to our attention in 1956, and his influence has grown over the years. He identified positive stress as *eustress*, typically related to good events in the client's life—doing well in an exam, winning an athletic contest, riding a roller coaster, or getting a promotion. When you bring out pleasant stories from a client's life, you help the client focus on positive stressors from the past. But too much positive stress can also wear on the body—and the client's body cannot tell the difference between eustress and negative stress. You will find some clients who actually "suffer" from eustress. They are often

TABLE 13.1	Positive and Destructive Effects of Stress
Positive Stress	**Destructive Stress**
Neurogenesis and learning	Excess cortisol and neural loss
Frontal cortex strengthened	Limbic system and negative feelings take control
Gray and white matter increase	Neural loss and brain shrinkage
Reduction in size of amygdala	Enlargement of amygdala
Longer, healthier life	May lead to mental and physical illness

© Cengage Learning

busy, multitasking people who enjoy their work and are constantly succeeding, but always asking to do more. Eventually, they and their body wear out. These clients can also benefit from stress management.

Table 13.1 summarizes some of the potential life benefits and challenges that stress provides us.

Figure 13.1 shows how a high level of stress activates the brain. This figure appeared in Chapter 1 (Figure 1.3), but we repeat it here because it so clearly illustrates the importance of stress management. Continuing stress over time from bullying, work pressure, or interpersonal difficulties can damage the brain and affect client functioning. Under abnormal levels of stress, the brain is turned on forcefully, is lit up, and eventually can burn out. The brain and body can endure only so much stress.

The centrality of stress in so many counseling and therapy issues means that counselors and psychotherapists need to be competent in the strategies of stress management, which are easily combined with virtually all theories of helping.

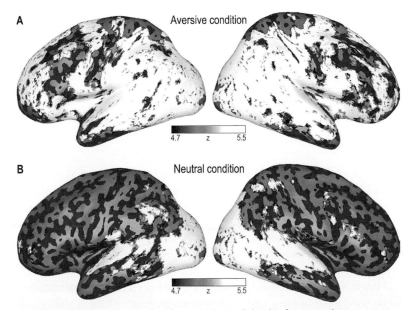

FIGURE 13.1 Effect of abnormal (aversive) and normal (neutral) levels of stress on brain activation.

Source: Hermans, E., van Marle, H., Ossewaarde, L., Henckens, A., Qin, S., Kesteren, M., Schoots, V., Cousijn, H., Rijpkema, M., Oostenveld, R., & Fernández, G. (2012). Stress-related noradregenic activity prompts large-scale neural network configuration. Science, 334, 1151–1153. Reprinted with permission from AAAS.

▶Stress Management and Therapeutic Lifestyle Changes

Treatment is based on managing stress.

—Patrick McGorry, M.D.

Stress management is an organized series of instructional strategies to enable clients to live with and deal with stress effectively. Most therapists and counselors draw on stress management strategies to help meet client needs and interests. Stress management is commonly thought of as a part of cognitive behavior therapy (CBT), but it stands on its own as a system and is used widely by therapists and counselors of many theoretical orientations and beliefs. In today's overly busy world, our brain circuits become overloaded and we may underperform (Hallowell, 2005). As a result of stress, we may make poor decisions, fail to decide what is most important or what to do next, or even suffer emotional breakdown.

Therapeutic lifestyle changes (TLC) are also instructional strategies that are oriented to physical and mental health, bringing together neuroscience, medicine, and counseling. TLC is a term that likely originated through early preventive work with heart attack survivors (Freidman et al., 1986). The Recurrent Heart Attack study cut the number of second heart attacks in half through behavioral counseling with Type A hostile and angry patients. Behavioral counseling emphasized a more relaxed lifestyle, provided specifics for behavioral change, and enabled the heart attack survivor to have a lengthened life. Since then an avalanche of primarily medical studies have reported successful applications of various TLC concepts with cancer survivors, people with diabetes, people with Alzheimer's disease, and many others.

STRESS MANAGEMENT AND THERAPEUTIC LIFESTYLE CHANGES	ANTICIPATED RESULT
These instructional strategies are designed to improve the physical and mental health of clients. Relaxation, meditation, irrational beliefs, thought stopping, and time management are some of the skills and strategies used to manage stress and change lifestyle.	Clients will use this information to practice managing their stress and change their lifestyle to improve physical and mental health. Physical and mental symptoms will improve over time.

Stress management traditionally has served as a *remedial* treatment for already stressed and needy clients. TLCs use stress management strategies and many others, but their focus is on *primary and secondary prevention* of stress, although TLCs are increasingly used as stress management treatments as well. Both stress management and TLCs have a positive impact on mental and physical health and are bringing medicine, counseling, and psychotherapy together through understanding of the brain and brain research. Considerable research has demonstrated their effectiveness and value.

Box 13.2 presents a list of common stress management and TLC instructional strategies that can be used in counseling and clinical practice. Some of these will be elaborated briefly in this chapter. For more detailed discussion of these strategies, plus many others, we suggest *The Relaxation and Stress Reduction Workbook* (Davis, Eshelman, & McKay, 2008), which is updated periodically.

BOX 13.2 Stress Management and TLC Instructional Strategies

*Associated Primarily
With Stress Management*

Logical consequences
Social skills training
Assertiveness training
Thought stopping
Imagery, guided imagery
Positive reframing
Time management
Conflict resolution
Gestalt exercises
Biofeedback
Neurofeedback
Relaxation training
Cognitive behavior therapy (CBT)
Reframing thoughts, emotions, and
   behaviors
Homework

Associated Primarily With Therapeutic Lifestyle Changes

The Big Six TLCs

1. Exercise (avoid sitting)
2. Sleep
3. Social relations
4. Nutrition
5. Cognitive challenge and education
6. Meditation

Other highly useful TLCs

Control screen time for TV, games, and
   computers
Prayer
Positive thinking/optimism
No drugs/limited alcohol
No smoking

Religion/spirituality/strong value
   system
Taking a nature break rather than a
   coffee break
Enjoyable hobbies of any type: art,
   music, collecting, cards, reading,
   etc.
Helping others, social justice action
Careful use of medications and
   supplements
Homework

▶Concepts Into Action: Psychoeducational Stress Management Strategies

Psychoeducation is instruction that has both an educational and a therapeutic purpose. One common style of psychoeducation involves life skills training; it can involve role-playing and enactment of scenes from the client's here-and-now life or anticipated future situations. This type of psychoeducation, which may involve several sessions, planned homework, and readings, is frequently found in group and family work. Communication skills training, assertiveness training, and thought stopping are three examples.

Communication Skills Training

Teaching attending behavior and the listening skills to clients and groups has been part of the microskill paradigm since its inception. Skills training is particularly effective with clients who are shy or have mild depression. You will recall that we showed how the teaching of microskills to college students and to depressed and even schizophrenic psychiatric patients proved an effective treatment supplement. At this time, it may be helpful to return to page 79 of Chapter 3. We consider the teaching of skills so important that we spoke of this in Chapter 3 and continue it throughout the book as part of competency education or training as treatment. Teaching communication skills can be a useful part of your strategies for counseling and therapy.

Microskill psychoeducation as treatment is similar to what you have encountered in your own practice sessions, but the emphasis is almost totally on attending, observation, and the basic listening sequence. The difference is that instead of counseling, you now teach client communication skills to help clients develop better interpersonal relationships. The specifics of teaching communication microskills summarized below will be quite familiar to you by this point.

The "how" of teaching skills to clients is straightforward and step-by-step. However, note that steps 1 through 4 and 8 through 9 below are particularly important for maintaining empathy and trust throughout instruction and psychoeducation (and other influencing strategies as well).

1. A solid empathic relationship is essential.
2. Hear the client's story and identify strengths. Look for specific areas of broken communication with parents, spouse, children, roommates, business associates, or the boss. Skills as basic as attending, paraphrasing, summarizing, and careful acknowledgment of feelings can help resolve serious difficulties and conflict with others. Teaching just attending alone often conveys the basic point of the importance of listening to those who are open.
3. Check out your clients' interest in learning listening or other skills—what are their goals? Depressed clients may have difficulty in setting goals, given their feelings of defeat and inadequacy. They often have difficulty communicating with others. Note their frequent lack of eye contact and "down" body language. A good way to help them and many other clients is to teach attending behavior and questioning skills. When one is attending to someone else, focus on oneself is difficult. And the response we get if we listen to others is a great reinforcement for our self-esteem.
4. Verbally give them information on the content of the skill (e.g., the three V's + B).
5. Role-play an ineffective style of listening, and discuss what is unproductive.
6. If possible, video or audio record that session. Watch or listen to the recording and encourage clients to decide what behaviors that they might want to change.
7. Role-play a positive session using good listening skills.
8. Contract for action in the real world via homework assignments that are meaningful for the client.
9. Follow up to check effectiveness and next steps.

Use the same framework for teaching other communication skills or to practice unique behavior changes that you and the client identify. For example, Allen used this framework successfully with seriously disturbed and hospitalized war veterans. He would videotape the patient in a session, review the tape with the patient, and then let the patient decide what behaviors he wanted to change. Highly distressed patients often learned best if only one component of attending behavior (e.g., eye contact, verbal following) was taught at a time.

Assertiveness Training

Assertiveness training follows a model similar to that of communication skills training via microskills. As an example, a client might be in a work situation where others treat her or him unfairly and the client simply sits, takes it, and does not speak up. Again, you need a good working alliance and relationship followed by hearing the client's story fully. Be sure to draw out strengths and resources; find places where the client has been effective. A community genogram in which you identify strengths will often be useful to further support the client in the here and now of the session. The client goal is obviously becoming more assertive and in control of one's own behavior.

Assertiveness training has long been a treatment of choice for women who find themselves in challenging situations. At the same time, be very careful, because a woman who is an emotionally abusive relationship may escalate a bad situation by becoming more assertive. The nature of counseling in this situation may be for the woman to think seriously about alternatives and her desire to continue the relationship. If the situation involves physical abuse, your first responsibility is keeping the woman safe—skills training for survival can come later. Know the location of a safe house.

During the restorying phase, carefully sort out thoughts and feelings and look for client illogical thinking and ineffective behaviors. Here cognitive behavioral strategies such as examination of automatic thoughts will be helpful (see Chapter 15). Once the goals are clear, help the client practice more assertive and appropriate behaviors for responding in the workplace or elsewhere. The video practice model of microskills above can be a powerful addition to assertiveness training. And, of course, contract for action and follow up with your client, providing more support and further assertiveness practice as needed. Some clients may be speaking up too loudly for their own beliefs and desires, disregarding others. The same system of assertiveness training can be used to educate them in more appropriate and useful modes of social interaction.

Thought Stopping

This strategy is so simple that it is easy to ignore or even forget the many things that go on during a session. But we have consistently found that it is one of the most effective interventions a counselor or therapist can use. Please give thought stopping your serious attention.

Many of us have a pattern of internalized negative self-talk. These stressful thoughts may be said to yourself, perhaps even several times a day, as in the following examples:

"I can't do anything right."
"I am not a good person."
"I always foul up."
"I should have done better."

Other areas for negative thoughts include guilty feelings, procrastination in which you might "overthink" the situation, worry over details, fear that things will only get worse, anger that others "never get it right," repetitively thinking about past failures, and always needing the approval of others.

The best way to understand the potential of thought stopping is to apply it to yourself. Start by thinking of some of your negative personal thoughts that are, in effect, self put-downs. For example, some people think that they are failures and can't complete a task. Often clients (and perhaps you as well) ruminate over something that was done wrong or something that someone said. These can repeat in your mind until they become part of your self-concept. Particularly important, start noticing negative thoughts that may come into your head. STOP THEM!

Having identified these areas, try one or more of the following:

1. Relax, close your eyes, and imagine a situation in which you are likely to make the negative self-statement. Take time and let the situation evolve. When the thought comes, observe what happens and how you feel. Then tell yourself silently, "STOP." In emotional thought stopping, say it loudly—even shouting—at first.
2. Place a rubber band around your wrist, and every time during the day that you find yourself thinking negatively, snap the rubber band and say, "STOP!"
3. Once you have developed some understanding of how often you use negative thinking, after you say "STOP," substitute a more positive statement such as the following, which are suggested in place of the negative self-statements.

   "I can do lots of things right."
   "I am a capable person."
   "I sometimes mess up—no one's perfect."
   "I did the best I could."

4. Do this a minimum of one full day.

What are some positive self-statements that you can make to replace your own negative self-talk?

Positive Guided Imagery

Closely allied with psychoeducation, imagery is a popular technique to help clients relax and discover positive resources (Utay & Miller, 2006). All of us have experiences that are memorable for us—maybe a lakeside or mountain scene, a snowy setting, or a special quiet place. The image can become a positive resource to use when you feel challenged or tense. For example, we feel tension in our bodies when we encounter conflict. If we learn to notice our internal body tension, then we can immediately and briefly focus on a relaxing scene, take a deep breath, and deal with the challenging situation more effectively. When giving guided imagery instructions, time your presentation with your observations of the client.

> "Imagine yourself in a pleasant and relaxing situation, where you feel totally yourself. Relax, then close your eyes and enjoy that feeling. (long pause) What are you seeing, hearing, feeling?" (This itself may produce the relaxation response. Visualizing positive situations when under stress is calming.)

> "What is your image of your ideal day/job/life partner?"

> "Close your eyes and visualize someone who was helpful and supportive to you in the past—a family member, a friend, a teacher, or someone else. Think just briefly of the issues that you face with that person. Now visualize that person listening carefully to you. What is that person saying to you? What are that person's feelings when saying this, and how do you feel inside right now?"

> "Please tell me about a time when you were able to express yourself in a challenging situation and hold your opinion." (The counselor draws out the strength story.) "Now imagine you are back in that situation. Close your eyes and tell me about it. What are you seeing, hearing, feeling?" (Better than the previous.)

Imagery directives are often powerful and must be used with care. They need to be followed by a debriefing and follow-up action with the client.

Exploration of negative images is usually inappropriate and can be unethical. The counselor needs to be fully qualified and the time and situation appropriate for the client. False memories can easily occur and harm the client.

Homework

Stress management and therapeutic lifestyle changes are irrelevant unless taken home, practiced, and used by your clients. The fifth stage is action, and that means action outside the session. Thus, we need to negotiate with our clients to ensure that they are interested, willing, and able to do something in the next day or week to change behavior and try something new. Concreteness and a contract for action are essential.

Homework assignments can include following up on a stress management strategy or a lifestyle change. Homework can be playing basketball with friends, starting a program of walking or running, or keeping a diary of foods eaten. If thought stopping is the issue, the homework may be to journal the level of success each day.

With some clients, just observing themselves leads to changes in their behavior. The client can also record what happened just before the thought and what happened afterward. Such records can be valuable in changing faulty thinking patterns. There are endless ways to involve clients. A couple with difficulties may be asked to observe and count the number of arguments they have. They record the before, during, and after aspects of the argument to discuss with the counselor. No change is expected because the client is simply observing and recording what is going on.

▶Therapeutic Lifestyle Changes

I don't smoke, I eat well, I exercise, I sleep. I have 80% less chance of chronic disease. There is no pill. Lifestyle is the medicine.

—David Katz, MD

Three quarters of health costs are related to lifestyle.

—Robert Jacobs

Therapeutic lifestyle changes (TLCs) are nearly perfect supplements for those of us who practice with a wellness orientation. Reviews of research show that they often are as effective as or more effective than medication (Ivey, 2012). TLCs are so effective as preventive agents for mental and physical health that they are now being recognized as bringing medical, counseling, and psychological practice together. The major difficulty in using TLCs is getting clients to try them and stay with them. Naturally, they will be most effective if they are included as part of treatment sessions coupled with supportive counseling to ensure action and generalization of lifestyle changes to daily life.

Bringing TLCs to counseling and psychotherapy represents a radical change for our profession. TLCs such as exercise, nutrition, and meditation are mentioned only briefly, if at all, in our training programs, although there has been a marked increase in interest in recent years. We suggest that you consider adding TLC instruction to your work in stress management and with many types of client issues.

A major challenge for counselors and therapists who wish to include TLCs as part of their treatment program is compliance—getting clients to take home and actually engage in desired behavior. The Big Six discussed below (e.g., exercise, sleep, nutrition) are likely well known to most clients, but many simply do not do what they know they should do. An assessment or a discussion of optimism and wellness from Chapter 2 may be useful in getting clients to think seriously about lifestyle changes and then actually moving to action. Setting up meaningful life goals (e.g., discernment) can be part of the process. TLCs will often be easier to use with clients who have serious physical challenges, those who fear Alzheimer's or diabetes, or clients recovering from a hospital stay, particularly after a stroke or heart attack. The survival rate for clients who have cancer is better for those who engage in TLCs. Mental health, of course, improves with all the TLCs.

Some 20 million new neural connections (synapses) are made and lost daily. We need to create new ones to compensate for those we lose. TLCs are an important part of the process of helping develop new connections and fostering mental health. All of them increase self-esteem, improve mental and physical health, are cost effective, and are backed up by extensive research in neuroscience, counseling, and medical research. *Counseling and medicine converge around both mental and physical health issues.*

While all of the TLC strategies summarized here can be therapeutic and support physical and mental health, six stand out as central: exercise, sleep, nutrition, social relations, cognitive challenge, and meditation. Termed the *Big Six*, they are particularly useful additions to counseling and psychotherapy with most clients. Many other TLC strategies should also be assessed and included as part of treatment. Individuals will have varying responses to different TLCs.

Exercise

Evidence is mounting for the benefits of exercise, yet psychologists (and counselors and physicians) don't often use exercise as part of their treatment arsenal.

—Kirsten Weir

TABLE 13.2 Therapeutic Lifestyle Changes: Exercise

Enriched	Destructive
Any exercise one likes and will do	Sitting reduces lifespan
Enhances sleep	Wakefulness
Produces dopamine, BDNF	Obesity
Treats depression, anxiety	Mental health issues
Increased gray matter	Loss of gray matter
Stand up for computer, reading	4 hours TV = 80% increase in heart death, 40% increase in death from all causes
Longer life	

© Cengage Learning

A central goal of stress management is to get blood flowing to your brain and body. Exercise increases brain volume, has been found as effective as meds for mild depression, may prevent cancer, and may slow the development of Alzheimer's disease. The best summary of exercise research and the implications for our practice may be found in John Ratey's (2008) *Spark: The Revolutionary New Science of Exercise and the Brain.*

Table 13.2 presents the major results of an exercise-enriched lifestyle as contrasted with one that is sedentary. All tables in this section are from Allen Ivey's (2012) review of the research literature.

We love and work more effectively if we are comfortable in our bodies. A sound body is fundamental to mental health. It is recommended that adults get at least 150 minutes of exercise each week—20 minutes a day, which is often 20 minutes more than people claim that they do. Just 15 minutes of daily exercise resulted in a 10% decrease in cancer death and 14% decrease in death overall. This means that exercise can give you, on average, an additional 3 years of life (Wen et al., 2011).

The central challenge in instructing clients in TLCs is compliance—encouraging them to take home what they know and do something about it. Obviously we can't tell a client what to do, as it would likely build resentment. Helping clients change their behavior involves subtle and skilled use of instructional strategies in which the client is a full co-participant in the process and "buys in." Box 13.3 presents evidence on the benefits of exercise.

The most powerful study on exercise to date found that exercise can overcome genetic issues that lead to brain atrophy and depression. Moreover, aerobic exercise can increase beneficial BNDF (brain-derived neurotrophic factor). BDNF has been referred to as "Miracle-Gro for the brain." It is critical in the growth of new neurons and synapses, especially in the hippocampus, cortex, and basal forebrain, which are all central to learning, memory, and higher thinking. This increase in BDNF leads to neurogenesis, a larger hippocampus, and less depression. This has been shown true in both mice and humans (Erickson, Miller, & Roecklin, 2012). The authors summarize their research findings with the following conclusions:

▲ Aging and lower BDNF produce lower hippocampal volume and thus poorer memory function. This, in turns, leads to increased likelihood of depression and possibly Alzheimer's.

▲ Exercise increases BDNF, serotonin, and hippocampus volume through neurogenesis. This results in improved memory and reduction or elimination of depression.

The Benefits of Exercise

Any physician (counselor or therapist) who does not prescribe exercise is unethical.

—A president of the American
Psychiatric Association
(Quoted by John Ratey, personal communication)

Exercise and mental health was the focus of the cover story of the American Psychological Association's *Monitor on Psychology* for December 2011 (Weir, 2011). The following points were made in summarizing key research:

▲ About 25% of the U.S. population gets no exercise at all.
▲ Exercise enhances mood. In some studies, clients with depression often do better with exercise than they do with medication. Those who continue exercise do better than those on medication.
▲ Exercise is beneficial to people with cancer, diabetes, multiple sclerosis, and other physical challenges.
▲ Exercise is useful in the treatment of anxiety and panic disorders.

▲ Smoking cessation can be more successful with exercise.
▲ Mice and humans share similar reactions to stress. Researchers found that "bullied" mice without the opportunity to exercise in enriched cages and hassled by dominant alpha mice hid in shadows and showed signs of anxiety and depression. However, mice in enriched cages with considerable exercise were able to "shrug" off bullying and social defeat, did not show the negative signs, and handled mazes competently.

Other research studies with older adults clearly show the benefits of exercise. One found increased plasticity in older adults who exercise (aerobic training and walking). They were examined using fMRI and were found to have increased executive functioning and better connections in higher-level networks (Voss et al., 2010, 2012). A study of African American women at risk for cognitive loss found the same results, with increased activity in the prefrontal cortex (Carlson et al., 2009). These findings have relevance for your clients, regardless of age.

▲ Highlighted areas in Figure 13.2 show increased brain gray matter as a result of exercise. Walking a few blocks weekly is useful, but 72 blocks (6–9 miles) is more effective.

Exercise alters the effect of HDL cholesterol genes. An increasing number of studies show that healthy behavior reaches down even to DNA and modifies genes in a positive fashion. (Other research shows that stress and other environmental factors do the opposite.) One study, involving 22,939 women, found that nine genes were associated with cholesterol level and that an increase in exercise affected the genes that relate to positive HDL levels (Ahmad et al., 2011). Dr. Brendan Everett, one of the study's authors, comments, "If someone has high cholesterol, the first thing I tell them is to get out and exercise."

Sexual activity is one form of exercise that needs more attention from the counseling profession. Research evidence shows that it lowers blood pressure, boosts immunity, burns calories, improves cardiovascular health, boosts self-esteem, improves intimacy, reduces pain, and helps a person sleep better (Stöppler, 2011). You may want to review her summary of research at www.medicinenet.com/sexual_health_pictures_slideshow/article.htm.

Sleep

Called by some #1 of the Big Six TLCs, sleep serves as a foundation of mental and physical health. Consider the summary in Table 13.3 and the importance of assessing sleep patterns in counseling and therapy. Difficulty with sleeping is one of the key signs of increasing depression. Assessing sleep patterns is important as so many mental health issues relate to sleep health. William Dement's 2000 book *The Promise of Sleep* remains the gold standard and is the major reference for the material presented here. More recent research elaborates and supports Dement's research and his comprehensive view of the field.

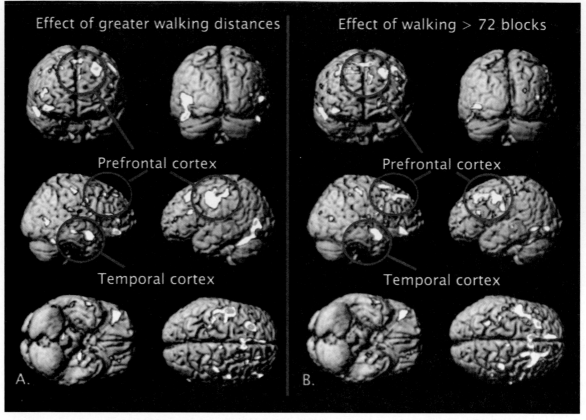

FIGURE 13.2 The effect of exercise on the brain.

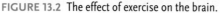

Source: Erickson, K., Miller, D., & Roecklein, K. (2012). The aging hippocampus: Interactions between exercise, depression, and BDNF. *The Neuroscientist*, 18, 82–97. Copyright 2012. Reprinted by permission of SAGE Publications.

A full rest is critical for brain functioning and development of new neural networks. With especially significant decisions, you may want to suggest that clients "sleep on it" and allow themselves time to reflect and not rush into a quick decision. Taking time allows for consolidation of memory and taking multiple factors into account for the final decision.

Table 13.3 summarizes the major issues and results of an enriched versus a destructive sleep environment.

TABLE 13.3   Therapeutic Lifestyle Changes: Sleep

Enriched	*Destructive*
7–9 hours of sleep	Sleep deprivation, jet lag
Reading, meditating, quiet, no TV	Eating late at night
Increases metabolism, hormones	Late time for school start
Consolidates learning	Parts of brain turn off during the day, effect similar to alcohol
Decreased attention	Increased risk of accidents
Improves mood	

Lack of sleep results in irritability, cognitive impairment, reduced moral judgment, symptoms similar to ADHD and depression, impaired immune system, increased heart attack risk, growth suppression, and diabetes. In addition, the other Big Six TLCs are impaired, as well as the ability to manage stress. Beyond lack of sleep, we find sleep disorders, narcolepsy, and restless leg syndrome.

Schoolchildren (ages 5–10) require 10 to 11 hours, adolescents 9, and adults 7 to 9, while pregnant women need more than 8 hours. The vast majority of North Americans do not reach these standards. Staying up late watching television or other forms of screen time is then followed by an early wakeup for school or a rushed commute to work.

Sleep is related to memory consolidation and strengthened synaptic connections, particularly in working memory. Increasingly, Freud's view of dreams is being validated, although his interpretation of symbols remains in doubt, particularly when multicultural considerations are added.

Help your clients plan systematically for sleep with the following strategies: (1) Exercise in the late afternoon can sometimes be beneficial, but not after the evening meal. (2) Never eat or drink alcohol after 7:00 P.M. (3) Plan a restful evening, but keep screen time to 90 minutes or less and stand up frequently. (4) Read before going to sleep, perhaps 30–90 minutes, but avoid the most stimulating novel and today's bad news. (5) Meditate for 15–20 minutes. (6) Sleep in a totally dark room with minimal noise. Sexual experience can also be helpful. Of course, each client is unique, and you will want to work to find the sleep methods that are most effective for each person.

Nutrition

You are what you eat.

—Victor Lindlahr

The evidence is now sufficient that all counselors and therapists need to add consideration of nutritional issues to their list of instructional strategies. The data are too clear to continue to omit assessment of nutrition as part of the helping process. Like exercise, sleep, meditation, and sexuality, this is a topic that seldom appears in counseling and therapy literature. You are not expected to be a nutrition expert, but some basic knowledge of proper eating habits and referral sources is now necessary. Vegetables, fruits, and local organic foods make a difference not only in physical condition, but also in mental health. Good mental health is difficult when physical health is compromised.

Obesity facilitates Alzheimer's development and creates addiction through loss of dopamine receptors. High fat activates genes to release RNA, causing apoptosis and cell death. Counselors should have available a variety of resources to help clients with weight issues and should encourage clients to write down a resolution and share it with others openly. To help clients lose weight and adopt better eating habits, recommend avoiding the "whites," such as most pasta, sugar, and white breads, and limiting snacks, because frequent snacking can be a critical route to eating issues and weight gain.

Many people might object to these weight loss recommendations. The fat acceptance movement talks about the "oppression of fatism." Some clients cannot lose weight. Your client may be aware of the dangers of obesity and the accompanying possibility of a shorter life, but still decide not to change. Work with such clients on the importance of other TLCs, particularly exercise, because being in good physical condition makes a significant difference to their health. Help them toward self-acceptance, and help them deal with name harassment associated with weight as fatism. Also, search YouTube using the key terms "fat acceptance" and "fatism" to see many passionate, personal statements.

Counselors need to develop an awareness of nutritional supplements and the constantly changing data from research. For example, omega 3 and fish oil are currently used widely

TABLE 13.4 Therapeutic Lifestyle Changes: Nutrition

Enriched	Destructive
Increased myelin	Inflammation
Low-fat, simple-carbohydrate diet	Diet high in sugar, pasta, complex carbs
Olive oil, canola oil	Palm oil
Richly colored fruits and vegetables	"Dirty dozen" fruits and vegetables (with most pesticide residue)
Vitamin D3, omega 3/salmon/fish oil	Meat hormones and antibiotics
Pure water	BHP plastic water bottles
Organic food	Junk food
Walnuts and other nuts	

© Cengage Learning

and appear to be successful supplements, even for ADHD and teenage schizophrenia prevention (Amminger et al., 2010). Vitamin D3 clears amyloids that appear to be causal in Alzheimer's disease ("Scientists Pinpoint," 2012). It's important that counselors be aware of such nutritional supplements, but clients should check with their physicians before including supplements in their diets.

We do not recommend that you advise clients to use supplements; this is a matter for consultation with a physician. However, factual information on the dangers of junk food, the "whites," the value of good nutrition, and medical referrals need to become part of the counselor's practice. Informing clients of research on nutrition can be a valuable part of counseling.

The summary of nutritional issues in Table 13.4 may be helpful.

Evidence is increasing that fish oil, omega 3, and eating fish can be helpful in mental health (Raji et al., 2011). A 5-year study found that those who ate fish at least once a week increased the amount of gray matter in the brain while reducing the risk of Alzheimer's fivefold. Baked and broiled fish made the difference; fried fish was ineffective. Fish and chips is not part of a healthy diet.

Omega 3 is one of the treatments used for teenage youth at ultra-high risk (UHR) for depression, bipolar diagnosis, and schizophrenia. The Orygen Youth Health program is focused on mental health preventive treatment for highly distressed youth (McGorry, 2007, 2009; McGorry, Nelson, Goldstone, & Yung, 2010). It includes psychoeducation, family work, and CBT, with minimal medication. Originator and researcher Dr. Patrick McGorry was named Australian of the Year in 2010 for Orygen's preventive mental health efforts, now implemented throughout the country. Treatment and prevention become one in this program. Research clearly shows that the Orygen program works to prevent schizophrenia, depression, and bipolar diagnosis.

A University of Cincinnati study found that reducing carbohydrates to 10% of the diet facilitated cognition in patients diagnosed with minimal cognitive impairment (MCI). Controlling stress and eating properly optimizes brain function and increases insulin supply (Jameson, 2012).

Vitamin D3 (5,000 units), plus the supplement curcumin (turmeric), was found to clear dangerous amyloids from the brain. Amyloids are most likely a major cause of Alzheimer's disease, the often expected result of MCI (Masoumi et al., 2012). MCI, far more common than Alzheimer's, has received little attention from the medical, psychological, or counseling community. Identifying cognitive changes may even prevent the movement to dementia. Many physicians diagnose Alzheimer's much too soon and are not aware of the term MCI. You can help both individuals and families by using the term MCI and helping them become fully aware of how TLCs and stress management can slow the progression of cognitive problems.

"Coffee May Ward Off Progression to Dementia" is the arresting title of an article (Anderson, 2012) reporting a study in which people with MCI who drank five cups of coffee daily avoided developing dementia over the next 2 to 4 years (Cao et al., 2012). A reasonable amount of alcohol has also been found beneficial (two drinks for men, one for women). Red wine is most recommended.

Given the increasing information on the relationship of nutrition to mental health, you will want to include an assessment of nutrition in your assessment of client needs.

Social Relations

Why can't we all get along?

—Rodney King

Evidence is clear that those who are alone and alienated have vastly increased mental and physical health issues. Counseling and psychotherapy are about social relationships. This book is founded on the idea that human relationships are central to our well-being, and we have endless research attesting to this idea, ranging from the child's need for a loving, supporting family to teens and their peer groups and on into adulthood. People with cancer and other diseases tend to live longer and healthier lives if they have solid social relationships.

The basic research and clinical findings are summarized in Table 13.5.

Assisting shy or isolated clients to get out and communicate more effectively with others is a useful route to better social relations. Those who are in close relationships are more likely to be in better health. Consider microskills training in listening skills as a way to assist clients.

Counseling and therapy were not evaluated in this study, but the findings remain powerful for us. Much of our work is supporting our clients and helping them improve their relationships.

Cognitive Challenge

There is not as much money in prevention as there is in medication and long-term therapy.

If we don't use our minds actively, we are likely to slow down and lose neurons and neural nets over time. Take a course, learn a language, learn a new instrument—basically do something different and challenging. Doing what we already do well may be useful, but not as growth producing as the creation of the new. Uncertainty can be growth producing. "Educators say that, for adults, one way to nudge neurons in the right direction is to challenge the very assumptions they have worked so hard to accumulate while young. With a brain already full of well-connected pathways, adult learners should 'jiggle their synapses a bit' by confronting thoughts that are contrary to their own" (Strauch, 2010). Some of these ideas are summarized in Table 13.6.

TABLE 13.5 **Therapeutic Lifestyle Changes: Social Relations**

Enriched	Destructive
Love, sex	Living alone
Joyful relationships	Negativism, criticism
Extended lifespan	Brain cell death
Higher levels of oxytocin	Stress
Helping others	Self-focus

© Cengage Learning

TABLE 13.6	Therapeutic Lifestyle Changes: Cognitive Challenge
Enriched	**Destructive**
Any cognitive challenge	TV
Change of any type	Repetitious routine
Learning a musical instrument	Being alone
Learning a new language	Vegetating, sitting
Playing bridge, crossword puzzles	Boredom
Computer time for elders	Too much screen time
Travel	

Two research-tested training programs are well worth consideration by you and your clients. Luminosity (www.lumosity.com) is challenging and enjoyable, a favorite of Mary. Evidence is that with about 15 minutes daily, one can improve attention, flexibility, and other cognitive dimensions. Even more thoroughly researched and more highly recommended are Posit Science programs, including general brain fitness, visual acuity, and alertness in driving (www.positscience.com). These Posit Science programs have more solid research, but take more time and commitment.

All of us can profit from challenging our brains. Older clients who may be concerned with preventing dementia should be encouraged to consider participating in many types of programs that engage the mind. The Mayo Clinic has found that both computer time and exercise relate to better cognitive functioning in adults diagnosed with mild cognitive impairment (Geda et al., 2012). On the other hand, too much screen time is considered unwise for children and teens, and probably for all of us (Rose, 2012), because screen time reduces exercise and socialization and can interfere with sleep.

Mindfulness Meditation and the Body Scan

Jon Kabat-Zinn (2005a, 2005b) has researched and promoted a deep-seated form of relaxation to help people cope with stress, pain, and anxiety. He calls his technique the "body scan," which he uses along with mindfulness meditation. The body scan is basically a focused relaxation system, useful in preparation for meditation.

One of Allen and Mary's finest life experiences was several weeks when they participated in Kabat-Zinn's systematic program and learned mindfulness meditation. Experienced practitioners usually prefer this technique over the body scan and systematic relaxation, which are more easily learned and taught. Counselors and therapists should not teach mindfulness unless they have sufficient training and have practiced it themselves. Mindfulness may require a lifestyle change for many. Kabat-Zinn (1990) has worked closely with the Dalai Lama and has participated in basic research showing that the brain does change with the positive open approach of mindfulness.

The benefits of mindfulness have been reviewed by Davis and Hayes (2012). These include decreased stress and anxiety, emotional regulation and reduced negative rumination and thinking, better working memory, increased response flexibility, and an increase in compassion.

Mindfulness meditation is derived primarily from Buddhist thought and practice. There is no "goal" except perhaps to live as much as possible in the immediate here and now.

Similar in some ways to relaxation, practitioners usually lie comfortably on the floor or sit in a suitable chair, then close their eyes. The focus becomes the *Now* and paying special attention to breathing, noting how the breath comes in and out. You may want to breathe in with one nostril and out with the other, as this tends to help one focus on the *Now*. Focusing on your breath and the up and down movement of your stomach can achieve the same effect. Thoughts and feelings will likely start running through your mind. Do not fight them; let them come, but as they enter your Now awareness, let them drift off. This part of meditation is the most difficult for beginners, and too often they stop at this point.

After practicing meditation, usually for several weeks, you may find a near perfect "stillness" and awareness of the present moment. There is clear evidence that this state alone allows new neural connections to develop in positive areas of the brain. If you keep this up, you will eventually notice the here and now more fully throughout the day. You'll notice the beauty of the world in new ways. Your partner or lover will appear very differently to you because you are in the moment.

We recommend that you refer clients to honest experts who truly know mindfulness. There are charlatans out there. A safe but secondary alternative for learning meditation is to refer them to the website www.mindfulnesstapes.com, where they can purchase materials and learn on their own.

Research supporting meditation is impressive. It has been demonstrated that meditation reduces stress and increases positive gray matter changes in the amygdala. The more seriously a person practices meditation, the greater the benefit (Hölzel et al., 2011). Serious study of yoga can also make a difference. Cancer survivors experienced a decrease in anxiety, perceived stress, and fatigue, and improvement in emotions. Damaging cortisol was reduced (Mulcahy, 2011).

Many now believe that quieting the brain through prayer, saying the Rosary, and other quiet religious practices have a similar effect as meditation.

We now move from the Big Six to a brief discussion of other useful programs for brain and physical health.

REFLECTION *How Are You Doing in Your Own Life With the Big Six and Other TLCs?*

How are you doing with stress management? Do you have an exercise plan? Do you get enough sleep? What are your eating patterns—are you avoiding the whites of sugar and flour? How are you doing socially—are you spending enough time with friends? At this point, you likely have enough cognitive challenge, but what might you plan for the future? Have you given any thought to a regular plan of meditation?

Review the other TLCs listed on pages 331–338 and in the Tree of Life. What do you notice and learn from the rest of the list? How much screen time with phones, computers, games, and TV do you have?

In the past, counseling clients on exercise, meditation, nutrition, and other health-oriented issues has not been central to our training and professional work. Research has shown that they are now a necessary part of competent practice. What do you think of these new developments, and are you prepared to add them to your set of counseling strategies and theories?

The Tree of Contemplative Practices

Figure 13.3 illustrates the many therapeutic lifestyle changes that are available to us as counselors and therapists. Enjoy looking at the many possibilities that The Center for Contemplative Mind in Society has kindly shared with us. Visit their website at www.contemplativemind.com.

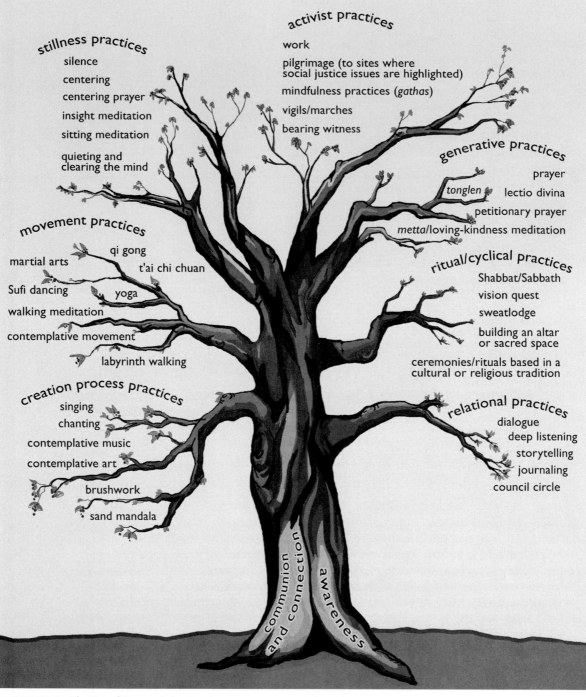

FIGURE 13.3   The Tree of Contemplative Practices.

Source: The Center for Contemplative Mind in Society, www.contemplativemind.org. Reproduced by permission.

▶Summary: Using the Concrete Action Strategies for Client Change

The skills and strategies of this chapter almost always belong in the restorying and action phases of the session. Along with the more listening-oriented, empathic confrontation, they are powerful tools for change and action in the client's daily life.

Logical consequences is a critical skill in decision counseling, which will be presented in some detail in Chapter 14, followed by a transcript of a complete decision-making example session. Instruction and psychoeducation strategies involve teaching strategies that require client interest and involvement. The multiple strategies of stress management and therapeutic lifestyle changes also involve psychoeducation, but they face a special challenge of compliance. Even though clients know that they "should" engage in these behaviors, it can be difficult to get them involved. An optimism and wellness assessment from Chapter 2 may be the first step in increasing client interest. If clients can see the lifetime benefits of a significant change in lifestyle, they are more likely to listen to you. However, forcing these ideas on clients is not appropriate—they must want them.

▼▼▼▼	Key Points
Logical consequences	Ideally, this is a gentle skill used to help people sort through issues when a decision needs to be made. Decisions can have both negative and positive consequences. The focus is on potential outcomes, and the task is to assist clients to foresee consequences as they review alternatives for action. A common statement used here is "If you _____, then _____ will possibly result." This skill predicts the probable results of a client's action, in five steps: 1. Listen to make sure you understand the situation and how the client understands what is occurring and its implications. 2. Encourage the client to think about positive and negative consequences of a decision. 3. If necessary, comment on the positive and negative consequences of a decision in a nonjudgmental manner. 4. Summarize the positives and negatives. 5. Let the client decide what action to take.
Instruction and psychoeducation	Instruction and psychoeducation are closely related. Instruction, providing information or advice, is brief, consisting of relatively short comments to facilitate action in the real world. Psychoeducation is more comprehensive. Many times clients need the counselor's knowledge and expertise around key life issues. The counselor knows the community and the resources available. He or she also knows the likely pattern and key issues of a divorce, the death of a family member, or other life issue. Psychoeducation is a more systematic way to teach clients about new life possibilities; this may range from training in communication skills to developing a successful wellness plan. Before you give instruction or engage in psychoeducation: 1. A solid working relationship is essential. 2. Hear the client's story and identify strengths. 3. Check out your clients' interest and readiness for receiving information. 4. Be clear and concise, and encourage client participation and feedback.

Psychoeducation includes teaching communication microskills	Teaching the communication skills of living can be helpful to a person who is depressed, a client who has difficulty in the work setting, or a couple in difficulty. The four points above are central, but add the following when you teach specific behavioral skills to clients or engage in assertiveness training.
	5. Obtain a clear understanding, with examples, of the specific places where the client has difficulty.
	6. Role-play the situation, with you playing the other person. Where possible, video record the session, thus providing specific feedback.
	7. View the role-play, note behaviors, and help the client identify places where he or she might be effective. Where useful, teach the specific microskills. Attending and open questions are often the most helpful at the beginning, while working on emotions can be especially useful as well. Sharing alternative behavioral change methods or suggesting a specific direction for solving an issue may help your client if done in a respectful and culturally sensitive manner.
	8. Work collaboratively and involve your client in the choice of directives. Appropriately attending behavior via assertive body language, vocal tone, and eye contact remain essential, as are clear, concrete verbal expressions and checking out the degree of client participation. Directives are usually applied in assertiveness training and social skills training, and in specific exercises such as imagery, thought stopping, journaling, or relaxation training.
Stress management	Changing the stressors or changing your reactions to them are key goals of stress management. Several strategies can help you achieve these goals, including relaxation, meditation, irrational beliefs, thought stopping, time management, and many other techniques mentioned in this chapter.
Therapeutic lifestyle changes (TLCs)	The TLCs are often well known to your clients. At issue is helping clients take them seriously and act on them. Exercise, sleep, nutrition, social relations, cognitive challenge, and meditation are known as the Big Six, although there are many other critical aspects of this approach to wellness and prevention. The Tree of Life provides an excellent overview of the many possibilities.
Brain and body protection	Sustained, chronic, or extreme stress accelerates the normal wearing and tearing of our body and mind. Psychoeducation, stress management, and TLC strategies are key to protecting our physical and mental health,
What else?	Most of the influencing skills require you to be concrete and specific. Remember to involve your client as a co-participant as you utilize influencing skills.

▶Competency Practice Exercises and Portfolio of Competence

 Go to CengageBrain.com to access Counseling CourseMate, where you will find an interactive ebook, quizzes, videos, interactive counseling and psychotherapy exercises, the Portfolio of Competence, case studies, a practice test, and more.

Individual Practice

Exercise 1. Writing Logical Consequence Statements   Using the five steps of the logical consequences strategy, briefly indicate to a client what might be the logical consequences of one of the following: staying in an abusive relationship; smoking while pregnant; moving from marijuana to cocaine.

a. Summarize the client's concern in your own words, using "if, then" language.
b. Ask specific questions about the positive and negative consequences of continuing the behavior.
c. Provide the client with your own feedback on the probable consequences of continuing the behavior. Use "if, then" language.
d. Summarize the differences between the feedback just given and the client's view when the client says she/he doesn't want to change (this implies the use of confrontation).
e. Encourage the client to make her/his own decision.

Exercise 2. Logical Consequence Using Listening Skills   By using questioning skills, you can encourage clients to think through the possible consequences of their actions. ("What result might you anticipate if you did that?" "What results are you obtaining right now while you continue to engage in that behavior?") However, questioning and paraphrasing the situation may not always be enough to make clients fully aware of the logical consequences of their actions. For each of the following clients and situations, write logical consequences statements that can help the client understand the situation more fully.

a. A student who is contemplating taking drugs for the first time.
b. A young woman contemplating an abortion.
c. A student considering taking out a loan for college.
d. An executive in danger of being fired because of poor interpersonal relationships.
e. A client who is consistently late in meeting you and is often uncooperative.

Exercise 3. Practicing Strategies   This chapter includes many different possibilities for practice. Try out each skill or strategy as time permits. Work through each of the strategies before using them with a client.

a. Try the strategy on yourself. Most can be done alone if you take time to really do it and study the process. What happens? What occurs for you? What did you learn? Would you like to continue and practice that method further?
b. Find a classmate or friend, get the person's permission, and work together using various strategies in practice sessions. What happens? What occurs for you? What did you learn? Would you like to continue and practice that method further?

Group Practice

Exercise 4. Logical Consequences and Instruction/Psychoeducation   Group work with these three influencing skills requires practice with each if you are to develop competence. The general model of group work is suggested, but only one strategy should be used at a time.

Remember to include the Client Feedback Form from Chapter 1 as part of the practice session. This is a particularly important place to practice group supervision, sharing, and feedback.

Step 1: Divide into practice groups.

Step 2: Select a leader for the group.

Step 3: Assign roles for each practice session.

▲ Client
▲ Counselor, who will begin by drawing out the client story or issue using listening skills and then attempt one of the influencing skills and strategies from this chapter

BOX 13.4 Feedback Form: Logical Consequences, Instruction/Psychoeducation, Stress Management, and TLCs

You can download this form from Counseling CourseMate at CengageBrain.com.

_____ (DATE)

_____   _____
(NAME OF COUNSELOR)   (NAME OF PERSON COMPLETING FORM)

Instructions: The two observers will complete the form and then discuss their observations with the practicing counselor and the volunteer client.

1. Did the counselor use the basic listening sequence to draw out and clarify the client's story or concern?
   How effectively?

2. Provide nonjudgmental, factual, and specific feedback for the counselor on the use of the specific influencing skill
   (logical consequences, instruction/psychoeducation, stress management, or TLC).

3. As you view the totality of the session, where was the client at the beginning on the Client Change Scale? Where
   was he or she at the conclusion? What aspects of the skill or strategy impressed you as most useful and effective?

- ▲ Observer 1, who will complete the Feedback Form in Box 13.4
- ▲ Observer 2, who will complete the feedback form and also the CCS Rating Form (see Chapter 10), deciding how much of an impact the counselor's influencing skills have made

Step 4: Plan.　In using influencing skills, the acid test of mastery is whether the client actually does what is expected (for example, does the client follow the directive given?) or responds to the feedback, self-disclosure, and so on, in a positive way. For each skill, different topics are likely to be most useful. State goals you want to accomplish in each instance. Some ideas follow:

- ▲ *Logical consequences.* A member of the group may present a decision he or she is about to make. The counselor can explore the negative and positive consequences of that decision.
- ▲ *Instruction/psychoeducation.* The counselor may provide instruction (information) or psychoeducation about a particular issue to the individual or group, such as the value of a wellness plan or dealing with a death in the family. The group gives feedback on whether the counselor was able to give information in a way that was clear, specific, interesting, and helpful. We suggest that you consider teaching microskills as communication skills to your individual or group.
- ▲ *Stress management and TLCs.* Select one of the strategies presented in this chapter and work through the specific steps. Involve your client in the process; the two of you together can select the strategy that you would like to try. As part of the practice session, be sure to tell the client what to expect and the likely results.

Step 5: Conduct a 5- to 15-minute practice session using the strategy.　Use listening skills along with the selected strategy. Is the client connected and involved?

Step 6: Review the practice session and provide feedback for 10 to 12 minutes.　Remember to stop the recording to provide adequate feedback for the counselor.

Step 7: Rotate roles.

Portfolio of Competence

This chapter is about multiple interpersonal influence strategies, and it covers considerable material. You cannot be expected to master these concepts until you have a fair amount of practice and experience. At this point, however, it will be helpful if you think about the major ideas presented in this chapter and where you stand currently. Also, where would you like to go in terms of next steps?

Use the following chart to evaluate your present level of mastery. As you review the items below, ask yourself, "Can I do this?" Check those dimensions that you currently feel able to do. Those that remain unchecked can serve as future goals. Do not expect to attain intentional competence on every dimension as you work through this book. You will find, however, that you will improve your competencies with repetition and practice.

Skill/Strategy	Level 1: Identify and discuss the skill or strategy	Level 2: Demonstrate the skill or strategy in action	Level 3: Use the skill or strategy with actual impact on the client	Level 4: Teach the skill or strategy to others
Logical consequences				
Instruction and psychoeducation				
Stress Management				
Teaching interpersonal communication via microskills				
Assertiveness training				
Thought stopping				
Positive imagery				
Therapeutic Lifestyle Changes (TLCs)				
Exercise				
Sleep				
Nutrition				
Social relations				
Cognitive challenge				
Meditation				
Spirituality/prayer				
Social justice action				
Other competencies you have mastered from the Tree of Contemplative Practices (Figure 13.3)				

▶Determining Your Own Style and Theory: Critical Self-Reflection on Influencing Skills

You have encountered the most active set of microskills and strategies in Section IV, and have had the opportunity for at least a brief introduction to each. With which of these skills and strategies do you feel most comfortable? Which might you seek to use? Which might you avoid? How do you feel about the idea of consciously influencing the direction of the session?

What single idea stands out for you among all those presented in this chapter, in class, or through informal learning? What stands out for you is likely to be important as a guide toward your next steps. How would you use these skills to help clients (and maybe yourself) manage their stress? What are your thoughts on multicultural issues and the use of this skill? What other points in this chapter strike you as important? How might you use ideas in this chapter to begin the process of establishing your own style and theory? Given the complexity of this chapter and the many possible goals you might set for yourself, list three specific goals you would like to attain in the use of influencing skills and strategies within the next month.

Skill Integration, Theory Into Practice, and Determining Personal Style

Whaat is your preferred style for counseling and psychotherapy? This section provides a framework to help you integrate the many skills and concepts of this book. Central to this process is Chapter 14, where we examine one counseling session in detail. Then we recommend that you again record a full session, develop a transcript, and analyze how your personal integration of skills affects the client. Competence and mastery of counseling begins to show when you can anticipate and evaluate the impact of your style on client growth and development.

Chapter 14. Skill Integration, Decisional Counseling, Treatment Planning, and Relapse Prevention There are several important concepts in this chapter. Decisional counseling, likely the most used theory of all, is introduced, followed by the transcript of a complete decisional counseling session with Allen and Mary illustrating the microskills and strategies in action. After an analysis of this session, the popular CBT strategy of relapse prevention (RP) is introduced to help ensure that client action happens after the session.

Most important, however, is the assignment at the very end of this chapter, where we suggest that you again record a session and compare it with your first recorded session. The transcript of Allen and Mary's decisional interview illustrates in considerable detail the complex working of a session. When you develop your own transcript, we hope you will give additional thought and attention to the session and the accompanying analysis. *This is what will make the most difference for you and help ensure that learning from the course and this book stays with you for your entire career.*

Chapter 15. How to Use Microskills and the Five Stages With Theories of Counseling and Psychotherapy   Here you will see brief summaries of decisional counseling, person-centered counseling, logotherapy, and multicultural counseling and therapy. This is followed by a more detailed presentation of crisis counseling and cognitive behavioral therapy. In addition, the online CourseMate provides transcripts and discussion of brief counseling, motivational interviewing, and the new field of counseling/coaching.

Chapter 16. Determining Personal Style and Future Theoretical/Practical Integration   This final chapter provides an opportunity to review your work with microskills, the five-stage structure, and your thoughts about counseling and therapy theories. You will be asked to think about your own natural style and plan for the future. You will be encouraged to use your knowledge and skills to build your own culturally intentional, culturally appropriate counseling style.

We are nearing the end of our journey through the basics of counseling and psychotherapy. You now have competencies that can be used in many settings, as the microskills and five stages are foundational units of all communication—counseling, psychotherapy, business, sales, law, medicine, peer helping, and many others.

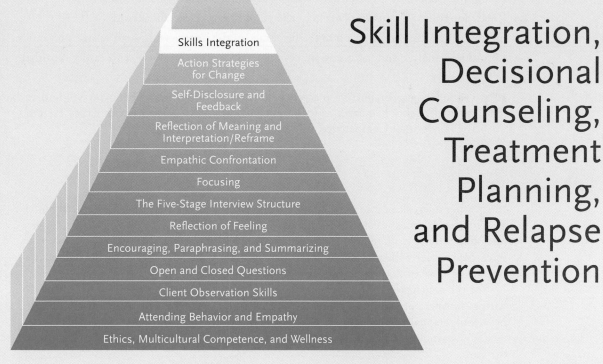

Skill Integration, Decisional Counseling, Treatment Planning, and Relapse Prevention

Skills Integration

Action Strategies for Change

Self-Disclosure and Feedback

Reflection of Meaning and Interpretation/Reframe

Empathic Confrontation

Focusing

The Five-Stage Interview Structure

Reflection of Feeling

Encouraging, Paraphrasing, and Summarizing

Open and Closed Questions

Client Observation Skills

Attending Behavior and Empathy

Ethics, Multicultural Competence, and Wellness

I have learned that people will forget what you said, people will forget what you did, but people will never forget how you made them feel.

—Maya Angelou

Mission of "Skill Integration, Decisional Counseling, Treatment Planning, and Relapse Prevention"

You and your clients will greatly benefit from a naturally flowing session and treatment plan using a smooth integration of the skills, strategies, and concepts of intentional counseling and psychotherapy. This is a concept-dense chapter explaining several key issues, but the most valuable of these is that you conduct a complete counseling session and analyze what happens between you and the client. Treatment planning, decisional counseling, and relapse prevention are three practices that can be invaluable in counseling and psychotherapy.

Chapter Goals and Competency Objectives Awareness, knowledge, skills, and actions developed through the concepts of this chapter will enable you to

▲ Integrate concepts, skills, and strategies learned in previous chapters.

▲ Understand the basics of decisional counseling as an integrative theory as well as how the five-stage decisional structure relates to other theories of counseling and psychotherapy.

▲ Plan for the complexity of the first session and use a checklist to help ensure that critical points are covered in the first session.

▲ View a transcript of a complete decisional counseling session and examine specifics for analyzing microskill usage, the Client Change Scale, and empathic constructs.

▲ Create long-term treatment plans for a client, and keep systematic records.

▲ Increase client take-home action in the fifth stage of the interview through relapse prevention.

▲ Record and analyze your own session; compare your counseling style with earlier recorded interviews.

 Go to CengageBrain.com to access Counseling CourseMate, where you will find an interactive ebook, quizzes, videos, interactive counseling and psychotherapy exercises, the Portfolio of Competence, case studies, a practice test, and more.

Integrating microskills into a meaningful, coherent session is what this chapter is about. You will see Allen conduct a decisional counseling session with Mary using the five-stage session as a framework to help her examine the implications of a career change. Thus, we start this chapter with a review of decisional counseling preplanning. After the demonstration session, we explore issues of treatment planning.

At the conclusion of this chapter, we suggest that you conduct a full counseling session with a classmate or volunteer, transcribe it, and then classify your own style and outline how you personally have integrated the skills and strategies of this book.

We begin with a definition of skill integration and its anticipated personal result for you.

SKILL INTEGRATION	ANTICIPATED RESULT
Integrate the microskills into a well-formed counseling session and generalize the skills to situations beyond the classroom.	Developing counselors will integrate skills as part of their natural style. Each of us will vary in our choices, but increasingly we will know what we are doing, how to intentionally flex when what we are doing is ineffective, and what to expect in the session as a result of our efforts.

▶Decisional Counseling: Overview of a Practical Theory

Not to decide is to decide.

—Harvey Cox

Decisions require creativity, a disciplined freedom, an openness to change, and the capacity to envision the potential consequences of given actions. There is an old Zen fable that goes something like this, updated for today.

A woman is hiking along a California Sierra trail along the edge of a 15-foot drop. As she rounds a bend, she sees a bear, who starts to charge. Surprised but still able, she grabs a wild vine and swings over the edge. As she thankfully hangs and looks for a safe place to jump, she sees another bear below! There are summer strawberries growing on the vine, so she decides to hold on with one hand and reaches for a few berries with the other. How sweet they taste!

Your clients face bears of decisions. One bear promises one thing, while the other may bring something else. We can help clients taste the sweetness of strawberries and the importance of the moment before they jump. Let us hope that their decisions are friendlier than bears.

Clients come to us with pieces of their lives literally "all over the place." The "magic" of creativity comes from the spontaneous generation of something new out of what already exists. Decision making is a creative practice.

Decisions are the stuff of life. Decisions are both our challenges and our opportunities. Decisional counseling (Ivey & Ivey, 1987; Ivey, Ivey, & Zalaquett, 2010) may be described as a practical model that recognizes decision making and the microskills as a system that will help you develop increased competence in the major theories of counseling and psychotherapy.

▶The History of Pragmatic Decisional Counseling: Trait-and-Factor Theory

Decisional counseling was introduced in Chapter 8 with a brief discussion of the historical contributions of Benjamin Franklin, who outlined the basics of the decisional process: (1) define the problem; (2) generate alterative solutions; (3) decide for action. This approach is "very American," focusing on what is pragmatic with no attention to theory. What is critical is to find something that "works" to resolve issues and move on with life.

The goal in decisional counseling theory and practice is to facilitate clients' decision making and to consider the many *traits and factors* underlying any single decision. Trait-and-factor theory has a long history in the counseling field, dating back to Frank Parsons's development of the Boston Vocational Bureau in 1908. Parsons expanded the Franklin framework and pointed out that in making a vocational decision the client needs to (1) consider personal traits, abilities, skills, and interests; (2) examine the environmental factors (opportunities, job availability, location, and so on); and (3) develop "true reasoning on the relations of these two groups of facts" (Parsons, 1909/1967, p. 5). Since that time, proponents of trait-and-factor theory have searched for the many dimensions that underlie "true reasoning" and decision making.

Gradually trait-and-factor theory came to be seen as incomplete, and new decisional and problem-solving models have arisen (Brammer & MacDonald, 2002; D'Zurilla, 1996; D'Zurilla & Nezu, 2007; Egan, 2010; Janis & Mann, 1977). All of these models can be described as modern reformulations of Benjamin Franklin's original model and trait-and-factor theory.

▶The Place of Decisional Counseling in Modern Practice

Decision making is a lifelong issue. Clients are always seeking to solve their life issues and make decisions. Young adults must choose a college or a career or decide whether to continue a relationship, get married, or have a child. Later they will make decisions about how to succeed in a work setting, deal with difficult colleagues, and plan finances for their children's education and their own retirement. They will find that decisions don't end with retirement; the first question faced by many retirees is "What shall I do with all this time?" Difficult decisions around health issues, wills, and plans for their own funeral often require counseling.

The many theories of counseling and therapy will, of course, focus on different content and use varying methods to help clients reach these decisions. Counseling and psychotherapy are clearly moving to more of a health and wellness orientation. Along with this comes the need to help clients make informed and wise decisions about taking care of themselves and their loved ones. For us, this means increased awareness of stress management, as well as enabling clients to make good use of therapeutic lifestyle changes.

In short, competence in decisional counseling is critical in virtually all issues and theories in counseling and therapy. You may be helping a child or adolescent decide how to deal with friends, cope with parent and teacher demands for behavior and academic performance, or choose between the army and college. Or imagine an adolescent or adult with mild, or even severe, depression—while cognitive behavioral therapy may be the treatment of choice, helping these clients solve interpersonal and job challenges is central to the treatment plan, including the decision to change lifestyle and/or possibly a referral for medication. For all of these clients and situations, decisions about managing stress and healthy lifestyle changes will be significant.

The practical basis of decisional counseling is found in the five stages of the counseling session (*empathic relationship—story and strengths—goals—restory—action*). This five-stage model provides a solid structure for creative and intentional decision making. Recall that the five-stage structure using only listening skills and skillful confrontation is often sufficient for clients to make major decisions and effect change. When you use minimal influencing skills, you are moving close to the practice of person-centered counseling, but still you are enabling better client decisions. Whether a client participates in person-centered counseling, cognitive behavioral therapy, or even long-term psychoanalysis, decisions will be part of every client meeting.

Decisional counseling can be equally useful with difficult personal decisions. As such, decisional counseling becomes integrative and eclectic in that it feels free to draw on other theories as they meet the immediate pragmatic needs of the client. Thus the client may start with career choice, but mention an issue/concern with alcohol along the way—or you discover underlying anxiety and depression. The decisional counselor may draw on motivational interviewing and brief counseling to help the client deal with the immediate career and job issues. With the anxiety and depression, the therapists may decide to bring in stress management and cognitive behavioral therapy.

▶Decisional Counseling and Emotional Understanding

Emotions and the feelings are not a luxury, they are a means of communicating our states of mind to others. But they are also a way of guiding our own judgments and decisions. Emotions bring the body into the loop of reason.

—Anthony Damasio

There is a common limitation in the problem-solving models, particularly those close to the original trait-and-factor framework, in that insufficient attention is paid to the *emotional* aspects of the decision. Very few of us will be satisfied if our decisions reflect only rational cognitive processes. Decisions require emotional energy, and this is a critical part of establishing newly created decisions in our long-term memory. Think about bringing more feeling and emotion to the session when you work with client decisions, regardless of your personal style and theories of choice.

How clients feel emotionally about a decision heavily determines whether or not they are going to act on that decision. Thus, the fourth and fifth stages of the counseling session need to give special attention to reflection of feeling and how emotions relate to the actions a client might take after the interview in the real world. The first route toward this, of course, is exploring the feelings and emotions associated with each alternative. For example, a client may not be fully aware of how small but continuing hurtful behaviors with a partner will ultimately destroy the relationship. How does this client feel emotionally with the situation as it is—what is the present story? Then, how might the client feel emotionally as he or she thinks ahead to the results of significant behavioral change?

With especially significant life decisions, a more systematic exploration of alternative futures with emotional consequences can be explored. The emotional balance sheet described below is an illustration of how we can explore in more complete detail the role of emotion in decisional counseling.

Cognitive and Emotional Balance Sheet

The balance sheet, created by the Australian Leon Mann, is an extension of logical consequences. Each alternative is written down in a list of gains and losses. Writing down the multiple aspects of a decision clarifies issues and helps ensure that the most critical matters are considered. As an example, Table 14.1 shows a decisional balance sheet created with a

TABLE 14.1   The Cognitive and Emotional Balance Sheet*

List below the positive and negative factual and emotional results for each of the possible alternatives. If there is more than one alternative, make a separate Cognitive and Emotional Balance Sheet for each one.

What is the decision? *What happens if I leave my abusing partner?*

What are the possible positive gains for me?	*What are the emotional gains for me?*	*What are the possible positive gains for others?*	*What are the emotional gains for others?*
Abuse will stop and I won't get hurt.	I won't be so scared.	My Mom won't have to talk to me on the phone constantly.	Mom will be so relieved that it's over.
I'll be able to move on with my life.	Perhaps I can return to feeling good about myself.	My Mom would like to help.	She'd feel that she is important to me again.
I can be myself.	I used to feel OK, and that would be a relief.		

What are the possible losses for me?	*What are the emotional losses I might face?*	*What are the possible losses for others?*	*What are the emotional losses for others?*
I'll be on my own.	This frightens me as much as staying.	None that I can think of.	Again, none that I can think of. They'll be happy to see him gone.
How can I finance things by myself?	This terrifies me.	My parents may have to support me for a while.	They aren't that well off, and they told me not to go out with him. They may be angry, even though they'll help.
I still love that man, despite it all.	I'll be lonely.	My friends will be there for me.	They'll be glad for me and listen.
He might follow me, and that might make it worse.	I'll have no future and be totally alone.	My counselor is there to advise and support me.	I can sense that I'm not as alone as I might think I am. I feel supported and cared for.

*Adapted from Leon Mann (Mann, 2001; Mann, Beswick, Allouche, & Ivey, 1989); also see Miller and Rollnick (2002).

woman who has experienced abuse. Notice how powerful the arguments are for staying in an abusive relationship. This is why your support in such cases is significant.

As another example, we can help substance abusers use the **emotional balance sheet** to look at issues around drinking or using drugs. Adding focus concepts to the balance sheet helps alcoholics see the broader implications of their drinking on the lives of others and makes possible the exploration of emotional issues. The balance sheet lists the positives—what drinking does for them—and then lists the negatives. All this is done on a balance sheet created jointly by the helper and the client. Motivational interviewing has adapted Leon Mann's strategies and uses the balance sheet with addiction (Miller & Rollnick, 2002).

Emotional balancing gives special attention to how the client might feel after each possible decision. Although decision making is a cognitive activity, real satisfaction comes with feelings of pleasure and satisfaction about the result. With this activity, emotions are stressed throughout the logical consequences strategy. For example, "What do you feel and enjoy about drinking?" "What does cocaine do for you?" "Imagine yourself not drinking; how would you feel about yourself?" "What would your family feel?" Use both here-and-now emotions and those of the past and the anticipated future. Look forward to the longer-term benefits that will come with change.

The Future Diary as a Further Emotional Support

A useful homework assignment is asking clients to think through and write what life will be like in the coming months, even years, if they make a particular decision one way or another. The future diary provides the client time to journal and think over the implications of difficult decisions. The journal helps the client to consider alternative decisions and their potential emotional consequences. The future diary can be completed with the client in the here and now of the session. Guided imagery helps clients anticipate both the result and the accompanying emotions associated with each decisional possibility. Decision making is generally considered a rational process, but an intellectual decision is not enough—emotional balance is critical for satisfactory decisions.

▶Importance of Case Conceptualization and Working Formulation

Foundational to your work is your understanding of your clients, their issues, and the decisions they want or need to make. A case conceptualization is an individualized application of your theoretical model that takes into consideration the antecedents of the case and your observations and inferences. The data to be considered include what you may find in the intake file, such as demographics, personal history, and presenting issues or concerns. In the ongoing sessions, many other factors—client verbal content and style, nonverbals, emotional experience, your experience of the client, what happens during your interactions, test results, diagnoses, strengths and assets, weaknesses and gaps, supporting materials, inferences and assumptions, working hypothesis, goals of treatment, treatment plans, possible barriers, interventions, and evaluation of outcomes—will result in constant changes to your case formulation for each client.

Based on all this information, you produce a case formulation that constantly changes but becomes more useful and precise as the relationship develops. Many counselors and psychotherapists suggest sharing your formulation with the client. It improves fit and accuracy, strengthens the relationship, and enhances collaboration and the possibility of clients' actually taking action based on their learning and decisions in the sessions. Case formulations can assist in creating behavioral, thought, and emotional change.

This formulation will be the working hypothesis used to guide treatment. It will evolve over sessions with the client and will help in making decisions on treatment plans, selecting interventions, assigning homework, coping with setbacks, assessing outcome, determining paths to action, and completing treatment. Allen's work with Mary, presented in detail below, illustrates many of these concepts.

▶Planning the First Session and Using a Checklist

The decisional counseling session presented in this chapter is a role-play conducted by Allen and Mary Ivey, based on Mary's real-life career planning.

In her intake information file, completed before counseling, Mary stated, "I find myself bored and stymied in my present job as a physical education teacher. I think it's time to look at something new. Possibly I should think about business. Sometimes I find myself a bit depressed by it all." This initial meeting illustrates that career counseling is closely related to personal counseling. The personal issues arise along with the career issues further into the session. Additionally, gender is a multicultural issue that needs to be considered constantly.

Given the five-stage structure, you can write and think through an advance plan for the session. Later this can also be used as a framework for writing up notes. Before the first meeting, study the client file; anticipate challenging issues and how you might handle them. The advance plan does not mean that you impose your views or concepts on the client; rather, you think through approaches that may enable the client to achieve whatever goal he or she may have. Just because you have a plan, don't expect things to always work out as you anticipate. Clients will bring up issues that cause you to rethink where to go next; you may even need to scrap your plan entirely. As always, intentionality—the ability to be open to alternatives and flex with the here and now—is critical.

The following plan shows Allen's assessment of his forthcoming meeting with Mary. He created the plan from his study of Mary's file, consisting of a pre-session questionnaire. Mary stated in her intake form, "I'd like to do something new with my career. I'm ready for something new—but what?" Note that the plan is oriented to helping the client create her own unique career plan and facilitating the discussion of personal issues as well. The plan is structured to help the client achieve her objectives and make her own decisions, but remains flexible enough to adapt as the session progresses and new issues are brought up.

STAGE/DIMENSION, KEY QUESTIONS	COUNSELOR PREPARATION
Empathic relationship. Initiate the session, develop rapport/structuring. *What structure do you have for this session? Do you plan to use a specific theory? What special issues do you anticipate with regard to rapport development?*	Mary appears to be a verbal and active person. I note she likes swimming and physical activity. I like to run, and that may be a common bond to discuss. I'll be open about structure but keep the five stages in mind. It seems she may be unhappy in her current job and want to look into another career choice. Another personal issue is her divorce. I'll need to listen to her stories, using mainly questions and reflective listening skills, and follow a decision-making model.

continued on next page

STAGE/DIMENSION, KEY QUESTIONS	COUNSELOR PREPARATION
Story and strength. Gather data, draw out stories, concerns, or issues. *What are anticipated issues? Strengths? How do you plan to define the issues with the client? Will you emphasize behavior, thoughts, feelings, meanings?*	I'll use Mary's wellness strengths early and focus on finding out what she *can* do. I'll use the basic listening sequence to bring out issues from her point of view and learn about her thoughts about her job and her thoughts about the future. I'll be interested in her personal life as well. How are things going since the divorce? What is it like to be a woman in a changing world? Mary may well bring up several issues. I'll summarize them toward the end of this phase, and we may have to list them and set priorities if there are too many issues. Mainly, however, I expect to talk about career choice.
Goals. Set goals mutually, establish outcomes. *What is the ideal outcome? How will you elicit the client's idealized self or world?*	I'll ask her what her fantasies and ideas are for an ideal resolution and follow up with the basic listening sequence. I'll end by confronting and summarizing the real and the ideal. As for outcome, I'd like to see Mary define her own direction from a range of alternatives.
Restory. Explore and create alternatives—confront client incongruities and conflict. *What theories would you probably use here? What specific incongruities have you noted or do you anticipate in the client? How will you generate alternatives?*	Working from the decisional model, I hope to begin this stage by summarizing her positive strengths and wellness assets. I'd like to see several new alternative possibilities considered. Counseling and business are indicated in her pre-session form as two good possibilities. Are there other possibilities? The main incongruity will probably be between where she is and where she wants to go. I expect to ask her questions and create some concrete alternatives even in the first session. I hope she will act on some of them following our meeting. I think career testing may be useful.
Action. Generalize and act on new stories. *What specific plans do you have for transfer of training? What will enable you to feel that the session was worthwhile?*	I'll feel satisfied if we have generated some new possibilities and can do some exploration of career alternatives. We can plan from there. I'd like it if we could generate at least one thing Mary can do for homework before our next session.

Given the complexity of relationships, particularly professional relationships, Atul Gawande has written *The Checklist Manifesto* (2009). Focusing first on medicine, he found that a surgical checklist of basic and often obvious factors significantly reduced dangerous errors during operations. He goes on to point out that thinking ahead about what one is going to do improves performance regardless of the field.

Box 14.1 provides a checklist for the first session. Even the most experienced and confident counselor or therapist is likely to forget some of the items in the checklist. Review the list before you talk with your client; then review it again afterwards, checking to see what might have been missed. There are items here that need to be considered in every session. For you, personally, what might you add to or delete from this checklist? Adapting the checklist so that it fits you and your agency is essential.

The first session presents many challenges as you create a relationship, structure the meeting, and get to know a new client. The session plan and checklist may be helpful in any session, but particularly that first session in which there are so many issues to cover. Experience reveals that even the most seasoned professional forgets and omits essential items on this list and in the treatment plan as well.

BOX 14.1 | Checklist for the First Session*

Before the Session

☐ Are you familiar with HIPAA, the policies of your agency, and key state laws? Are key policies posted in the agency waiting area? These need to be shared early with the client.

☐ Is there a file, and have you read it? Do you need notes from the file for a refresher?

☐ Do the room and setup ensure confidentiality? If you are working in an open setting, how will you maintain privacy?

☐ Does the room provide adequate silence? Do you need a sound machine working outside your door?

☐ Is there an inviting atmosphere where you will meet the client? Is it neutral, or do you have interesting art and objects relevant to those who may come to this office? Are chairs placed in a position where the power is relatively equalized?

☐ If an informal setting such as counseling on the street or gym, again is the situation as comfortable as possible?

Stage 1: Empathic Relationship

Initiate the session. Develop rapport and structuring. "Hello, what would you like to talk about today?" Did you:

☐ Plan ahead flexibly to ensure an empathic relationship and connect with this unique client?

☐ Discuss the client's rights and responsibilities? The counseling and therapy relationship works in part because of clearly defined rights and responsibilities of each person involved.

☐ Provide an explanation of what might happen in the session and/or how the conversation is likely to be structured?

☐ Review HIPAA, agency policies, and key legal issues? If your agency requires diagnostic labels, did you explain that to the client and offer to share that diagnosis if he or she wishes?

☐ Discuss confidentiality and its limits?

☐ Obtain the client's permission to take notes and/or record the session? Was the client informed that these notes and the recording are available if he or she wishes to review them?

☐ If working with an underage client, obtain the appropriate parental permission as required by your agency and/or state law?

☐ Provide an opportunity for the client to ask you questions before you started? Were issues of multicultural differences addressed?

☐ Work with the client to establish an early preliminary goal or objective for the session?

☐ Come prepared if the client immediately started talking about issues and concerns? Did you listen carefully and return later to cover those matters that you may not have had time to attend to?

Stage 2: Story and Strengths

Gather data. Use the BLS to draw out client stories, concerns, problems, or issues. "What's your concern?" "What are your strengths and resources?" Did you:

☐ Allow and encourage the client to present the story fully? Did you reframe the word *problem* into a more positive, change-oriented perspective using words such as *issue*, *challenge*, *concern*, or *opening for change*?

☐ Bring out the key facts, thoughts, feelings, and behaviors related to the story? Did you perhaps also look for underlying deeper meanings behind the story?

☐ Avoid becoming enmeshed in the client's story by becoming a voyeur (endless fascination and searching for details about the client's interesting issues) or by unconsciously putting a "negative spin" on what the client said? To paraphrase Eldridge Cleaver, "Is the counselor part of the problem or part of the solution?"

☐ Bring out stories and concrete examples of client personal strengths and external resources? Did you search for specific images within these stories and perhaps anchor these positive images in specific areas of the body?

☐ Ask the critical questions, "What else relates to what we've talked about so far?" "What else is going on in your life?" and "Is there anything else I should have asked you but didn't?"

Stage 3: Goals

Set goals mutually. The BLS will help define goals. "What do you want to happen?" "How would you feel emotionally if you achieved this goal?" Did you:

☐ Review the early goals set by the client and revise them in accordance with new information about the story and strengths?

☐ Jointly make these goals as specific and observable as possible?

☐ When necessary, break down large goals into manageable step-by-step objectives that can be reached over time? Did you prioritize these goals?

(continued)

---

*This checklist was inspired by the book by A. Gawande, *The Checklist Manifesto* (New York: Holt, 2009). Gawande talks specifically about the importance of a checklist for successful surgery and suggests that the idea be taken into other areas as well. This counseling checklist was authored by Allen Ivey. However, he gives permission for anyone to photocopy this checklist with the request that he be given appropriate credit. Copyright © 2014 Allen Ivey.

BOX 14.1 (continued)

❏ Remind the client of the strengths and resources that he or she brings to achieve these goals?

Stage 4: Restory

Explore alternatives via the BLS. Confront client incongruities and conflict, restory. "What are we going to do about it?" "Can we generate new ways of thinking, feeling, and behaving?" Did you:

❏ Include brainstorming without a theoretical orientation? Confront with a supportive challenge, summarizing the goal and the issue? ("On one hand the goal is _____, but on the other hand the main challenges you face are _____. Now what occurs to you as a solution?) Often clients with your support will come up with their own unique and workable answers, often ones that you did not think of.

❏ Use appropriate theories and strategies with this client?

❏ Use a variety of listening and influencing strategies to facilitate client reworking and restorying of issues? What were they?

❏ Use identified positive strengths and resources to remind clients during low points of their own capabilities?

❏ Agree on homework or personal experiments to be completed after the session?

❏ Develop a clear definition of a more workable story that can lead to action and transfer to the real world?

Stage 5: Action

Plan for generalizing session learning to "real life." "Will you do it?" Use BLS to assess client commitment to taking action after the session. Did you:

❏ Build on the new story, or start of a new story, and work with the client to take specific action and learning to the "real world"?

❏ Agree on a plan for transfer of learning that is clear and doable?

❏ Work with the client to develop a systematic and work-able relapse prevention plan? (See pages 388–390 of this chapter for CBT's approach to ensure that a client takes action outside of the interview.)

❏ Contract with the client to do at least one thing differently during the week, or even tomorrow?

❏ Agree to plans to look at this homework during the next session?

❏ Check how it was for the client? Does the client think he or she could work with you? Did you agree to work together?

❏ Set a date for next session and/or follow-up sessions?

❏ Write interview notes as soon as possible and seek consultation from supervisors or colleagues as necessary?

▶A Full Transcript of a Counseling Session: I'd Like to Find a New Career

Why do we present a full transcript and analysis at this point? We are aware that reading a transcript line by line is not always fascinating, but we recommend careful reading and analysis. You have seen many specific examples of counseling throughout the book. At this point, you already know a great deal about the counseling session. You are aware that counseling and therapy is a complex professional field. Reviewing this transcript in full detail will solidify your understanding of what occurs between counselor and client.

While the microskills have broken down the session into specific concrete observable steps, putting it all together is a different matter; here your integration of skills, strategies, and concepts becomes art form. What is presented here serves several purposes:

1. You will see the "flow" of a real session and observe that decisions and change move more slowly and deliberately than suggested in our other, briefer transcript examples.
2. The skill classification and process notes provide information on what is happening in the session.
3. The transcript serves as a model that you can use when you conduct your own sessions and analyze your behavior and its impact on the client. We recommend that you analyze your own sessions in this fashion, even when you are an experienced professional.

Periodically record sessions on video, and review what happened. Occasionally bring the client in with you for the review to provide additional feedback.

In this decisional counseling session, Mary plays the role of a 36-year-old divorced client with two children. For purposes of analysis, we present the session in three columns. The first column is the counselor and client verbatim transcript. The second column classifies counselor skills and client responses. The third column analyzes the effectiveness of skills throughout the session, with special attention to empathy and the Client Change Scale. The ability to classify and analyze your behavior and its effect on the client is a most useful competency that some experienced professionals may not have mastered.

As you read the transcript, evaluate and assess Allen's personal style. What responses make sense to you? Are his interventions appropriate? What might you do differently? While you want to define your own natural style, reflect on your style choices and how various client might think and react to your choices. You will find that some responses and strategies are less effective than others. We all make errors; it is our ability to learn from them and change that enables us to become more effective.

The first stage of the session focuses on building a working relationship. As you read through the first stage, please respond to the following questions.

1. How did Allen attempt to put Mary at ease? Do you feel it was effective?
2. Putting it into practice for yourself, please list at least one comment/response that you would have used to help Mary feel at ease.
3. Mary is a high functioning client. Please list at least three ways you would have to restructure the beginning of a session with a child, an emotionally distraught client, or an involuntary client.

COUNSELOR AND CLIENT CONVERSATION	SKILL CLASSIFICATIONS, INCLUDING FOCUS AND CONFRONTATION	PROCESS COMMENTS, INCLUDING EMPATHY AND CLIENT CHANGE SCALE
STAGE 1. EMPATHIC RELATIONSHIP Initiate the session. develop rapport and structuring. "Hello, what would you like to talk about today?"		
1. *Allen*: Hi, Mary. How are you today?	Open question Focus on client	Allen walks to the reception area, smiles and invites her in.
2. *Mary*: Ah . . . just fine. . . . How are you?	Self-disclosure (SD), returns focus to the counselor	As Mary walks in, note that she opens with two speech hesitations.
3. *Allen*: Good, just fine. Nice to see you. . . . Hey, I noted in your file that you've done a lot of swimming, even swim meets.	Self-disclosure, paraphrase from reading file Focus on counselor, client	Allen notices these breaks and senses some nervousness on her part. Consequently, he decides to take a little time to develop rapport, seeking to put Mary at ease. Note that he focuses on positive wellness as one of her strengths. Recognize client's strengths, even this early in the session. (Additive)

continued on next page

COUNSELOR AND CLIENT CONVERSATION	SKILL CLASSIFICATIONS, INCLUDING FOCUS AND CONFRONTATION	PROCESS COMMENTS, INCLUDING EMPATHY AND CLIENT CHANGE SCALE
4. *Mary*: Oh, yeah, (smiling) . . . I like swimming; I enjoy swimming a lot.	Self-disclosure of positive strength	
5. *Allen*: With this hot weather, I've been getting out some myself. Have you been able to lately?	Self-disclosure, closed question Focus on counselor, client	First sessions often begin with informal talk, designed to both relax and get to know the client a bit before the meeting really starts.
6. *Mary*: Yes, I enjoy the exercise. It's good relaxation.	Self-disclosure	
7. *Allen*: I also saw you won quite a few awards along the way. (*Mary*: Um-hmm.) . . . You must feel awfully good about that.	Paraphrase from file, reflection of feeling Client focus	Mary's nonverbal behavior is becoming more relaxed. Client and counselor now are beginning to show some body language symmetry.
8. *Mary*: I do. I do feel very good about that. It's been lots of fun.	Self-disclosure, focuses on wellness strength and enjoyment	
9. *Allen*: It's great that you get that exercise and have fun with it. (pause) Before we begin, I'd like to ask if I can tape-record this talk. I'll need your written permission, too. Do you mind? You can review all written notes at any time. And sometimes people find it helpful to take home the recording.	Feedback, information giving	Obtaining permission to record the session is essential. If the request is presented in a comfortable, easy way, most clients are glad to give permission. At times it may be useful to give the tapes to clients to take home and listen to again.
10. *Mary*: No, that's okay with me. (signs form permitting use of tape for *Intentional Interviewing and Counseling*) Perhaps I will want to listen to that recording at some time.	Self-disclosure	Mary is now ready to begin the session, but more structuring is needed. Some clients may be so anxious to start talking about their issues that some parts of structuring will have to be inserted during the session itself. But, it is best to clarify these important details before you start.
11. *Allen*: As we start, Mary, there are some things to discuss as we begin. We'll have about an hour today and then we can plan for the future together. Today, I'd like to get to know you and I'll try to focus mainly on listening to your concerns. At the	Information giving, self-disclosure Client, counselor, cultural/environmental context	Allen provides some additional structure so that Mary knows what she might expect. He introduces gender differences and provides an opportunity for Mary to react and ask questions. Note "*we* can plan." This leads to a more mutual approach. (Additive)

COUNSELOR AND CLIENT CONVERSATION	SKILL CLASSIFICATIONS, INCLUDING FOCUS AND CONFRONTATION	PROCESS COMMENTS, INCLUDING EMPATHY AND CLIENT CHANGE SCALE
same time, from your file, I know that some of your issues relate to women's issues. Obviously, I'm a man, and I think it is important to bring this up so that you will be more likely to feel free to let me know if I seem to be "off-target" or misunderstand something. Feel free to ask me any questions you'd like around this or other matters. [At this point Allen discusses confidentiality, HIPPA, and other key agency policies.]		
12. *Mary*: I feel comfortable with you already. But, a couple questions. One is that I'm interested in the counseling field as a possibility—and the other is around the issue of living with divorce and being a single parent. What can you say about those?	Client, counselor As virtually all of Mary's comments involve self-disclosure, this classification of her comments will not be included in the rest of this transcript.	Mary gives the OK, but then asks two questions. She leans forward when she asks them. This question is a surprise to the counselor and should be noted, as divorce and relationship issues may show themselves later to need considerable attention.
13. *Allen*: Well, first I'm divorced and have one child living with me while the other is in college. Of course, I'd be glad to talk about the counseling career and share some of my thoughts. What thoughts occur to you around divorce and counseling?	Self-disclosure, open question Counselor, client	Keep self-disclosures brief and return the focus to the client. But be comfortable and open in that process.
14. *Mary*: That helps. Going through my divorce was the worst thing of my life. My children are so important to me. Perhaps your experience with divorce will help you understand where I am coming from.	Client, main theme	Mary smiles, sits back, and appears to have the information she was wondering about.

The second stage is story and strengths. As you read through Mary's story, please consider the following questions and provide your responses.

1. Make a list of the strengths that Mary and Allen identify.
2. What stands out to you as key points of Mary's story?
3. Allen provides a paraphrase in line 19. Please write your own paraphrase for what you have heard so far.

COUNSELOR AND CLIENT CONVERSATION	SKILL CLASSIFICATIONS, INCLUDING FOCUS AND CONFRONTATION	PROCESS COMMENTS, INCLUDING EMPATHY AND CLIENT CHANGE SCALE
STAGE 2. STORY AND STRENGTHS Gather data, draw out stories, concerns, issues.		
15. *Allen*: You've talked about quite a few things. Could you tell me what you'd like to start with?	Open question Client	In this series of leads, you'll find that Allen uses the basic listening sequence of open question, encourager, paraphrase, reflection of feeling, and summary, in that order. Many counselors in different settings will use the sequence or a variation to define client issues.
16. *Mary*: Well . . . ah . . . I guess there's a lot that I'd like to talk about. You know, I went through . . . ah . . . a difficult divorce and it was hard on the kids and myself and . . . ah . . . we've done pretty well. We've pulled together. The kids are doing better in school and I'm doing better. I've . . . ah . . . got a new friend. (breaks eye contact) But, you know, I've been teaching for 13 years and really feel kind of bored with it. It's the same old thing over and over every day; you know . . . parts of it are okay, but lots of it I'm bored with.	Client, issue/concern, others	As many clients do, Mary starts the session with a "laundry list" of issues. Though the last thing in a laundry list is often what a client wants to talk about, the break in eye contact at mention of her "new friend" raises an issue that should be watched for in the future. As the session moves along, it becomes apparent that more than the career issue needs to be looked at. Mary discusses a pattern of boredom. This is indicative of an abstract client who is able to reflect on herself and see patterns of behavior.
17. *Allen*: You say you're *bored* with it?	Encourage Client	The key word *bored* is emphasized.
18. *Mary*: Well, I'm bored, I guess . . . teaching field hockey and . . . ah . . . basketball and softball, certain of those team sports. There are certain things I like about it, though. You know, I like the dance, and you know, I like swimming—I like that. Ah . . . but . . . you know . . . I get tired of the same thing all the time. I guess I'd like to do some different things with my life.	Client, issue/concern	Note that Mary elaborates in more detail on the word *bored*. Allen used verbal underlining and gave emphasis to that word, and Mary did as most clients would: She elaborated on the meaning of the key word to her. Many times short encouragers and restatements have the effect of encouraging client exploration of meaning and elaboration on a topic. "I'd like to do some different things" is a more positive "I" statement.
19. *Allen*: So, Mary, if I hear you correctly, sounds like change and variety are important instead of doing the same thing all the time.	Paraphrase Client	Note that this paraphrase has some dimensions of an interpretation in that Mary did not use the words *change* and *variety*. These words are the opposite of

COUNSELOR AND CLIENT CONVERSATION	SKILL CLASSIFICATIONS, INCLUDING FOCUS AND CONFRONTATION	PROCESS COMMENTS, INCLUDING EMPATHY AND CLIENT CHANGE SCALE
		boredom and doing "the same things all the time." This paraphrase takes a small risk and is slightly additive to Mary's understanding. This is an example of the positive asset search, in that it would have been possible to hear only the negative "bored." Working on the positive suggests what *can* be done. Note her next response.
20. *Mary*: Yeah . . . I'd like to be able to do something different. But, you know, ah . . . teaching's a very secure field, and I have tenure. You know, I'm the sole support of my two daughters, but I think, I don't know what else I can do exactly. Do you see what I'm saying?	Client, family, issue/ concern	Mary, being heard, is able to move to a deeper discussion of her issues. Note that Mary tends to be abstract and discusses patterns and generalizations. If she were primarily concrete, she would give many more linear details and tell specific stories about her issues. She continues for most of the session in this mode of expression. She has equated "something different" with a lack of security. As the session progresses, you will note that she associates change with risk. These basic meaning constructs, already apparent, lie under many of her issues.
21. *Allen*: Looks like the security of teaching makes you feel good, but it's the boredom you associate with that security that makes you feel uncomfortable. Is that correct?	Reflection of feeling with confrontation of conflicting emotions followed by checkout Client, issue	This reflection of feeling contains elements of a confrontation as well, in that the good feelings of security are contrasted with the boredom associated with teaching. (Potentially additive empathy)
22. *Mary*: Yeah, you know, it's that security. I feel good being . . . you know . . . having a steady income and I have a place to be, but it's boring at the same time. You know, ah . . . I wish I knew how to go about doing something else.	Client, issue	Note that Mary often responds with a "Yeah" to the reflections and paraphrases before going on. Here she is wrestling with Allen's confrontation in line 21. She adds new data, as well, in the last sentence. On the CCS, this would be acceptance and recognition (Level 3).
23. *Allen*: So, Mary, let me see if I can summarize what I've heard. Ah . . . it's been tough since the divorce, but you've gotten things together. You mentioned the kids are doing pretty well. You talked about a new relationship. *I heard you mention that.*	Summary, checkout Client, family, issue/ concern	This summarization concludes the first attempt at definition of key issues. Allen uses Mary's own words for the main things and attempts to distill what has been said. The positive asset search has been used briefly ("You've gotten things together . . . kids . . . doing well"). See other leads

continued on next page

COUNSELOR AND CLIENT CONVERSATION	SKILL CLASSIFICATIONS, INCLUDING FOCUS AND CONFRONTATION	PROCESS COMMENTS, INCLUDING EMPATHY AND CLIENT CHANGE SCALE
(*Mary*: Yeah.) But the issue that you'd like to talk about now is . . . this feeling of boredom (*Mary*: Umm . . .) on the job, and yet you like the security of it. But maybe you'd like to try something new. Is that the essence of it?		that emphasize client strength. Mary sits forward and nods with approval throughout this summary. The confrontation of the old job with "maybe you'd like to try something new" concludes the summary. Note the checkout at the end of the summary to encourage Mary to react. (Interchangeable)
24. *Mary*: That's right. That's it.	Client, issues/concerns	Mary again responds at Level 3 on the CCS, acceptance and recognition.

The third stage of the session is setting goals. As you read through goal setting, please consider the following questions and provide your responses.

1. This stage is goal setting, yet Mary does not seem to establish goals. Reading through the session and her responses to Allen's questions, what could you ask Mary to help clarify her goals for counseling?
2. How does this stage differ from Stage 2, Story and Strengths?
3. Allen provides a response to Mary's identified issue/concern in line 41. Please formulate your own response. How is it different from Allen's response? Why did you respond in this way?

COUNSELOR AND CLIENT CONVERSATION	SKILL CLASSIFICATIONS, INCLUDING FOCUS AND CONFRONTATION	PROCESS COMMENTS, INCLUDING EMPATHY AND CLIENT CHANGE SCALE
STAGE 3. GOALS Set goals mutually, establish outcomes.		
25. *Allen*: I think it might be helpful if you could specifically define what some things are that might represent a more ideal situation.	Open question in the form of a statement Main theme	In Stage 3, seek to discover where the client wants to go. Note that the basic listening sequence remains central in this stage. Counselor statements oriented to goals have the potential to be additive.
26. *Mary*: Ummm. I'm not sure. There are some things I like about my job. I certainly like interacting with the other professional people on the staff. I enjoy working with the kids. I enjoy talking with the kids. That's kind of	Client, issue/concern, others	Mary associates interacting with people as a positive aspect of her job. When she says "enjoy working with kids," her tone changes, suggesting that she doesn't enjoy it that much. But the spontaneous tone returns when she mentions "talking with

COUNSELOR AND CLIENT CONVERSATION	SKILL CLASSIFICATIONS, INCLUDING FOCUS AND CONFRONTATION	PROCESS COMMENTS, INCLUDING EMPATHY AND CLIENT CHANGE SCALE
fun. You know, it's the stuff I have to teach I'm bored with. I have done some teaching of human sexuality and drug education.		them" and talks about teaching subjects other than team sports.
27. *Allen*: So, would it be correct to say that some of the teaching, where you have worked with kids on content of interest to you, has been fun? What else have you enjoyed about your job?	Paraphrase, open question Client, main theme	The search here is for positive assets and things that Mary enjoys. Note the "what else?" (Potentially additive)
28. *Mary*: Well, I must say I enjoy having the same summer vacations the kids have. That's a plus in the teaching field. (pause)	Client, family, issue/ concern	
29. *Allen*: Yeah . . .	Encourage	Mary found only one plus in the job. Allen probes for more data via an encourager. This type of encourager can't be classified in terms of focus.
30. *Mary*: You see, I like being able to . . . Oh, I know, one time I was able to do a workshop for our own teachers and that was really . . . I really felt good being able to share some of my ideas with some people on the staff. I felt that was kind of neat, being able to teach other adults.	Client, others	Mary brings out new data that support her earlier comment that she likes to teach when the content is of interest to her. The "I" statements here are more positive and the adjective descriptors indicate more self-assurance.
31. *Allen*: Do you involve yourself very much in counseling the students you have?	Closed question Client, others	A closed question with a change of topic to explore other areas.
32. *Mary*: Well, the kids . . . you know, teaching them is a nice, comfortable environment, and kids stop in before class and after class and they talk about their boyfriends and the movies; I find I like that part . . . about their concerns.	Client, others	Mary responds to the word *counseling* again with discussion of interactions with people. Mary wants work that keeps her in contact with others.

continued on next page

COUNSELOR AND CLIENT CONVERSATION	SKILL CLASSIFICATIONS, INCLUDING FOCUS AND CONFRONTATION	PROCESS COMMENTS, INCLUDING EMPATHY AND CLIENT CHANGE SCALE
33. *Allen*: So, as we've been reviewing your current job, it's the training, the drug education, some of the teaching you've done with kids on topics other than phys. ed. (*Mary*: That's right.) And getting out and doing training and other stuff with teachers . . . ah, sharing some of your expertise there. And the counseling relationships. (*Mary*: Ummm.) Out of those things, what are some fields you've thought of transferring to?	Summary, open question Client, issue/concern, others	This summary attempts to bring out the main strands of the positive aspects of Mary's job. Note, however, that the counselor still directs the flow with the open question and has changed topic somewhat. Potentially additive, but would it have been more useful to have explored the interpersonal contacts Mary just mentioned? This is a good example of how questions can focus and direct the conversation, but is this the time for such direction? What do you think?
34. *Mary*: Well, a lot of people in physical education go into counseling. That seems like a natural second thing. Ah . . . of course, that would require some more going to school. Umm . . . I've also thought about doing some management training for a business. Sometimes I think about moving into business . . . entirely away from education. Or even working in a college as opposed to working here in the high school. I've thought about those things, too. But I'm just not sure which one seems best for me.	Client, issue/concern	Mary talks with only moderate enthusiasm about counseling. In discussing training and business, she appears more involved. Mary appears to have assets and abilities, makes many positive "I" statements, is aware of key incongruities in her life, and seems to be internally directed. She is clearly an abstract, formal operational client. For career success, she also needs to become more concrete and action oriented.
35. *Allen*: So, the counseling field, the training field. You've thought about staying in schools and perhaps in management as well. (*Mary*: Um-hm, um-hm.) Anything else that occurs to you?	Paraphrase, open question Client, issue/concern	This brief paraphrase distills Mary's ideas in her own words. (Interchangeable)
36. *Mary*: No, I think that seems about it.	Issue/concern	
37. *Allen*: Before we go further, you've talked about teaching and the security it offers. But at the same time you talk about *boredom*. You talk with excitement about business and training. How do you put this together? What does it *mean* to you?	Summary, confrontation, open question, eliciting meaning Client, issue/concern	This summary includes confrontation and catches both content and feeling. The question at the end is directed toward issues of meaning. The word *boredom* was underlined with extra vocal emphasis. Meaning brings in emotional issues that are often helpful in decision making. (Interchangeable)

COUNSELOR AND CLIENT CONVERSATION	SKILL CLASSIFICATIONS, INCLUDING FOCUS AND CONFRONTATION	PROCESS COMMENTS, INCLUDING EMPATHY AND CLIENT CHANGE SCALE
38. *Mary*: Uhhh . . . Ah . . . If I stay in the same place, it's just more of the same. I see older teachers, and I don't want to be like them. Oh, a few have fun; most seem just *tired* to me. I don't want to end up like that.	Client, others	Mary elaborates on the meaning and underlying structure of *why* she might want to avoid the occasional boredom of her job. When she talks about "ending up like that," we see deeper meanings. On the CCS, the client may again be rated at acceptance and recognition (Level 3). Though considerable depth of understanding and clarity are being developed, no large change has occurred. You will find that developmental movement often is slow and arduous. Nonetheless, each confrontation moves toward more complete understanding.
39. *Allen*: You don't want to end up like that.	Encourage/restatement Client	The key words are repeated.
40. *Mary*: Yeah, I want to do something new, more exciting. Yet my life has been so confused in the past, and it is just settling down. I'm not sure I want to risk it.	Client	Mary moves on to talk about what she wants, and a new element—risk—is introduced. Risk may be considered Mary's opposing construct to security.
41. *Allen*: So, Mary, risk frightens you?	Reflection of feeling with an interpretive overtone Client	This reflection of feeling is tentative and said in a questioning tone. This provides an implied checkout and gives Mary room to accept it or suggest changes to clarify the feeling. (Interchangeable, possibly additive)
42. *Mary*: Well, not really, but it does seem scary to give up all this security and stability just when I've started putting it together. It just feels strange. Yet I do want something new so that life doesn't seem so routine . . . and . . . ah . . . I think maybe I have more talent and ability than I used to think I did.	Client, issue/concern	The statement above was additive. Mary responds as might be predicted with a deeper exploration of feelings of fear of change. At the same time, she draws on her personal strengths to cope with all this. This positive self-statement is a step toward Level 4 on the CCS.

continued on next page

COUNSELOR AND CLIENT CONVERSATION	SKILL CLASSIFICATIONS, INCLUDING FOCUS AND CONFRONTATION	PROCESS COMMENTS, INCLUDING EMPATHY AND CLIENT CHANGE SCALE
43. *Allen*: So you've felt the meaning in this possible job change as an opportunity to use your *talent* and take risks in something new. This may be contrasted with the feelings of stability and certainty where you are now. But *now* means you may end up tired and burned out like some co-workers you have observed. Am I reaching the sense of things? How does that sound?	Reflection of meaning, checkout Client, issue/concern	This reflection of meaning also confronts underlying issues that impinge on Mary's decision. It contains elements of the positive asset search or positive regard as Allen verbally stresses the word *talent*. (Interchangeable)
44. *Mary*: Exactly! But I hadn't touched on it that way before. I do want stability and security, but not at the price of boredom and feeling down as I have lately. Maybe I do have what it takes to risk more.	Client, issue/concern	Mary is reinterpreting her situation from a more positive frame of reference. Allen could have said the same thing via an interpretation, but reflection of meaning lets Mary come up with her own definition. This reinterpretation of Mary's meaning represents generation of a new solution (CCS Level 4). She has a new frame of reference with which to look at herself. But this newly integrated frame is *not* issue/concern resolution; it is a *step* toward a new way of thinking and acting. Allen decides to move to Stage 4 of the session. It would be possible to explore definition of key concerns and detail the goals more precisely, but we can take up these matters in later contacts.

The fourth stage is restorying, which helps clients to think through alternatives and confront discrepancies in their thinking and plans. As you read through restorying, please consider the following questions and provide your responses.

1. Mary seems pretty certain that she wants a change. What evidence do you have that she wants to make a change? What is holding her back from making changes?
2. How does Allen successfully walk her through these challenges? Is there anything you would have done differently or other areas you would have addressed?
3. Mary's issues really begin to come to the surface at the end when the time for the session is almost over. This is common as clients begin to feel more comfortable with the counselor and have had a chance to explore their issues. Although Mary's concerns are important, Allen can suggest that they explore these issues in the next meeting. How might your response in line 85 differ if your client was in an intense emotional state?

COUNSELOR AND CLIENT CONVERSATION	SKILL CLASSIFICATIONS, INCLUDING FOCUS AND CONFRONTATION	PROCESS COMMENTS, INCLUDING EMPATHY AND CLIENT CHANGE SCALE
STAGE 4. RESTORY Explore and create alternatives, confront client incongruities and conflict.		
45. *Allen*: Mary, from listening to you, I get the sense that you do have considerable ability. Specifically, you can be together in a warm, involved way with those you work with. You can describe what is most meaningful to you. You come across to me as a thoughtful, able, sensitive person. (pause)	Feedback Client	Allen combines feedback on positive assets with some self-disclosure here and uses this lead as a transition to explore alternative actions. The emphasis here is on the positive side of Mary's experience. Allen's vocal tone communicates warmth, and he leans toward Mary in a genuine manner. (Additive)
46. *Mary*: Ummm . . .		During the feedback, Mary at first shows signs of surprise. She sits up, then relaxes a bit, smiles, and sits back in her chair as if to absorb what Allen is saying more completely. There are elements of praise in Allen's comment.
47. *Allen*: Other job ideas may develop as we talk . . . ah . . . I think it might be appropriate at this point to explore some alternatives you've talked about. (*Mary*: Um-hm.) The first thing you talked about was you liked teaching drug education and sexuality. What else have you taught kids?	Directive, paraphrase, open question Client, main theme	Allen starts exploring alternatives a little more concretely and in depth. The systematic issue/concern-solving model—define the issue/concern, generate alternatives, and set priorities for solutions—is in his mind throughout this section. He begins with a mild directive. "What else?" keeps the discussion open. (Additive)
48. *Mary*: Let's see . . . The general areas I liked were human sexuality, drug education, family life, and those kinds of things. Ah . . . sometimes communication skills.	Client, issue/concern	
49. *Allen*: Have you attended workshops on any of these topics?	Closed question Issue/concern	Closed questions oriented toward concreteness can be helpful in determining specific experiences that may be useful in her decision making.
50. *Mary*: I've attended a few. I've enjoyed them . . . I really did. You know, I've gone to the university and taken workshops in values clarification and communication skills. I liked the people I met.	Client, issue/concern, others	Note that virtually all counselor and client comments have focused on the client and her issues and concerns. Remember to consider the client in each of your responses; too heavy an emphasis on difficulties and challenges may cause you to miss the unique person before you. At the same time, a broader focus

continued on next page

COUNSELOR AND CLIENT CONVERSATION	SKILL CLASSIFICATIONS, INCLUDING FOCUS AND CONFRONTATION	PROCESS COMMENTS, INCLUDING EMPATHY AND CLIENT CHANGE SCALE
		might expand the issue and provide more understanding. Social work, for example, might emphasize the family and social context.
51. *Allen*: Sounds like you've really enjoyed these sessions. One of the major roles in counseling, education, and business is training—for example, psychological education through teaching others skills of living and communication. How does that type of work sound to you?	Reflection of feeling, information, checkout Client, issue/concern	Allen briefly reflects her positive feelings and then shares a short piece of occupational information. This is followed by a checkout returning the focus to Mary. (Additive)
52. *Mary*: I think I would enjoy that sort of thing. Um-hmmm . . . It sounds interesting.	Client, issue/concern	
53. *Allen*: Sounds like you have also given a good deal of thought to . . . ah . . . extending that to training in general. How aware are you of the business field as a place to train and teach employees?	Paraphrase, open question Client, issue/concern	Mary's background and interest in a second alternative are explored.
54. *Mary*: I don't know that much about it. You know, I worked one summer in my dad's office, so I do have an exposure to business. That's about it. They all have been saying that a lot of teachers are moving into the business field. Teaching is not too lucrative, and with all the things happening here in California and all the cutbacks, business is a better long-term possibility for teachers these days. It just seems like an intriguing possibility for me to investigate or look into. The latest business cutbacks are scary, too.	Client, issue/concern, environmental context	Mary talks in considerably greater depth and with more enthusiasm when she talks about business. The most frequent descriptive words she has used with teaching include *boring*, *security*, and *interpersonal interactions*, while *interest* and *excitement* were used for training and teaching psychologically oriented subjects as opposed to physical education. Now she mentions cutbacks. Business has been described with more enthusiasm and as more lucrative. We may anticipate that she will eventually associate the potential excitement of business with the negative construct of risk and the lack of summer vacations and time to be with her children.
55. *Allen*: Mm-hmm, . . . so you've thought about it . . . looking into business, but you've not done too much about it yet. Neither teaching nor business is really promising now, and that's a little scary.	Paraphrase, reflection of feeling, including confrontation of discrepancies in behavior Client, issue/concern	This paraphrase is somewhat subtractive. Mary did indicate that she had summer experience with her father. How much and how did she like it? Allen missed that. The paraphrase involves a confrontation between what Mary says and her lack of doing anything extensive in terms of a search. The reflection of feeling acknowledges emotion.

COUNSELOR AND CLIENT CONVERSATION	SKILL CLASSIFICATIONS, INCLUDING FOCUS AND CONFRONTATION	PROCESS COMMENTS, INCLUDING EMPATHY AND CLIENT CHANGE SCALE
56. *Mary*: That's right. I've thought about it, but . . . ah . . . I've done very little about it. That's all . . .	Client	Mary feels a little apologetic. She talks a bit more rapidly, breaks eye contact, and her body leans back a little. Mary's response is at Level 2 on the CCS. She is only partially able to work with the issues of the confrontation.
57. *Allen*: And, finally, you mentioned that you have considered the counseling field as an alternative. Ah . . . what about that?	Interpretation Issue/concern	Allen omitted further exploration of business. If Allen had focused on positive aspects of Mary's experience and learned more about her summer experience, the confrontation (of thinking without action) probably would have been received more easily. As this was a demonstration based on real issues, Allen sought to move through the stages perhaps a little too fast. Also, the counseling field is an alternative, but it seems to come more from Allen than from Mary. An advantage of transcripts such as this is that one can see errors. Many of our errors arise from our own constructs and needs. This intended paraphrase is classified as an interpretation, as it comes more from Allen's frame of reference than from Mary's. Hard to classify in terms of empathy, as the comment is both a little subtractive and potentially additive as it elaborates information in more detail.
58. *Mary*: Well, I've always been interested, like I said, in talking with people. People like to talk with me about all kinds of things. And *that* would be interesting . . . ah . . . I think, too. (pause)	Issue/concern, others	Mary starts with some enthusiasm on this topic, but as she talks her speech rate slows and she demonstrates less energy.
59. *Allen*: Um-hmmm.	Encourage	
60. *Mary*: You know, to explore that. (pause)	Issue/concern	Said even more slowly.
61. *Allen*: Um-hmmm. (pause)	Encourage	Allen senses her change of enthusiasm, is a bit puzzled, and sits silently, encouraging her to *talk more*. When you have made an error and the client doesn't respond as you expect, return to attending skills.

continued on next page

COUNSELOR AND CLIENT CONVERSATION	SKILL CLASSIFICATIONS, INCLUDING FOCUS AND CONFRONTATION	PROCESS COMMENTS, INCLUDING EMPATHY AND CLIENT CHANGE SCALE
62. *Mary*: But . . . I'd have to take some *courses* . . . if I really wanted to get into it.	Issue/concern	One reason for Mary's hesitation appears. It is becoming clearer that Allen's comments at 57 are a slight distraction, but the information obtained still is useful.
63. *Allen*: So putting those three things together, it seems that you want people-oriented occupations. They are particularly interesting to you.	Interpretation/ reframe Client, issue/concern	This is a mild interpretation, as it labels common elements in the three jobs. It could also be classified as a paraphrase. Not all skill distinctions are clear. (Interchangeable, but Allen is more back on track)

This would have been a good point to present the Balance Sheet to Mary. Together Allen and Mary could look at the positives and negatives of changing career and the ultimate logical consequences of the decision. Allen did work through the balance sheet later, but it would have been more timely to have introduced it at this time. Thinking through the balance sheet and future diary would be a useful homework assignment for her.

COUNSELOR AND CLIENT CONVERSATION	SKILL CLASSIFICATIONS, INCLUDING FOCUS AND CONFRONTATION	PROCESS COMMENTS, INCLUDING EMPATHY AND CLIENT CHANGE SCALE
64. *Mary*: Definitely . . . and that's where I am most happy.	Client	Mary has returned to a Level 3 on the CCS.
65. *Allen*: And, Mary, as I talk I see you . . . ah . . . coming across with a lot of enthusiasm and interest as we talk about these alternatives. I do feel you are a little less enthusiastic about returning to school. (*Mary*: Right!) I might contrast your enthusiasm about the possibilities of business and training with your feelings about education. There you talk a little more slowly and almost seem bored as you talk about it. You seem lively when you talk about business possibilities.	Feedback with confrontation of where she is and where she'd like to be Client, issue/concern	Allen gives Mary specific and concrete feedback about how she comes across in the session. There is a confrontation as he contrasts her behavior when discussing two topics. Confrontation— the presentation of discrepancies or incongruity—may appear with virtually all skills. It may be used to summarize past conversation and stimulate further discussion, leading toward a resolution of the incongruity. (Additive)
66. *Mary*: Well, they sound kind of exciting to me, Allen. But I just don't know how to go about getting into those fields or what my next steps might be. They sound very exciting to me, and I think I may have some talents in those areas I haven't even discovered yet.	Client, issue/concern	Mary talks rapidly, her face flushes slightly, and she gestures with enthusiasm. She meets the confrontation and seems to be willing to risk more. This may still be considered a Level 3 on the CCS, but there may be movement ahead.
67. *Allen*: Um-hmmm. Well, Mary, I can say one thing. Your enthusiasm and ability to be open will be helpful to you in your search. Ah . . . at the same time, business and schools	Feedback, information, logical consequences, confrontation Client, issue/concern	This statement combines mild feedback with logical consequences. A warning about the consequences of client action or inaction is spelled out. Mary is also confronted with some consequences of choice. (Additive)

COUNSELOR AND CLIENT CONVERSATION	SKILL CLASSIFICATIONS, INCLUDING FOCUS AND CONFRONTATION	PROCESS COMMENTS, INCLUDING EMPATHY AND CLIENT CHANGE SCALE
represent different types of lifestyles. I think I should give you a warning that if you go into the business area you're going to lose those summer vacations.		
68. *Mary:* Yeah, I know that . . . and you know, that special friend in my life—he's in education—I don't think he would like it if I was, you know, working all summer long. But business does pay a lot more, and it might have some interesting possibilities. (*Allen:* Um-hmm.) . . . It's a difficult situation.	Client, issue/concern others	Confrontations often result in clients' presenting new information and thoughts that have not been discussed previously. A new issue has emerged that may need definition and exploration. Mary is still responding at Level 3 on the CCS, but Allen is obtaining a more complete picture of the issues and of the client.
69. *Allen:* A difficult situation?	Encourage/restatement Issue/concern	Again, the encourager is used to find deeper meanings and more information.
70. *Mary:* Um-hmm. I guess I'm saying that . . . I'm . . . ah . . . you know, my friend . . . I don't think he would approve or like the idea of me having two weeks' vacation. (*Allen:* Uh-huh.) He wants me to stay in some field where I have the same vacation time I have now so we can spend that time together.	Client, issue/concern, others	Mary has more speech hesitations and difficulties in completing a sentence here than she has anywhere in the session. This suggests that her relationship is meaningful to her, and her friend's attitude may be a part of the final career decision. Much career counseling involves personal issues as well as career choice. Both require resolution for true client satisfaction.
71. *Allen:* I hear you saying that your friend has a lot to say about your future. How does that strike you as an independent woman who has been on your own successfully for quite a while?	Interpretation/reframe, open question Client, others, cultural/environmental context	Here we see the introduction of gender relations as a cultural/environmental/contextual issue. Allen's reframing of the situation offers Mary a chance to explore her relationship with her friend from a different contextual perspective. (Additive)
72. *Mary:* It really is . . . well, Bo's a special person . . .	Others	Mary's eyes brighten.
73. *Allen:* And, I sense you have some reactions to his . . .	Interpretation Client, others	Allen interrupts, perhaps unnecessarily. It might have been wise to allow Mary to talk about her positive feelings toward Bo.
74. *Mary:* Yeah, I'd like to be able to explore some of my own potential without having those restraints put on me right from the beginning.	Client, issue/concern	Mary talks slowly and deliberately, with some sadness in her voice. Feelings are often expressed through intonation. Here we see the beginning of a critical gender

continued on next page

COUNSELOR AND CLIENT CONVERSATION	SKILL CLASSIFICATIONS, INCLUDING FOCUS AND CONFRONTATION	PROCESS COMMENTS, INCLUDING EMPATHY AND CLIENT CHANGE SCALE
		issue. Women often feel constraints in career or personal choices, and men in this culture often place implicit or explicit restraints on critical decisions. Feminist counseling theorists argue that a male helper may be less effective with these types of issues/concerns. What are your thoughts on this issue?
75. *Allen*: Um-hmm . . . In a sense he's almost placing similar constraints on you that you feel in the job in physical education. There are certain things you have to do. Is that right?	Interpretation/reframe, confrontation, checkout Client, issue/concern, others, cultural/environmental context	This interpretation relates the construct of boredom and the implicit constraint of being held down to the constraints of Bo. The interpretation clearly comes from Allen's frame of reference. With interpretations or helping leads from your frame of reference, the checkout of client reactions is even more essential. The drawing of parallels is abstract, formal operational in nature. (Potentially additive)
76. *Mary*: Yes, probably so. He's putting some limits on me . . . setting limits on the fields I can explore and the job possibilities I can possibly have. Setting some limits so that my schedule matches his schedule.	Client, issue/concern, others	Mary answers quickly. It seems the interpretation was relatively accurate and helpful. One measure of the function and value of a skill is what the client does with it. Mary changes the word *constraints* to the more powerful word *limits*. Mary remains at Level 3 on the CCS, as she is still expanding on aspects of the issue/concern.
77. *Allen*: In response to that you feel . . . ? (deliberate pause, waiting for Mary to supply the feeling)	Open question, oriented to feeling Client	Research shows that *some* use of questions facilitates emotional expression. (Additive)
78. *Mary*: Ah . . . I feel I'm not at a point where I want to *limit things*. I want to see what's open, and I would like to keep things open and see what all the alternatives are. I don't want to shut off any possibility that might be really exciting for me. (*Allen*: Um-hmm.) A total lifetime of careers.	Client, issue/concern	Mary determinedly emphasizes that she does not want limits.
79. *Allen*: So you'd like to have a life of exciting opportunity, and you sense some limiting . . .	Reflection of feeling, paraphrase, confrontation Client, issue/concern	A brief but useful confrontation of Bo versus career.

COUNSELOR AND CLIENT CONVERSATION	SKILL CLASSIFICATIONS, INCLUDING FOCUS AND CONFRONTATION	PROCESS COMMENTS, INCLUDING EMPATHY AND CLIENT CHANGE SCALE
80. *Mary*: He reminds me of my relationship with my first husband. You know, I think the reason that all fell apart was my going back to work. You know, assuming a more nontraditional role as a woman and exploring my potential as a woman rather than staying home with the children . . . ah . . . you know, sort of a similar thing happened there.	Client, issue/concern, others, cultural/ environmental context	Again, the confrontation brings out new information about Mary's present and past. Is she repeating old relationship patterns in this new relationship? The counselor should consider issues of cultural sexism as an environmental aspect of Mary's planning. This does not appear in this meeting, but a broader focus on issues in the next session seems imperative. Other focus issues of possible importance include Mary's parental models, others in her life, a women's support group, the present economic climate, the attitudes of the counselor, and "we"—the immediate relationship of Mary and Allen. So far he has assumed a typical Western "I" form of counseling in which the emphasis is on the client. With the development of new, more integrated data, this could move toward a more inclusive construct (Level 5 response on the CCS).
81. *Allen*: There really are a variety of issues that . . . you're looking at. One of these is the whole business of a job. Another is your relationship with Bo and your desire to find your own space as an independent woman.	Summary Client, issue/concern, others, cultural/ environmental context	The time is waning, and Allen must plan a smooth ending and plan for the next session. He catches the incongruity that Mary faces between work and relationship. Allen fails to pick up fully on the cultural/environmental context focus. Many of Mary's issues relate to women's issues in a sometimes sexist world.
82. *Mary*: (slowly) Um-hmmm . . .		Mary looks down, relaxes, and seems to go into herself.
83. *Allen*: You look a little sad as I say that.	Reflection of feeling Client	This reflection of feeling comes from nonverbal observations and picks up on her facial reactions. (Interchangeable)
84. *Mary*: It would be nice if the two would mesh together, but it seems difficult to have both things fit together nicely.	Issue/concern	Mary is describing her ideal resolution. Here the session could recycle back to Stages 2 and 3, with more careful delineation of the issue/concern between job and personal relationships and defining the ideal resolution more fully. *An issue/concern exists only if there is a difference between what is actually happening and what you desire to have happen.* This sentence illustrates the importance of issue/concern definition

continued on next page

COUNSELOR AND CLIENT CONVERSATION	SKILL CLASSIFICATIONS, INCLUDING FOCUS AND CONFRONTATION	PROCESS COMMENTS, INCLUDING EMPATHY AND CLIENT CHANGE SCALE
		and goal setting. Mary's response to the confrontation is 4 on the CCS. We have an important new insight, but insight is not action. She also needs to act on this awareness.
85. *Allen*: Well, that's something we can explore a little bit further. It seems this is an important part of the puzzle. Let's work on that next week. Would that be okay? I see our time is about up now. But it might be useful if we can think of some actions we can take between now and the next time we get together.	Information, directive Issue/concern	Many clients bring up central issues just as the meeting is about to end. Allen makes the decision, difficult though it is, to stop for now and plan for more discussion later. Note that Mary is still talking about her relationship mainly from an abstract, formal operational orientation. Clients often bring up central issues late in the session.

The fifth stage is action, giving your client areas to work on between sessions as they begin working on creating new stories for their life. As you read through the action stage, please consider the following questions and provide your responses.

1. What action steps does Allen ask Mary to complete? What steps does Mary come up with on her own? Can you think of anything else? How might you use the Balance Sheet as a homework assignment?
2. After Mary leaves, what notes would you write for yourself to act on or follow up on?

COUNSELOR AND CLIENT CONVERSATION	SKILL CLASSIFICATIONS, INCLUDING FOCUS AND CONFRONTATION	PROCESS COMMENTS, INCLUDING EMPATHY AND CLIENT CHANGE SCALE
STAGE 5. ACTION Generalize and act on new stories.		
86. *Allen*: We have come up so far with three things that seem to be logical: business, counseling, and training. I think it would be useful, though, if you were to take a set of career tests. (*Mary*: Uh-huh.) That will give us some additional things to check out to see if there are any additional alternatives for us to consider. How do you feel about taking tests?	Summary, open question Client, issue/concern	Allen continues his statement and moves to Stage 5. He summarizes the career alternatives generated so far and raises the possibility of taking a test. Note that he provides a checkout to give Mary an opportunity to make her own decision about testing. (Additive)

COUNSELOR AND CLIENT CONVERSATION	SKILL CLASSIFICATIONS, INCLUDING FOCUS AND CONFRONTATION	PROCESS COMMENTS, INCLUDING EMPATHY AND CLIENT CHANGE SCALE
87. *Mary*: I think that's a good idea. I'm at the stage where I want to check all alternatives. I don't want *anything* to be limited. I want to think about a lot of alternatives at this stage. And I think it would be good to take some tests.	Client, issue/concern	Mary approves of testing and views this as a chance to open alternatives. She verbally emphasizes the word *anything*, which may be coupled with her desire to avoid limits to her potential. Some women would argue that a female counselor is needed at this stage. A male counselor may not be sufficiently aware of women's needs to grow. Allen could unconsciously respond to Mary in the same ways she views Bo as responding to her.
88. *Allen*: Then another thing we can do . . . ah . . . is helpful. I have a friend at a local firm who originally used to be a coach. She's moved into personnel and training at Jones. (*Mary*: Ummm.) I can arrange an appointment for you to see her. Would you like to go down and look at the possibilities there?	Information Client, issue/concern, counselor	Allen suggests a concrete and specific alternative for action. Mary is predominantly abstract formal operational; she has tended to talk about issues and avoid action. This avoidance of action is also indicative of Level 3 on the Client Change Scale. Until Mary takes some form of concrete action or resolves the issue in her mind, she will remain at Level 2 or 3 on the CCS. If some action is taken on her concerns during the coming week, then she will have moved at least partially to Level 4 on the CCS. (Additive)
89. *Mary*: Oh, I would like to do that. I'd get kind of a feel for what it's like being in a business world. I think talking with someone would be a good way to check it out.	Client, issue/concern	Stated with enthusiasm. The proof of the helpfulness of the suggestion will be determined by whether she does indeed have a talk with the friend at Jones and finds it helpful in her thinking.
90. *Allen*: You're a person with a lot of assets. I don't have to tell you all the things that might be helpful. What other ideas do you think you might want to try during the week?	Feedback, open question Client, issue/concern	Allen recognizes he may be taking charge too much and pulls back a little. Although he is encouraging Mary to act, he is now using her ideas. (Additive)
91. *Mary*: What about checking into the university and ah . . . advanced degree programs? I have a bachelor's degree, but . . . maybe I should check into school and look into what it means to do more coursework.	Client, issue/concern	Mary, on her own, decides to look into the university alternative. This is particularly noteworthy, as earlier indications were that she was not all that interested. Note that real generalization is usually concrete *action*.

continued on next page

COUNSELOR AND CLIENT CONVERSATION	SKILL CLASSIFICATIONS, INCLUDING FOCUS AND CONFRONTATION	PROCESS COMMENTS, INCLUDING EMPATHY AND CLIENT CHANGE SCALE
92. *Allen*: Okay, that's something else you could look into as well. (*Mary*: Uh-huh.) So let's arrange for you then to follow up on that. I'd like to see you doing that. (*Mary*: Um-hmmmm.) And . . . ah . . . we can get together and talk again next week. You did express some concern about your relationship with your friend, Bo, ah . . . would you like to talk about that as well next week? And, as I look back on this session, one theme we haven't discussed yet is how being a woman with family responsibilities relates to all this. Maybe this is something to be explored next week as well?	Summary Client, issue/concern, cultural/environmental context	Allen is preparing to conclude. Fortunately, he does consider the women's issues. Probably this should have been done sooner in the session. Is this an issue with which he can help, or would you recommend referral? (Basically additive as it opens possible directions for the future)
93. *Mary*: I think so, they sort of all . . . one decision influences another. You know. It all sort of needs to be discussed. And thanks for bringing up women's issues and my children. That's important to me.	Client, cultural/ environmental context	An important insight at the end. Mary realizes her career issue is more complex than she originally believed. If you were Allen's supervisor, would you recommend a primary emphasis on career counseling or on personal counseling in the next session? Or perhaps some combination of them both? What else would you advise him to do?
94. *Allen*: Okay. I look forward to seeing you next week, then.	Self-disclosure Client, counselor	
95. *Mary*: Thank you.		

At the completion of the session, Allen walks out with Mary and they arrange for the next appointment. He also shows Mary the career information available in the counseling office.

▶Transcript Analysis

Through the way you listen and the topics you select to reinforce by attending, you influence what happens in the session. Effective listening will increase the control clients have and allow them to become partners or "co-constructors" of what happens in the session. Examine your behavior and become aware of your impact on your client. If you plan a career session, you most likely will talk about careers. If you decide to "let the client talk and see what happens," the sessions may lack direction, but "what happens will happen."

Of course, any session will not completely follow your plan, but your personal decisions influence what happens. This is a critical reason to use the checklist, develop plans, examine notes, and reexamine your own style. We recommend a detailed analysis of your counseling style and behavior continuously throughout your career. We also suggest appropriate sharing of your thoughts and analysis with your clients. Ask colleagues and supervisors to review your work, so you can grow and improve.

▶Skills and Their Impact on the Client

Let us turn to a microskill analysis of the session. Table 14.2 presents a skill summary of Allen's meeting with Mary. You will see that each stage involved different patterns of microskill usage.

Stage 1: Empathic Relationship   The session began with Allen using both listening and influencing skills. We see open questions, a combined paraphrase and reflection of feeling, information giving, and self-disclosure. He focused immediately on Mary's wellness

TABLE 14.2   **Types of Skills the Counselor Used in Each Stage of the Session**

Stage	Skills Used	Focus
Stage 1. *Empathic Relationship*: Initiating the session	6 attending skills 7 influencing skills 0 confrontations	5 client 0 issue/concern 3 counselor 1 cultural/environmental context
Stage 2. *Story and Strengths*: Gathering data	7 attending skills 0 influencing skills 1 confrontation	5 client 2 issue/concern 1 family 1 cultural/environmental context
Stage 3. *Goals*: Mutual goal setting	15 attending skills 3 influencing skills 1 confrontation	7 client 5 issue/concern 2 significant other
Stage 4. *Restory*: Working	18 attending skills 14 influencing skills 5 confrontations	14 client 15 issue/concern 4 significant other 3 cultural/environmental context
Stage 5. *Action*: Terminating	4 attending skills 3 influencing skills 0 confrontations	5 client 4 issue/concern 2 counselor 1 cultural/environmental context
Total	50 attending skills 27 influencing skills 7 confrontations	46 client 24 issue/concern 1 family 6 significant other 5 counselor 6 cultural/environmental context

strengths in swimming, obtained permission to record the session, and offered the opportunity for Mary to ask him questions. Observation skills helped Allen decide when it was time to move on with the conversation.

Stage 2: Story and Strengths   Only listening skills were used to draw out Mary's story and concerns. The primary focus was on changing careers from physical education to either counseling or business.

Stage 3: Goals   Listening and influencing skills were used as Mary spoke about her goals in more detail. The issue of security in teaching versus risk in business was a central concern, as revealed in the reflection of meaning with a confrontation (line 43).

Stage 4: Restory   In this "working" phase, Allen used both influencing skills and confrontation of incongruity and discrepancies extensively. The cultural/contextual issue of gender is explored though interpretation/reframing, listening skills, and periodic summaries.

Stage 5: Action   The session closes with specific plans for generalization and homework for Mary. This stage begins with a summary of the session and ends with a discussion of plans for the future.

In terms of the overall balance of skill usage, Allen used a ratio of approximately two attending skills for every influencing skill. When you look at competence levels, Allen is able to identify and classify the several skills and stages of the session. He is able to identify the impact of his skills on the client. Allen also demonstrates his ability to use the basic listening sequence to structure a session in five stages and to employ intentional counseling skills. Note that Allen focused primarily on the client in the earlier phases of the session and only in the later portions increased emphasis on the career issues and challenges. This demonstrates that he can balance focus between the person and client issues, concerns, and challenges. An ineffective counselor might have focused early on what he or she might have called the "issue/concern" and missed Mary as a unique person.

In terms of focus dimensions, Allen's focus remained primarily on the client, although the majority of his focus dimensions were dual, combining focus on Mary with focus on the issues and concerns Mary brought to the session. Focus analysis points out that Allen did not focus extensively on others and the family (on Bo or on Mary's children, for example). The relationship with Bo appeared with greater clarity later. Allen brought in the cultural/environmental/contextual focus (lines 71, 75, and 81), enabling a beginning discussion of gender issues that clearly need further work.

Mary appeared to move a little deeper into personal insights concerning her present and future life following each of the five confrontations in Stage 4. She responded primarily at Level 3 on the Client Change Scale (CCS) each time. Note that Mary was led to discuss her personal life and relationship with Bo (line 71). At 79, Allen comments on Mary's desire to have a life of "exciting opportunity," but she senses that Bo is putting limits on her. Mary moves easily here to discuss her own personal wishes in more depth. Allen included a checkout at the end of a slightly inaccurate confrontation (75), and Mary was able to introduce her own constructs, substituting the word *limits* for Allen's *constraints*.

As the session ended, Mary appeared ready and willing to take action. On the Client Change Scale, she has moved to Level 4, or a new way of thinking about her issues. The CCS (introduced in Chapter 10) is a systematic way for you to monitor the progress of your clients toward their goals both inside and outside the session. The real proof of success, however, will have to wait until the next meeting so that we can determine whether the generalization plan was indeed acted on. Thoughts, feelings, and behaviors need to change for true generalization. *The work that clients do after the session is as valuable as or more valuable than what they do during the session.* You can also evaluate the effectiveness of the meeting in terms of the number of choices available to Mary. A good rule of thumb is "If you don't have at least three possibilities, you don't have a choice." Mary appears to have achieved that objective. In addition, the issue of her relationship with Bo has been unearthed, and this topic may open her to further counseling possibilities. The question must be raised whether Allen, as a man, is the most appropriate counselor for Mary to see. Answers to that question will vary with your personal worldview. Again, what are your evaluations? What would you do differently?

Your first counseling session with a client is always unique and gives you an opportunity to learn about diversity and the complexity of the world. Box 14.2 presents an international view of the work with clients and recommendation for practice.

BOX 14.2   ## National and International Perspectives on Counseling Skills

What's Happening With Your Client While You Are Counseling?

Robert Manthei, Christchurch University, New Zealand

There is more going on in counseling beyond what we see happening during the session. Clients are good observers of what you are doing, and they may not always tell you what they think and feel. Research shows that clients expect counseling to be shorter than do most counselors and therapists. Clients see counselors as more directive than counselors see themselves. And what the counselor sees as a good session may be seen otherwise by clients, and vice versa. Counselors and clients may vary in their perceptions of the effectiveness of their work together.

I conducted a study of client and counselor experience of counseling. Among the major findings are the following:

Clients Often Have Sought Help Before
Most people don't come for counseling immediately. Talking with friends and family and trying to work it out on their own were usually tried first. Reading self-help books, prayer, and alcohol and drugs are among other things tried. Some deny that they have difficulties until these become more serious.

Implications for Practice   Ask clients what they have tried before they came to you, and find out what aspects of prior efforts seemed to have helped. You may want to build on past successes. This is an axiom of brief solution-oriented counseling.

First Impressions Make a Difference
That first meeting sets the stage for the future, and the familiar words "relationship and rapport" are central. I found that clients generally had favorable impressions of

the first session and viewed what happened even more positively than counselors. Sometimes sharing experience helps. One client who did not feel positive about the first session commented, "Maybe if the counselor had gone through a similar experience of divorce and children, it would have helped."

Implications for Practice   Obviously, be ready for that first session. Cover the critical issues of confidentiality and legal issues in a comfortable way. Structure and let the client know what to expect. Some personal sharing, used carefully, can help. And empathic listening always remains central.

Counseling Helps, but So Do Events Outside of the Session
Resolution of their issues was attributed to counseling by 69% of clients, while 31% believed events outside the session made the difference. Among things that helped were talking and socializing more with family and friends, taking up new activities, learning relaxation, and involvement with church.

Implications for Practice   What you do in the session needs to be supplemented with plans for generalization of behavior and thought to daily life. Homework and specific ideas for using what is learned in the session help ensure that action follows the session.

Things That Clients Liked
Relationship variables such as warmth, understanding, and trust are essential. Clients liked being listened to and

(continued)

BOX 14.2 (continued)

being involved in making decisions about the course of counseling. Reframes and interpretations helped them see their situations in a new way; also valued were new skills such as imagery, relaxation training, and thought stopping to eliminate negative self-talk.

All of the above speaks to respecting the client's ability to participate in the change process. We need to disclose with clients the rationale for what we are doing, but also ask them to share their perceptions of the session(s) with us. We can learn much from the client by adopting an egalitarian approach.

▶Additional Considerations: Referral, Treatment Planning, and Case Management

What treatment, by whom, is most effective for this individual with that specific problem, and under what set of circumstances?

—Gordon Paul

Respect the client's ability to participate in the change process. We need to disclose with clients the rational for what we are doing, but also ask them to share their perceptions of the session(s) with us. We can learn much from the client by adopting an egalitarian approach.

—Robert Manthei

These two quotations provide a background for integrating and thinking through the many ideas, concepts, and theories discussed in this book. While we have broken down the counseling and therapy process into clear steps, it is the totality of the many basics of helping that enable us to realize the real complexity of this interactive process—we and the client learn and change with each other.

Whether you are involved with decisional, person-centered, crisis, or cognitive behavioral therapy, you need to consider what to do next. The following sections address three key issues in planning for the future.

Referral

No counselor has all the answers. In the case of Mary, although Allen did discuss women's issues with her, *referral* to a women's group might be helpful as many of her issues are common to women looking for a career change. In addition, she may need a referral to the university financial aid office. An important part of individual counseling is helping your clients find community resources that may facilitate their growth and development. The community genogram (presented in Chapter 9) helps counselors and clients think more broadly and consider appropriate referral sources.

Sometimes the counselor/client relationship simply doesn't work as well as we all would like. When you sense the relationship isn't doing well, avoid blaming either the client or yourself. Focus on client goals, and seek to hear the story completely and accurately. Ask the client for feedback on how you might be more helpful. Seek consultation and supervision, and most often these "difficult patches" can be resolved to the benefit of all. When an appropriate referral needs to be arranged, you do not want to leave clients "hanging" with no sense of direction or fearful that their problems are too difficult. Maintain contact with the client as the referral process evolves, sometimes even continuing for a session or two until arrangements are complete. It is critical that the client never feel rejected by you. Your understanding, empathic support during the referral process is essential.

Another key referral issue is whether counselor expertise and experience are sufficient to help the client. Even if you think that you are working effectively, this may not be enough; it may be a case in which supervision and case conferences can be helpful. Opening up your work to others' opinions is an important part of professional practice. Clients, of course, should be made aware that you as counselor or therapist are being supervised.

Treatment Planning

Treatment planning, in the form of specific written goals and objectives, is increasingly standard and often required by agencies and insurance companies. When possible, negotiate specific goals with the client and write them down for joint evaluation. They should be as concrete and clear as possible, including specific indicators of behavioral change and emotional satisfaction. The more structured counseling theories, such as cognitive behavioral, strongly urge counseling and treatment plans with specific goals developed for each issue. Their counseling and treatment plans are often more specific than the one presented in the example that follows.

Less structured counseling theories (Gestalt, psychodynamic, person-centered) tend to give less emphasis to treatment plans, preferring to work in the moment with the client. In short-term counseling and interviewing, the interview plan serves as the treatment plan. As you move toward longer-term counseling (five to ten sessions), a more detailed treatment plan with specific goals is often required. Many agencies also use case management as part of the treatment plan.

Following is Allen's plan for the second session with Mary. Developed from information gained in the first session, the plan organizes the central issues of the case, allowing for new input from Mary as the session progresses.

STAGE/DIMENSION, KEY QUESTIONS	COUNSELOR ASSESSMENT AND PLAN
Empathic relationship. Initiate the session, develop rapport and structuring. *What structure do you have for this interview? Do you plan to use a specific theory? What special issues do you anticipate with regard to rapport development?*	Mary and I have reasonable rapport. As I look at the first session, I note I did not focus enough on Mary's context nor did I attend to other things that might be going on in her life. It may be helpful to plan some time for general exploration *after* I follow up on the testing and her interviews with people during the week. Mary indicated an interest in talking about Bo. Two issues need to be considered at this session in addition to general exploration of her present state. I'll introduce the tests and follow that with discussion of Bo. For Bo, I think a person-centered method emphasizing listening skills may be helpful.
Story and strengths. Gather data, draw out stories, concerns, problems, or issues. *What are anticipated problems? Strengths? How do you plan to define the issues with the client? Will you emphasize behavior, thoughts, feelings, meanings?*	1. Check how Mary sees her career concerns defined now. Use basic listening sequence. 2. Later, and as appropriate, open up the issue of Bo with a question, then follow through with reflective listening skills. Be alert to a woman's perspective. 3. Mary has many assets. She is bright, verbal, and successful in her job. She has good insight and is willing to take reasonable risks and explore new alternatives. These assets should be noted in our future sessions. 4. Explore women's issues with her.

continued on next page

STAGE/DIMENSION, KEY QUESTIONS	COUNSELOR ASSESSMENT AND PLAN
Goals. Set goals mutually, establish outcomes. *What is the ideal outcome? How will you elicit the client's idealized self or world?*	We have already discussed her career goals, but they may need to be reconsidered in light of the tests, further discussion of Bo, and so on. It is possible that late in this session or in a following session we may need to define a new outcome in which careers and her relationships are both satisfied.
Restory. Explore and create alternatives, confront client incongruities and conflict. Working and restorying. *What theories would you probably use here? What specific incongruities have you noted or do you anticipate in the client? How will you generate alternatives?*	1. Check on results of tests and report them to Mary. 2. Explore her reactions and consider alternative occupations. 3. Use person-centered, Rogerian counseling and explore her issues with Bo. 4. Relate careers to the relationship with Bo. Give special attention to confronting the differences between her needs as a "person" and Bo's needs for her. Note and consider the issue of women in a changing world. Does Mary need referral to a woman or to a women's group for additional guidance? Would assertiveness training be useful?
Action. Generalize and act on new stories. *What specific plans do you have for transfer of training? What will enable you, the counselor, to personally feel that the interview was worthwhile?*	At the moment it seems clear that further exploration of careers outside the interview is needed. We will have to explore the relationship with Bo and determine her objectives more precisely.

Counselors and therapists working in community and hospital clinics are often required to have precise goals in their treatment plans, both for the good of the client and to fulfill the requirements of many insurance companies. A brief outline of what many agencies consider important in a treatment plan is shown in Figure 14.1. Note the emphasis on concrete goals, specificity of interventions, and a planned date for evaluation of goal achievement. Increasingly, you will find yourself working with some variation of this goal-oriented form, regardless of setting.

Case Management as Related to Treatment Planning

Although this book focuses on the session and counseling skills, treatment planning often needs to be extended to *case management*. For human service professionals, social workers, and school counselors, case management will be as or more important than treatment planning. Case management requires the professional helper to coordinate community services for the benefit of the client and, very often, the client's family as well. Let's look at the complexity of case management with an example.

> *A single mother and her 10-year-old boy are referred to a social worker in a family services agency. The family physician thinks that the child's social interaction issues may be a result of Asperger's syndrome. The child has few friends, but is doing satisfactorily in school. The social worker interviews the mother and reports the child's social and academic situation at school. The mother says that she has financial problems and difficulty in finding work. Meeting with the the child a few days later, the worker notes good cognitive and language competence, but that the child is unhappy and demonstrates some repetitive, almost compulsive behavioral patterns.*

COMMUNITY CLINIC
Behavioral Treatment Plan

This form will be reviewed again in no more than two months, and progress toward goals will be noted. Changes in interventions or goals should be noted immediately.

Patient's Name, Address, Phone, Email: _____

_____

Clinic Record Number _____   Insurance _____

Diagnosis: Summary of Patient's Original Concerns:

Axis I _____   _____

Axis II _____   _____

Axis III _____   _____

Axis IV _____   _____

Axis V _____   _____

Identified Patient Strengths and Resources (to be added to throughout therapy):

_____

_____

_____

_____

Interview Progress Narrative

Problem/Concern #1		
Goal	Interventions	Progress Toward Goal

Problem/Concern #2		

Problem/Concern #3		

Signature _____   Date _____

Patient signature _____   Date _____

If patient is a child:

Name of child _____   Age _____

Parent signature _____   Date _____

FIGURE 14.1 Treatment plan example.

The agency staff meets and starts to initiate a case management treatment plan. It is clear that the mother needs counseling and that the child needs psychological evaluation and likely treatment as well. At the staff meeting, the following plan is developed: The child is to be referred to a psychologist for evaluation; based on that evaluation, recommendations for treatment are likely to be made. The social worker is assigned to do supportive counseling with the mother and to take overall responsibility for case management. The social worker will eventually need a treatment plan for the mother and a case management plan for the family.

School performance is good, but the social worker contacts the elementary school counselor and finds that the counselor already has the child in counseling. In fact, the counselor has been working with the teacher to help the client develop better classroom and playground relationships. The school counselor and social worker discuss the situation and realize that the boy is often left alone without much to do. Staying after school for an extended day would likely be helpful. The school counselor, with the social worker's backing, contacts a source for funding through a local men's club.

This is but a beginning for many cases—case management involves multiple dimensions beyond this basic scenario. Counseling and therapy are an important part of case management, but only a part. The Division of Youth Services or Family and Children Services may need to be called to review the situation if the counselor or another mandated reporter suspects signs of abuse or neglect; the mother may need short-term financial assistance and career counseling. If the father has not been making child support payments, legal services may have to be called in.

Through all of the above, the social worker maintains awareness of all aspects of the case. Through individual counseling with the mother, the worker is in constant touch with all that is going on and remains in contact with key figures working toward the positive treatment of the child. But the child can only be successfully treated in the family, school, and community situation.

Advocacy action is important throughout, and it does not happen by chance. Throughout this process, the social worker will constantly advocate for the mother and the child by establishing connections with various agencies. The school counselor who goes to the local men's club for funding has to advocate for the child; in addition, encouraging a teacher to change teaching style to meet individual child needs requires advocacy. Elementary counselors spent a good deal of time in various activities advocating for students individually, and often advocating for fair school policies as well.

Social justice issues may appear. In many communities and agencies, all of the above services may not exist—and equally or more likely, they do not work together effectively. This may require organizing to produce change. The child may be bullied, but the school has no policy to prevent bullying. Social justice and fairness demand that each child be safe. Discrimination against those who are poor or are from minority backgrounds may exist. Individual and group education or actual forced change may be necessary.

▶Maintaining Change: Relapse Prevention

Change that is not planned and contracted with the client is less likely to occur. One approach used to reduce slips or relapses is the relapse prevention plan. This approach, originally developed for use in alcohol or drug abuse treatment, is now used in a variety of settings. Relapse prevention is now a standard part of most cognitive behavioral counseling, and the concepts presented here will enrich all theories of helping described in this chapter.

The Maintaining Change Worksheet: Self-Management Strategies for Skill Retention form (Box 14.3) helps the client plan to avoid relapse or slips into the old behavior. The counselor hands the client the worksheet and they work through it together, giving special emphasis to things that may come up to prevent treatment success. Research and clinical experience in counseling both reveal that this may be effective with clients and help ensure that they actually *do* something different as a result of their experience in the session or treatment series.

| **BOX 14.3** | Maintaining Change Worksheet: Self-Management Strategies for Skill Retention |

I. Choose an Appropriate Behavior, Thought, Feeling, or Skill to Increase or Change

Describe in detail what you intend to increase or change.

_____

_____

Why is it important for you to reach this goal?

_____

_____

What will you do specifically to make it happen?

_____

_____

II. Relapse Prevention Strategies

A. Strategies to help you anticipate and monitor potential difficulties: regulating stimuli

Strategy	*Assessing Your Situation*
Do you understand that a temporary slip may occur but it need not mean total failure?	_____
What are the differences between learning the behavioral skill or thought and using it in a difficult situation?	_____
Support network: Who can help you maintain the skill?	_____
High-risk situations: What kinds of people, places, or things will make retention or change especially difficult?	_____

B. Strategies to increase rational thinking: regulating thoughts and feelings

What might be an unreasonable emotional response to a temporary slip or relapse?	_____
What can you do to think more effectively in tempting situations or after a relapse?	_____

C. Strategies to diagnose and practice related support skills: regulating behaviors

What additional support skills do you need to retain the skill? Assertiveness? Relaxation? Microskills?	_____

Permission to use this adaptation of the Relapse Prevention Worksheet was given by Robert Marx, University of Massachusetts, Amherst.

(continued)

BOX 14.3 | (continued)

D. Strategies to provide appropriate outcomes for behaviors: regulating consequences

Can you identify some probable outcomes of succeed-   _____
ing with your new behavior?   _____
How can you reward yourself for a job well done?   _____
Generate specific rewards and satisfactions.   _____

III. Predicting the Circumstance of the First Possible Failure (Lapse)

Describe the details of how the first lapse might occur; include people, places, times, and emotional states.

_____

_____

_____

Remember that every client is unique and the relapse prevention plan needs to be tailored to the individual client and the characteristics of the issue. Every person's prevention plan should look somewhat different.

Include in the plan a list of people who can be counted on for support, the triggers to the unwanted behaviors, possible responses to those triggers, and rewarding alternatives client can do instead of engaging in the negative behavior.

People you can contact via telephone, texting, or email: Develop a list of three or more people who can provide immediate support during times of potential relapse or slip.

Triggers to undesired behavior, and possible responses for those triggers: Discuss and write down the biggest temptations or triggers. What are the things or situations that make you want to engage in the negative behavior (frequenting some people, visiting a particular place, the sight, smell, or sound of . . .)? What can you plan to do in response to each of these triggers?

Alternative and rewarding activities that do not involve the unwanted behavior: Inactivity and boredom are some of the greatest threats to change. Working with your client, develop a plan of activities that he or she enjoys and wants to do to prevent voids of time that can facilitate negative behaviors.

Rewards the client can give her- or himself for meeting change targets: What would the client do when reaching a week of progress, two weeks, and longer periods of change?

Remember that change takes time and effort. Maintaining an intentional effort to change will pay off at the end.

▶Summary: Meeting the Clients Where They Are

It's not the mistakes you make, but what you do to correct them that counts.

—Allen's plumber

This chapter has provided you with a model for transcribing and generating your own analysis of sessions. It has also provided an opportunity to examine, analyze, and note effective behaviors of Allen as he worked with Mary. *No interview is perfect.* What counts is your ability to be intentional and creatively change your behavior to be with the client, wherever he or she "is."

Knowledge of the basics of decisional counseling, case conceptualization, referral, treatment planning, and relapse prevention offers you more possibilities to improve your counseling and psychotherapy competencies.

Of course, what is by far the most crucial: How do you view your counseling and therapy style? What are your strengths? Where do you see the opportunity for further creative growth?

▼▼▼▼	Key Points
Decisional counseling	Decisional counseling is a practical model that assumes that most, perhaps all, clients come to us for help in making a wide array of decisions. These range from daily life decisions to extremely complex issues, such as deciding a college major, whether or not to end a relationship, or where to live after a major disaster. Clients and their families need to make decisions when they face diagnoses such as attention deficit disorder, anxiety or depression, or the possibility of schizophrenia. Decision making and the microskills are a foundation for most systems of counseling and most client concerns, issues, and challenges.
Decisional structure and additional strategies	The restorying model can be considered a basic decisional model underlying other theories of counseling and therapy. Once you have mastered the skills and strategies of *Intentional Interviewing and Counseling*, you will find that you can more easily develop competence in other theories of helping. Some strategies that are central to decisional counseling are confrontation, creative brainstorming, logical consequences, balance sheet, emotional balancing, and future diary.
Planning the first session and using the checklist	The first session presents many challenges; using the checklist helps you plan for the session and reduces the chances of missing the basics. You can systematically plan for the initial session using the five-stage structure. Be intentionally flexible and ready to change your plan if events in the session suggest that another approach is needed. The five stages provide a useful checklist to ensure that you cover all points, even if the session does not go as expected.
Session analysis	Using the constructs of this book, examine your own style and that of others for microskill usage, focus, structure of the session, and the resultant effect on a client's cognitive and emotional development style. The transcript provides a systematic analysis through process comments and behavioral counts of microskill usage, examines the five-stage structure, and evaluates client movement on the Client Change Scale. Counselors with varying theoretical orientations will approach the case in different ways. A person-centered counselor might focus more on Mary and her feelings, thoughts, and meanings about herself. A cognitive-behavioral counselor would be more interested in specific behavioral descriptions and aim for change in actions. These also will be strengthened with knowledge of decisional counseling.
Treatment plan	This is a long-term plan for conducting a course of counseling sessions. It should be made available to the client. Clients need to be part of this plan; be sure to discuss benchmarks for change and the achievement of clients' goals.
Maintaining change and preventing relapse	A relapse prevention plan helps ensure that the fifth stage of the interview is indeed accomplished and that slips are reduced. The form for Maintaining Change Worksheet: Self-Management Strategies for Skill Retention helps clients plan to avoid relapses or slips.

►Competency Practice Exercises and Portfolio of Competence

 Go to CengageBrain.com to access Counseling CourseMate, where you will find an interactive ebook, quizzes, videos, interactive counseling and psychotherapy exercises, the Portfolio of Competence, case studies, a practice test, and more.

Individual Practice

You have engaged in the systematic study of the counseling process and have experienced many ideas for analyzing your style and skill usage. These responses must be genuinely your own. If you use a skill or strategy simply because it is recommended, it could be ineffective for both you and your client. We hope that you will draw on the ideas presented here, but ultimately *you* will put the science together in your own art form.

You have practiced varying patterns of helping skills with diverse clientele. We hope that you have developed awareness and knowledge of individual and multicultural differences. Study and learn how to "flex" and be intentional when you encounter varied clients with differing needs. For example, you may be more comfortable with teenagers than with children or adults, or you may have special abilities with elders.

Exercise 1: Conducting and Transcribing a Full Session   Now that you are finishing this chapter, to help ensure your understanding of your own style, complete another audio or video session. Use the following guidelines.

1. *Find a volunteer client* willing to role-play a concern, issue, or opportunity.
2. *Counsel the volunteer client* for at least 15 minutes. Avoid sensitive topics. Feel free to go further to gain a sense of completion.
3. *Use your own natural communication style.*
4. *Ask the volunteer client,* "May I record this session?"
5. *Inform the client* that the recording device may be turned off any time he or she wishes.
6. *Select a topic.* You and the client may choose interpersonal conflict, a specific issue, or one of the elements from the RESPECTFUL model.
7. *Follow the ethical guidelines* from Chapter 2. Common sense demands ethical practice and respect for the client.
8. *Obtain feedback.* You will find it very helpful to get immediate feedback from your client. As you practice the microskills, use the Client Feedback Form (Box 1.4, page 24).
9. *Also obtain feedback from another student.* We have found that it is very helpful and clarifying if you and a student partner exchange transcripts and comment fully on each. This gives you important additional feedback.

Transcribe the session (see Box 14.4), and use what you have learned so far to fully analyze your work. A careful analysis of your behavior in the session will aid in identifying your natural style and its special qualities as well as your skill level. Use the ideas presented in this chapter to further examine and analyze your work in the session.

Remember that *you* are the person who will integrate what you have learned here into your own practice. Identifying the nature of your personal style and current skill level will set you on the road to continued growth and competence as a professional counselor or therapist. Please give special attention to your understanding and use of cultural/environmental/contextual issues. Look at your natural style of counseling, and evaluate your multicultural expertise.

Finally, go back to the transcript or recording of the session you recorded earlier, in Chapter 1, and note how your style has changed and evolved since then. What particular strengths do you note in your own work?

Organize the transcript in a format similar to the transcript presented in this chapter. A session transcript may be useful in demonstrating your competence and obtaining an internship or job. Consider the transcript a permanent part of your developing professional life and your portfolio.

Check off the following points to make sure you have included all the necessary information in the transcript:

❑ Describe the client briefly. Do not use the client's real name.

❑ Outline your session plan *before* the session begins.

❑ Be sure you obtain the client's permission before recording the session, and include a summary of this agreement in the transcript. The client should be free to withdraw at any time. Ethically, we protect the rights of the client.

❑ Number all interactions, and be sure to indicate who is speaking at the beginning of each interaction.

❑ Mark the focus of each interaction and note your use of attending and influencing skills.

❑ If you confront, note this in the Skill Classifications column; in the Process Comments, note how the client responds, using the Client Change Scale.

❑ Comment on your interactions as appropriate to the situation. Discuss what you feel was good or needs improvement in your skills. If you feel that you used a skill inappropriately, describe what you believe would have been a better approach. Note also what skills worked well!

❑ Indicate when you think that you have reached the end of a stage. Do not feel that you must cover all stages; in some cases, you may cycle back to an earlier stage or forward to a later stage.

❑ Write a commentary that summarizes what happened.

❑ Summarize your use of skills through a skill count.

❑ Assess your competence levels. What skills have you mastered, and what do you need to do next? This is also a summary of your strengths and the areas that need further development. What did you like and not like about your work? Your ability to understand and process "where you are" and discuss yourself is necessary for personal and professional growth.

❑ End the transcript with a treatment plan for a hypothetical future series of sessions.

For the exercise in this section, you don't have to transcribe a full session, although a full transcript likely will be most beneficial to you. A twenty-minute transcript from within a longer session is enough. But if you do such an excerpt, be sure to indicate what happened in the rest of the counseling session so that the context of the transcript is clear.

Group Practice

Exercise 2: Exchanging Feedback Work with one of your classmates to obtain feedback about your work. Share the audio, video, or transcript with others. Take turns providing feedback and enriching each other's skills. Don't forget to emphasize strengths.

Portfolio of Competence

Include your transcript and analysis in your Portfolio of Competence. Include written feedback from your colleague. Many students take their full portfolio to interviews for internships or even professional positions. The transcript analysis will provide a good sense of your understanding and competence in the counseling and psychotherapy field.

▶Determining Your Own Style and Theory: Critical Self-Reflection on Your Own Practice

We are not presenting a checklist of specific competencies for this chapter, but we would like you to reflect on your work and your changes during the semester. Please comment or put in your journal reflections on the process of completing a full session.

CHAPTER 15

How to Use Microskills and the Five Stages With Theories of Counseling and Psychotherapy

Skills Integration

Action Strategies
for Change

Self-Disclosure and
Feedback

Reflection of Meaning and
Interpretation/Reframe

Empathic Confrontation

Focusing

The Five-Stage Interview Structure

Reflection of Feeling

Encouraging, Paraphrasing, and Summarizing

Open and Closed Questions

Client Observation Skills

Attending Behavior and Empathy

Ethics, Multicultural Competence, and Wellness

There is nothing so practical as a good theory.

—Kurt Lewin

Mission of "How to Use Microskills and the Five Stages With Theories of Counseling and Psychotherapy"

The microskills and the five-stage model can be used to explain the structure and practice of the many theories you will encounter in your counseling practice. This chapter explains how you can use the concepts of this book in virtually all of your sessions, with almost any client.

Chapter Goals and Competency Objectives Awareness, knowledge, skills, and actions in applying microskills to counseling and psychotherapy will enable you to

▲ Observe how the microskills framework is used across multiple theories of counseling and psychotherapy.

▲ Briefly review four theories that have been featured in this text, their philosophy, and their strategies: person-centered counseling, decisional counseling, multicultural/feminist theory, and logotherapy.

▲ Read a transcript, practice, and engage in some of the basics of two additional major approaches: crisis counseling and cognitive behavioral therapy.

▲ Review transcripts of brief counseling, motivational interviewing, and counseling/coaching, which can be found on the accompanying CourseMate website.

This chapter is divided into three parts, each focusing on counseling theory in a different way.

Part I. Microskills, Five Stages, and Theory points out how varying theories use the microskills in varying patterns. You will find that the five-stage structure is applicable in all sessions and with all theories.

Part II. A Brief Summary of Theories Discussed in Earlier Chapters outlines some basics of decisional counseling, person-centered counseling, multicultural/feminist theory, and logotherapy. Discussed for each is their philosophy, key methods, and implications for your practice. Transcripts related to these theories are available in CourseMate.

Part III. Crisis Counseling and Cognitive Behavioral Therapy (CBT) again presents philosophy, key methods, and implications for your practice. It includes detailed transcripts with microskill analysis, illustrating how you can use the five stages and microskills in very different approaches to the issues clients face.

▶Part I: Microskills, Five Stages, and Theory

You will find that the competencies you have developed with this book will enable you to understand and work with each approach more quickly and competently. Although virtually all theoretical approaches use microskills and the five stages, each of them has a conceptual framework and worldview that differ from the others.

Table 15.1 reveals that each theory of counseling has a distinct pattern of microskill use. All use the basic listening sequence, but counselors using each system listen to stories from a different frame of reference. The person-centered system tends to use the listening skills most frequently and focuses most on the individual client, giving less attention to the cultural/environmental context (CEC). Brief counseling and counseling/coaching use many questions; CBT and Gestalt use many directives; and multicultural counseling and therapy (MCT) and feminist therapy pay more attention to the cultural/environmental context. Decisional counseling, MCT, and feminist theories are the most eclectic of the theories as they tend to use a variety of skills and focus dimensions.

Each theory, because of its particular philosophic orientation, emphasizes different aspects of life experience and gives different emphasis to each of the five stages. Pragmatic decisional counseling, for example, tends to work in the here and now on immediate life issues or facilitate life planning. In contrast, the person-centered counselor emphasizes relationship and self-actualization. Logotherapy focuses on life's meaning, and CBT on cognitions and behavior. MCT and feminist theory draw on all of the above, but always seek to situate the individual in the cultural/environmental context so that clients are well aware of the impact the surrounding world has on their cognitions and emotions.

Each theoretical system has considerable merit. This may help you to understand why theories have proliferated over the years. No one theory has all the answers, but resolving one major issue or problem inevitably has an impact on the total human being. For example, effective use of CBT will often result in more self-actualized clients who are better able to make decisions and may even have clearer discernment of their life direction.

TABLE 15.1 Microskills patterns of differing approaches to the interview

		Decisional counseling	Person-centered	Logotherapy	Multicultural and feminist therapy	Crisis counseling	Cognitive behavioral therapy	Brief counseling	Motivational interviewing	Counseling/coaching	Psychodynamic	Gestalt	Business problem solving	Medical diagnostic interview
BASIC LISTENING SKILLS	Open question	●	○	●	◐	◐	◐	●	●	●●	◐	●	◐	◐
	Closed question	◐	○	●	◐	●	●	◐	◐	◐	○	◐	◐	◐
	Encourager	●	◐	●	◐	◐	◐	◐	●	●	◐	◐	◐	◐
	Paraphrase	●	●	●	◐	◐	◐	●	●	●	◐	○	◐	◐
	Reflection of feeling	●	●	●	●	◐	◐	●	●	●	◐	●	○	○
	Summarization	◐	◐	●	◐	◐	◐	●	◐	●	◐	○	◐	◐
INFLUENCING SKILLS	Reflection of meaning	◐	●	●	●	○	○	○	◐	◐	◐	○	○	○
	Interpretation/reframe	◐	○	◐	◐	◐	○	◐	●	○	●	●	◐	◐
	Logical consequences	◐	○	○	◐	◐	◐	◐	○	◐	○	○	●	◐
	Self-disclosure	◐	◐	◐	◐	◐	◐	◐	◐	◐	○	◐	○	○
	Feedback	◐	◐	◐	◐	○	◐	◐	●	●	○	◐	◐	○
	Instruction/psychoeducation	●	○	○	●	◐	●	●	◐	◐	○	○	●	◐
	Directive	◐	○	◐	◐	◐	●	○	◐	◐	○	●	●	●
	CONFRONTATION (Combined skill)	◐	◐	◐	●	○	◐	◐	◐	●	◐	●	◐	◐
FOCUS	Client	●	●●	●	◐	◐	●	◐	◐	◐	◐	●	◐	◐
	Main theme/issue	●	○	◐	◐	◐	●	●	◐	●	◐	◐	●	●
	Others	◐	○	◐	◐	◐	◐	◐	◐	◐	◐	◐	○	○
	Family	◐	○	◐	◐	◐	◐	◐	◐	○	◐	○	○	○
	Mutuality	◐	◐	◐	◐	○	◐	◐	◐	●	○	○	○	○
	Counselor/therapist	○	◐	◐	◐	○	○	○	○	○	○	○	○	○
	Cultural/environmental/contextual	◐	○	◐	●●	●	◐	◐	◐	◐	○	○	◐	○
ISSUE OF MEANING (Topics, key words likely to be attended to and reinforced)		Problem solving	Self-actualization, relationship	Values, meaning vision for life	How CEC impacts client	Immediate action, meeting challenges	Thoughts, behavior	Problem solving	Change	Strengths and goals	Unconscious motivation	Here-and-now behavior	Problem solving	Diagnosis of illness
COUNSELOR ACTION AND TALK TIME		Medium	Low	Medium	Medium	High	High	Medium	Medium	Medium	Low	High	High	High

LEGEND

● Frequent use of skill   ◐ Common use of skill   ○ Occasional use of skill

© Cengage Learning

Perhaps Kurt Lewin's famous statement on theory could be rephrased: *There is nothing so practical as becoming competent in several theoretical approaches.*

▶Part II: A Brief Summary of Theories Discussed in Earlier Chapters

The following four approaches to counseling and psychotherapy have been discussed in this book. At this point, you likely have become familiar with and developed at least beginning competence in decisional counseling. We have referred frequently to Carl Rogers and person-centered counseling and the critical importance of the empathic relationship and working alliance. If you try to eliminate questioning and most of the influencing skills from your session, you will be close to using the person-centered style.

Viktor Frankl's logotherapy has been featured in relation to the skill of reflecting meaning and the discernment process. While there is much more to learn about logotherapy, the beginning here may be all there is until you start reading Frankl's work on your own. His innovative and pioneering work is often not stressed in our training programs. A focus on meaning, mission, and life direction, coupled with the five stages, will provide a foundation for using some of his concepts. The person who has a meaning and life direction typically can survive many hurts and traumas throughout life.

Multicultural counseling and therapy (MCT) issues have been stressed beginning in Chapter 1 and throughout the book, and attention has been paid to gender issues. Although you will be taking courses in this area, if you use the five stages and include a special focus on the cultural/environmental context, you have a good start toward mastery of MCT.

Decisional Counseling

The doors we open and close each day decide the lives we live.

—Flora Whittemore

Philosophy   Decisional counseling is representative of an underlying philosophy in the United States: Find something practical and pragmatic—something that "works." However, many other countries and cultures make decisions through some variety of the decision-making model. An early example is Benjamin Franklin's three-stage decision model, created around 1750. Franklin is the person we can credit with the basic three stages of defining the problem, generating alternatives, and deciding on action. In the 1890s, C. S. Peirce and William James established the philosophy of pragmatism, which again is very American and provides a rationale for Franklin's basic work, as well as decisional and problem-solving counseling. Their basic idea is that theory and its conceptions need to show themselves useful in practical matters, thought is to guide action, and "truth" shows itself by results. An updated version of pragmatism is "Walk the talk!"

Key Methods   The five-stage decisional structure, plus microskills, is actually the basic strategy. You have seen decisional counseling in practice throughout this book. Chapter 1 presented the five-stage session, which is a decisional framework, an expansion of Franklin's original three stages with an emphasis on empathic relationship. Stage 2 uses the microskills of the basic listening sequence (BLS), drawing out the client's story and strengths, which Franklin termed "the problem." Stage 3, specifying goals, adds direction to the Franklin model. Stage 4 uses influencing skills and the BLS to create alternatives for action. Decisional counseling adds Stage 5, action, which brings the realization that if something isn't done about what is decided, nothing will happen.

Sample Strategies The strategies of decisional counseling are integrated with the methods above. Feeling and emotion are given central attention in decisional counseling. Unless the client is emotionally satisfied with the decision, the chosen alternatives are less likely to be beneficial or even used. For this reason, the balance sheet and future diary are helpful strategies to organize the decisional process. Decisional counseling uses all the microskills and does not hesitate to draw on other theoretical approaches as appropriate.

Implications for Your Practice By this time, you have likely mastered decisional counseling and are ready to move on to other theoretical systems. But keep decisional counseling in mind, regardless of where you head. Your clients always need to make pragmatic decisions, particularly when they face crises. Moreover, whether you favor person-centered counseling, cognitive behavioral therapy, or some other system, all require decisions and, ultimately, client action in the real world. Thus, the five stages of decisional counseling can serve as a useful model and checklist for virtually any counseling and therapy theory.

Suggested Next Steps

Nezu, A. M., Nezu, C. M., & D'Zurilla, T. J. (2013). *Problem-solving therapy: A treatment manual.* New York: Springer.
  This is a detailed manual of how to treat a variety of health and mental health issues using PST. The book offers specific treatment guidelines, exercises, homework assignments, and case examples.

Haley, J. (1987). *Problem-solving therapy.* San Francisco: Jossey-Bass.
  This older volume is well worth ordering inexpensively as a used book. Haley is a well-known exponent of Ericksonian therapy, an interesting alternative to and expansion of decisional and problem-solving frameworks.

Person-Centered Counseling

> *When I look at the world I'm pessimistic, but when I look at people I am optimistic.*
>
> —Carl Rogers

Philosophy Carl Rogers believed in a person-centered approach over any theory or opinion of the counselor. Clients are the experts in their own life history and need the counselor primarily as a facilitator. Rogers (1961, p. 32) once commented:

> In my early professional years I was asking the question: How can I treat, or cure, or change this person? Now I would phrase the question in this way: How can I provide a relationship that this person may use for his [or her] own personal growth?

Out of this philosophy comes a belief in human dignity and personal self-actualization to reach one's full potential. The goal is individual decision making and determining one's own life path.

Key Methods Rogers was the ultimate listener. View his famous films, and you will see that he has perfect attending behavior and is superb at encouraging, paraphrasing, reflecting feelings (and meaning), and summarizing. In addition, many of his summaries are excellent examples of quiet confrontations in which discrepant elements of the client's story are examined. His focus is almost always on the client. Occasionally, you will find him using feedback, and very occasionally self-disclosure and interpretation.

Person-centered theorists are strongly opposed to the use of questions, considering them intrusive and a limit on the client's self-discovery process. However, careful examination of Rogers's work reveals an occasional question, usually associated with some form of "What do you want?" But these questions come in very careful relationship to what the client has said earlier in the session. Rogers would not use directives, as used in many theories, and often spoke strongly against them.

Carl Rogers was definitely opposed to much of what we see in current theoretical practice. He worried that other models, more goal centered, would take over the direction of the session. Decisional counseling, CBT, Gestalt, and other theories take a more active stance in the session, and we need to heed his warning. Rogers comments (1961, p. 186):

> The good life is a process, not a state of being. It is a direction, not a destination.

This is a very profound and attractive statement. Rogers was what we would call a process counselor, focusing on the here and now. Being in the moment is central. In contrast, current theories typically focus on outcome and accountability, ideas that might make Rogers sad. Evidence is quite clear that the more active theories bring about significant change and do so more quickly than person-centered therapy. However, they cannot be fully successful unless the Rogerian empathic relationship is established throughout counseling and therapy. Without a working alliance, success will be most difficult.

In summary, look at some key aspects of Rogers's lasting legacy: the importance of relationship and empathic understanding; the centrality of emotion; the emphasis on listening (although he did not label the skills). He showed us the importance of looking at what actually happens in the session through his courage in recording his sessions and sharing them, and he was ahead of his time when he encouraged and worked for world peace and multicultural understanding through his group work and international presentations.

Implications for Your Practice   You have seen here that Carl Rogers's influence is present throughout this book, even when we don't speak of him. Regardless of your chosen theory, listening will always be central in helping your client. Continue listening, seek empathic understanding, and keep your eye on client desires and goals, not your own wishes. Be patient and keep honing those attending skills. Make your own judgments on the value of questions and influencing skills.

Suggested Next Steps

Rogers, C. R. (1961). *On becoming a person.* Boston: Houghton Mifflin.
   Likely you can best absorb Carl Rogers's thinking by reading his book slowly and studying the brief transcripts. Beyond that, visit YouTube and view the long list of therapy examples and key lectures.

Logotherapy

Once at a therapy conference (a general, not logotherapy, conference) someone from the audience asked one of the speakers: "What do you call the spiritual encounter between two people?" The speaker answered "Viktor Frankl."

Philosophy   There is always meaning to life. Logotherapy aims to help the person become one's own best person through discerning meaning and purpose. Even in the most miserable circumstances, life still has meaning and can provide support. We can choose to be unhappy and live life without meaning or choose life with meaning. Paraphrasing Nietzsche's philosophy: *Those who have a why can find a how and bear any situation.*

Frankl's survival in the Nazi death camps and his influential life serve as a model for us all. If we can discern a meaning for our lives, we can "survive" the many challenges and difficult decisions we face in our life.

Key Methods   That word *meaning* and its relationship to counseling and psychotherapy may be Frankl's most significant contribution. No other theory speaks so well and completely to the meaning of life. You will find logotherapy discussed in very few textbooks, yet we have found his thinking so powerful that it permeates our being. Discerning what we care for, what we are about, and our life's meaning and purpose seems to transcend all other issues. Whether prison or palace, what meaning does the person have to live?

Questioning and reflecting meaning are the microskills that we identify with Frankl. He was a good listener. Used effectively, a focus on personal meaning can turn a session around and change a negative direction to a positive and fulfilling discussion. Clients can hyperreflect on problems to the point of self-destruction. Dereflection is an excellent strategy to help clients reframe and think in new and more positive ways about their issues, and even their entire life plan.

Discernment is not specifically a Frankl strategy, but it is clearly a meaning-oriented strategy inspired by him. Discernment, finding a life vision, is a decisional process. "Everything can be taken from . . . [us] . . . but the last of the human freedoms—to choose one's attitude in any given set of circumstances, to choose one's own way" (p.104).

Many see Frankl as the first and original cognitive behavioral theorist, and he identified himself as such. How can that be? Simply put, skilled use of logotherapy is a powerful way to enable clients to *reframe* cognitions and emotions and think in new ways. Most modern theories operate from his premise. Furthermore, Frankl always emphasized that thought and meaning need to be taken into action in the real world. Thinking is not enough.

Implications for Your Practice   Look for meaning issues as you listen to the client story. Working in depth with meaning and discernment of life's direction takes a bit more time than brief therapy, CBT, and others, but can be life changing. Other theories and practices sometimes ignore underlying meaning issues. Exploring meaning can be particularly useful in crisis situations. Once clients reframe the trauma or tragedy from the meaning frame of reference, they can more readily calm down and think through concrete action plans with your help.

Beyond that, Frankl emphasized that we must live and act on our meaning. He once commented, "It's not hard to make decisions when you know what your values are." You can easily bring basic concepts of logotherapy into other theories by making meaning a focus of some of your sessions.

Suggested Next Steps
Frankl, V. (1959) *Man's search for meaning.* New York: Beacon.

This is Frankl's most famous work, discussing his experiences in a German concentration camp during World War II. Widely available in paperback and inexpensive, this is the book to read when you face serious life challenges; it will inspire and help you keep going. We frequently give a copy of this book to clients. As with Rogers, you will find many YouTube videos showing his style of work and his personal approach to his key ideas.

Multicultural Counseling and Therapy (MCT) and Feminist Therapy

All counseling and psychotherapy are multicultural.

—Paul Pedersen

Women belong in the house . . . and the Senate.

—Author unknown

Philosophy   Multicultural counseling and therapy (MCT) and gender and women's issues have both been emphasized in this book. But why are we presenting the two together? Both rest on a foundation of cultural/environmental context (CEC). Making CEC central in the session is seen as essential for effective counseling and therapy. If CEC is not considered, then the counseling and therapeutic work is incomplete.

Feminist therapy, also known as relational cultural therapy (RCT), is usually considered separate from MCT. It has created its own separate body of literature focusing on women's issues. At the same time, a commitment to feminist theory includes a respect for and inclusion of MCT. Similarly, MCT cannot truly exist without feminist concepts.

Key Methods   Both theories draw on all the other theories discussed in this book and more. In that sense, they are eclectic and integrative. All microskills will be used, as appropriate to the theoretical approach used, but CEC issues will always be considered and typically actively involved in the session. The issue is how does the cultural, environmental, and social context affect the client.

Cultural identity development (CID) is a critical part of both approaches. CID identifies five stages of awareness of self as a cultural person (Cross, 1991). CID speaks to all ethnicities and cultures, whether the person is of European or African descent, Asian, Latina/o, Native American, or First Nations.

Here are two examples of how cultural identity development theory may appear in the session. At each stage you will find the same person thinking a different way and requiring a different approach than before. Please keep in mind that the stages are not always linear. At stage 1, an African American naively denies or ignores racism and seeks to emulate White people; or a woman accepts and even supports the status quo. At stages 2 and 3, they encounter discrimination and begin to know themselves as cultural beings with pride in being African American or a woman. Anger and action against the "system" may be a feature of this stage.

Reaching stage 4, individuals realize that they are identifying themselves as "against the oppressor," rather than really understanding the meaning of their unique cultures. African Americans or women often retire to reflect, study, and put themselves together in new ways, avoiding the mainstream culture when possible. Stage 5, termed *internalization*, is integrative in that the person values all earlier stages, except the first, and may behave and think within that stage, according to the situation. The person often makes a commitment to action.

MCT and feminist practitioners seek to identify the cultural identity level and stage of their clients. They may first meet and counsel the client at that level, but the goal is to increase consciousness of what it means to be a woman, an African American, a gay person, or other cultural identity. This occurs through listening to past and present experiences using the BLS, but questioning becomes useful as the counselor helps clients see how cultural issues and external oppression are part of their concerns and life challenges. A good deal of instruction about the nature of oppression and what can be done about it may be included. Of course, advocacy and social justice action are critical to help prevent oppression.

Implications for Your Practice   The central implication is to remember that clients come from a cultural background that affects their identity. We have spoken only of African Americans and women, but European Americans and those of all ethnicities and races, as well as men, also come from a CEC background with varying levels of identity development. White people, in general (not all), often are unaware of the fact that they enjoy privileges denied to others because of the color of their skin. Men throughout all societies enjoy gender entitlement and special privileges.

Think of CEC as referring to many groups. The RESPECTFUL model (D'Andrea & Daniels, 2001) is one way to outline cultural differences. The model proposes ten cultural issues: Religion/spirituality, Economic/class background, Sexual orientation/gender identity, Personal style and education, Ethnic/racial identity, Chronological challenges (age), Trauma, Family background, Unique physical characteristics, and Location of residence and language differences.

Perhaps the central implication is that "there is a lot to learn and experience." We cannot know all dimensions of culture, but we can develop an appreciation and learn to use these concepts in the session. Give special attention to the most prominent cultural groups in your community.

Suggested Next Steps

Sue, D. W., & Sue, D. (2013). *Counseling the culturally diverse: Theory and practice* (6th ed.). Hoboken, NJ: Wiley.

> This has been the basic textual overview of MCT and includes significant material on feminist issues.

Brown, L. (2009). *Feminist therapy*. Washington, DC: American Psychological Association.

> This book is a superb summary of feminist therapy, the how and why.

YouTube video demonstrations and presentations. Key words are *feminist therapy*, *cultural relational therapy*, *multicultural counseling*, and *multicultural therapy*.

▶Part III: Crisis Counseling and Cognitive Behavioral Therapy (CBT)

Crisis counseling and cognitive behavioral therapy have been chosen to illustrate how the five stages and microskills can be applied to theories other than those described earlier in this book. Crisis counseling is a basic model, but one that requires special training and experience as it involves real challenges for any counselor. Cognitive behavioral therapy (CBT) is currently the most used and researched method of professional helping. The stress management strategies of CBT have become a central part of many treatment protocols, particularly since research has shown that therapeutic lifestyle changes (TLCs) are effective and potentially life changing.

Three other systems of counseling and therapy are presented on the accompanying CourseMate website: brief therapy, motivational interviewing, and counseling/coaching. Brief therapy uses a framework similar to the five stages, but starts with an emphasis on goals. Motivational interviewing has been shown effective for treatment of alcohol and substance abuse. Counseling/coaching is relatively new as a term for counseling and therapy, but anytime we are using TLCs and encouraging our clients to find and follow their goals, we are touching on the central dimensions of counseling/coaching. It may be valuable for you to give this area special attention, as working as a professional coach with individuals, families, and organizations is a growing field.

Crisis Counseling

We want people to know that their emotions and reactions are completely normal.

—Anonymous crisis helper

Philosophy   Crisis counseling is the most pragmatic and action-oriented form of helping. The word *pragmatism* comes from the Greek word for deed, act, to practice and to achieve πράγμα (*pragma*). Even more than decisional counseling, crisis counseling is concerned with action and useful, pragmatic results for the client. However, pragmatism is embedded in a caring attitude that is fully aware that most responses to a crisis can be considered completely normal.

What is a crisis? *Crisis* is closely related to *trauma*; it means stress, stress hormones, and a rise in cortisol to the brain. It has been pointed out that virtually all the world's population experience one or more crises/traumas in their lifetime. In that sense, crisis is a normal life event, and "normalizing" the crisis is one foundational idea to keep in mind.

You may encounter many different types of crisis. Immediate here-and-now crises demanding rapid practical action include flood, fire, earthquake, rape, war, refugee status, a school or community shooting, personal assault (including abuse), being held hostage, a serious accident, and the sudden discovery or diagnosis of a major medical problem (e.g., heart attack, cancer, multiple sclerosis). These call for immediate practical action. But many crisis/trauma survivors will benefit from further support and counseling.

The second type of crisis is a more "normal" part of life. Many of your clients will come to you to talk about divorce or the breakup of a long-term relationship, foreclosure of their home, job and income loss, a home break-in and burglary, or the death of a loved one. Examples of children's crises include bullying, dealing with their own or a parent's illness, or leaving friends and moving to a new location. For some, not getting into a desired college or failing an exam will become a crisis situation.

All crisis counseling involves two phases: (1) working through the initial trauma and (2) appropriate follow-up and further counseling. Counseling skills are obviously needed in both phases, but the immediate crisis will demand more cognitive and emotional flexibility—and the ability to join the client in the here and now. The second phase usually gives you more time and elements to work with and will begin to look like more typical counseling.

A more comprehensive team approach is required in serious crises such as when a person with a gun approaches and attacks a school, university, bank, or office. The aftermath of a bombing, a fire, earthquake, or other disaster will typically need follow-up group and individual work.

As one example of the team approach, Mary Bradford Ivey, as a counselor in an elementary school, helped usher the excited elementary students into the school auditorium where they were to watch the liftoff of the NASA spaceship *Challenger*. Their excitement turned to fear and tears when the ship blew up in front of them. Mary and the teaching staff faced a major here-and-now crisis.

The teachers had worked with Mary to plan ahead if a school crisis should occur. It had been agreed that the teachers would take the children to their homerooms and encourage them to talk and ask questions, with awareness that each child would have her or his own unique reactions. The teacher needs to maintain composure and reassure children that they are safe. Encouraging children to ask questions helps give them control. Mary went from classroom to classroom, supporting each teacher and the students. If one or more children were particularly upset, Mary took them with her to the school counselor's office.

The next day and later that week, the debriefing continued. Some children wanted to express their feelings through art. Mary had had extensive experience in group work and set up several sessions for students who wanted or needed to explore issues in more detail.

The message here: *Prepare for crisis.* We never know when one will occur and what the crisis will be.

Key Methods   Although we are speaking here of the immediate survivors of a major crisis such as a fire or flood, the suggestions offered also hold for those you may meet after the event. Once again, you will find the five stages and the microskills a useful framework for thinking about how to help these clients. The transcript example beginning on page 406, which presents a woman the day after a frightening fire and the loss of her home, illustrates the five stages.

Normalizing.   Do we call those who experience trauma "victims" or "survivors"? The second term is more empowering for clients and puts them more in control. Thinking of people as victims tends to depersonalize them and put them in a helpless position, controlled by external forces.

Many crisis workers object to the term *posttraumatic stress disorder* (PTSD), pointing out that virtually any serious encounter with crisis will produce extreme stress and challenges to the whole physical and mental system. *Disorder* is an inappropriate term because the client has actually responded in a normal fashion to an insane situation. Many prefer to refer to a normal *posttraumatic stress reaction* (PTSR). Your clients have gone through a far from normal experience. Helping them see that their "problem" is not inside them, but the logical result of external stressors, is one step toward normalizing the situation.

Furthermore, labeling the client in any way, even with a term such as *stress reaction*, should not be central—all these people are facing serious issues and need individual support and respect. Thus, the cultural/environmental/contextual focus remains central as we want to avoid attributing any client response as being solely "in the person." All survivors of a crisis need to know that however they responded to severe challenges is OK and to be expected.

Calming and caring.   A second major concept is to "normalize" the situation for the client and provide some sense of calm and possibility. This means establishing an empathic relationship by indicating that you care and will listen. Often the first thing in a crisis is that you must be calm and certain. Your personal bearing will do much to meet this first criterion of effective crisis work—*calming the client.*

Don't say "Calm down, it will be OK" or " You're lucky you survived." Better calming language includes such comments as "It's safe now" (if that is true), "We will see that this situation is taken care of," "I feel bad myself; that was a terrible thing to go through," "Your reaction and what you did are common and make sense," and the critical "What would help you right now?" For those who are having flashbacks (and this occurs with all types of trauma), calming and normalizing what they are experiencing are essential.

In particular, do not minimize the crisis. In some situations, you will think and even know that the client is overreacting. Be aware of your thoughts and feelings, which may be valid, but join the client where he or she is. "Enter the client's shoes," as Carl Rogers might say.

Safety.   Crisis and trauma survivors need to know that they are safe from the danger they have gone through. With soldiers and many others suffering from serious posttraumatic stress (not PTSD), creating a sense of safety and calm may not be accomplished immediately. Offer verbal reassurance that the crisis is over and they are now safe—again, if they are indeed safe. However, more than words may be needed. A woman who has experienced spousal abuse or a homeless and hungry person needs to find a safety house or a place to stay and eat immediately. To some counseling and therapist supervisors, this is "violating boundaries." This type of thinking is a relic of the field's past, but still there are some with these attitudes and beliefs. Stand up for what is right, help clients find what they need, connect them with resources. And consider the statement, with appropriate timing, "I'll be there with you to help."

Action.   A good place to start is "What do you need now?" "What help do you want?" For yourself, what can you do that is *possible* in the here and now, and in the future? Do not overpromise. As noted above, some clients need a place to stay that night. Others need to know facts immediately. "Will I have to go through a vaginal exam after the rape?" "Are we going to be taken away by a bus?" "Where is the high ground in case the water comes again?" Answering these and other questions calmly and clearly will do much to alleviate anxiety.

The next step is to stay with the client and ensure that their needs are met. Crisis situations are often confusing. Following Katrina, many clients lost their helpers and thus experienced even more anxiety and tension. Volunteers in the Haiti earthquake went to help with good intentions and, indeed, did provide valuable assistance. But soon they had to return home, and often people were left "up in the air" with no knowledge of what to do next.

Debriefing the story.   Have you ever talked to family members or friends who have had a difficult and traumatic hospital operation? Have you noted that they often give you detail after detail? And then, the next time you see them, they tell you the same painful story . . . and perhaps even a third or fourth time. Freud called this "wearing away the trauma." People need to tell their stories, and many need to tell them again and again. Here the basic listening sequence becomes the treatment of choice. If you paraphrase, reflect emotions, and summarize what they have said authentically and accurately, they will know that someone has finally heard them.

Follow-up.   Concrete action in the immediacy of crisis is essential. Where possible, you want to arrange to meet the client again for debriefing and planning in more detail for the future. In some cases, longer-term counseling and therapy will be needed.

Watch for strengths and resilience. If given sufficient early support, most people work through their crises. They have internal strengths that will carry them through. Look for these strengths and external resources that will enable them to recover. At the same time, even the most resilient survivors need to debrief what has happened.

Implications for Your Practice   There is much more to crisis counseling than what is said here, but you will find that competence and expertise in the BLS and the five-stage structure will provide a map that will help carry you through some challenging situations.

All of us need to be ready to help in crisis situations. We may have to deal with immediate crises such as the ones we have focused on here. But we also need to understand the concepts underlying crisis counseling because so many clients will have experienced, or be experiencing right now, extremely difficult situations.

BOX 15.1 | Research and Related Evidence That You Can Use

Systematic Emergency Therapy for Sexual Assault, Personal Assault, and Accident Survivors

Ideas from this research program should be used only under appropriate supervision and after you have acquired sufficient knowledge and practice. There are important implications for your practice here, but considerable experience with severe distress is necessary before exploring this experimental treatment.

In a study conducted in Atlanta by Rothbaum and Keane (2012), 137 trauma survivors (about one-third had experienced rape, one-third assault, and one-third an automobile accident) were divided into two groups. The first group received standard trauma assessment; the second group received the assessment plus the experimental systematic treatment, which consisted of an initial treatment followed by two additional sessions one week apart.

Dr. Rothbaum summarized the first session:

We asked people to go back to the traumatic event, to go through it in their mind's eye and recount it out loud over and over. We tape-recorded it, and we gave them that tape to listen to. All of this happened very quickly, in about an hour, because they had

already been in the ER [emergency room] for a long time and just wanted to go home.

The patients were given cognitive behavioral therapy in which client negative thoughts were identified (e.g., "I never will feel safe again," "I'll never drive a car after that accident"). Strategies such as thought stopping and cognitive reframing were taught as ways to avoid harmful cognitions and emotions. Sessions 2 and 3 continued the process of debriefing, with continued emphasis on homework.

Follow-up at 12 weeks revealed that the sexual assault victims had substantially reduced their amount of posttraumatic stress; the personal assault and accident survivors had also improved, but not as much. "More are going to end up with PTSD at week 4 and week 12 if they don't get the intervention," said Dr. Rothbaum.

Rothbaum, B., & Keane, T. (2012, April 13). *Emergency therapy may prevent PTSD in trauma victims.* Anxiety Disorders Association of America (ADAA) 32nd Annual Conference, Session 318R.

Think of the need for counselors to debrief what people have witnessed at the scene of an accident, perhaps seeing a dead child with a bloody mother stuck in a seat belt and the father stunned and speechless. EMTs (and police and firefighters) don't forget experiences like this; they wear on them emotionally and frequently lead to depression. There is a real need to provide counseling and support after such traumatic experiences.

Counselors and therapists also experience trauma burnout. Counselors often suffer burnout when they work several days with a major disaster, listen to endless sad stories on a crisis line, or just do daily intervention work at a mental health center. The continual load of people in crisis wears on helpers, who may become traumatized themselves as they listen to horrific stories. Counselors need support when they work with these difficult situations. Counseling and therapy for the counselor need to be considered as part of this support process.

Microskills and the five stages give you a start in understanding crisis work, but you have much more to learn to be fully helpful in such situations. At the same time, some crisis situations require many helpers and counselors. Seek some training and offer yourself to others.

▶Crisis Counseling First Session Transcript

Each type of crisis is different; adapt your approach accordingly. Establish trust and the working relationship as quickly as possible. You will often have to act swiftly and sometimes decisively to help your clients reach the next stage beyond that first session.

Following is a sample transcript involving a family in a big city dealing with the loss of their apartment after a fire. The counselor, Angelina Knox, meets the mother, Dalisay Arroyo, in the office of one of the managers at the community center the day after the fire. Dalisay, 31, is employed as an aide in a nursing home and has two children. The father has only occasionally been involved since the children's birth. The fire department took Dalisay and her children to her parents' small apartment where they spent the night, but they obviously can't stay there more than a few days. Thus, as the counselor prepares to meet with the mother, first responders have already worked with early crisis safety and basic needs. The counselor may need to focus more on emotional reactions and planning for the future.

This transcript is an edited and condensed version of a half-hour session. This large city has a history of preparation for crisis; it is in a flood zone and subject to summer fires. In addition, Homeland Security has strengthened existing resources. Thus, the session occurs within an ideal support system, something that is not available in all but a few settings. Therefore, we will review the crisis situation again after the transcript, outlining what can be done in more difficult situations with inadequate support systems.

COUNSELOR AND CLIENT CONVERSATION	PROCESS COMMENTS
1. *Angelina*: (Walks to the secretary's office, smiles warmly, and invites Dalisay in.) Hello Ms. Arroyo, I'm Angelina Knox. You and your children have had a terrible night. I'm a community counselor here in town and want to see how things are going and how I might be helpful. But, before we start, are there any questions that you might want to ask me?	Angelina is ready to spend time on developing an empathic relationship, but like many trauma survivors, the client wants to start immediately.

COUNSELOR AND CLIENT CONVERSATION	PROCESS COMMENTS
2. *Dalisay*: Angelina, I'm not sure where I should go or what I should do. I don't want to stay with my parents. They are good people, but don't have room for us and they get impatient with the kids. All my furniture is gone. I don't know what to do. (Starts crying softly.)	This is a common type of statement in crisis. Clients are "all over the place," topic jumping. Other clients may be unable to talk coherently; still others angrily demand that action be taken immediately. Be ready for almost any reaction, and remember they are all normal and to be expected.
3. *Angelina*: It's hard . . . really hard. (She sits in silence for a minute until Dalisay looks up.) Your reactions make sense and are totally normal. It will take some time to sort things out, but we have some resources here that will help. But, before going on, could you tell me what happened?	Angelina acknowledges Dalisay's feelings and encourages her to tell her story, while seeking to normalize her thoughts and emotions. (Interchangeable empathy)

Clients reacting to trauma need to tell their stories. Some will tell them at length, while others may simply describe the bare facts. Emotions will vary from a loss of control to numbness without much feeling expressed. The following is a much shorter version of what was said over 5 minutes of interaction. More tears flowed, but Dalisay also felt relief that no one was hurt.

COUNSELOR AND CLIENT CONVERSATION	PROCESS COMMENTS
4. *Dalisay*: I was almost asleep and then I smelled something strange in the kitchen. I went in and there was a small fire in the wastebasket. I must not have put the cigarette out. Then, all of sudden, it went "poof" and I ran to get the children out. . . . It spread so fast, but the neighbors called the fire department right away and only our apartment is gone. But then we got out in the cold. Firemen wrapped their blankets around us, asked if we had any help, and then they took us to my parents, 10 blocks away. But during all that time, the children were frantic and I couldn't quiet them. Their dolls and toys are gone. They couldn't stop crying until Grandma held them. The fire chief called this morning and said that you could likely help me figure out what to do next. My father drove me here on the way to work, the children are with their grandmother, but I guess I'll have to walk back to them, but all my warm clothes are gone. I called the nursing home today and the shift supervisor said that I could have the rest of the week off, but that likely means no pay and I can hardly pay bills now. (Serious crying.) I don't have enough money to rent a new place, but I can't stay with my parents.	Throughout the longer story, Angelina offered solid attention and a fair amount of natural spontaneous body mirroring. Her comments were short and usually took the form of encouragers and restatement. She did acknowledge Dalisay's emotions, but did not reflect them, believing that would be more appropriate later. Angelina's session behavior represented interchangeable empathy.
5. *Angelina*: Dalisay, you and your daughters have had a terrible experience, but it is good that you got out in time and had your parents to stay with. I can sense the horror you must have experienced and felt, even though I wasn't there. And . . . then . . . the children. I can see that you worry about them. It's great that your parents are close, even though they can only do so much. As I listen to you, I get the feeling that you already have useful strengths and some clear ideas about what needs to be done. That contact with your shift supervisor was wise and will be helpful in the long run. Not everyone . . .	What we see here is a brief summary of Dalisay's situation and recognizing her emotions without pressing issues. Angelina then brings in a family focus, along with feedback supporting what Dalisay has done already to remedy her situation. Dalisay, like almost all people in crisis, has assets, strengths, and resources. (Interchangeable empathy with some additive dimensions)

continued on next page

COUNSELOR AND CLIENT CONVERSATION	PROCESS COMMENTS
6. *Dalisay*: (Interrupts anxiously) Thanks, but I can only be gone so long. As soon as everyone was safe, I started thinking how things could work out and I realized that I must hang onto my job. It's really scary. How am I to manage?	"What am I to do next?" While listening skills remain central, this is the time for Angelina to move to more direct influencing in the session. Dalisay needs listening and emotional support, but the real issues require action.
7. *Angelina*: I hear your worry. There are some things that our office can offer. We are lucky here in that we have a trauma relief center that provides much of what you need, including some limited financial help for a few days. You won't have to stay with your parents long, as I think we can arrange for a temporary furnished apartment for you. I've already contacted the Women's Center and they have clothes and some kitchen essentials that will help. If you are interested, I'll call and arrange for a time for you to meet with them. So, you see that there are several possibilities, but we don't want to do anything until it makes sense to you.	Angelina again acknowledges emotions and comes up with very specific directives and suggestions as to how she and her agency can help. Is Angelina offering too much material aid so soon? Certainly it would not have been too soon to offer hurricane or flood survivors clear statements of what actually could be done for them. Many felt lost in the vagueness of helping efforts, and sometimes more was promised than would ever be delivered. Here we see a large city well prepared for crisis—Homeland Security has done its job here. (Potential additive empathy with action)
8. *Dalisay*: (Seeming relieved) Wow, that is more than I expected. I thought I'd be hanging like Katrina survivors. It's terrific that I can get an apartment, and I'm amazed at the possibility of financial help. This will enable me to keep my job and take care of my children. But the next thing is, what about the children and school?	We see positive movement on the Client Change Scale. Dalisay is moving from "I can't" to "I think I can," with Angelina's help.
9. *Angelina*: Well, we need to talk about that. We will try to find an apartment near your old place so that they don't have to change, but that might not happen. We will do the best we can. I know that it isn't easy for you or them.	Try not to overpromise, but to say what you and your agency really can do in a straightforward fashion. Following information giving, we see a brief acknowledgment of feeling. (Lower-level interchangeable empathy)
10. *Dalisay*: I feel a little better, but still a lot anxious and worried.	More movement on the CCS.
11. *Angelina*: Clearly we need to talk over the fire in more detail, the fright you experienced, what it did to the children, and how you handled it. And then there is a lot of worry over what will happen next. If we can get together tomorrow or the next day, we can do some more serious debriefing of what happened. Would you like to do that?	As noted, this is an edited version of the longer session. The session has now gone on for about 20 minutes and provided the client with security about what will happen next and how Angelina will follow up. Debriefing of the story and the trauma needs to start as soon as possible. Note the open invitation to talk, rather than telling Dalisay that she has to return. This moves toward a more egalitarian relationship.
12. *Dalisay*: Angelina, you have been so much help and so understanding. Yes, I'd like to talk about what happened. All of us, the children and me, had nightmares last night. It wasn't good. But I know that we have to get settled right now, and I'll meet and talk with you later on this.	Dalisay is much calmer than she was at the beginning of the session. She has moved from a Level 1+ on the Client Change Scale to a beginning Level 3. Emotionally, at least in the moment, she may have reached Level 4, but don't expect this to hold unless further counseling and support are provided.

COUNSELOR AND CLIENT CONVERSATION	PROCESS COMMENTS
Like many crisis sessions, this one moved from topic to topic and stage to stage with a flow that did not follow the typical pattern. Those who have gone through trauma need (1) personal supportive contact; (2) understanding and clarification of the crisis trauma; (3) awareness of their own personal strengths, as well as what external resources are available to them; (4) some short-term achievable goals; and (5) an immediate, clear, concrete action plan, with arrangements for later personal follow-up and further discussion and debriefing as soon as possible.	
13. *Angelina*: Let's write down together where we are and what we can and need to do before we meet again. Where shall we start?	This directive brings in Dalisay as an egalitarian partner in finding solutions. Angelina could tell her client what to do, but success is much more likely if Dalisay is respected and fully involved. (Additive empathy)

14. *Dalisay*: I really appreciate that you could help us find housing. Could we begin with that?
    (Angelina brings out paper and pen for both of them, and they start to work.)
	It takes 10 minutes for Angelina and Dalisay to write down the action plan, who will do what and when. Dalisay occasionally starts to cry, but more easily regains self-control. Out of this come workable alternatives that can be implemented in stages.
	    Follow up on actions and debriefing of the trauma, but this has to be something that Dalisay wants. The children also need to tell their stories, and the school counselor needs to be consulted. At some point, it may be useful to bring the grandparents in.

Crisis counseling demands much from you, the counselor, but it also provides many rewards when you can provide concrete help and see relief start to come in for the client and the family. But imagine what it feels like in a major crisis when you don't have the resources described above and you meet with ten or more people who have just gone through a fire, flood, or earthquake the middle of the night in the rain. These clients have even more needs, and they could be hungry. You may only have 15 minutes and never see the person again. Thus, the calming and caring are needed continually, both for the client and yourself.

Again, burnout can be a problem for the crisis counselor. There is also your own emotional involvement. You may care for clients and their future, but follow-up to make sure that they have followed an action plan is often not possible, leaving you wondering how helpful you (and the crisis team) really were. Thus, crisis counseling can often turn into a developing crisis for the helper. This means that you need to take care of yourself throughout each day. Take breaks, seek to get enough sleep, try to get a little exercise, and make sure that you debrief your experiences with understanding colleagues and/or supervisors.

Suggested Next Steps
Miller, G. (2012). *Fundamentals of crisis counseling.* Hoboken, NJ: Wiley.
Kanel, K. (2011). *A guide to crisis intervention.* Belmont, CA: Cengage.
    These two textbooks are commonly used in the field, and you will find it beneficial to have one or both on your bookshelf. A number of certification programs are available on the Internet. Some will be helpful, but study each with real care.

Cognitive Behavioral Therapy

[People] are disturbed not by things, but by the view that they take of them.

—Epictetus

The key to change . . . is doing.

—Carlos Zalaquett

Philosophy   Cognitive behavioral therapy (CBT) originated in two different philosophic traditions. The cognitive portion of CBT is rooted in Epictetus' famous Stoic statement above. Viktor Frankl's logotherapy is often seen as the first cognitive theory because of his emphasis on reframing cognitions to more positive thought patterns, though Frankl also stressed the importance of taking thought into action. But it took Albert Ellis to bring cognitive work to center stage with what he first called rational emotive therapy (RET). He soon changed the name of his theory to rational emotive behavioral therapy (REBT), emphasizing the importance of making ideas and cognitions real through behavioral change and homework. Aaron Beck has since become central to cognitive therapy, and Donald Meichenbaum is known for his integrative CBT model, with a strong emphasis on behavioral change and stress management. CBT is now eclectic and related in some way to all the skills and strategies of this text. Ultimately, we can view CBT as pragmatic and practical, searching constantly for "what works" with each client.

Key Methods   The National Association of Cognitive-Behavioral Therapists defines CBT as follows (www.nacbt.org/basics-of-cbt.aspx):

> Cognitive-behavioral therapy does not exist as a distinct therapeutic technique. The term "cognitive-behavioral therapy (CBT)" is a very general term for a classification of therapies with similarities. There are several approaches to cognitive-behavioral therapy, including Rational Emotive Behavior Therapy, Rational Behavior Therapy, Rational Living Therapy, Cognitive Therapy, and Dialectic Behavior Therapy.

Frame of reference.   CBT is an information processing system in which thoughts influence our feelings and actions. The purpose of CBT is to explore thought patterns, help the client see that they are ineffective or irrational, and enable the client to "think different." Some specifics of CBT, as outlined by the national association, include the following:

1. A base of cognitive response as being key to change
2. Time limited with specific goals
3. A sound relationship is needed, but is not central
4. A collaborative venture between counselor and client that uses a Socratic question and answer style
5. Based on aspects of Stoic philosophy ("It is not things, but what one thinks of things that counts.")
6. Structured and directive
7. Based on an educational model *(psychoeducation)*
8. Relies on induction (encourages clients to look at their thoughts and draw their own conclusions, although the counselor will make serious challenges and confrontation)
9. Homework is essential

CBT also encourages self-healing and aims to increase clients' competency and provide coping skills they can use when facing new concerns and challenges.

Key CBT propositions (Dobson, 2009, p. 4):

▲ Cognitive activity affects behavior.

▲ Cognitive activity may be monitored and altered.

▲ Desired behavior change may be affected through cognitive change.

As you can see, much of what we have emphasized in this book is in accord with CBT tenets. However, we believe that attention to relationship, feelings, and meanings is more essential than CBT typically suggests. An excellent illustration of why CBT also needs to consider emotional experience can be seen in the session transcript that follows. Neuroscience findings indicate that thoughts come first from feelings. In fact, there are many CBT specialists who agree with neuroscience that thoughts come first from feelings and emotions. Albert Ellis's rational emotive behavioral therapy (REBT), of course, includes emotional issues as a basic factor.

▶Cognitive Behavioral Session Transcript

Cognitive behavior therapy and its relation to microskills and the five stages will be presented through a demonstration by Carlos Zalaquett with a client, Renée, who is anxious about her performance in her internship. She was referred by her site supervisor, who was concerned about her level of anxiety, which seemed to affect her performance.

Session 1 Carlos went through the necessary introductory steps such as discussing confidentiality, HIPPA, and talking about the session structure and process. He spent time establishing the relationship, and Renée spoke about her concerns and anxiety around being an effective counselor. She showed a fair amount of anxiety during the first 10 minutes of the session, but gradually relaxed. Having heard the general details of her story, Carlos turned to examining strengths. Renée was an excellent student and had worked as a peer counselor the previous year on campus, but graduate school and the internship felt very different. Carlos summarized these and other past strengths and then asked her to describe a successful case. At this point, she started smiling. He also found that Renée had a good program of exercise and several friends. The anxiety seemed to focus on her desire to be perfect immediately and her concern that she might not help clients as much as she wished.

Session 2 The relationship had been established in the first session, and Carlos asked about her week, searching for positives and what might be "new and good." She said that she felt more optimistic, had continued exercising, and had been getting out with her friends. Carlos summarized the strengths from the previous session and then turned to her stage 3 concerns. Before starting CBT, he wanted to know what thoughts, feelings, and behaviors might represent a more positive Renée. She stated that she'd like to relax more, learn how to work better with clients, and become more confident. As a result, she hoped and anticipated that her counseling behavior would be more effective.

Stages 4 and 5 of Session 2 As CBT moves to stage 4, expect the counselor, Carlos, to increase his talk time and use more questions. Why? Active interpersonal influence strategies are an essential part of CBT. Having listened to the story and identified strengths, Carlos will now turn to examining and challenging the negative parts of her story. At this stage, look for the following counseling style in the transcript:

1. He will ask for more detail and for concrete evidence of Renée's self-damaging beliefs.
2. The client and counselor will look for other, more useful ways to understand and reframe the evidence and the story. Can we look at the concern in new and more positive ways?

3. What are the consequences if the client continues to maintain the current negative set of cognitive beliefs? What will be the consequences of a more positive frame of reference?

4. Expect considerable use of the skills of instruction/psychoeducation to help the client contract for specific changes in thoughts, feelings, and behaviors.

We will join the conversation about 15 minutes into the second session. The transcript has been shortened and edited, but the main aspects of changing cognitions and beliefs are presented in the transcript.

COUNSELOR AND CLIENT CONVERSATION *STAGE 4. RESTORYING*	PROCESS COMMENTS
1. *Carlos*: Well, to review—this is an interesting situation because your internship site supervisor referred you. I understand that you did well in the program, but perhaps could use some help as you prepare for future sessions with clients in your clinic. Is that right? Is that your goal?	Carlos briefly summarizes stages 1, 2, and 3, and we see the reason for Renée's consultation. Notice how he uses a questioning tone at the end to check out the accuracy of his statement.
2. *Renée*: I've been feeling some anxiety with my new job as a counselor. A client that I really like didn't come back. I just recently graduated, so every day I'm feeling insecure about seeing clients. I just feel nervous all the time.	Renée's body language indicates tension. She is sitting back in her chair and her legs are jiggling. Renée seems to start the session at Level 2 on the Client Change Scale.
3. *Carlos*: I can see some discomfort as you speak about it. Are you hesitant about what the client's going to do, or you worry because you feel that you may not be as competent with them?	Acknowledgment of emotions followed by an interesting question on her nervousness coupled with a mild interpretation. Using a question for an interpretation at times softens the impact and allows the client to reflect on what you say. (Potentially additive)
4. *Renée*: That's exactly it. I feel that I'm second guessing myself. Am I doing the right thing? Is this what's best for them? You know, I just graduated. So I just have all this insecurity that I'm trying to deal with every day. It's becoming increasingly difficult.	The additive questioning interpretation brings out both cognitions and the emotions associated with her difficulties.
5. *Carlos*: I see. So, share with me some situations in which you felt that way.	This could be best described as a question formed as an encouraging statement. The focus is on the main theme or issue. (Potentially additive)
6. *Renée*: Before I see a client, I always feel this way.	Renée sees her behavior as a pattern (abstract formal operational thought).
7. *Carlos*: So, as you are going to see a client, you feel anxious about how you will do, how you will perform?	Reflection of feeling, combined with a question. Carlos seeks to confirm the experience of the client. (Interchangeable)

COUNSELOR AND CLIENT CONVERSATION	PROCESS COMMENTS
STAGE 4. RESTORYING	
8. *Renée*: Am I good enough?	Here we see one of Renée's central cognitions, which in turn leads to feelings of anxiety.
9. *Carlos*: How good you are and how competent, I see. Well, I think I understand the reason for the referral. I'm going to focus on what we call cognitive behavior therapy, also known as cognitive therapy, to help you cope with your worrying thoughts. In cognitive therapy we believe that your thoughts affect your behavior and your mood. For example, a person may say "I'm going to see a client and I feel anxious," suggesting that the situation is what triggers the anxiety. However, what we have learned in CT is that it's not the situation that triggers the reaction. There is something in between. And this is your thoughts or images. The question is, do you understand what happens between seeing a client and your emotional response?	Carlos restates the central cognition and lets the client know he has understood her concern. He then structures using instruction/psychoeducation to introduce the basics of cognitive behavioral therapy. The question at the end seeks to check out if Renée has an understanding of what he just said.
10. *Renée*: I haven't really thought about it because I just feel so overwhelmed with emotions that I can't really put the thing into perspective.	Renée's emotions are leading her cognitions. Reversing this pattern is a major goal of CBT. The cognitive way we think about things affects the nature of things and how we feel about them.
11. *Carlos*: So let me help you by engaging in a process of discovery of "what is in between." What's the first thing that happens when you picture in your mind that you are going to see a client?	Cognitive therapists see clients as practical scientists and help them to engage in self-discovery. Imagery is used to help the client gain a different understanding of her issue. (A key additive strategy)
12. *Renée*: I just get nervous.	
13. *Carlos*: You get nervous. Exactly. It's such an automatic connection. It is hard to ask what else could be there.	Restatement of feelings followed by instruction/psychoeducation to explain that immediacy of emotional reactions precludes clear thinking. (Continues the additive strategy)
14. *Renée*: Right.	
15. *Carlos*: Now, let me ask you then to stop for a second and notice what happens when you are on the brink of seeing a client. . . . Now, let me ask what crossed your mind?	Imagery directive to help Renée observe her thoughts. One key to CBT is helping clients observe their own thoughts, feelings, and behaviors. The question is similar to free association strategies and is a useful directive in CBT. (Potentially additive)

continued on next page

COUNSELOR AND CLIENT CONVERSATION	PROCESS COMMENTS
STAGE 4. RESTORYING	
16. *Renée*: Um, am I gonna use the appropriate counseling techniques for this client? Will I be able to help this client? I'm never sure.	Renée thinks back and identifies the negative cognition. She is starting self-observation. This represents the awareness and the beginning of Level 3 on the CCS.
17. *Carlos*: Um-hum. And what do we call this thought that goes through your mind?	Carlos uses a minimal encourager and questions to draw out the client's label for the cognition. Use the client's key words and everyday vocabulary.
18. *Renée*: Negative thinking.	Voila! Renée shows clear understanding and thus is now at Level 3 on the CCS. Many/most clients will not be able to identify and name their style of thinking so quickly.
19. *Carlos*: Negative thinking. Very well, this is exactly how we call these in our work, negative thinking or negative thoughts. And the reason, as you see, is that they are negative in nature. Okay.	Restatement and instruction/psychoeducation helps Renée feel understood and provides her with additional CBT information. (Interchangeable and somewhat additive)
20. *Renée*: Okay.	
21. *Carlos*: So the sequence in our view is that when you are facing an event, something goes through your mind, and that is really what triggers your reaction, your emotional reaction. Are you with me?	Carlos provides further psychoeducation, but checks out to see if Renée understands. (Additive)
22. *Renée*: Yeah. Right now, I'm thinking about other things that go through my mind before I get ready to see a client. Even just talking about it now makes me feel a little tense.	Renée "gets it" and demonstrates a good level of understanding by providing further examples of thoughts and physical reaction. Clear Level 3 on the CCS.
23. *Carlos*: Very good. It's interesting you say this because I was at the brink of asking you to go back to the original situation to see if you could discover some other thoughts. So let's look at your situation again. Very good, educate me. Help me know what else goes through your mind as you are facing these types of situations.	Notice the frequent use of encouragers. The counselor engages the client as a co-collaborator, as she is the expert on her own experiences. Renée can help Carlos understand her better, promoting an active process of self-discovery. (Additive)
24. *Renée*: Is the client gonna want to come back?	Our greatest fear! And 25% of clients don't come back, more if the client is culturally different from you. There is always some truth behind cognitions that lead us to fear and inaction.
25. *Carlos*: Um-hum…	Use of encouragers to further promote self-discovery.

COUNSELOR AND CLIENT CONVERSATION	PROCESS COMMENTS
STAGE 4. RESTORYING	
26. *Renée*: Am I good enough for that client? Am I gonna be successful? These types of thoughts.	Notice how Renée's responses address previous questions raised in the session. She is engaged.
27. *Carlos*: Good, sometimes these thoughts, that we call negative thoughts, are part of what we call a core belief. These are long-held thoughts that affect our behavior and emotions. It's not easy to identify core beliefs, but they are central to whatever we do. I mention this because automatic thoughts spring from core beliefs, which should be the ultimate focus of our attention. But for the time being, we can focus on the negative thoughts. So this is a two-step process. Step one is working with the current thoughts. Step two is discovering the core beliefs and then dealing with these over time. But let's go back to your current situation because I understand you want to do something about the negative thoughts. Since you understand this very well, we will continue to analyze your situations from a CBT therapy point of view. Is this okay?	Instruction/psychoeducation to further advance the CBT model. Carlos ends with a brief summary of Renée's concerns and, in the spirit of collaboration, checks to see if he is heading in the right direction and she concurs. What you see here is a brief summary of an hour session. Remember that CBT does not typically move this fast, nor do we always have a client who grasps the purpose of counseling this quickly.
28. *Renée*: Okay.	Psychoeducation will be ineffective unless the client is ready, willing, and able.
29. *Carlos*: I have a chart that I share with my clients to record their automatic thoughts. As you can see, it's a chart with three columns. The first is used to report a specific event; okay, then here in the third column we look at the emotional or behavioral reaction; and then we spend time in the second column identifying the thought or image that may be involved in the situation.	Carlos introduces a CBT chart to record automatic thoughts. This chart in its simplest form displays three columns: Event—Thought—Response. See Figure 15.1.
30. *Renée*: Okay.	

THOUGHT RECORD		
Event	**Thought**	**Response** (Emotions/Behaviors)
Meeting a client	*Am not good enough.*	*Anxious, worry, and feel less effective in the meeting, although somehow I survive. But, I wonder about the client.*

FIGURE 15.1 Basic thought record sheet.

COUNSELOR AND CLIENT CONVERSATION	PROCESS COMMENTS
STAGE 4. RESTORYING	
31. *Carlos*: What we did before actually followed these three columns (points to chart), so let's look at it from this chart's point of view. Can you see it well?	Carlos has laid the foundation for the work with this chart and builds on what has transpired in the session so far.
32. *Renée*: Yes, I see it.	
33. *Carlos*: Think about your situation right now; you are going to see a client. Okay. And you're feeling anxious. As you represent this in your mind, pay close attention to your thoughts. What thoughts did you identify?	Carlos instructs Renée in how to use the three-column chart. The event (seeing a client) is first. Her anxiety about the session is in the third column.
34. *Renée*: There were several of them: "I'm not sure if I'm going to be proficient." "I'm not sure I'm going to do well." "I'm not sure if the client will like to come back or if the client will be successful."	Here Renée has been asked to explore the cognitions that will be recorded in the second column. Her responses are Level 3 on the CCS scale as Renée shows awareness of what is happening to her, but already in a more optimistic framework.
35. *Carlos*: Good job. I'm going to put these thoughts down here. Now, we have identified the specific parts of this chart, including the negative thoughts that are creating the anxiety in your relationships with clients. The situation, the thoughts, and then the emotional reaction. This column representing the thoughts is essential. In our view, it's not the situation itself that upsets you. It's what you think or how you perceive the situation that affects you. That's the reason why the focus is on the thoughts. Sounds reasonable?	Carlos demonstrates the use of the chart and helps familiarize Renée with CBT's view of the importance of thought processes. Renée has just been given a lot of information. In the actual interview, more time was spent in back and forth discussion and obtaining Renée's thoughts and feelings about the chart and how it would apply to her. Don't just pass the chart to clients. Work with them as they think it through.
36. *Renée*: Yes, it does.	
37. *Carlos*: Now we are going to take one step forward. Think about past experiences. How many times have you been, for a lack of a better word, incompetent?	Carlos challenges the automatic thought of incompetence by asking for evidence in its support. CBT practitioners usually ask, "What's the evidence that the thought is true? That it's not true?" Notice the client's response. (Potentially additive)
38. *Renée*: I don't think I've ever been incompetent. I always try my best.	Anxiety rides on the shoulders of expectations, not of actual experiences.
39. *Carlos*: Earlier we talked of your many competencies. Recall them? Um-hum. So let me ask you the opposite then. What evidence do you have that suggests that you may be competent?	Carlos refers back to earlier discussions of strength and asks the client to reflect on her competence. (Additive)

COUNSELOR AND CLIENT CONVERSATION	PROCESS COMMENTS
STAGE 4. RESTORYING	
40. *Renée*: Well, in my work throughout practicum and internship, I did well. I completed all my counseling courses successfully. I graduated, so that must mean that I've received the proper training and that I should be competent to perform these new skills that I've been given.	A further demonstration of CBT: Negative thoughts have less actual support than positive thoughts, but the negatives influence behavior and feeling more. This is a truism reinforced by neuroscience research.
41. *Carlos*: Help me understand this. So you don't think that you may be very competent, but then there are these facts that, if I heard you correctly, demonstrate that you have been competent. How do you reconcile these?	Carlos uses gentle confrontation to help Renée restory. (Additive)
42. *Renée*: How am I competent?	Notice how Renée's response demonstrates a shift to a positive self-perception. Early step to Level 4 of the CCS and real change.
43. *Carlos*: Yeah, how do you know that you have been competent?	Open question to promote further restorying and drawing of competence evidence. (Additive)
44. *Renée*: Like I said, probably my graduate coursework. That to me is evidence that I've been competent. Perhaps—I guess I learned in my classes that my first sessions would not be perfect and that some clients do not return regardless.	
45. *Carlos*: How fascinating. You are right about what happens to all of us. I was curious about how you knew that you have been competent in the past. Now let me ask you a different question. Let's say that you see a client and that you don't do your very best. What's the worst that could happen?	Carlos continues to work with Renée to challenge her negative belief. He introduces a CBT technique called *worst case scenario*.
46. *Renée*: Maybe the client won't come back. One didn't.	We've heard that before, but now we are working on the negative condition on a basis of positive assets and strengths.
47. *Carlos*: Maybe they won't come back. Uh-huh. And how bad would that be?	Restatement followed by question to further explore client's worst expectations. (Potentially additive)
48. *Renée*: I guess it isn't so terrible. Would just give me a chance to maybe practice more, work on my skills more. I guess it wouldn't be the end of the world or anything.	Renée reveals the catastrophic thought underlying her fears but demonstrates she can challenge that thought.

continued on next page

COUNSELOR AND CLIENT CONVERSATION	PROCESS COMMENTS
STAGE 4. RESTORYING	
49. *Carlos*: It wouldn't be . . . ?	Minimal encourager is offered to get Renée to repeat her statement and reinforce thought change.
50. *Renée*: It wouldn't be the end of the world.	Another step to Level 4 change in cognitions.
51. *Carlos*: Oh, it wouldn't be the end of the world. I see. Sometimes when my clients say something like this, I ask them if they could restate their thoughts in positive terms. Can you do this?	Reframing. Carlos assists Renée in transforming her thought into a positive statement. (Additive)
52. *Renée*: Absolutely. It will be an opportunity to improve.	
53. *Carlos*: I see. Is there other evidence that suggests you have the competence to do your work?	Positive asset search. Carlos continues searching for strengths and positives. (Additive)
54. *Renée*: I have a few clients that I've seen that always come back, and they look forward to scheduling the appointments with me and seeing me again, so I guess that's good news. I must be doing something right.	Note that a negative experience can lead to fears and ineffective cognitions, even when the evidence suggests otherwise.
55. *Carlos*: Uh-hum. You know, I always wonder when people say I'm doing well, because doing well for one person may have a different meaning than it has for you, so when you say doing well, in your case that clients are coming back, what helps you to do well?	Open question to help Renée own her skills. (Additive)
56. *Renée*: Um, I guess being confident in the techniques that I use with my clients. They even want to come back and see me, and it seems like I have been using the appropriate ones and that they've been working well with the client, because they want to come back and continue to work on their issues with me.	Notice Renée's active involvement in cognitive restructuring and restorying.

Stage 5. Action: generalizing new cognitions and behaviors to the real world

57. *Carlos*: Very well. So let me go back to our initial situation. We were talking about seeing clients and the fact that you were feeling anxious about it, and we made a connection that demonstrates that negative thoughts trigger our negative emotions. I would like you to think about seeing a client, a new client right now, what will go through your mind as you work to prepare to do that?	Carlos moves to prepare Renée for the action phase of counseling. (Additive)

COUNSELOR AND CLIENT CONVERSATION	PROCESS COMMENTS
STAGE 5. ACTION	
58. *Renée*: I will think about the positive things that I'm doing and probably how I've been successful so far, so there's no reason for me to think that I'm going to fail or am incompetent.	The basis for change and eventual maintenance of new patterns of thought.
59. *Carlos*: Very good. In the past, when you were at the brink of seeing a new client, you felt anxious. Concern was about not doing it well, not having the client returning, not having all the confidence to do your best work, and all that was defined as the words or thoughts that triggered your emotional reaction. Now, I see that when you look at these situations, you have more positive thoughts.	Carlos summarizes CBT model and session work.
60. *Renée*: Yes, I would be more confident and enjoy my work even if the client doesn't want to come back; I realize that it's not the end of the world; it will give me a chance to improve.	Level 4 cognitive change has been achieved, but this is still not the real world. Homework and follow-up will be necessary to achieve lasting change.
61. *Carlos*: Good. As you can see, with this cognitive approach we not only begin to address your situation, but also learn about its practice. What I would like you to do is to use this very same chart to monitor your thoughts. Monitor what happens when you are going to see a client. So that will be the event, and then we will see how you feel emotionally about it. Then, spend some time paying attention to identify the thoughts and the dreaded consequences or results that crossed your mind in that situation. Okay. And I have a date here because I'm going to ask you to do this throughout the week. This will give us a chance to identify more clearly the negative thoughts and to use those as a foundation to stop and challenge them with your strengths, where you feel competent and good about yourself. There are a lot of things you do well.	Carlos adds more information about the model, assigns homework, and encourages action.
62. *Renée*: Okay.	
63. *Carlos*: Any questions, any comments about this particular homework?	Carlos uses questions to find out if Renée has doubts or concerns.
64. *Renée*: No. It is something I can do. It will be very helpful.	

continued on next page

COUNSELOR AND CLIENT CONVERSATION *STAGE 5. ACTION*	PROCESS COMMENTS
65. *Carlos*: Very good. One last question. How did it go? How do you feel?	Final checkouts to determine client's satisfaction and feelings.
66. *Renée*: Good. I feel better.	Moving to Level 4 on the CCS.

As demonstrated in Renée's experience, automatic thoughts can trigger feelings and affect behavior. Discovering negative automatic thoughts helped the client replace them with more appropriate thoughts and improve her situation. Renée has learned a new cognitive technique in this session. Clients like her can learn to detect automatic thoughts by

▲ Learning about the CBT conceptualization
▲ Discovering negative automatic thoughts using thought recording charts
▲ Finding ways to replace these thoughts with more appropriate ones
▲ Applying these techniques in their everyday situations to effect positive changes in their lives

Change doesn't come from the sky. It comes from human action.

—Tenzin Gyatso, the 14th Dalai Lama

Suggested Next Steps

Beck, J., & Beck, A. (2011). *Cognitive behavioral therapy* (2nd ed.). New York: Guilford Press.
Ellis, A., & Dryden, W. (2007). *The practice of rational emotive behavior therapy.* New York: Springer. (Originally published 1997)
  Put these two books together and you will have a solid introduction to the basics of CBT from two rather different and even competing perspectives.

BOX 15.2 Research and Related Evidence That You Can Use

fMRI Predicts Response to CBT

Functional magnetic resonance imaging (fMRI) of battered women revealed that certain patterns of brain activity predicted better response to cognitive behavioral therapy (Aupperle & Hunt, 2012). More specifically, greater anterior cingulate activity and less posterior insula activity were found to be critical in the different responses to treatment. The anterior cingulate monitors conflicts in information processing and guides decision making, while the insula has been found to relate to pain and the six basic feelings of Chapter 7.

With some studies showing only a 50% response rate to CBT in the domestic violence population, "there is room for improvement," noted lead investigator

Robin Aupperle. "If we can find techniques to target these areas specifically to enhance CBT and future treatments, that is important."

We have said that "counseling changes the brain." But, in a way, this is not new. Sigmund Freud in 1895 theorized that experience would change the brain (Centonzea et al., 2004). We now can help clients change memories, and thus their thinking, feeling, and behavior. With this new knowledge, clients can find new and more powerful meanings for their lives. We are likely nearing the time where we can anticipate with increasing precision the power of effective counseling and therapy.

▶Three Additional Theories

The introduction to this chapter mentioned three additional theories that also serve as examples of how you can use and transfer the ideas of the five stages to other systems and methods of counseling and psychotherapy. The CourseMate website presents information and transcripts and gives ideas for using brief counseling, motivational interviewing, and counseling/coaching. Here are some brief suggestions for their use.

Brief Counseling

Chapter 1 provided data showing that many clients are satisfied with just one session and most counseling is completed by the tenth. This suggests that you will want to give special attention to brief counseling. You can enter into brief counseling easily through microskills and the five stages. The major adaptation is moving goal setting to early in the session. Then, all that follows is oriented to the client. We have suggested that early goal setting is a useful part of decisional counseling, and all counseling and therapy, with a review of goals in stage 3.

Motivational Interviewing

Motivational interviewing is a thoroughly researched system that began with a focus on alcohol and substance abuse. You will find that it is closely similar to the five-stage model and decisional counseling, except for the clear focus on what works with these difficult clients.

Counseling/Coaching

We have added the word *counseling* to coaching to separate it from professional coaching. However, now the National Board of Certified Counselors (NBCC) offers supplementary training for the coaching profession. The structure of the coaching interview again closely follows the five-stage model and teaches the same microskills featured here. The content of training, of course, is different, but the ideas of coaching do have much to offer all counselors and psychotherapists.

▶Summary: Practice and Integration Promote Personal Theory Development

This chapter is designed to facilitate your own evolving theory of how you want to practice counseling and therapy. The skills and the five-stage structure emphasized in this book, plus the theories presented here, give you a practical map of the multiple approaches.

You will gain beginning competence in many of these approaches to counseling if you can take time to deliberately practice each one. This may not be possible at this time, but it is something for you to think about. After all, you now have the basic skills and concepts to start practice in virtually all counseling and therapy theories. But be careful, this is a beginning level of competence; real mastery requires further study and supervision.

Your ultimate goal is to use this newly acquired knowledge with your clients. Now you can access a variety of counseling styles to provide them with more alternatives, and thus deliver more effective help. We encourage you to study the theories presented here and to engage in the intentional use of the tools they provide.

Multiple approaches	Table 15.1 summarizes microskill use in many different counseling approaches, including decisional counseling, person-centered counseling, logotherapy, multicultural counseling and therapy, feminist therapy, crisis counseling, cognitive behavioral therapy, brief counseling, motivational interviewing, and counseling/coaching. Though all these approaches may be explained and understood in terms of their use of microskills and how the session is structured, note that their emphases are quite different.
	For example, consider the differences between decisional counseling, which emphasizes careful listening to the story/concern/challenge of the client before acting, and brief counseling and counseling/coaching, which emphasize working on issues as quickly as possible. Person-centered helping stresses listening to the client's feelings and story in detail, while CBT actively encourages the client to change and adopt new thoughts and behaviors, while avoiding questioning and directives.
Multicultural issues	Each theory requires different adaptations to be meaningful in multicultural situations. Particularly helpful in this regard is the concept of focus (Chapter 9). By focusing on the cultural/environmental/contextual dimensions, you can bring in these issues fairly easily to all helping approaches. However, you still must recognize that the aims of each approach may not be fully compatible with varying cultures. This same point, of course, should be made with regard to the client regardless of cultural background. Some clients may prefer the Rogerian person-centered approach; others may want solutions and cognitive behavioral action. Avoid stereotyping any client with prior expectations.
Cultural intentionality	We are suggesting that the intentional counselor or therapist will have more than one theoretical alternative available. At the same time, you need to select those approaches to helping that are most comfortable for you. Balancing your knowledge, skills, and interests as you counsel varying clients entails a lifetime process of learning for any helping professional.

►Competency Practice Exercises and Portfolio of Competence

 Go to CengageBrain.com to access Counseling CourseMate, where you will find an interactive ebook, quizzes, videos, interactive counseling and psychotherapy exercises, the Portfolio of Competence, case studies, a practice test, and more.

Individual Practice

Exercise 1. Theory Practice   Choose one of the theories and audio or video record a brief session with a volunteer.

▲ Plan for a minimum session of 15 minutes, but be flexible so you can cover the basics of the theory.
▲ Select a concern for the role-play. Be specific, and use a mild concern.
▲ Record the session on audio or video.
▲ Review the session using the feedback form provided in Box 15.3.

You can download this form from Counseling CourseMate at CengageBrain.com.

Theoretical system selected for practice_____

_____ (DATE)

_____   _____
(NAME OF COUNSELOR)   (NAME OF PERSON COMPLETING FORM)

Empathic relationship: Initiating the session, rapport and structuring ("Hello; this is what might happen in this session"). How well did the counselor establish rapport, and how did he or she accomplish this objective? Were preliminary goals identified? According to the theory, was goal setting carried to more specificity?

Story and strengths: Gathering data, drawing out stories, concerns, problems, or issues ("What's your concern? What are your strengths or resources?"). Was at least one positive asset or strength of the client identified? How completely did the counselor draw out the story and/or issues? Were the strengths and resources adequately explored?

Goals: Mutual goal setting ("What do you want to happen?"). Was it effective? Were the original goals of the session reviewed, and were the desired outcomes of the client really clear? With brief counseling, review of goals can also be helpful at this point.

Restory: Working. How was this approached? Were thoughts, feelings, behaviors, or meanings a primary focus? What specifics did the counselor use to encourage creation of new ways of thinking and being?

(continued)

BOX 15.3 (continued)

Action: How did the counselor go about helping the client create a concrete plan for action? Was homework agreed to by counselor and client?

Did the counselor help the client plan for generalization to daily life?

General comments on the counselor and skill usage:

Group Practice

Exercise 2. Practice With Crisis Counseling and CBT   Select one theory and build from there.

▲   Work with a partner, switching the roles of client and counselor. Plan for a minimum session of 15 minutes, as this is likely enough to cover basics. But be flexible as more time is often needed.

▲   Select a concern for the role-play. This time the issues need to be very specific—for example, dealing with something meaningful to you such as a past or present conflict on the job, in the family, or with a friend or partner. Consider issues of life goals and vision. Aim for concreteness and clarity throughout the storytelling.

▲   Record each of the sessions on audio or video, perhaps using your computer, cell phone, or video-equipped digital camera to provide some instant feedback.

▲   With your partner, search through the discussion of each theory and create a tentative treatment plan in accord with the basic tenet of the theory, structure, and microskill usage.

▲   Review both sessions with your partner, using the feedback form in Box 15.3. Determine strengths and areas for improvement, and share suggestions for achieving desired outcomes.

Portfolio of Competence

Developing and evaluating your skills and competence using each of the theories practiced should be included in your Portfolio of Competence.

▶Determining Your Own Style and Theory: Critical Self-Reflection on Theoretical Orientations

How does the concept of theoretical orientation relate to your own developing style and theory? Which of the approaches presented most appeals to you? Do you agree with us that decisional counseling and the five-stage structure underlie most other approaches as a basic model?

We will not ask you to assess your competence in any of these approaches as it is far too early and you will want to work further with each one. Rather, please focus your attention on your early impressions and where you think you might go next in building competence in these or other theoretical orientations.

What single idea stands out for you among all those presented in this chapter, in class, or through informal learning? What stands out for you can be a guide toward your next step. What are your thoughts on multicultural issues and the various theoretical approaches? What other points in this chapter strike you as useful? How might you use ideas in this chapter to begin the process of establishing your own style and theory?

CHAPTER 16

Determining Personal Style and Future Theoretical/ Practical Integration

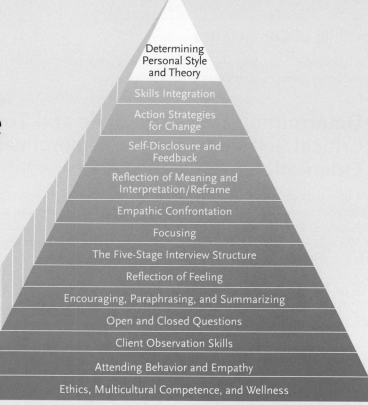

Determining
Personal Style
and Theory

Skills Integration

Action Strategies
for Change

Self-Disclosure and
Feedback

Reflection of Meaning and
Interpretation/Reframe

Empathic Confrontation

Focusing

The Five-Stage Interview Structure

Reflection of Feeling

Encouraging, Paraphrasing, and Summarizing

Open and Closed Questions

Client Observation Skills

Attending Behavior and Empathy

Ethics, Multicultural Competence, and Wellness

Now is the time when your action is practice.

—The Dalai Lama

Mission of "Determining Personal Style and Future Theoretical/Practical Integration"

Developing your own personal approach to counseling and psychotherapy involves a multiplicity of factors. At this moment we are asking you to reflect on yourself, your values and personal meanings, and your skills. Where are your strengths? What areas need further development? Where would you like to go? What are your competencies for working with clients different from you in terms of personal style, values, type of concern presented, and cultural difference?

Chapter Goals and Competency Objectives Awareness, knowledge, skills, and actions in the skills and strategies of this book will enable you to

▲ Utilize the skills and concepts of this book as an effective framework to continue to increase competence in the counseling and therapy process. You can meet the many challenges ahead from a solid base.

▲ Review the concepts of this book and evaluate your ability to master the basic competencies of the microskills hierarchy and the counseling session.

▲ Continue the process of determining which theory or theories are most relevant to you and how you currently integrate skills and strategies.

▲ Consider how your helping style and possible theoretical orientation will be relevant to the many differing types of clients you will encounter.

▶Introduction: Identifying an Authentic Style That Relates to Clients

Some counselors and therapists have developed individual styles of helping that require their clients to join them in their orientation to the world. Such counselors believe they have found the one true and correct formula for counseling and therapy; clients who have difficulty with that formula are often termed "resistant" or "not ready" for counseling. These counselors do indeed have their own style, but their methods tend to be rigid and dogmatic. They can and do produce effective change, but they may be unable to serve many client populations. Missing from their orientation is a broader understanding of the complexity of humanity and the helping process.

As you generate your own approach to the field, remember that clients may have widely different values and experiences from yours. They may wish to head in many directions different from yours. Remain aware that the skills, strategies, and theories you favor may not be preferred by your client. If so, then you may wish to expand your skills and knowledge in areas where you are now less comfortable. It is their life, not yours.

As you expand your competence, maintain your authenticity as a person. With study, patience, and experience, you will increase your abilities to work with those who are unlike you. The opportunity to learn from clients different from you is one of the special privileges of being a counselor or a therapist. You will want to expand your understanding of cultural differences including race/ethnicity, gender, spiritual/religious orientation, disability, sexual orientation, age, and socioeconomic status.

This brief chapter returns to the beginning, but the focus is totally on you. In Chapter 1 we saw your natural style. Is it the same? How might it have changed? Table 15.1 (page 396) provided a brief map showing how the microskills and concepts play themselves out in different models of counseling and therapy, which should aid you in the search for your personal style and your progress.

▶The Microskills Hierarchy: Assessing Your Competencies

The preceding 15 chapters have presented 39 major concepts and skill categories. Within those major divisions are more than 100 specific methods, theories, and strategies. Ideally, you will commit them all to memory and be able to draw on them immediately in practice to facilitate your clients' development and progress in the session.

These skills, concepts, and understandings are but a beginning. We have not considered here details of personality development, testing, and the many other theories of counseling and therapy you will encounter in the future. Although these microskills and training concepts are used in multiple fields and by people around the world, only a brief introduction to the multitude of applications has been provided in this book.

How can you manage to retain and use all of this book's concepts and the theories presented? Most likely, you cannot at this point. But recall the story of the samurai (page 81). With time and experience, you will develop increased understanding and expertise. As you grow as a counselor or therapist, you will find the ideas expressed here becoming increasingly clear, and your mastery of skills and theories will likewise continue to increase. Use this book as a resource that can help you recall key concepts that you will use with your varied clients.

Retaining and mastering the concepts of the microskills hierarchy may be facilitated by what is termed *chunking*. We do not learn information just in bits and pieces; we learn it best by organizing it into patterns. The microskills hierarchy is a pattern that can be visualized and experienced. For example, at this moment you can probably immediately recall that attending behavior has certain major concepts "chunked" under it (the "three V's + B" of culturally appropriate visuals, vocal tone, verbal following, and body language). The basic listening sequence (BLS) will be easy to recall and essentially covers the first half of this book. You can probably also recall the purpose of open questions and perhaps which questions lead to which likely outcomes (for example, *how* questions lead to process and feelings).

With periodic review and experience, the many concepts will become increasingly familiar to you. If you complete a transcript examining and classifying your own counseling style similar to the one presented in Chapter 14, the ideas of this book will become especially clear. As you must have noticed, the skills you have practiced instead of just reading are the ones that you understand best and that have the most relevance for you. Reading is a useful introduction to counseling and psychotherapy, but the results of practice and experience will stick with you far into the future.

Table 16.1 summarizes the major concepts of the book. Take some time to review your own competence levels in each of the 39 major areas and enter the results in the table. What is your present level of mastery and competence in each area?

TABLE 16.1   Self-Assessment Summary

	Identification and Classification	Basic Competence	Intentional Competence	Teaching Competence	Evidence of Achieving Competence Level
Skills and Concepts					
1. Attending behavior					
2. Questioning					
3. Observation skills					
4. Encouraging					
5. Paraphrasing					
6. Summarizing					
7. Reflecting feelings					
8. Basic listening sequence					
9. Positive psychology and wellness: strengths, resources, positive assets					
10. Empathy					

11.	Empathic relationship—story and strengths—goals—restory—action				
12.	Focusing				
13.	Empathic confrontation				
14.	Client Change Scale				
15.	Reflection of meaning				
16.	Interpretation/reframe				
17.	Self-disclosure				
18.	Feedback				
19.	Logical consequences				
20.	Instruction/psychoeducation				
21.	Directives				
22.	Stress management				
23.	Therapeutic lifestyle changes				
24.	Analysis of the counseling session (Chapter 14)				
Theoretical/Practical Strategies					
25.	Wellness assessment				
26.	Ethics				
27.	Multicultural competence				
28.	Community genogram				
29.	Decisional counseling				
30.	Person-centered counseling				
31.	Logotherapy				
32.	Multicultural counseling and therapy, feminist therapy				
33.	Crisis counseling				
34.	Cognitive behavioral counseling				
35.	Brief solution-focused interviewing and counseling and psychotherapy (see the optional CourseMate website)				
36.	Motivational interviewing (see the optional CourseMate website)				
37.	Counseling/coaching (see the optional CourseMate website)				
38.	Neuroscience and how it may relate to counseling and psychotherapy				
39.	Defining personal style and theory				

1. *Identify and classify the concept.* If the skill or concept is present in a counseling session, can you label it? These concepts provide a vocabulary and communication tool with which to understand and analyze your counseling and psychotherapy behavior and that of others. The ability to identify concepts means most likely that you have chunked most of the major points of the skill together in your mind. You may not immediately recall all seven types of focus, but when you see a session in progress you will probably recall which one is being used.

2. *Demonstrate basic competence.* At this level, you will be able to understand and practice the concept. Continued practice and experience form the foundation for later intentional mastery.

3. *Demonstrate intentional competence.* Skilled counselors and therapists can use the microskills and the concepts of this book to produce specific, concrete effects with their clients. Please visit the Ivey Taxonomy (Appendix A) and review the specifics of the results you may anticipate. If you reflect feelings, do clients actually talk more about their emotions? If you provide an interpretation, does your client see her or his situation from a new perspective? If you work through some variation of the positive asset search, does your client actually view the situation more hopefully? If you conduct a well-formed, five-stage counseling session, how does the client change on the Client Change Scale? Do behavior and thinking change? Intentional competence shows up not in your use of the skills and concepts but rather in what your client does. To demonstrate this level of skill, you need to be able to produce *results* due to your efforts in helping.

4. *Demonstrate teaching competence.* You are not expected to master teaching all the skills and concepts of this book at this point. However, you should have learned that you can teach attending behavior and the basic listening sequence to your clients during the counseling session. Some of you may have had the opportunity to take these skills and conduct teaching workshops with community volunteers, church groups, and peer counselor training programs. For the longer term, we suggest that you continue to think of the possibility of teaching skills to clients and their families. And if you become a professional, you most likely will find yourself teaching listening skill workshops at some point in your career. Box 16.1 provides a personal account of using the microskills throughout one's career.

BOX 16.1 National and International Perspectives on Counseling Skills

Using Microskills Throughout My Professional Career

Mary Rue Brodhead, Executive, Canadian Food Inspection Agency

My first encounter with the microskills program was through my doctoral program in teacher education. I wanted to learn about counseling so that I could better reach students, particularly those at risk who often were the least likely to approach a counselor. What I learned very rapidly was that teachers often fail to listen to their students—in fact, one research study found that out of nearly 2,000 teacher comments, there was only one reflection of feeling and, of course, most comments focused on providing information. Often, even when teachers did listen, they weren't able to recognize the verbal and nonverbal cues that were reflecting the reality of their students.

After completing my degree, I entered teacher education and found that microskills lead to a more student-centered teaching. I also found that teachers who matched their students' cognitive/emotional style were the more effective. If a student is concrete, the teacher needs to provide specific examples and use concrete questions. If the student is more reflective, then formal operational strategies can be used. Bringing in emotional involvement via sensorimotor strategies enriched teaching.

I next lived on an island off Vancouver Island where my husband Dal and I worked with members of the Kwakiutl Nation. I trained teachers, but I was involved in counseling and established an alternative program, as part of the local School Board, for youth who had dropped out of high school. I actually taught them the same interviewing skills that you are learning in this book. Needless to say, the multicultural orientation here helped sensitize me to differences among people.

One of our first activities was a trip in the Chief's fishing boat to gather Christmas trees for the old people, a cherished tradition in the community. This was the first time most of these young people had participated in the ritual. As they delivered the trees, the recipients, in an expression of gratitude, invited them in for something to eat and began to tell stories. These old people, thrilled to have an audience, would shower attention on these alienated youth who, in response, would listen with respect. And so an upward cycle of communication began. Eventually we raised money to buy tape recorders and tried to capture these tales on tape (a commonplace activity now, but quite new in those days), leading to further strengthening of the students' listening and questioning skills. In this case, attending behaviors led to an increased sense of value and respect on both sides.

After three years in British Columbia, my husband and I returned to Ottawa where I worked in a federal government employment equity program. The task centered around providing culturally sensitive counseling and opportunities to members of the Canadian four "equity groups" (visible minorities, Aboriginal peoples, persons with disabilities, and women in nontraditional occupations) to develop the skills needed for career development and success in the Federal Public Service.

As part of my work, I used microskills to train government officials to listen and to really hear and understand the variety of perspectives, strengths, and styles of working found within members of our multicultural workforce. Here, I also learned the importance of language and eventually became reasonably fluent in French, an essential skill for success in bilingual, multicultural Canada. Our team developed a multicultural counseling course, used—all or in part—within 15 universities in Canada, that included many ideas presented here.

The agency where I now work is responsible for animal health, plant protection, and food safety. It is one of the "science departments," involving agriculture, fisheries, scientific research, and many other issues. Given the diversity of my workforce, my first task with my team has been to build a "culture of learning" where vast amounts of information and knowledge can be shared effectively and efficiently. Microskills are a key element in management training, and a good communication skills workshop can be vital in team building. And, of course, all our managers need to listen to and motivate those with whom they serve.

Looking back over my career, I'm amazed to find that the basic microskills have been useful in my teaching, counseling, multicultural work, and governmental leadership positions. "Training as treatment" and "teaching competence" in microskills can help us all make a difference throughout our careers.

▶Your Personal Style and Future Theoretical/ Practical Integration

We make a living by what we get. We make a life by what we give.

—Winston Churchill

The fact that you took this course in counseling and psychotherapy skills and read this book suggests that you have a strong interest in working with and serving others. The helping fields are rich in opportunity for personal joy and satisfaction. You can make a difference in other people's lives. As you think about your personal style and future theoretical/practical orientation, what do you want to give to your clients and the world?

There are three major factors to consider as you move toward identifying your own personal style and integrating the many available theories: your own personal authenticity; the needs and style of the client; and your own life goals, values, and vision. Unless a skill or theory harmonizes with who you are and your sense of meaning, it will be false and less effective. *Competence, caring, and a sense of direction and purpose are essential.*

Remember that you are one of a kind, as are those whom you would serve. We all come from unique life experiences, varying families, differing communities, and distinct views of gender, ethnic/racial, spiritual, and other multicultural issues. It is obvious that modifying

The following issues are presented for you to consider as you continue to identify your natural style and future theoretical/practical integration of skills and theory.

Goals and Values   What do you want to happen for your clients as a result of their working with you? What would you *desire* for them? How would you serve? How are these goals similar to or different from those of decisional, solution, person-centered, cognitive-behavior therapy, or other theory?

Skills and Strategies   What microskills and strategies do you personally favor? What do you see as your special strengths? What are some of your needs for further development in the future? What else?

Cultural Intentionality   With what cultural groups and special populations do you feel capable of working? What knowledge do you need to gain in the future? How aware are you of your own multicultural background? What else?

Theoretical/Practical Issues   What theoretical/practical story would you provide now that summarizes how you view the world of counseling and psychotherapy? How do the several theoretical orientations of Chapter 15 relate to your present style? Where next would you like to focus your efforts and interests? What else?

natural style and theoretical orientation will be necessary if you are to be helpful to the endlessly varying, challenging, and interesting clients you will meet.

If you have presented and analyzed a transcript of a counseling session as recommended in Chapter 14, you have an excellent beginning for understanding yourself and how clients relate to you. If you are competent in microskills and structure of the session, you have a foundation in the critical basics of decisional and person-centered counseling. With additional thought and experience, you can fairly rapidly become competent in logotherapy, multicultural and feminist theories, and stress management. Crisis counseling and the several other theories are now readily accessible to you.

At this point it may be useful to summarize your own story of counseling and psychotherapy. Box 16.2 asks you to review your goals, your special skills, and your plans for the future. Where are you going next?

Go to CengageBrain.com to access Counseling CourseMate, where you will find an interactive ebook, quizzes, videos, interactive counseling and psychotherapy exercises, the Portfolio of Competence, case studies, a practice test, and more.

▶Summary—As We End: Thanks, Farewell, and Success in Your Future Growth and Professional Journey

We have come to the end of this phase of your counseling and psychotherapy journey. You have had the chance to learn the foundation skills and how they are structured in a variety of theoretical/practical approaches. Skills that may have seemed awkward and unfamiliar are often now automatic and natural. As in the Samurai effect, you now do not need to think of them constantly. The basic listening sequence is likely part of your being at this point. Moreover, expect that the *empathic relationship—story and strengths—goals—restory—action* framework will become part of your practice, regardless of your final theoretical orientation(s).

The next steps are yours. Many of you will be moving on to individual theories of counseling, exploring issues of family counseling and therapy, becoming involved in the community, perhaps becoming a coach, and learning the many aspects of professional practice. We have designed this book as a clear summary of the basics; a naturally skilled person can use the information here for many effective and useful helping sessions.

The relationship is forever. . . . Find joy in helping.

—Benjamin Zander

We have enjoyed sharing this time with you. We have come a long way together, and we appreciate your commitment and dedication. Many of the ideas presented in this book have come from students. We hope you will take a moment to provide us with your feedback and suggestions for the future. These pages will be constantly updated with new ideas and information. You have joined a never-ending time of growth and development. and positively affecting the lives of others.

Allen, Mary, and Carlos
allenivey@gmail.com
mary.b.ivey@gmail.com
carlosz@usf.edu

▶Suggested Supplementary Readings for Follow-up on Microskills and Multicultural Issues

Chapter 15 presented key books as suggested next steps for the six theories summarized there. The literature of the field is extensive, and you will want to sample it on your own. Here we would like to share a few books that we find helpful as next steps to follow up on microskills and specifics of multicultural counseling and psychotherapy.

Microskills

www.emicrotraining.com
Visit this website for up-to-date information on microcounseling, microskills, and multicultural counseling and therapy.

Daniels, T., & Ivey, A. (2007). *Microcounseling* (3rd ed.). Springfield, IL: Charles C Thomas.
The theoretical and research background of microskills is presented here in detail.

Evans, D., Hearn, M., Uhlemann, M., & Ivey, A. (2011). *Essential interviewing* (8th ed.). Belmont, CA: Brooks/Cole.
Microskills in a programmed text format.

Ivey, A., Ivey, M., Gluckstern-Packard, N., Butler, K., & Zalaquett, C. (2012). *Basic influencing skills* (4th ed.) [DVD]. Alexandria, VA: Microtraining/Alexander Street Press.
Updated video demonstrations of the skills of the second half of this book. Included are the videos of Allen Ivey and Nelida Zamora (reflection of meaning and integration of skills).

Ivey, A., Ivey, M., Gluckstern-Packard, N., Butler, K., & Zalaquett, C. (2011). *Basic stress management* [DVD]. Alexandria, VA: Microtraining/Alexander Street Press.
Comprehensive videos illustrating main aspects of stress management and several CBT strategies. Demonstrations include meditation, spiritual imagery, automatic thoughts, psychoeducation, and Gestalt.

Ivey, A., Ivey, M., & Zalaquett, C. (2011). *Essentials of intentional interviewing: Counseling in a multicultural world* (2nd ed.). Belmont, CA: Brooks/Cole.

Zalaquett, C. P. (2008). *Las habilidades atencionales básicas: Pilares fundamentales de la comunicación efectiva* [DVD]. Alexandria, VA: Microtraining/Alexander Street Press.
Each of the basic attending skills are role-played and discussed in Spanish with Latina/o participants in this video. A full interview using the BLS is also demonstrated. www.emicrotraining.com

Zalaquett, C. P., Ivey, A. E., Gluckstern-Packard, N., & Ivey, M. B. (2008). *Las habilidades atencionales básicas: Pilares fundamentales de la comunicación efectiva.* Alexandria, VA: Microtraining/Alexander Street Press.
A book with a Spanish presentation of the microskills for Latina/o practitioners and professionals interested in learning or practicing their language and counseling skills. The book is filled with practical information, exercises, and feedback forms. www.emicrotraining.com

Theories of Counseling and Psychotherapy With a Multicultural Orientation

Ivey, A., D'Andrea, M., & Ivey, M. (2012). *Theories of counseling and psychotherapy: A multicultural perspective* (7th ed.). Thousand Oaks, CA: Sage.

The major theories are reviewed, with special attention to multicultural issues and neuroscience. Includes many applied exercises to take theory into practice.

Sue, D.,W., & Sue, D. (2013). *Counseling the culturally diverse* (6th ed.). Hoboken, NJ: Wiley.

The original and classic text on most types of multicultural issues.

Thomas, R. (2000). *Multicultural counseling and human development theories: 25 theoretical perspectives.* Springfield, IL: Charles C Thomas.

Multiple orientations are presented in a comprehensive fashion.

The Ivey Taxonomy

Definitions and Anticipated Results

Skill, Concept, or Strategy	Results You Can Anticipate When Using Skill, Concept, or Strategy
Ethics Observe and follow professional standards, and practice ethically. Particularly important issues for beginning counselors are *competence*, *informed consent*, *confidentiality*, *power*, and *social justice*.	Client trust and understanding of the counseling process will increase. Clients will feel empowered in a more egalitarian session. When you work toward social justice, you contribute to problem prevention in addition to session healing work.
Multicultural Issues Base counselor and therapist's behavior on an ethical approach with respect and an awareness of the many issues of diversity. Include the multiple dimensions described in Chapter 2. All of us have many intersecting multicultural identities.	Anticipate that both you and your clients will appreciate, gain respect, and learn from increasing knowledge in intersecting identities and multicultural competence. You, the counselor, will have a solid foundation for a lifetime of personal and professional growth.
A Positive Strength and Wellness Approach Help clients discover and rediscover their strengths by listening carefully for present strengths and resources. In addition, consider a wellness assessment. Find strengths and positive assets in the clients and in their support system. Identify multiple dimensions of wellness.	Clients who are aware of their strengths and resources can face their difficulties and discuss problem resolution from a positive foundation. Also, effective and positive counseling and psychotherapy can be anticipated to strengthen the frontal cortex and hippocampus, while potentially resulting in a smaller amygdala.
Attending Behavior Support your client with individually and culturally appropriate visuals, vocal quality, verbal tracking, and body language.	Clients will talk more freely and respond openly, particularly about topics to which attention is given. Depending on the individual client and culture, anticipate fewer breaks in eye contact, a smoother vocal tone, a more complete story (with fewer topic jumps), and a more comfortable body language.
Empathy Experiencing the client's world and story as if you were that client; understanding his or her key issues and saying them back accurately, without adding your own thoughts, feelings, or meanings. This requires attending and observation skills plus using the important key words of the client, but distilling and shortening the main ideas.	Clients will feel understood and engage in more depth in exploring their issues. Empathy is best assessed by the client's reaction to a statement and his or her ability to continue the discussion in more depth and, eventually, with better self-understanding.

continued on next page

Skill, Concept, or Strategy	Results You Can Anticipate When Using Skill, Concept, or Strategy
Additive Empathy Interviewer adds meaning and feelings beyond those originally expressed by the client.	Clients reach a better understanding of their own issues and engage in more depth in exploring of these issues.
Subtractive Empathy Interviewer response gives back to the client less than what the client said and perhaps even distorts what has been said. In this case, the listening or influencing skills are used inappropriately.	Inappropriate use of skills subtracts from the client's experience. Client doesn't feel understood.
Observation Skills Observe your own and the client's verbal and nonverbal behavior. Anticipate individual and multicultural differences in nonverbal and verbal behavior. Carefully and selectively feed back some here-and-now observations to the client as topics for exploration.	Observations provide specific data validating or invalidating what is happening in the session. Also, they provide guidance for the use of various microskills and strategies. The smoothly flowing session will often demonstrate movement symmetry or complementarity. Movement dissynchrony provides a clear clue that you are not "in tune" with the client.
Questions Questions can be open or closed.	Effective questions encourage more focused client conversations with more pertinent detail and less wandering.
Open Questions Begin open questions with the often useful *who, what, when, where,* and *why. Could, can,* or *would* questions are considered open but have the additional advantage of being somewhat closed, thus giving more power to the client, who can more easily say that he or she doesn't want to respond.	Clients will give more detail and talk more in response to open questions. *Could, would,* and *can* questions are often the most open of all, because they give clients the choice to respond briefly ("No, I can't") or, much more likely, explore their issues in an open fashion.
Closed Questions Closed questions may start with *do, is,* or *are.*	Closed questions may provide specific information but may close off client talk.
Encouraging Give short responses that help clients keep talking. They may be verbal restatements (repeating key words and short statements) or nonverbal actions (head nods and smiling).	Clients elaborate on the topic, particularly when encouragers and restatements are used in a questioning tone of voice.
Paraphrasing (also known as reflection of content) Shorten, clarify the essence of what has just been said, but be sure to use the client's main words when you paraphrase. Paraphrases are often fed back to the client in a questioning tone of voice.	Clients will feel heard. They tend to give more detail without repeating the exact same story. If a paraphrase is inaccurate, the client has an opportunity to correct the counselor.
Summarizing Summarize client comments and integrate thoughts, emotions, and behaviors. This technique is similar to paraphrase but used over a longer time span. Important in the summary is that you seek to find strengths and resources that support the client.	Clients will feel heard and often learn how the many parts of important stories are integrated. The summary tends to facilitate a more centered and focused discussion. The summary also provides a more coherent transition from one topic to the next or a way to begin or end a full session.

Skill, Concept, or Strategy	Results You Can Anticipate When Using Skill, Concept, or Strategy
Basic Listening Sequence (BLS) The BLS consists in the microskills of using open and closed questions, encouraging, paraphrasing, reflecting feelings, and summarizing. These are supplemented by attending behavior and client observation skills. Select and practice all elements of the basic listening sequence.	Clients will discuss their stories, issues, or concerns, including the key facts, thoughts, feelings, and behaviors. Clients will feel that their stories have been heard.
Reflection of Feeling Identify the key emotions of a client and feed them back to clarify affective experience. With some clients, the brief acknowledgment of feeling may be more appropriate. Often combined with paraphrasing and summarizing.	Clients will experience and understand their emotional state more fully and talk in more depth about emotions and feelings. They may correct the counselor's reflection with a more accurate descriptor.

The Five-Stage Interview Structure

Skill, Concept, or Strategy	Results You Can Anticipate When Using Skill, Concept, or Strategy
1. *Empathic relationship:* Initiate the session. Develop rapport and structuring. "Hello, what would you like to talk about today?"	The client feels at ease with an understanding of the key ethical issues and the purpose of the session. The client may also know you more completely as a person and professional.
2. *Story and strengths:* Gather data. Use the BLS to draw out client stories, concerns, problems, or issues. "What's your concern?" "What are your strengths and resources?"	The client shares thoughts, feelings, and behaviors; tells the story in detail; presents strengths and resources.
3. *Goals:* Set goals mutually. The BLS will help define goals "What do you want to happen?" "How would you feel emotionally if you achieved this goal?"	The client will discuss directions in which he or she might want to go, new ways of thinking, desired feeling states, and behaviors that might be changed. The client might also seek to learn how to live more effectively with situations or events that cannot be changed at this point (rape, death, an accident, an illness). A more ideal story ending might be defined.
4. *Restory:* Explore alternatives via the BLS. Confront client incongruities and conflict, restory. "What are we going to do about it?" "Can we generate new ways of thinking, feeling, and behaving?"	The client may reexamine individual goals in new ways, solve problems from at least those alternatives, and start the move toward new stories and actions.
5. *Action:* Plan for generalizing session learning to "real life." "Will you do it?" Use BLS to assess client commitment to taking action after the session.	The client demonstrates changes in behavior, thoughts, and feelings in daily life outside of the session.

Skill, Concept, or Strategy	Results You Can Anticipate When Using Skill, Concept, or Strategy
Focusing Use selective attention and focus the counseling session on the client, theme/concern/issue, significant others (partner/spouse, family, friends), a mutual "we" focus, the counselor, or the cultural/environmental context. You may also focus on what is going on in the here and now of the session.	Clients tend to focus their conversation or story on the areas that the counselor responds to. As the counselor brings in new focuses, the story is elaborated from multiple perspectives. If you selectively attend only to the individual, the broader dimensions of the social context are likely to be missed.
Empathic Confrontation Supportively challenge the client to address observed discrepancies and conflicts. 1. Listen, observe, and note client conflict, mixed messages, and discrepancies in verbal and nonverbal behavior. 2. Summarize and clarify internal and external discrepancies by feeding them back to the client, usually through summarization. 3. Evaluate how the client responds and whether the confrontation leads to client movement or change. If the client does not change, flex intentionally and try another skill.	Clients will respond to the confrontation of discrepancies and conflict by creating new ideas, thoughts, feelings, and behaviors, and these will be measurable on the five-point Client Change Scale. Again, if no change occurs, *listen*. Then try an alternative style of confrontation.
Client Change Scale (CCS) The CCS helps you evaluate where the client is in the change process. Level 1. Denial. Level 2. Partial examination. Level 3. Acceptance and recognition, but no change. Level 4. Creation of a new solution. Level 5. Transcendence.	The CCS can help you determine the impact of your use of skills. This assessment may suggest other skills and strategies that you can use to clarify and support the change process. You will find it invaluable to have a system that enables you to (1) assess the value and impact of what you just said; (2) observe whether the client is changing in response to a single intervention; or (3) use the CCS as a method for examining behavior change over a series of sessions.
Reflection of Meaning Meanings are close to core client experiencing. Encourage clients to explore their own meanings and values in more depth from their own perspective. Questions to elicit meaning are often a vital first step. A reflection of meaning looks very much like a paraphrase but focuses on going beyond what the client says. Often the words *meaning, values, vision,* and *goals* appear in the discussion.	The client will discuss stories, issues, and concerns in more depth with a special emphasis on deeper meanings, values, and understandings. Clients may be enabled to discern their life goals and vision for the future.
Interpretation/Reframe Interpretation and reframing can provide the client with a new meaning or perspective, frame of reference, or way of thinking about issues. Interpretations/reframes may come from your observations; they may be based on varying theoretical orientations to the helping field; or they may link critical ideas together.	The client will find another perspective or way of thinking about a story, issue, or problem. The new perspective could be generated by a theory used by the counselor, by linking ideas or information, or by simply looking at the situation afresh.

Skill, Concept, or Strategy	Results You Can Anticipate When Using Skill, Concept, or Strategy
Self-Disclosure As the counselor, share your own related past personal life experience, here-and-now observations or feelings toward the client, or opinions about the future. Self-disclosure often starts with an "I" statement. Here-and-now feelings toward the client can be powerful and should be used carefully.	When used appropriately and empathically, the client is encouraged to self-disclose in more depth and may develop a more egalitarian relationship in the session. The client may feel more comfortable in the relationship and find a new solution relating to the counselor's self-disclosure.
Feedback Present clients with clear, nonjudgmental information on how the counselor believes they are thinking, feeling, or behaving and how significant others may view them or their performance.	Clients may improve or change their thoughts, feelings, and behaviors based on the therapist's feedback.
Logical Consequences Explore specific alternatives with the client and the concrete positive and negative consequences that would logically follow from each one. "If you do _____, then _____ this may result."	Clients will change thoughts, feelings, and behaviors through better anticipation of the consequences of their actions. When you explore the positives and negatives of each possibility, clients will be more involved in the process of decision making.
Instruction and Psychoeducation Share specific information with the client, such as career information, choice of major, or where to go for community assistance and services. Offer advice or opinions on how to resolve issues and provide useful suggestions for personal change. Teach clients specifics that may be useful, such as helping them develop a wellness plan, teaching them how to use microskills in interpersonal relationships, and educating them on multicultural issues and discrimination.	If information and ideas are given sparingly and effectively, the client will use them to act in new, more positive ways. Psychoeducation that is provided in a timely way and involves the client in the process can be a powerful motivator for change.
Stress Management and Therapeutic Lifestyle Changes These instructional strategies are designed to improve the physical and mental health of clients. Relaxation, meditation, irrational beliefs, thought stopping, and time management are some of the skills and strategies used to manage stress and change lifestyle.	Clients will use this information to practice managing their stress and change their lifestyles to improve physical and mental health. Physical and mental symptoms will improve over time.
Directives Direct clients to follow specific actions. Directives are important in broader strategies such as assertiveness or social skills training or specific exercises such imagery, thought stopping, journaling, or relaxation training. They are often important when assigning homework for the client.	Clients will make positive progress when they listen to and follow the directives and engage in new, more positive thinking, feeling, or behaving.

continued on next page

Skill, Concept, or Strategy	Results You Can Anticipate When Using Skill, Concept, or Strategy
Skill Integration Integrate the microskills into a well-formed counseling session and generalize the skills to situations beyond the classroom.	Developing counselors will integrate skills as part of their natural style. Each of us will vary in our choices, but increasingly we will know what we are doing, how to intentionally flex when what we are doing is ineffective, and what to expect in the session as a result of our efforts.
Determining Personal Style and Theory As you work with clients, identify your natural style, add to it, and think through your approach to counseling and psychotherapy. Examine your own preferred skill usage and what you do in the session. Integrate learning from theory and practice in counseling and psychotherapy into your own skill set.	You, as a developing counselor or psychotherapist, will identify and build on your natural style. You will commit to a lifelong process of constantly learning about theory and practice while evaluating and examining your behavior, thoughts, feelings, and deeply held meanings.

The Family Genogram

Along with the community genogram, the family genogram brings additional information about all-important family history. We frequently use both strategies with clients and often hang the family and community genograms on the wall in our office during the session, thus indicating to clients that they are not alone in the interview. Many clients find themselves comforted by our awareness of their strengths and social context.

The family genogram is one of the most fascinating exercises that you can undertake. You and your client can learn much about how family history affects the way individuals behave in the here and now. The classic source for family genogram information is McGoldrick and Gerson (1985). You will find the book *Ethnicity and Family Therapy* a most valuable and enjoyable tool for helping to expand your awareness of racial/ethnic issues (McGoldrick, Giordano, & Garcia-Preto, 2005).

Many of us have important family stories that are passed down through the generations. These can be sources of strength (such as a story of a favorite grandparent or ancestor who endured hardship successfully). Family stories are real sources of pride and can be central in the positive asset search. There is a tendency among most counselors and therapists to look for problems in the family history and, of course, this is appropriate. But be sure to search for positive family stories as well as problems.

Children often enjoy the family genogram, and a simple adaptation called the "family tree" makes it work for them. The children are encouraged to draw a tree and put their family members on the branches, wherever they wish.

We present briefly here major "how's" of developing a family genogram. Specific symbols and conventions have been developed that are widely accepted and help professionals communicate with one another.

This brief overview will not make you an expert in developing or working with genograms, but it will provide a useful beginning with a helpful assessment and treatment technique. First go through this exercise using your own family; then you may want to interview another individual for practice.

1. List the names of family members for at least three generations (four are preferred) with ages and dates of birth and death. List occupations, significant illnesses, and cause of death, as appropriate. Note any issues with alcoholism or drugs.

2. List important cultural/environmental/contextual issues. These may include ethnic identity, religion, economic, and social class considerations. In addition, pay special attention to significant life events such as trauma or environmental issues (e.g., divorce, economic depression, major illness).

3. Basic relationship symbols for a genogram are shown on the right, and an example of a genogram is shown on the following page.

4. As you develop the genogram with a client, use the basic listening sequence to draw out information, thoughts, and feelings. You will find that considerable insight into one's personal life issues may be generated in this way.

Close	═══
Enmeshed	═══
Estranged	—//—
Distant	------
Conflictual	MWW
Separated	—/—

Basic relationship symbols

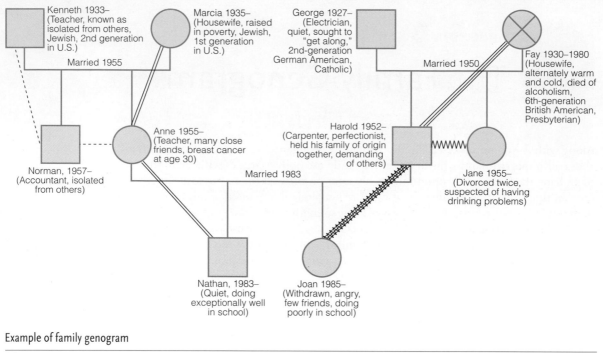

Example of family genogram

© Cengage Learning

Developing a genogram with your clients and learning some of the main facts of family developmental history will often help you understand the context of individual issues. For example, what might be going on at home that results in Joan's problems at school? Why is Nathan doing so well? How might intergenerational alcoholism problems play themselves out in this family tree? What other patterns do you observe? What are the implications of the ethnic background of this family? The Jewish and Anglo backgrounds represent a bicultural history. Change the ethnic background and consider how this would impact counseling. Four-generation genograms can complicate and enrich your observations. Note: The clients here have defined their ethnic identities as shown. Different clients will use different wording to define their ethnic identities. It is important to use the client's definitions rather than your own.

Counseling, Neuroscience, and Microskills

Experiences, thoughts, actions, and emotions actually change the structure of our brains. . . . Indeed, once we understand how the brain develops, we can train our brains for health, vibrancy, and longevity.

—John Ratey

Mission of "Counseling, Neuroscience, and Microskills"

Neuroscience will become an increasingly important part of counseling and therapy education, practice, and research. Appendix C is designed to provide an integrated summary of necessary knowledge.

However, for many of us, the basics of how neuroscience informs our practice may be challenging—and thus more than one reading is recommended. In addition, visit the Internet for further explanation of key concepts. For those with a science background, your task is to take these beginnings and move farther along the path.

Appendix C Goals and Competency Objectives Awareness, knowledge, skills, and actions developed through this appendix will enable you to

▲ Understand some basics of brain physiology that are especially relevant to the counseling and therapy process.

▲ Explain some neuroscience concepts to your clients, thus encouraging compliance and more willingness to take learning in the session to the real world.

▲ Utilize microskills, counseling theory, stress management, and therapeutic lifestyle changes to plan specific interventions that benefit your clients' brains as well as their bodies.

Counseling and psychotherapy can build new brain networks. You are entering our field at what likely will be its most exciting and productive time. The bridge between biological and psychological processes is erasing the old distinction between mind and body, between mind and brain—*the mind is the brain.* We believe it is time to embrace a broader view that integrates counseling and psychotherapy, neuroscience, neuroimaging, chemistry, molecular biology, and the cognitive sciences.

Neuroscience research and theory lend strong scientific support to what we have long been doing in counseling and psychotherapy. Moreover, developing knowledge of the brain will continually enable us to become more precise and effective in our work with clients. As noted throughout this book, each microskill, used effectively, makes a difference. Combined with the wellness and positive asset search and the multiple strategies of varying theoretical approaches, these skills will give you an increasingly effective approach to counseling and psychotherapy.

As early as 1989, Kandel argued that because psychotherapy involves learning new ways of functioning, structural changes occurring in client brains would soon be detectable by neuroimaging machines that identify specifically what is going on inside the brain (Kandel, 2007). Helping prove that prediction today are positron emission tomography (PET) scans and functional magnetic resonance imaging (fMRI). These highly sophisticated methods have found that behavioral, cognitive, and interpersonal therapy can change the structure of the brain (Colozino, 2010; Goldapple et al., 2004; Martin, Martin, Rai, Richardson, & Royall, 2001). Clients, with the help of counseling (or medication at times), are capable of functionally "rewiring" the brain.

▶The Holistic Brain/Body and the Possibility of Change

The more we pulverize matter, the more it insists on its fundamental unity.

—Teilhard de Chardin

The whole brain is greater than the sum of its parts. The brain is a constantly interacting system within itself and in relation to the cultural/environmental context. Each component affects the total system of the holistic brain. Knowledge of the brain and awareness of the new knowledge being constantly developed will lead you to more effective intentional counseling and psychotherapy. Of necessity, the following discussion breaks down the brain into specific parts, which are critical for you to know if you are to communicate with other professionals in the near future.

The brain is, simultaneously, a localized and distributed system. While some of its functions are associated with specific brain structures and regions, these regions act in concert with other, sometimes distant, brain regions. What we come to experience as "mind" is the result of this intense connectivity. Each of our 100 billion neurons connects through even more synapses with an almost infinite number of receptors. Brain interactivity was highlighted by Freed and Mann (2007), who reviewed 22 studies examining sadness and the brain. There is evidence that sadness causes reactions in at least 77 different brain regions.

The human connectome project. Scientists are currently developing a detailed map of all the neural connection paths within the brain (Seung, 2012). Each of us, as a result of genetics and environmental experience, will have unique connections and pathways—in effect, *you are your connectome.* The connectome studies will ultimately provide us with a clear map of how distinct parts of our brains are joined via neural networks. This is a key example of the new future we face and the importance of maintaining an interest in the constant new developments in neuroscience, particularly as they relate to counseling and clinical practice.

Synchrony along the pathways and neural nets appears to be essential for mental health, and likely physical health as well (the later as yet unproven). The nature of each individual's default mode network (DMN), or the brain in the "resting state," will be an essential clue leading to treatment of serious mental challenges (also known as psychopathology) (Hoffman, 2012). The DMN and its implications are discussed in more detail later in this appendix.

Neuroplasticity. The key term for this new future is *neuroplasticity*—the brain's ability to change and reorganize itself throughout life. For counseling and therapy, this means the brain can change—it is not fixed, but responds to external environmental events and actions or initiations by the individual. The old idea that the brain does not change is simply wrong. Neuroplasticity means that even in old age, new neurons, new connections, and new neural networks are born and can continue development—a brain can rewire itself.

Therapeutic lifestyle changes appear to enhance neuroplasticity. For example, we have discussed the effectiveness of exercise, likely through increased blood flow to the brain. Four weeks of serious meditation measurably builds brain gray matter (Tang, Lu, Fan, Yang, & Posner, 2012). Video games have been shown to increase eye-hand coordination, and even cognition. Medications such as fluoxetine (Prozac) and Aricept also show promise to enhance neuroplasticity (Vogel, 2012).

Particularly fascinating is *neurogenesis*, the development of completely new neurons, even in the aged. There is evidence that this occurs in adults primarily in the hippocampus, the main seat of memory, discussed later in this appendix (Siegel, 2007). This is where effective counseling can affect the generation of new neural connections, leading to client change.

Exercise is particularly relevant as a lifetime process to ensure brain and physical health (Ratey, 2008b, 2012). Exercise increases blood flow and the release of positive neurotransmitters such as serotonin. Many of you reading this have experienced the serotonin "high" of a beautiful sunset, the here-and-now immediacy of a close relationship, or running or other physical activity. This positive high through exercise needs to be part of your treatment regime. Exercise is particularly helpful for depression because of serotonin release. If you are sad—walk or run! If you can't run, meditate and use relaxation training.

▶Some Basic Brain Structures

This introductory summary of some key aspects of the brain will enable you to understand and converse with physicians, neuropsychologists, and others who will be important for your career.

The frontal lobe (see Figure C.1) is our chief executive officer (CEO); it is associated with executive functioning, abstract reasoning, and decision making. Critical for long-term memory, it is also the focus of attentional processes and much of our motor behavior, enabling us to function effectively in social systems. Emotional regulation is located here through connections

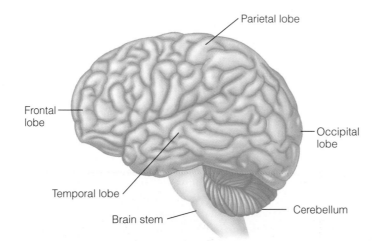

FIGURE C.1   Basic areas of the brain.

with the limbic system. However, in situations of danger, emergency, or mental distress, or under the influence of drugs or alcohol, the limbic system may take over. Clients with frontal lobe issues may show poor emotional control, language problems, personality changes, apathy, or inability to plan. In addition, moral and value decisions (good/bad) rest here.

The *parietal lobe* gives us our spatial sense, but it also serves as a critical integrating force for the senses (see/hear/feel/taste/touch) and our motor abilities. Synthesizing, putting things together, is a function of the parietal. Problems in the parietal lobe may show in personality change, lack of self-care or dressing, making things, and drawing ability. Any failure to integrate may involve the parietal lobe. Difficulties with these functions are often associated with Alzheimer's disease.

The *temporal lobe* is concerned with auditory processing, language and speech production, aspects of sexuality, and memory. The limbic system (shown in Figure C.2), which includes the hippocampus, center of long-term memory, is located in the temporal lobe.

The *occipital lobe* is for visual processing and color recognition.

The *cerebellum* is approximately 10% of the brain's volume, but contains more than 50% of the total number of neurons. Not so long ago, it was ignored as a vestige of the past, but recent research has revealed its centrality and significance in brain functioning—with more research to come. It is a vital part of smooth, coordinated body movement and balance. It also has a role in several cognitive functions, including attention, language processing, and the sensory modalities. A common test for healthy cerebellum motor control is to ask the person to move the fingertip in a rapid straight trajectory; a person with damage will move slowly and erratically.

The *brain stem* connects the brain to the rest of the body via the spinal cord. It is a conduit for integrating the whole brain and is critical for central nervous system functioning such as heart rate, respiration, attention, and consciousness. It also regulates the sleep cycle.

Central and peripheral nervous systems (CNS and PNS).   The CNS consists of the brain and spinal cord; it integrates information and coordinates the body and mind. The PNS connects the CNS to the sensory organs, muscles, glands, and blood vessels. In addition, the PNS is where we find the neurotransmitters that are so important in understanding the counseling process and treatment. The PNS is divided into the somatic nervous system and the autonomic nervous system (ANS), which also includes some sensory systems. The ANS includes the sympathetic and the parasympathetic nervous system. The sympathetic nervous system operates through a series of interconnected neurons and activates the fight-or-flight response. Effects of this activation include pupil dilation, increased sweating, increased heart rate, and increased blood pressure. The parasympathetic nervous system, also called the rest-and-digest system, serves to conserve energy. Effects of its activation include slowing the heart rate, increasing intestinal and glandular activity, and relaxing sphincter muscles in the gastrointestinal tract. Both systems play a pivotal role in the stress response.

A note on the triune brain.   A useful way to think about the brain and its central functions is the concept of the triune brain (McLean, 1985). We can think of the brain as divided into three key parts: the brain stem or reptilian brain, the limbic complex, and the cerebral cortex.

Drawing from theories of evolution, we find that the reptilian complex is primarily focused on basic body functions and pleasure/pain. Think of the spinal cord, brain stem, and the "floor" of the brain such as the basal ganglia. For a simple visual illustration, make a fist, but put your thumb inside the fist. The wrist represents the spinal cord and the lower hand the brain stem. The limbic system is the thumb powering emotion and making memory possible, while the outer core is the cortex, the highest evolutionary point.

The whole is greater than the sum of its parts. The complex, holistic, interacting brain needs all parts working in harmony. A developing idea, only discussed informally so far, is that dyssynchrony of the brain is a major contributor to mental and physical ill health. Effective counseling works toward synchronizing the triune brain.

▶Basics of Feeling and Emotion

The *limbic system* (Figure C.2) is of prime importance for us as counselors and therapists, as it helps us to understand issues of emotion, feeling, and memory. Through understanding feelings, we can help clients improve TAP executive emotional regulation. As stressed throughout the book, the *amygdala* is the energizer of emotive strength. It is the power of feelings that place information in memory in the hippocampus; thus, the interrelationship of the amygdala and hippocampus is central. Drawing sensory information (what is seen, heard, felt, tasted, smelled) from other parts of the brain, the amygdala signals intensity. Nothing happens in other parts of the brain or body (or memory) unless there is sufficient external or internal stimulation.

The *hippocampus* is our memory "organ" and works closely with the amygdala and cerebral cortex. Energy from the amygdala tells the hippocampus which information should be remembered. When there is not enough interest or energy, no memory is produced. In contrast, a highly stressful event (such as war or rape) can overwhelm the whole system like a lightning bolt and result in destruction of neurons and distressed memory. New, negative neural networks take over. The research discussed earlier shows us the importance of wellness and positive assets as we seek to develop and strengthen positive memories in the hippocampus. Again, *effective counseling can affect the brain in positive ways.*

The amygdala is recognized as the central area for feelings, particularly the four negative feelings of fear, anger, sadness, and disgust. Surprise, of course, can be negative or positive, depending on the situation and context. All of these five feelings are protective, evolved over time to keep us from danger. The sixth, gladness and its varieties, ranging from satiation through happiness, joy, and contentment, are believed to have evolved later, and

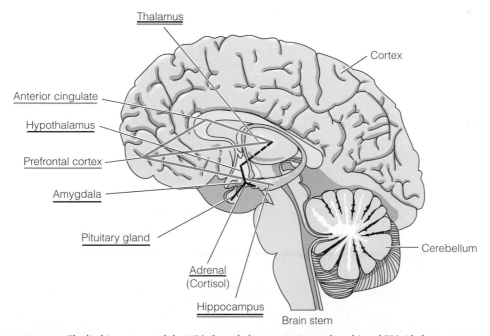

FIGURE C.2 The limbic system and the HPA (hypothalamus, pituitary, adrenals) and TAP (thalamus, anterior cingulate cortex, prefrontal cortex) axes.

thus the prefrontal cortex becomes central. The social emotions, such as guilt and shame, are blends of feelings as we cognitively respond to what occurs in our social-emotional environment and relations with others. These require more sophisticated cognitive processes than the basic feelings. Interestingly, this view has been recently introduced and supported by theory and research presented by Lindquist and colleagues (2012), who specify that emotions are produced through our interaction with family, culture, and environment.

It can be argued that the six basic feelings developed through evolution can be reduced to only two fundamental behavioral reactions—approach and avoidance. At all levels of development, from small organisms through snakes and mammals to humans, survival depended on knowing when and how to obtain food and reproductive opportunities and avoiding danger of all types. Drawing from this, it is helpful to think again of the HPA and TAP axes (Figure C.2). The HPA is the location of the amygdala and the seat of protective feelings, while the evolutionarily more recent TAP is deeply involved in basic positives, as well as defining the social emotions, as outlined above.

Accepting the above explanation, you can see the critical importance of a positive approach, stress management, and therapeutic lifestyle changes (TLC) in the counseling and therapy process. Building on this foundation, neuroscience research has offered exciting findings. For example, we can now identify specific neurons in the amygdala that affect anxiety, depression, and posttraumatic stress. These harmful networks remain in place unless treated effectively. Using classic behavioral methods derived from Pavlov's work, evidence is that we can change the power of these neurons "through presenting the feared object in the absence of danger." Medications can do the same thing as counseling and therapy if targeted to specific "intercalated neurons" (Ekaterina, Popa, Apergis-Schoute, Fidacaro, & Paré, 2008). These are very clear examples of the approach/avoidance hypothesis. The implications for counseling and therapy is one that we have frequently made—a positive wellness approach is the most effective way to help clients deal with their issues.

The amygdala has complex responses to our social environment, enlarging with some experiences and decreasing in size with others. For example, the broader your social environment, the larger your amygdala will be (Bickart, Wright, Dautoff, Dickerson, & Barrett, 2011). Trauma has the opposite effect. When 24 traumatized women diagnosed with borderline personality disorder were compared with 25 healthy controls, it was found that their amygdala was reduced in size by 22% and the hippocampus by 11%. In some cases, significantly impaired cognitive performance was also noted (Weniger, Lange, Sachsee, & Irle, 2009).

The Nobel Prize–winning psychologist Daniel Kahneman (2011) carries this discussion a bit further. He comments that our emotional likes and dislikes determine what we believe about the world—politics, irradiated food, global warming, motorcycles, tattoos. Once we are settled emotionally, it is difficult for us or our clients to change. *Many times, perhaps most of the time, our feelings and emotions guide our cognitions.* Again, this reinforces the importance of exploring and reflecting feelings and emotions. When we see a picture of a scary spider or view blood and gore, the amygdala and negative feelings are activated—both verbally and nonverbally. With things that we like or have a deep interest in, our pupils dilate with positive feelings.

▶The HPA and TAP and Other Structures

No cognition without emotion. No emotion without cognition.

—Jean Piaget

We have made a strong case in this book for the importance of limbic HPA emotion in counseling and therapy. At the same time, we also emphasize the TAP as our control station and the prefrontal cortex as the CEO making decisions and monitoring the more capricious HPA. We can also think of the executive TAP as the critical factor in emotional regulation.

Think of the *hypothalamus* as a "switching station" for the HPA in which messages from inside and outside are transferred; it controls hormones that affect sex, hunger, sleep, aggression, and other biological factors. The *pituitary* is a "control" gland that relates to the hypothalamus. It also influences growth, blood pressure, sexual functioning, the thyroid, and metabolism. The *adrenal glands* produce the all-important *corticosteroids*, including *cortisol* (potentially damaging due to stress, but also necessary for stimulating memory). The neurotransmitter *epinephrine* (also known as adrenaline or norepinephrine) regulates heart rate and the fight or flight response.

The *thalamus* is also a switching station focusing more on the TAP, relaying sensorimotor signals to the cerebral cortex. It also regulates consciousness, sleep, and alertness. It is central in emotional regulation as well, because of its connections with the limbic HPA.

The *anterior cingulate cortex* is a "collar" around the corpus callosum relaying neural signals between the left and right brain (discussed below). It regulates important cognitive functions, including decision making, empathy, and emotion. In addition, it plays a part in blood pressure and heart rate. It "links the body, brainstem, limbic, cortical, and social processes into one functional whole" (Siegel, 2007, p. 38).

Mirror neurons, located primarily in the motor and parietal cortexes (Rizzolatti & Craighero, 2004), are of particular interest for counseling and human change as they enable us to understand the actions of others and thus learn from them. (We discuss mirror neurons and counseling in Chapter 3.) You can get a better sense of how mirror neurons work if you notice what occurs in your body when you see an exciting ballgame or an involving movie. Many of us find ourselves tensing up and clenching our fists in close or exciting situations. We may even sway as the pass receiver grabs the ball and heads down the field. In good movies that touch you emotionally in some way, the same thing happens. Your heart rate goes up and you may duck a swing from the villain as you sit on the edge of your seat.

The distinction between the left and right hemispheres of the brain is oversimplified in popular discussion. The *corpus callosum* connects the two sides, while the switching station thalamus is folded in. It is clear that the two sides work together, and their differences and similarities go far beyond the common generalization of the linear left brain and the supposedly more interesting, intuitive, and creative right brain. Recall that the executive left brain is associated with positive emotions while the right is more associated with the negative. The left brain modulates the negative, and the cognitive CEO decides what to do about challenging issues. LaCombe and McGraw (2010) sensibly summarize the issues:

> Note that both concepts are fictions—there is no such thing as a right brain, separate and independent of a left one. They are not in conflict, nor is one "superior" to the other. For example, when you're having a conversation, your left brain will focus on the meaning of the words exchanged, while your right brain will observe whether the other person gets what you're saying.
>
> We use the terms "left brain" and "right brain" to refer to the two basic ways the brain processes information. Thus, when we say that someone is "in their right brain" it simply means that they are processing their experience in a holistic, "big picture" way.
>
> In fact, no one functions on one side or the other all the time. It's just that distinguishing these two brain functions helps you to understand how some counseling techniques work and why they are effective.

The *nucleus accumbens* (not shown in Figure C.2) is our pleasure center; it plays a part in reward, laughter, addiction, aggression, fear, and the placebo effect. GABA, the inhibitory neurotransmitter, is produced here, as is acetylcholine, which transmits information throughout the brain and the peripheral nervous system. The nucleus accumbens is significant in sexual functioning and the "high" from certain recreational drugs, which increase the supply of pleasurable dopamine. Key in understanding addiction, the nucleus accumbens is particularly responsive to marijuana, alcohol, and related chemicals. It also provides a partial explanation of the challenges we face in counseling around sex addiction and, speculatively, with those who engage in stalking behavior. One of our great challenges is helping these clients examine and rewrite their stories and find new actions through healthy alternative highs to replace the rewards of addiction. When you find these clients developing new life satisfactions and interests (wellness), you are influencing them toward behavior that can result in new positive responses in the nucleus accumbens and other parts of their brain.

You will find clients who tend to operate spontaneously, on the spur of the moment. They may be impulsive, and it gets them in trouble. They may be creative, but have difficulty organizing their many ideas. They may be out of control, with overmedication or drugs. First, it is important to join them in their stories and understand their emotions. Then, more linear and cognitive theories such as cognitive behavioral therapy or motivation interviewing may be useful. For those caught in obsessive cognitive thinking and rumination, emotion-focused therapies may be helpful. The surprise of an effective confrontation or providing useful factual information via psychoeducation, stress management, and therapeutic lifestyle changes may make the difference.

Emotion may underlie cognition, but new cognitions change emotional experience as the executive TAP regulates our feelings. We believe that cognitive behavioral therapy will be most effective when relationship and emotional issues are also considered. Rogerian person-centered therapy needs a balance of feelings and cognitions. Decisional counseling at first glance is very cognitive, but, as we saw in Chapter 14, decisions made without consideration of emotion are likely to be unsatisfactory. There is a need for a balance between the limbic HPA and the executive TAP.

▶Neurons, Neural Networks, and Neurotransmitters

Neurons that fire together wire together.

—Donald Hebb

Sigmund Freud was a young medical student at the University of Vienna when he realized that the brain was composed of cells. He then predicted the future and the reality of neurotransmitters when he called the gap between cells *contact barriers*. In 1895, in his "Project for a Scientific Psychology," Freud presented his model of the brain and mind, even describing neurons responsible for consciousness, memory, and perception (Freud, 1895/1953).

Modern research estimates indicate that there are between 85 and 100 billion *neurons*, or nerve cells, in the brain. Neurons are connected through *synapses* to other neurons in *neural networks*. A neural network and its connections with others is shown in Figure C.3, followed by an enlarged representation of the end of the neuron connecting to another neuron via the synapse and transmission of neurotransmitters. At the center of the neuron is DNA. Neurotransmitters are chemical molecules that transmit signals from one neuron to another. Without neurotransmission, nothing happens in the brain or body—no movement, no learning. The interaction in counseling and therapy affects the transmission of neural impulses just like medication such as Prozac or drugs such as alcohol, marijuana, and cocaine. Both counseling and medication can increase neurogenesis, while there is evidence that drugs destroy neurons.

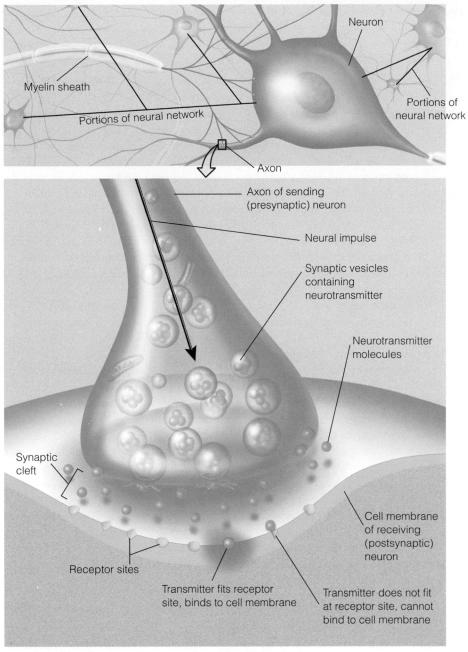

FIGURE C.3 Neural network, neurons, and neurotransmitters.

From W. Weiten, *Psychology*, 7th ed. © 2008 Wadsworth, a part of Cengage Learning. Reproduced by permission. www.cengage .com/permissions

As in all our discussions of the brain, neurotransmission is more complex than indicated here. For a more elaborate presentation, we recommend viewing one or more of the following YouTube videos:

How Neurotransmission Works (90 seconds,
    www.youtube.com/watch?v=p5zFgT4a0fA)

Neurotransmission—3D Medical Animation (38 seconds;
    www.youtube.com/watch?v=cNaFnRKwpFk)
2.1 Neurotransmission (more detailed, 61 minutes;
    www.youtube.com/watch?v=KEuJFb_mVUw)

"Neurons that fire together wire together." It takes more than one neuron to produce significant change. The *neural network* related to a single neuron is shown at the top of Figure C.3. However, this is a simplified picture as neurons are involved in multiple, complex, interacting networks. Neurons fire when we have any type of experience or stimulus, including the counseling interview. The neurons and neural net are where learning from counseling and therapy ultimately takes place through the transmission of neurotransmitters. If strong or frequent enough, this information becomes part of memory in the hippocampus. You can have a large influence on the developing brain through neuroplasticity. Your counseling skills and strategies can facilitate the movement of neurotransmitters and encourage strengthened neural connections. In our language, we call that learning or change. We also can measure this through positive development on the Client Change Scale. As a person learns, we can now see changes in brain scans such as the PET and fMRI. As research evolves and becomes more precise, scans may become key diagnostic instruments and even show that your work has actually affected specific areas of the brain.

The nerve impulse travels through the axon of the sending neuron and, if of sufficient strength, impels the *synaptic vesicles* to release the transmitter. The chemical neurotransmitter molecules then enter the synapse or synaptic cleft, where they then seek to bind with their unique *receptor sites* in the next neuron. There are more than 100 identified neurotransmitters, each with its own unique set of receptors.

The receptor sites can be fooled if a foreign chemical enters the bloodstream. For example, alcohol influences several transmitters including pleasurable dopamine, enhances inhibitory GABA, inhibits the excitatory learning glutamate, and promotes an endorphin high. Along with these good feelings come less control, less attention to consequences, and reduced effectiveness of motor control and cognition.

Psychiatric medications focus on and influence the action of transmitters and thus neural nets and on to memory and behavior changes. As an example, consider selective serotonin reuptake inhibitors (SSRIs, such as Prozac) used for depression. These medications, which mimic serotonin neurotransmitters, can often alleviate major depression, although evidence for SSRI effectiveness with moderate of minor depression is more equivocal (Fournier et al., 2010). Many people consider medications a mixed blessing. For example, a study of depression among 7,696 pregnant women found that untreated women had babies with slower rates of head and body growth. Women treated with SSRIs had reduced depression, but child head growth was again delayed and they were at higher risk for preterm babies (Marroun et al., 2012). At another level, severe mental and emotional distress such as schizophrenia has been targeted with varying success with an increasing number of antipsychotics (first-generation Haldol and Thorazine, atypical second- and third-generation Risperdal, Zyprexa, and Abilify). The primary focus of these powerful antipsychotics is dopamine, but some also affect serotonin, noradrenaline, and acetylcholine. However, antipsychotics have been found to reduce gray matter and brain volume (Ho, Andreasen, Ziebell, Pierson, & Magnotta, 2011; Lewis, 2011). We recommend against these drugs wherever possible, particularly now as pharmaceutical companies are starting to market them extensively for both children and the elderly. Both of these groups are particularly vulnerable to harm.

Given the mixed findings on medications, we need to consider whether effective therapy will influence outcome. Cognitive behavioral therapy (CBT) has often been found as effective or more effective with depression than medication, including more success at 6-month follow-up (e.g., Fava et al., 1998). A study of posttraumatic stress survivors found that 12 weeks of CBT were more effective than medications, which had no impact (Shalev et al., 2012). Research with teens who were in danger of psychosis found that CBT coupled with a broad array of supplementary preventions (including nutrition, family counseling, and social skills education) significantly reduced the number who actually become psychotic. Several studies of this type have been conducted by Patrick McGorry and his staff at Royal Melbourne Hospital in Australia and, by extension, throughout the world. In these programs, an attempt is made to limit medications and to avoid antipsychotics if at all possible. Research consistently attests to effective improvement in at-risk teens (McGorry, 2012). However, the fifth edition of the *Diagnostic and Statistical Manual* uses the words "attenuated psychosis syndrome" rather than high risk (c.f. Woods, Walsh, Saksa, & McGlashan, 2010). This nomenclature tends to pathologize teens and suggests that antipsychotics are much more likely to be used than prevention strategies and counseling.

While this book has given attention to broad issues of neuroscience and the impact of counseling on the brain, here we will take a risk and say that affecting neurotransmitters through *effective and quality* counseling and therapy is where the "rubber hits the road." As we have said, new neural networks through counseling is change—the creation of the *New*. Science and the art of counseling come together at this point.

Consider Table C.1 as a beginning presentation showing how your practice can influence neurotransmitters and produce change and create the *New*. Art becomes science and science becomes art.

TABLE C.1 Neurotransmitters and Possible Treatment Strategies

Neurotransmitter	Possible Impact of Counseling and Therapy
Glutamate—most important brain excitatory neurotransmitter, vital for neuroplasticity, movement, memory, and learning. Moderates neural firing. Monosodium glutamate (MSG), chemically close to glutamate, can cause problems for some people.	Generally, we want to increase this central neurotransmitter. Exercise facilitates glutamate production. Stress management and wellness activities are useful for balancing. Preliminary evidence of glutamate abnormalities in depression and schizophrenia. (Medications—glutamate uptake inhibitors)
GABA (gamma-aminobutric acid)—inhibitory, prevents neurons from becoming too active and regulates neuron firing. Important in limbic system and amygdala. Alcohol and barbiturates increase GABA, which results in lowered sensitivity to stimuli, plus cognitive and sensorimotor issues.	Calming strategies of CBT stress counseling, meditation, and the here-and-now emphasis are likely to be useful and increase the release of GABA. The basic listening sequence will help clients as you listen to their stories. (Medications—minor tranquilizers, antianxiety medications, lithium for low GABA)
Dopamine—attentional processes, pleasure, memory, reward system, fine motor movement. Addictive substances increase release. Low dopamine common in depression.	Seek to help clients find joy in their lives. Therapeutic lifestyle changes and counseling focused on stories of strengths and positive narratives should help dopamine production. All effective restorying should improve dopamine release as we move away from depression and ineffective behavior. (Medications—dopamine reuptake inhibitors [NDRIs] act as antidepressants.)
Serotonin—vital to mood, sleep, anxiety control, and self-esteem. Implicated in depression, impulsiveness, and anger/aggression.	Think of the serotonin "high" of running. Get clients moving. It is hard to be depressed when one is exercising. Wellness, meditation, cognitive behavioral counseling, and finding clear visions and meaning for life should be helpful. Positive restorying and action following the interview are important. (Medications—SSRIs permit more transmission. Ketamine, also known as the dangerous hallucinogenic street drug Special K, has been shown experimentally to improve depression rapidly.)

(continued)

TABLE C.1 (continued)

Neurotransmitter	Possible Impact of Counseling and Therapy
Norepinephrine (closely related to epinephrine, also known as adrenaline)—released immediately in stress, but also makes one cognitively and physically more aware and active. Involved in heart rate and helps new information transfer to long-term memory. With too much, damaging cortisol is released. Related to anxiety, depression, and bipolar diagnosis.	Again, get clients active and moving. When needed, use stress management, decisional counseling, and CBT to lessen stress. As always, telling one's story in a relationship of caring is calming. People can become addicted to an adrenaline high—you may have seen this in runners and even people overinvolved and excited at work. Finding meaning should help clients meet the challenges of life more effectively. (Medications—SSRIs, sometimes coupled with dopamine as an antidepressant)
Anandamide—impacts cannabinoid receptors (yes, that is what they are called), marijuana, affects nucleus accumbens, the brain's pleasure center. Involved in addictive behavior. Tetrahydrocannabinol (THC)—the active ingredient of marijuana activates receptors.	Motivational interviewing likely the most effective strategy as it attacks addictive issues directly (visit the accompanying website for more specifics). The client has enjoyed the "highs" of drugs and needs alternative approaches to find positives and strengths in life. Referral to Alcoholics Anonymous or support groups focusing on other issues (e.g., sexual, drugs, and other addictions) likely to be helpful. Marijuana appears to be helpful with many medical issues and Alzheimer's disease, but it has also been shown to increase teen suicide and potential for psychosis. (Medications—none available, but some in trials)
Acetylocholine (ACH)—first neurotransmitter to be discovered. Affects memory, cognitive functioning, emotion, aggression, central nervous system. Loss of ACH is a central indicator of Alzheimer's disease.	Exercise, meditation, social relationships, positive activities can slow Alzheimer's. Your work with families will be central to help make decisions and support the client appropriately. Work with clients to help them deal with this increasingly common challenge of life. (Medications—Aricept™, cholestine inhibitors; many new medications in advanced testing stages)
Enkephalins and endorphins—Endogenous morphine-like peptides such as enkephalins and beta-endorphins are present within the central nervous system. Endorphins are released in response to pain or sustained exertion. They serve as internal analgesics and seem to have a role in appetite control.	Pain management has an increasing role in counseling and therapy through therapeutic lifestyle changes such as relaxation, exercise, and meditation, which modulate pain, reduce stress, and produce a sensation of calm. Meditation and mindfulness training and related counseling strategies have been found very useful for pain relief and are considered preferable to potentially addictive pain relievers, which usually have side effects. (Medications—an array of over-the-counter and prescription pain and headache relievers, including codeine and morphine)

►Microskills and Their Potential Impact on Change

The microskills of attending, observation, and the basic listening sequence are vital for the communication of empathy. We start with the biological possibility of "feeling the feelings" of others because of mirror neurons. Through our childhood and later developmental experiences, we become more or less attuned to others. Neuronal structures of empathic understanding can pass away if not nourished. In turn, the teaching of empathy, particularly through the listening skills, may be helpful in human change. Moreover, if you are empathic with a client, you are helping that person become more understanding of others.

A classic study by Restak (2003, p. 9) found that training volunteers in movement sequences produced sequential changes in activity patterns of the brain as the movements became more thoroughly learned and automatic. Systematic step-by-step learning, such as that emphasized in this book, is an efficient learning system also used in ballet, music, golf, and many other settings. If there is sufficient skill practice, changes in the brain may be expected, and increased ability in demonstrating these skills will appear in areas ranging from finger movements to dance—and from the golf swing to counseling skills. Table C.2 presents a summary of how various microskills relate to the learning process involved in counseling and psychotherapy.

TABLE C.2 Key Microskills Concepts and Neuroscience	
Microskills Concept	**Some Issues Related to Neuroscience and Neuropsychology**
Attending behavior	Attention is measurable through brain imagining and is now recognized as the central process within the brain by many. When client and counselor attend to the story, the brains of both counselor and client become involved. Factors in attention are arousal and focus. Arousal involves the brain's core, which transmits stimuli to the cortex and activates neurons firing throughout many areas. Selective attention "is brought about by . . . a part of the thalamus, which operates rather like a spotlight, turning to shine on the stimulus" (Carter, 1999, p. 186). If you listen with energy and interest, and this is communicated effectively, expect your client to receive that attention as a positive resource in itself. Attention is central to functioning of the CEO prefrontal cortex. Where our attending is directed influences not only cognitions, but also emotions and feelings. Attention to positives reduces the effect of negative issues in one's life.
Questions	New histories and stories are written in the counseling session. The very asking of questions affects old memories stored in the hippocampus. Creating "new history is influenced by current determinants of neural experience, and such factors are usually very different from those that affected the original experience a long time ago" (Grawe, 2007, p. 67).
Observation	As you learn to observe your client more effectively, your brain is likely developing new connections. Expect your multicultural learning to become one of those new connections. Japanese have been found to be more holistic thinkers than Westerners. Expect different cognitive/emotional styles when you work with people who are culturally different from you—but never stereotype!
	Blacks and Whites both exhibit greater brain activation when they view same race faces and less when race is different. We tend to feel more comfortable when people are like us. This suggests that discussing racial and other cultural differences early in the session can be a helpful way to build trust. Interestingly, similar findings exist for political persuasions.
	Expressions can transmit emotions to others. If you smile, the world does indeed smile with you (up to a point). Experiments in which tiny sensors are attached to the "smile" muscles of people looking at faces have shown that the sight of another person smiling triggers automatic mimicry—albeit so slight that it may not be visible. The brain concludes that something good is happening out there, which creates a feeling of pleasure.

(continued)

TABLE C.2 (continued)

Microskills Concept	Some Issues Related to Neuroscience and Neuropsychology
Encouraging, paraphrasing, and summarizing	Active listening is a key aspect of relationship—consider the importance of listening to wellness strengths as well as client challenges. Similarly, if you listen to problems only, expect the nerve cells to communicate that as well.
Reflection of feeling	Traditional categories of feeling (sad, mad, glad, fear) appear in brain imaging. The limbic system organizes bodily emotions and includes the amygdala, hypothalamus, thalamus, hippocampal formation, and cortex. The cortex receives this information and determines how to name feelings and what can be done about them, thus regulating emotions. The central feelings of fear are located in the amygdala, which also transmits the intensity of emotions. In times of emergency or impulse, the limbic system can and will overcome the judgments of the cortex. Reflection of feeling is basic to communicating empathy. The counselor's mirror neurons "light up" when hearing the emotions and stories of the client.
Confrontation	What some call "creativity" may be located in the connections between the intuitive right brain and the linear left brain, with the participation of the mainly unconscious limbic system. The executive left brain provides the final spark integrating creativity and the creation of the New. While not yet final, many believe that new learning (neuroplasticity) occurs when the two hemispheres synchronize their activity (Goodwin, Lee, Puig, & Sherrard, 2005; Puig, Lee, Goodwin, & Sherrard, 2006). A confrontation that points out incongruities in a person's life is, by definition, an effort to open the person to a new way of thinking. Gentle and supportive confrontations often can reach underlying emotional structures as the empathic atmosphere provides the setting for creative new learning.
Focusing	Client selective attention is guided by existing patterns in the mind; focusing is an intentional skill that can open up more possibilities for client thoughts, feelings, and actions. A number of regions of the prefrontal cortex are activated during attentional task preparation and execution. Self-regulation and understanding of others (empathy) are deeply affected by attentional systems. Focusing will also help clients learn new behaviors.
Reflection of meaning	Meaning is closely allied to emotion. Depression is marked by wide-ranging symptoms, but its cardinal feature is the draining of meaning from life. By contrast, those in a state of mania see life as a gloriously ordered, integrated whole. The area of the brain that is most noticeably affected in both depression and mania is an area on the lower part of the internal surface of the prefrontal cortex, the brain's emotional control center. It is exceptionally active in bouts of mania and inactive (along with other prefrontal areas) during depression. Ratey (2008a, p. 41) has commented: "You have to find the right mission, you have to find something that's organic, that's growing, that keeps you focused on and continues to provide meaning and growth and development for yourself. . . . Spirituality even lights up key centers in the brain. Meaning drives the lower centers and is connected to emotions and motivations areas. . . . If you can get people into a situation where they have the meaning direction provided by their mission or their job or their goal, they don't need medicine."

Microskills Concept	Some Issues Related to Neuroscience and Neuropsychology
Interpretation/reframe, logical consequences, instruction/ psychoeducation, directives	Some clients enter therapy with negative emotions and amygdalas on over-drive, anxiously fearing what the therapy relationship will hold. In the language of neurobiology, the aim is to reduce this hyperactivity and bolster the activity of the nucleus accumbens, a brain area associated with pleasure.
	We need to activate cortical functions where positive thoughts and feelings are generated so that we can deal effectively with issues and concerns. For example, cognitive therapy can encourage left brain activity to gain control over negative emotions. The influencing skills and strategies, used effectively, provide clients with specific things they can do to build more positive thoughts, feelings, and behaviors. In this way, clients can deal more effectively with their issues.
	Under times of severe stress or panic, the amygdala can take over. Thus, you will find many clients who fail to use more positive memories and personal skills to counteract negativity. We need to build positive emotions to cope with the negative. Building on wellness and strengths will enable clients to cope with major challenges. Some people even speculate that practitioners in the not too distant future will be able to tailor specific treatments to modify brain circuits through counseling, medication, meditation, or other positive interventions.

▶The Brain's Default Mode Network: What's Happening When the Brain Is at Rest?

A great deal of meaningful activity is occurring in the brain when a person is sitting back and doing nothing at all.

—Marcus E. Raichle

Active brain, also called the *task positive brain* (TPB), has been the focus of this book and this brief summary of key brain structures. The TPB is concerned with daily life, doing and acting, thinking, feeling, and behavior. Surprisingly, the resting brain, the default mode network (DMN), is even more active. It was not until 2001 that the brain at rest was studied seriously (Raichle & Snyder, 2007). Our active TPB uses only 10% of the brain's potential energy, while the DMN consumes many times more than that. Virtually anything we do, from swatting a fly to writing a complex paper or hitting a tennis ball, requires less effort from the brain than what occurs at the default level.

The integrative pathways of the connectome, discussed in the early paragraphs of this appendix, provide important clues as to the nature and importance of the default mode network. Increasingly, neuroscientists believe that dyssynchrony among the paths and neural networks leads to mental (and physical) ill health. Hoffman (2012) has stated that "Changing our brains is about changing our connectome." He believes that the connectome and the default network represent a turning point in human history. Experience changes our brain, as do genes. Rewiring is what we do in counseling; therapy, neurofeedback, stimulation are all about rewiring.

Typical activities of the brain at rest are self-referential, such as introspection, mind wandering, daydreaming, memories, and thinking about the perspectives of others. Any thoughts not directly associated with the immediate here and now are believed to be basic to the DMN. It is believed that this is how our brain consolidates information, because so much information comes into our brain that we cannot deal consciously with it all. We suggest that you visit a 2-minute introduction to the DMN at http://www.youtube.com /watch?v=6A-RqZzd2JU; you can also search www.youtube.com using the words "resting brain" or "default mode network."

Many people believe that the DMN represents much of unconscious thought validating much of what Freud said about the mind (e.g., Viamontes & Beitman, 2007). A child's developmental experiences are imprinted in long-term memory and organized in the wandering mind of the default mode. Throughout our lives as adults, these and new experiences constantly change memories in our hippocampus and are part of the neurogenesis process. This complex array of sensory information is abstracted and forms the substrate of the unconscious. Only a small amount of subconscious data ever appear in conscious thought in the task positive brain. Yet it is always there at some level in the DMN, ultimately influencing conscious decisions. Some people go so far as to suggest that consciousness and individual agency may not exist (Ivey, 1986/2000; Lacan, 1977).

There is evidence that the default networks of people diagnosed with autism, schizophrenia, depression, ADHD, and Alzheimer's are distinctly different from others (Hoffman, 2012; Othmer, 2012). People with these challenges appear to have an overactive DMN. Some theorists believe that assessment of the DMN can lead to therapeutic change, and they consider neurofeedback a promising therapeutic mode (see page 310). For example, one study found that 40 sessions of neurofeedback served a calming function, smoothing out the brain waves of ADHD children as seen in fMRI evaluation. ADHD children are constantly active and changing what they are doing. The calming from neurofeedback also resulted in significant differences in behavior at school and at home (Russell-Chapin et al., in press).

Research on the DMN is expanding rapidly, and it is anticipated that it will have a powerful impact on theories of personality and human development, as well as on counseling and therapeutic practice. There is evidence through fMRI study that different personality types have varying patterns of activity in the DMN. Five personality styles (extraversion, neuroticism, openness/intellect, agreeableness, and conscientiousness) were identified (Sampson, Soares, Coutinho, Sousa, & Gonçalves, 2012):

> Extraversion and agreeableness were positively correlated with activity in the midline core of the DMN, whereas neuroticism, openness, and conscientiousness were correlated with the parietal cortex system. Activity of the anterior cingulate cortex (ACC) was positively correlated with extraversion and negatively with conscientiousness. Regions of the parietal lobe were differentially associated with each personality dimension.

Thus, it may be possible to assess personality via fMRI! This is powerful preliminary evidence supporting the earlier ideas around the unconscious of Viamontes and Beitman outlined above.

▶Social Justice, Stress Management, and Therapeutic Lifestyle Changes

Poverty in early childhood poisons the brain. . . . neuroscientists have found that many children growing up in very poor families with low social status experience unhealthy levels of stress hormones, which impair their neural development. The effect is to impair language development and memory—and hence the ability to escape poverty—for the rest of the child's life.

—Paul Krugman

Stress is an issue in virtually all client issues and problems. It may not be the presenting issue, but be prepared to assess stress levels and provide education and treatment as needed, using some of the strategies of Chapters 13 and 15. Stress will show in body tension and nonverbal behavior. Cognitive/emotional stress is demonstrated in vocal hesitations, emotional difficulties, and the conflicts/discrepancies clients face in their lives.

A moderate amount of stress, if not prolonged, is required for development and for physical health. For example, repeated stressing of a muscle through weight lifting breaks down muscle fibers, but after rest the rebuilding muscle gains extra strength. A similar pattern occurs for running and other physical exercise. If there is no stress, neither physical development nor learning will occur. Stress should also be seen as a motivator for focus and as a condition for change. The amygdala needs to be energized. For example, the skill of confrontation can result in stress for the client. But used as a supportive challenge, this stressor becomes the basis for cognitive and emotional change. In all of the above, note the word *moderate* and the need for rest between stressors.

Toxic and long-term stress is damaging. *"Cortisol* is the long-acting stress hormone that helps to mobilize fuel, cue attention and memory, and prepare the body and brain to battle challenges to equilibrium. Cortisol oversees the stockpiling of fuel, in the form of fat, for future stresses. Its action is critical for our survival. At high or unrelenting concentrations such as post-traumatic stress, cortisol has a toxic effect on neurons, eroding their connections between them and breaking down muscles and nerve cells to provide an immediate fuel source" (Ratey, 2008b, p. 277).

An article entitled "Excessive Stress Disrupts the Architecture of the Developing Brain" (National Scientific Council on the Developing Child, 2005) includes the following useful points:

1. In the uterus, the unborn child responds to stress in the mother, while alcohol, drugs, and other stimulants can be extremely damaging.
2. For the developing child neural circuits are especially plastic and amenable to growth and change, but again excessive stress results in lesser brain development and in adulthood that child is more likely to have depression, an anxiety disorder, alcoholism, cardiovascular problems, and diabetes.
3. Positive experiences in pregnancy seem to facilitate child development.
4. Caregivers are critical to the development of the healthy child.
5. Children of poverty or who have been neglected tend to have elevated cortisol levels.

If you review the sections above, particularly those talking about the frontal cortex and limbic system, you can obtain some sense of what poverty and challenges such as racism and oppression do to the brain. Incidents of racism place the brain on hypervigilance, thus producing significant stress, with accompanying hyperfunctioning of the amygdala and interference with memory and other areas of the brain. We need to be aware that many environmental issues, ranging from poverty to toxic environments to a dangerous community, all work against neurogenesis and the development of full potential. And let us expand this to include trauma.

Recall that the infant, child, and adolescent brain can only pay attention to what is happening in the immediate environment. Again, think of the varying positive and negative environments that your clients come from. One of the purposes of the community geno-gram is to help you and the client understand how we as individuals relate and are related to individuals, family, groups, and institutions around us. The church that welcomes you helps produce positive development; the bank that refuses your parents a loan or peers that tease and harass you harm development.

Clients need to be informed about how social systems affect personal growth and individual development. Our work here is to help clients understand that the problem does not lie in them but in a social system or life experience that treated them unfairly and did not allow an opportunity for growth.

Finally, there is social action. What are you doing in your community and society to work against social forces that bring about poverty, war, and other types of oppression?

Are you teaching your clients how they can work toward social justice themselves? A social justice approach includes helping clients find outlets to prevent oppression and work with schools, community action groups, and others for change.

▶Looking to the Future

Neuroscience research provides a biological foundation for understanding the impact of our work. *The very act of counseling and therapy produces changes in client memory (and your own).* Always be aware that new ideas and learning are being constructed in the session. We suggest that you continue to study and learn about brain structures and functions, as new findings may provide further support for our work and suggest specific guidelines for practice.

Brain research is not in opposition to the cognitive, emotional, behavioral, and meaning emphasis of counseling and therapy. Rather, it will help us pinpoint types of interventions that are most helpful to the client. In fact, one of the clearest findings is that the brain needs environmental stimulation to grow and develop. You can offer a healthy atmosphere for client growth and development. We advocate the integration of counseling, psychotherapy, neuroscience, molecular biology, and neuroimaging and the infusion of knowledge from such integrated fields of study to practice, training, and research.

▶Recommended for Further Study

Listed below are materials that have informed and shaped our integration of neuroscience and counseling. The first five listed include three researchers (Sapolsky, Decety, and Jackson) and two theorist/clinicians (Grawe and Ratey). All of them present neuroscience in an accessible and interesting manner, although Klaus Grawe is more of a challenge. Several other key scholars and researchers are included later in this section, and many others are well worth examining.

Sapolsky, R. (2012). *Biology and human behavior: The neurological origins of individuality* (2nd ed.) [Audio]. Chantilly, VA: The Teaching Company. (www.thegreatcourses .com; en.wikipedia.org/wiki/Robert_Sapolsky)
This is where Allen began his study of neuroscience. Sapolsky is an outstanding presenter and provides the basics in a clear and understandable format. Updated frequently, so check for most recent edition.

Grawe, K. (2007). *Neuropsychotherapy: How the neurosciences inform effective psychotherapy.* Mahway, NJ: Erlbaum. Grawe, sadly, died before this book was published, but read his obituary at www.psychotherapyresearch.org/displaycommon .cfm?an=1&subarticlenbr=50.
This is *the* book that best provides specifics of connections between neuroscience and therapy. Allen has read this book three times. In neuroscience, 2007 is a long time ago, but Grawe was very far ahead of our field.

Decety, J., & Jackson, P. (2004). The functional architecture of human empathy. *Behavioral and Cognitive Neuroscience Reviews, 3*(2), 71–100. (Available at www .herasaga.com/wp-content/uploads/2009/12/Decety-Jackson-Empathy.pdf; see also en.wikipedia.org/wiki/Jean_Decety.)
Decety and Jackson remains what we consider the best article on empathy and neuroscience. They know counseling and therapy research and tie it clearly to workings of the brain.

Ratey, J. (2008) *Neuroscience and the brain: Implications for counseling and therapy* [Video]. Framingham, MA: Microtraining Associates.

This video provides the "guts" of brain structure and science.

Ivey, A. (2009). *Counseling and neuroscience: Implications for microskills and practice* [Video]. Framingham, MA: Microtraining Associates. (www.emicrotraining.com) Allen discusses his integration of neuroscience and counseling with a focus on microskills.

Following is a list of other key authors you should investigate for theory, research, and practice. Please use Google or another search engine to find their most recent works and references. They will be updating and adding to their works often in the next few years. There is some debate regarding the use of Wikipedia as a source for scholarship. However, the websites listed below serve two useful purposes: (1) they will acquaint you with the life work of the scholar, and (2) you will find key references for further reading beyond the specific suggestions here. Knowing the life story and narratives of those whom many consider the main influencers will provide you with a historical and contextual sense of neuroscience. These brief web presentations will also point you to future developments in the field. You also can find YouTube presentations for all these authors.

Richard Davidson—meditation and the brain
     (en.wikipedia.org/wiki/Richard_J._Davidson)
Antonio Damasio—excellent on emotion and meaning
     (en.wikipedia.org/wiki/António_Damásio)
Jon Kabat-Zinn—brought mindfulness meditation to center stage
     (en.wikipedia.org/wiki/Jon_Kabat-Zinn)
Eric Kandel—a pioneer in neuroscience and practice
     (en.wikipedia.org/wiki/Eric_Kandel)
Michael Merzenich—foremost researcher on brain plasticity and change
     (en.wikipedia.org/wiki/Michael_Merzenich)
Daniel Siegel— interpersonal neurobiology and mindfulness
     (en.wikipedia.org/wiki/Daniel_Siegel)

REFERENCES

Abe, N., Okuda, J., Suzuki, M., Matsuda, T., Mori, E., Minoru, T., et al. (2008). Neural correlates of true memory, false memory, and deception. *Cerebral Cortex, 18*, 2811–2819.

Ahmad, T., Chasman, D., Buring, J., Lee, I., Ridker, P., & Everett, B. (2011). Physical activity modifies the effect of LPL, LIPC, and CETP polymorphisms on HDL-C levels and the risk of myocardial infarction in women of European ancestry. *Circulation: Cardiovascular Genetics, 4*, 74–80.

American Counseling Association. (2005). *Code of ethics.* Alexandria, VA: Author. Retrieved from www.counseling.org /files/fd.ashx?guid=ab7c1272-71c4-46cf-848c-f98489937dda

American Psychiatric Association. (2000). *Diagnostic and statistical manual of mental disorders* (4th ed., Text Revision). Washington, DC: Author.

American Psychological Association. (2002). *Ethical principles of psychologists and code of conduct.* Washington, DC: Author.

American Psychological Association. (2010). *Ethical principles of psychologists and code of conduct.* Washington, DC: Author. Retrieved from www.apa.org/ethics/code/index.aspx

Ammentorp, J., Sabroe, S., Kofoed, P., & Mainz, J. (2007). The effect of training in communication skills on medical doctors' and nurses' self-efficacy: A randomized controlled trial. *Patient Education and Counseling, 66*, 270–277.

Amminger, G., Schäfer, M., Papageorgiou, K., Klier, C., Cotton, Harrigan, S., Mackinnon, A., McGorry, P., & Berger, G. (2010). Long-chain ω-3 fatty acids for indicated prevention of psychotic disorders: A randomized, placebo-controlled trial. *Archives of General Psychiatry, 67*, 146–154.

Anderson, P. (2012). Coffee may ward off progression to dementia. *Medscape Medical News.* Retrieved from www.medscape.com/viewarticle/765781

Arehart-Treichel, J. (2001). Evidence is in: Psychotherapy changes the brain. *Psychiatric News, 36*, 33.

Asbell, B., & Wynn, K. (1991). *Touching.* New York: Random House.

Aupperle, R. L. (2012, April). *Neural system function predicts response to cognitive trauma therapy in women with domestic violence-related PTSD.* Paper presented at the Anxiety Disorders Association of America (ADAA) 32nd Annual Conference.

Aupperle, R., & Hunt, A. (2012, April). *fMRI may predict response to cognitive behavioral therapy.* Paper presented at the Anxiety Disorders Association of America (ADAA) 32nd Annual Conference.

Back, A., Arnold, R., Baile, W., Fryer-Edwards, K., Alexander, S., Barley, G., et al. (2007). Efficacy of communication skills training for giving bad news and discussing transitions to palliative care. *Archives of Internal Medicine, 167*, 453–460.

Barrett, M., & Berman, J. (2001). Is psychotherapy more effective when therapists disclose information about themselves? *Journal of Consulting and Clinical Psychology, 69*, 597–603.

Barrett-Lennard, G. (1962). Dimensions of therapist response as causal factors in therapeutic change. *Psychological Monographs, 76*, 43 (Ms. No. 562).

Bensing, J. (1999a). *Doctor–patient communication and the quality of care.* Utrecht, Netherlands: Nivel.

Bensing, J. (1999b). The role of affective behavior. *Communication,* 1188–1199.

Bensing, J., & Verheul, W. (2009). Towards a better understanding of the dynamics of patient provider interaction: The use of sequence analysis. *Patient Education and Counseling, 75*(2), 145–146.

Bensing, J., & Verheul, W. (2010). The silent healer: The role of communication in placebo effects. *Patient Education and Counseling, 80*(3), 293–299.

Bickart, K., Wright, C., Dautoff, R., Dickerson, B., & Barrett, L. (2011). Amygdala volume and social network size in humans. *Nature Neuroscience, 14*, 163–164.

Blumenfeld, W., Joshi, K., & Fairchild, E. (Eds.). (2009). *Investigating Christian privilege and religious oppression in the United States.* Rotterdam, Netherlands: Sense Publishers.

Brammer, L., & MacDonald, G. (2002). *The helping relationship* (8th ed.). Boston: Allyn and Bacon.

Brauser, D. (2012). Junk food linked to depression. *Medscape Medical News.* Retrieved from www.medscape.com /viewarticle/762655

Bryner, J. (2012, May). 2-hour therapy cures spider phobia by rewiring the brain. *Scientific American.* Retrieved from www.scientificamerican.com/article.cfm?id=spider-phobia -cured-with-2-hour-therapy&page=2

Bylund, C., Brown, R., Lubrano di Ciccone, B., Levin, T., Gueguen, J., Hill, C., et al. (2008). Training faculty to facilitate communication skills training: Development and evaluation of a workshop. *Patient Education and Counseling, 70*, 430–436.

Camus, A. (1955). *The myth of Sisyphus.* New York: Vintage Books.

Canadian Counselling and Psychotherapy Association. (2009). *Code of ethics.* Ottawa: Author. Retrieved from www.ccacc.ca/_documents/CodeofEthics_en_new.pdf

Cao, C., Loewenstein, D., Lin, X., Zhang, C., Wang, L., Ranjan Duara, R., et al. (2012). High blood caffeine levels in MCI linked to lack of progression to dementia. *Journal of Alzheimer's Disease, 30,* 559–572.

Carkhuff, R. (1969). *Helping and human relations: Practice and research.* New York: Holt, Rinehart, Winston.

Carkhuff, R. (2000). *The art of helping in the 21st century.* Amherst, MA: Human Resources Development Press.

Carlson, M., Erickson, K., Kramer, A., Voss, M., Bolea, N., Mielke, M., et al. (2009). Evidence for neurocognitive plasticity in at-risk older adults: The Experience Corps program. *Journals of Gerontology Series A: Biological Sciences and Medical Sciences, 64*(12), 1275–1282.

Carlstedt, R. (2011). *Handbook of integrative clinical psychology, psychiatry, and behavioral medicine: Perspectives, practices, and research.* New York: Springer.

Carstensen, L., Pasupathi, M., Mayr, U., & Nesselroade, J. (2000). Emotional experience in everyday life across the life span. *Journal of Personality and Social Psychology, 79,* 644–655.

Carter, R. (1999). *Mapping the mind.* Berkeley: University of California Press.

Carter, R. (2010). *Mapping the mind* (Rev. ed.). Berkeley: University of California Press.

Centonzea, D., Siracusanoc, A., Calabresia, P., & Bernardia, G. (2004). The project for a scientific psychology (1895): A Freudian anticipation of LTP-memory connection theory. *Brain Research Reviews, 46,* 310–314.

Colozino, L. (2010). *The neuroscience of psychotherapy: Healing the social brain* (2nd ed.). New York: Norton.

Contrada, R., Goyal, T., Cather, C., Rafalson, L., Idler, E., & Krause, T. (2004). Psychosocial factors in outcomes of heart surgery: The impact of religious involvement and depressive symptoms. *Health Psychology, 23,* 227–238.

Corey, G., Corey, M., & Callanan, P. (2011). *Issues and ethics in the helping professions* (8th ed.). Belmont, CA: Brooks/Cole.

Croce, A. 2003. *Non-verbal communication causes cultural misconceptions.* Retrieved December 25, 2008, from www.ramcigar.com/media/storage/paper366/news/2003/02/28/Campus/NonVerbal.Communication.Causes.Cultural.Misconceptions-382215-page2.shtml

Cross, W. (1971). The Negro-to-Black conversion experience. *Black World, 20,* 9, 13–27.

Cross, W. (1991). *Shades of black: Diversity in African-American identity.* Philadelphia: Temple University Press.

D'Andrea, M., & Daniels, J. (2001). RESPECTFUL counseling: An integrative model for counselors. In D. Pope-Davis & H. Coleman (Eds.), *The interface of class, culture and gender in counseling* (pp. 417–466). Thousand Oaks, CA: Sage.

D'Zurilla, T. (1996). *Problem-solving therapy.* New York: Springer.

D'Zurilla, T., & Nezu, A. (2007). *Problem-solving therapy: A positive approach to clinical interventions* (3rd ed.). New York: Springer.

Damasio, A. (2003). *Looking for Spinoza: Joy, sorrow, and the feeling brain.* New York: Harcourt.

Daniels, T. (2014). A review of research on microcounseling: 1967–present. In A. Ivey, M. Ivey, & T. Daniels, *Intentional interviewing and counseling: CourseMate interactive website.* Belmont, CA: Brooks/Cole.

Daniels, T., & Ivey, A. (2007). *Microcounseling: Making skills training work in a multicultural world* (3rd ed.). Springfield, IL: Thomas.

Danner, D., Snowdon, D., & Friesen, W. (2001). Positive emotion in early life and longevity. *Journal of Personality and Social Psychology, 80,* 804–813.

Davidson, R. (2001). The neural circuitry of emotion and affective style: Prefrontal cortex and amygdala contributions. *Social Science Information, 40.*

Davidson, R. (2004). Well-being and affective style: Neural substrates and biobehavioral correlates. *The Philosophical Transactions of the Royal Society, 359,* 1395–1411.

Davidson, R., & McEwen, B. (2012). Social influences on neuroplasticity: Stress and interventions to promote well-being. *Nature Neuroscience, 15,* 689–695.

Davidson, R., Pizzagalli, D., Nitschke, J., & Putnam, K. (2002). Depression: Perspectives from affective neuroscience. *Annual Review of Psychology, 53,* 545–574.

Davis, D., & Hayes, J. (2012, July/August). What are the benefits of mindfulness? *Monitor on Psychology,* pp. 64, 66–70.

Davis, M., Eshelman, E., & McKay, M. (2008). *The relaxation and stress reduction workbook* (6th ed.). Oakland, CA: New Harbinger.

Decety, J., & Jackson, P. (2004). The functional architecture of human empathy. *Behavioral and Cognitive Neuroscience Reviews, 3,* 71–100.

Decety, J., & Jackson, P. (2006). A social-neuroscience perspective on empathy. *Journal of Association for Psychological Science, 15*(2), 54–58.

Dement, W. (2000). *The promise of sleep.* New York: Dell.

deWaal, E. (1997). *Living with contradiction: An introduction to Benedictine spirituality.* Harrisburg, PA: Morehouse.

Dietrich, A., & Kanso, R. (2010). A review of EEG, ERP, and neuroimaging studies of creativity and insight. *Psychological Bulletin, 136,* 822–848.

Dixon, D. (2011). Exercise psychotherapy. In R. Carlstedt (Ed.), *Handbook of integrative clinical psychology, psychiatry, and behavioral medicine: Perspectives, practices, and research* (pp. 737–754). New York: Springer.

Dobson, K. (Ed.). (2009). *Handbook of cognitive behavioral therapies* (3rd ed.). New York: Guilford Press.

Donk, L. (1972). Attending behavior in mental patients. *Dissertation Abstracts International, 33* (Ord. No. 72-22 569).

Draganski, B., Gaser, C., Busch, V., Schuierer, G., Bogdahn, U., & May, A. (2004). Neuroplasticity: Changes in grey matter induced by training. *Nature, 427,* 311–312.

Duncan, B., Miller, S., & Sparks, J. (2004). *The heroic client.* San Francisco: Jossey-Bass.

Eberhardt, J. L. (2005). Imaging race. *American Psychologist, 60*(2), 181–190.

Egan, G. (2010). *The skilled helper* (9th ed.). Belmont, CA: Brooks/Cole.

Ekaterina, L., Popa, D., Apergis-Schoute, J., Fidacaro, G., & Paré, J. (2008). Amygdala intercalated neurons are required for expression of fear extinction. *Nature, 454,* 642–645.

Ekman, P. (1999). Basic emotions. In T. Dalgleish & M. Power (Eds.), *Handbook of cognition and emotion.* Sussex, UK: Wiley.

Ekman, P. (2007). *Emotions revealed* (2nd ed.). New York: Henry Holt.

Erickson, K., Miller, D., & Roecklein, K. (2012). The aging hippocampus: Interactions between exercise, depression, and BDNF. *The Neuroscientist, 18,* 82–97.

Ericsson, A., Charness, N., Feltovich, P., & Hoffman, R. (2006). *Cambridge handbook on expertise and expert performance.* Cambridge, UK: Cambridge University Press.

Farnham, S., Gill, J., McLean, R., & Ward, S. (1991). *Listening hearts.* Harrisburg, PA: Morehouse.

Fava, G., Rafanelli, M, Grandi, S., Conti, S., & Belluardo, P. (1998). Prevention of recurrent depression with cognitive behavioral therapy. *Archives of General Psychiatry, 55,* 816–820.

Fiedler, F. (1950). A comparison of therapeutic relationships in psychoanalytic, nondirective, and Adlerian therapy. *Journal of Consulting Psychology, 14,* 435–436.

Fournier, J., DeRubeis, R., Hollon, S., Dimidjian, S., Amsterdam, J., Shelton, R., et al. (2010). Antidepressant drug effects and depression severity: A patient-level meta-analysis. *Journal of the American Medical Association, 303,* 47–53.

Fox, E. (2012). *Rainy brain, sunny brain: How to retrain your brain to overcome pessimism and achieve a more positive outlook.* New York: Basic Books.

Frankl, V. (1959). *Man's search for meaning.* New York: Simon and Schuster.

Frankl, V. (1978). *The unheard cry for meaning.* New York: Touchstone.

Fredrickson, B., Tugade, M., Waugh, C., & Larkin, G. (2003). A prospective study of resilience and emotion following the terrorist attacks on the United States on September 11, 2001. *Journal of Personality and Social Psychology, 84,* 365–376.

Freed, P., & Mann, J. (2007). Sadness and loss: Toward a neurobiopsychosocial model. *American Journal of Psychiatry, 164,* 28–34.

Freud, S. (1953). Project for a scientific psychology. *Complete psychological works of Sigmund Freud* (Vol. 1, pp. 283–397). London: Hogarth Press. (Original work published 1895)

Friedman, M., Thoresen, C., Gill, J., Ulmer, D., Powell, L., Price, V., et al. (1986). Alteration of type A behavior and its effect on cardiac recurrences in post myocardial infarction patients: Summary results of the recurrent coronary prevention project. *American Heart Journal, 112,* 653–665.

Fukuyama, M. (1990, March). *Multicultural and spiritual issues in counseling.* Workshop presentation for the American Counseling Association Convention, Cincinnati.

Gawande, A. (2009). *The checklist manifesto: How to get things right.* New York: Henry Holt.

Gazzaniga, M. (2000). Cerebral specialization and interhemispheric communication: Does the corpus callosum enable the human condition? *Brain, 123,* 1293–1326.

Gearhart, C., & Bodie, G. (2011). Active-empathic listening as a general social skill: Evidence from bivariate and canonical correlations. *Communication Reports, 24*(2), 86–98.

Geda, Y. E., Silber, T. C., Roberts, R. O., Knopman, D. S., Christianson, T. J. H., Pankratz, V. S., Boeve, B. F., Tangalos, E. G., & Petersen, R. C. (2012). Computer activities, physical exercise, aging, and mild cognitive impairment: A population-based study. *Mayo Clinic Proceedings, 87,* 437–442.

Gendlin, E., & Henricks, M. (n.d.). *Rap manual* [Mimeographed]. Cited in E. Gendlin, *Focusing.* New York: Everest House.

Gergen, K., & Gergen, M. (2005, February). The power of positive emotions. *The Positive Aging Newsletter.* Retrieved from www.healthandage.com

Golby, A., Gabrelli, J., Chiao, J., & Eberhardt, J. (2001). Differential responses in the fusiform region to same-race and other-race faces. *Nature Neuropsychology, 4,* 845–850.

Goldapple, K., Segal, Z., Garson, C., Lau, M., Bieling, P., Kennedy, S., & Mayberg, H. (2004). Modulation of cortical-limbic pathways in major depression: Treatment-specific effects of cognitive behavior therapy. *Archives of General Psychiatry, 61,* 34–41.

Goodwin, L., Lee, S., Puig, A., & Sherrard, P. (2005). Guided imagery and relaxation for women with early stage breast cancer. *Journal of Creativity in Mental Health, 1*(2), 53–66.

Gottman, J. (2011). *The science of trust: Emotional attunement for couples.* New York: Norton.

Gould, E., Beylin, A., Tanapat, P., Reeves, A., & Shors, T. (1999). Learning enhances adult neurogenesis in the hippocampal formation. *Nature Neuroscience, 2*(3), 260–265.

Grawe, K. (2007). *Neuropsychotherapy: How the neurosciences inform psychotherapy.* London: Erlbaum.

Greene, D., & Stewart, F. (2011). African American students' reactions to Benjamin Cooke's "Nonverbal communication among Afro-Americans: An initial classification." *Journal of Black Studies, 42,* 389–401.

Hall, E. (1959). *The silent language.* New York: Doubleday.

Hall, J., & Schmid Mast, M. (2007). Sources of accuracy in the empathic accuracy paradigm. *Emotion, 7,* 438–446.

Hall, R. (2007). Racism as a health risk for African Americans. *Journal of African American Studies, 11,* 204–213.

Hallowell, E. (2005, January). Overloaded circuits: Why smart people underperform. *Harvard Business Review,* pp. 1–10.

Hargie, O., Dickson, D., & Tourish, D. (2004). *Communication skills for effective management.* New York: Palgrave Macmillan.

Haskard, K., Williams, S., DiMatteo, M., Heritage, J., & Rosenthal, R. (2008). The provider's voice: Patient satisfaction and the content-filtered speech of nurses and physicians in primary medical care. *Journal of Nonverbal Behavior, 32,* 1–20.

Henretty, J., & Levitt, H. (2010). The role of therapist self-disclosure in psychotherapy: A qualitative review. *Clinical Psychology Review, 30,* 63–77.

Hermans, E., van Marle, H., Ossewaarde, L., Henckens, A., Qin, S., Kesteren, M., et al. (2012). Stress-related noradrenergic activity prompts large-scale neural network configuration. *Science, 334,* 1151–1153.

Hill, C. (2004). *Helping skills.* Washington, DC: American Psychological Association.

Hill, C. (2009). *Helping skills* (3rd ed.). Washington, DC: American Psychological Association.

Hill, C., & O'Brien, K. (1999). *Helping skills.* Washington, DC: American Psychological Association.

Hill, C., & O'Brien, K. (2004). *Helping skills: Facilitating exploration, insight, and action.* Washington, DC: American Psychological Association.

Hillman, C., Erickson, K., & Kramer, A. (2008). Be smart, exercise your heart: Exercise effects on brain and cognition. *Nature Reviews Neuroscience, 9,* 58–65.

Ho, B., Andreasen, N., Ziebell, S., Pierson, R., & Magnotta, V. (2011). Long-term antipsychotic treatment and brain volumes: A longitudinal study of first-episode schizophrenia. *Archives of General Psychiatry, 68,* 128–137.

Hoffman, M. (2012, November 12). *The resting state brain.* Seminar presentation at the Roskamp Institute, Sarasota, FL.

Holland, J., Neimeyer, R., & Currier, J. (2007). The efficacy of personal construct therapy: A comprehensive review. *Journal of Clinical Psychology, 63,* 93–107.

Hölzel, B., Carmody, J., Vangel, M., Congletona, C., Yerramsetti, S., Gard, T., et al. (2011). Mindfulness practice leads to increases in regional brain gray matter density. *Psychiatry Research: Neuroimaging, 191,* 36–43.

Hunter, W. (1984). *Teaching schizophrenics communication skills: A comparative analysis of two microcounseling learning environments.* Unpublished doctoral dissertation, University of Massachusetts, Amherst.

Imel, Z., & Wampold, B. (2008). The importance of treatment and the science of common factors in psychotherapy. In S. D. Brown & R. W. Lent (Eds.), *Handbook of counseling psychology* (4th ed., pp. 249–266). Hoboken, NJ: Wiley.

Ishiyama, I. (2006). *Anti-discrimination response training (A.R.T.) program.* Framingham, MA: Microtraining Associates.

Ivey, A. (1971). *Microcounseling: Innovations in interviewing training.* Springfield, IL: Thomas.

Ivey, A. (1973). Media therapy: Educational change planning for psychiatric patients. *Journal of Counseling Psychology, 20,* 338–343.

Ivey, A. (2000). *Developmental therapy: Theory into practice.* North Amherst, MA: Microtraining Associates. (Originally published 1986)

Ivey, A. (2012, May). *Neuroscience: Implications for counseling and medical practice.* Presentation to Oakland University Medical School and Counseling Program.

Ivey, A., D'Andrea, M., & Ivey, M. (2012). *Theories of counseling and psychotherapy: A multicultural perspective* (7th ed.). Thousand Oaks, CA: Sage.

Ivey, A., & Ivey, M. (1987). Decisional counseling. In A. E. Ivey, M. B. Ivey, & L. Simek-Downing, *Counseling and psychotherapy* (pp. 25–48). Englewood Cliffs, NJ: Prentice-Hall.

Ivey, A., Ivey, M., & Daniels, T. (2014). *Intentional interviewing and counseling: CourseMate interactive website.* Belmont, CA: Brooks/Cole.

Ivey, A., Ivey, M., Gluckstern-Packard, N., Butler, K., & Zalaquett, C. (2011). *Basic stress management* [DVD]. Alexandria, Virginia: Microtraining/Alexander Street Press.

Ivey, A., Ivey, M., Gluckstern-Packard, N., Butler, K., & Zalaquett, C. (2012). *Basic influencing skills* (4th ed.) [DVD]. Alexandria, VA: Microtraining/Alexander Street Press.

Ivey, A., Ivey, M., Myers, J., & Sweeney, T. (2005). *Developmental counseling and therapy: Promoting wellness over the lifespan.* Boston: Lahaska/Houghton Mifflin.

Ivey, A., Ivey, M., & Zalaquett, C. (2010). *Intentional interviewing and counseling: Facilitating client development in a multicultural society* (7th ed.). Belmont, CA: Brooks/Cole.

Ivey, A., Ivey, M., & Zalaquett, C. (2011). *Essentials of intentional interviewing: Counseling in a multicultural world* (2nd ed.). Pacific Grove, CA: Brooks/Cole.

Ivey, A., Normington, C., Miller, C., Morrill, W., & Haase, R. (1968). Microcounseling and attending behavior: An approach to pre-practicum counselor training [Monograph]. *Journal of Counseling Psychology, 15,* Part II, 1–12.

Ivey, A., Pedersen, P., & Ivey, M. (2001). *Intentional group counseling: A microskills approach.* Belmont, CA: Brooks/Cole.

Jacobs, R. (1991). *Be an outrageous older woman.* Manchester, CT: Knowledge, Trends, and Ideas.

Jameson, M. (2012, February). Fear dementia? Your diet, weight more important than genes, experts say. *Orlando Sentinel.*

Janis, I., & Mann, L. (1977). *Decision making: A psychological analysis of conflict, choice, and commitment.* New York: Free Press.

Johnson, B. (2012). Psychology's paradigm shift: From a mental health to a health profession? *Monitor on Psychology, 43,* 72.

Kabat-Zinn, J. (1990). *Full catastrophe living: Using the wisdom of your mind to face stress, pain and illness.* New York: Dell.

Kabat-Zinn, J. (2005a). *Coming to our senses: Healing ourselves and the world through mindfulness.* New York: Hyperion.

Kabat-Zinn, J. (2005b). *Wherever you go, there you are: Mindfulness meditation in everyday life.* New York: Hyperion.

Kahnemann, D. (2011). *Thinking fast and slow.* New York: Farrar, Straus, & Giroux.

Kaiser, C., Drury, B., Malahy, L., & King, K. (2011). Nonverbal asymmetry in interracial interactions: Strongly identified Blacks display friendliness, but Whites respond negatively. *Social Psychology and Personality Science, 2*, 554–559.

Kandel, E. (2007). *In search of memory: The emergence of a new science of mind.* New York: Norton.

Karno, M., & Longabaugh, R. (2005). Less directiveness by therapists improves drinking outcomes of reactant clients in alcoholism treatment. *Journal of Consulting and Clinical Psychology, 73*(2), 262–267.

Kim, E., Park, N., & Peterson, C. (2011). Dispositional optimism protects older adults from stroke: The Health and Retirement Study. *Stroke.* Retrieved from stroke. ahajournals.org/content/early/2011/07/21/STROKEAHA.111.613448.full.pdf?ijkey=EgeC0lK195rBJDH&keytype=ref

Kolb, B., & Whishaw, I. (2009). *Fundamentals of human neuropsychology* (6th ed.). New York: Worth.

Lacan, J. (1977). *The four fundamental concepts of psychoanalysis: The seminar of Jacques Lacan, Book XI* (Jacques-Alain Miller, Ed.; Alan Sheridan, Trans.). New York: Norton.

Lacey, S., Stilla, R., & Sathian, K. (2012). Metaphorically feeling: Comprehending textural metaphors activates somatosensory cortex. *Brain and Language.* Retrieved from www.sciencedirect.com/science/article/pii/S0093934X12000028

LaCombe, S., & McGraw, T. (2010). *Right brain functioning.* Retrieved from www.myshrink.com/counseling-theory.php?t_id=62

LaFrance, M., & Woodzicka, J. (1998). No laughing matter: Women's verbal and nonverbal reactions to sexist humor. In J. Swim & C. Stangor (Eds.), *Prejudice: The target's perspective.* San Diego, CA: Academic Press.

Lane, P., & McWhirter, J. (1992). A peer mediation model: Conflict resolution for elementary and middle school children. *Elementary School Guidance and Counseling, 27*, 15–23.

Lane, R. (2008). Neural substrates of implicit and explicit emotional processes: A unifying framework for psychosomatic medicine. *Psychosomatic Medicine, 70*, 214–231.

Lee, C. (1992). *Empowering young black males.* Ann Arbor, MI: ERIC.

Lewis, D. (2011). Antipsychotic medications and brain volume: Do we have cause for concern? *Archives of General Psychiatry, 68*, 126–127.

Li, J., & Lambert, V. (2008). Job satisfaction among intensive care nurses from the People's Republic of China. *International Nursing Review, 55*, 34–39.

Libert, Y., Marckaert, I., Reynaert, C., Delvaux, N., Marchal, S., Etienne, A., et al. (2007). Physicians are different when they learn communication skills: Influence of the locus of control. *Psycho-Oncology, 16*, 553–562.

Lindquist, K., Wager, T., Kober, H., Bliss-Moreau, E., & Barrett, L. (2012). The brain basis of emotion: A meta-analytic review. *Behavioral and Brain Sciences, 35*, 121–143.

Loftus, E. (1997, September). Creating false memories. *Scientific American*, pp. 51–55.

Loftus, E. (2003). Our changeable memories: Legal and practical implications. *Nature Reviews: Neuroscience, 4*, 31–34.

Loftus, E. (2011). We live in perilous times for science. *Skeptical Inquirer, 35*, 13.

Logothetis, N. (2008). What we can do and what we cannot do with fMRI. *Nature, 453*, 869–878.

Lucas, L. (2007/2008). The pain of attachment—"You have to put a little wedge in there": How vicarious trauma affects child/teacher attachment. *Childhood Education, 84*, 85–91.

Luthar, S., Cicchetti, D., & Becker, B. (2000). The construct of resilience: A critical evaluation and guidelines for future work. *Child Development, 71*(3), 543–562.

MacLean, P. (1985). Brain evolution relating to family, play, and the separation call. *Archives of General Psychiatry, 42*, 405–417.

Mahoney, M., & Freeman, A. (Eds.). (1985). *Cognition and psychotherapy.* New York: Springer.

Mann, L. (2001). Naturalistic decision making. *Journal of Behavioural Decision Making, 14*, 375–377.

Mann, L., Beswick, G., Allouache, P., & Ivey, M. (1989). Decision workshops for the improvement of decision making skills. *Journal of Counseling and Development, 67*, 237–243.

Marci, C., Ham, J., Moran, E., & Orr, S. (2007). Physiologic correlates of perceived therapist empathy and social-emotional process during psychotherapy. *Journal of Nervous and Mental Disease, 195*, 103–111.

Marroun, H., Jaddoe, V., Hudziak, J., Roza, S., Steegers, E., Hofman, A., et al. (2012). Maternal use of selective serotonin reuptake inhibitors, fetal growth, and risk of adverse birth outcomes. *Archives of General Psychiatry, 69*, 706–714.

Marshall, J. (2010, February 22). Poverty during early childhood may last a lifetime. *Discovery News.* Retrieved May 10, 2010, from news.discovery.com/human/poverty-children-income-adults.html

Martin, S., Martin, E., Rai, S., Richardson, M., & Royall, R. (2001). Brain blood flow changes in depressed patients treated with interpersonal psychotherapy or venlafaxine hydrochloride: Preliminary findings. *Archives of General Psychiatry, 58*, 641–648.

Martin Luther King Jr. Center. (1989). *Six steps of nonviolent social change.* Retrieved from www.thekingcenter.org/king-philosophy#sub3

Maslow, A., Frager, R., & Fadiman, J. (1987). *Motivation and personality.* New York: Harper.

Masoumi, A., Goldenson, B., Ghirmai, S., Avagyan, H., Zaghi, J., Abel, K., et al. (2009). 1α,25-dihydroxyvitamin D3 interacts with curcuminoids to stimulate amyloid-β clearance by macrophages of Alzheimer's disease patients. *Journal of Alzheimer's Disease, 17*, 703–717.

Masuda, T., & Nisbett, R. (2001). Attending holistically versus analytically: Comparing the context sensitivity of Japanese and Americans. *Journal of Personality and Social Psychology, 81,* 922–934.

Matsumoto, D., Hwang, H., Skinner, L., & Frank, M. (2011, June). Evaluating truthfulness and detecting deception. *FBI Law Enforcement Bulletin.*

Mayo, C., & LaFrance, M. (1973). *Gaze direction in interracial dyadic communication.* Paper presented at the Eastern Psychological Association meeting, Washington, DC.

McGoldrick, M., & Gerson, R. (1985). *Genograms in family assessment.* New York: Norton.

McGoldrick, M., Giordano, J., & Garcia-Preto, N. (2005). *Ethnicity and family therapy* (3rd ed.). New York: Norton.

McGorry, P. (2007). Early intervention in psychotic disorders: Detection and treatment of the first episode psychosis in the critical early stages. *Medical Journal of Australia, 187,* 8–10.

McGorry, P. (2009). Intervention in individuals at ultra-high risk for psychosis: A review and future directions. *Journal of Clinical Psychiatry, 70,* 1206–1212.

McGorry, P. (Ed.). (2012). *Early intervention in psychiatry.* Retrieved from onlinelibrary.wiley.com/journal/10.1111/(ISSN)1751-7893/issues

McGorry, P., Nelson, B., Goldstone, S., & Yung, A. (2010). Clinical staging: A heuristic and practical strategy for new research and better health and social outcomes for psychotic and related mood disorders. *Canadian Journal of Psychiatry, 55,* 486–497.

McIntosh, P. (1988). *White privilege and male privilege: A personal account of coming to see correspondences through work in women's studies.* Wellesley, MA: Wellesley College Center for Research on Women.

Meara, N., Pepinsky, H., Shannon, J., & Murray, W. (1981). Semantic communication and expectation for counseling across three theoretical orientations. *Journal of Counseling Psychology, 28,* 110–118.

Meara, N., Shannon, J., & Pepinsky, H. (1979). Comparisons of stylistic complexity of the language of counselor and client across three theoretical orientations. *Journal of Counseling Psychology, 26,* 181–189.

Mikecz, R. (2011). Interviewing elites: Addressing methodological issues. *Qualitative Inquiry, 18,* 482–493.

Miller, G. (1956). The magical number 7 plus or minus 2: Some limits on our ability for processing information. *Psychological Review, 63,* 81–87.

Miller, K. (2007). Compassionate communication in the workplace: Exploring processes of noticing, connecting, and responding. *Journal of Applied Communication Research, 35,* 223–245.

Miller, S., Duncan, B., & Hubble, M. (2005). Outcome-informed clinical work. In J. Norcross & M. Goldfried (Eds.), *Handbook of psychotherapy integration* (pp. 84–104). Oxford, UK: Oxford University Press.

Miller, W., & Rollnick, S. (2002). *Motivational interviewing: Preparing people for change.* New York: Guilford Press.

Monk, G., Winslade, J., Crocket, K., & Epston, D. (1997). *Narrative theory in practice: The archaeology of hope.* San Francisco: Jossey-Bass.

Moos, R. (2001, August). *The contextual framework.* Presentation at the American Psychological Association, San Francisco.

Mulcahy, N. (2011). *Randomized trial of yoga in metastatic breast cancer.* Study presented at the 34th Annual San Antonio Breast Cancer Symposium.

Myers, D. (2013). *Psychology* (10th ed.). New York: Worth.

Myers, J., & Sweeney, T. (2004). The indivisible self: An evidenced-based model of wellness. *Journal of Individual Psychology, 60,* 234–245.

Myers, J., & Sweeney, T. (Eds.). (2005). *Counseling for wellness: Theory, research, and practice.* Alexandria, VA: American Counseling Association.

National Association of Social Workers. (2008). *Code of ethics.* Washington, DC: Author. Retrieved from www.naswdc.org/pubs/code/code.asp

National Organization for Human Services. (1996). *Ethical standards for human service professionals.* Canton, GA: Author. Retrieved from www.nationalhumanservices.org/ethical-standards-for-hs-professionals

National Scientific Council on the Developing Child. (2005). *Excessive stress disrupts the architecture of the developing brain* (Working Paper No. 3). Retrieved from www.developingchild.harvard.edu

Nes, L., & Segerstrom, S. (2006). Dispositional optimism and coping: A meta-analytic review. *Personality and Social Psychology Review, 10,* 235–251.

Nwachuku, U., & Ivey, A. (1991). Culture-specific counseling: An alternative approach. *Journal of Counseling and Development, 70,* 106–151.

Nwachuku, U., & Ivey, A. (1992). Teaching culture-specific counseling use in microtraining technology. *International Journal for the Advancement of Counseling, 15,* 151–161.

Ogbonnaya, O. (1994). Person as community: An African understanding of the person as intrapsychic community. *Journal of Black Psychology, 20,* 75–87.

Oliveira-Silva, P., & Gonçalves, Ó. (2011). Responding empathically: A question of heart, not a question of skin. *Applied Psychophysiology and Biofeedback, 36,* 201–207.

Othmer, S. (2012, September). *Resting state networks and ILF neurofeedback.* Presentation at Clinical Summit, Chicago.

Parsons, F. (1967). *Choosing a vocation.* New York: Agathon. (Originally published 1909)

Pfiffner, L., & McBurnett, K. (1997). Social skills training with parent generalization: Treatment effects for children with attention deficit disorder. *Journal of Consulting and Clinical Psychology, 65,* 749–757.

Pos, A., Greenberg, L., Goldman, R., & Korman, L. (2003). Emotional processing during experiential treatment of depression. *Journal of Clinical and Consulting Psychology, 73,* 1007–1016.

Posner, M. (Ed.). (2004). *Cognitive neuropsychology of attention*. New York: Guilford Press.

Power, S., & Lopez, R. (1985). Perceptual, motor, and verbal skills of monolingual and bilingual Hispanic children: A discrimination analysis. *Perceptual and Motor Skills, 60*, 1001–1109.

Probst, R. (1996). Cognitive-behavioral therapy and the religious person. In E. Shafranski (Ed.), *Religion and the clinical practice of psychology* (pp. 391–408). Washington, DC: American Psychological Association.

Puig, A., Lee, S., Goodwin, L., & Sherrard, P. (2006). The efficacy of creative arts therapies to enhance emotional expression, spirituality, and psychological well-being of newly diagnosed Stage I and Stage II breast cancer patients: A preliminary study. *The Arts in Psychotherapy, 33*, 218–228.

Raichle, M. E., & Snyder, A. (2007). A default mode of brain function: A brief history of an evolving idea. *NeuroImage, 37*(4), 1083–1090.

Raji, C., Ericson, K., Lopez, O., Kuller, L., Gach, M., Thompson, P., et al. (2011, December). *Regular fish consumption is associated with larger gray matter*. Paper presented at the Radiological Society of North America, Chicago.

Ratey, J. (2008a). *Neuroscience and the brain* [Transcript from video interview]. Framingham, MA: Microtraining Associates.

Ratey, J. (2008b). *Spark: The revolutionary new science of exercise and the brain*. New York: Little, Brown.

Ratey, J. (2012, April). *Exercise as a key to mental and physical health*. Presentation to Trends in Neuroscience Conference, Bradley University, Peoria, IL.

Restak, R. (2003). *The new brain: How the modern age is rewiring your mind*. New York: Rodale Press.

Rigazio-DiGilio, S., Ivey, A., Grady, L., & Kunkler-Peck, K. (2005). *The community genogram*. New York: Teachers College Press.

Rizzolatti, G., & Craighero, L. (2004). The mirror-neuron system. *Annual Review of Neuroscience, 27*, 169–192.

Rogers, C. (1957). The necessary and sufficient conditions of therapeutic personality change. *Journal of Consulting Psychology, 21*, 95–103.

Rogers, C. (1961). *On becoming a person*. Boston: Houghton Mifflin.

Rose, L. (2012, May 29). Too much screen time, too little playtime for Canadian children. *Vancouver Sun,*. Retrieved from www.vancouversun.com/mobile/life/parenting/much+screen+time+little+playtime+Canadian+kids+report+card+finds/6694913/story.html

Rothbaum, B., & Keane, T. (2012, April). *Emergency therapy may prevent PTSD in trauma victims*. Paper presented at the Anxiety Disorders Association of America (ADAA) 32nd Annual Conference.

Roysircar, G., Arredondo, P., Fuertes, J., Ponterotto, J., & Toporek, R. (2003). *Multicultural competencies*. Washington, DC: Association for Multicultural Counseling and Development.

Russell-Chapin, L., Kemmerly, T., Liu, W., Zagardo, M., Chapin, T., Dailey, D., et al. (in press). A pilot study of neurofeedback, fMRI, and the default mode network: Implications for counseling treatment of attending deficit disorder. *Journal of Neuropathy & Experimental Neurology*.

Sampson, A., Soares, J., Coutinho, J., Sousa, M., & Gonçalves, Ó. (2012). *The big five default brain: Functional evidence*. Unpublished manuscript, University of Minho, Portugal.

Scheier, M., Carver, C., & Bridges, M. (1994). Distinguishing optimism from neuroticism (and trait anxiety, self-mastery, and self-esteem): A reevaluation of the Life Orientation Test. *Journal of Personality and Social Psychology, 67*(6), 1063–1078.

Schwartz, J., & Begley, S. (2003). *The mind and the brain: Neuroplasticity and the power of mental force*. New York: Regan.

Scientists pinpoint how vitamin D may help clear amyloid plaques found in Alzheimer's. (2012, March 6). *Science Daily*. Retrieved from www.sciencedaily.com/releases/2012/03/120306131845.htm

Seki, T., Sawamoto, K., Parent, J., & Alvarez-Buylla, A. (2011). *Neurogenesis in the adult brain I: Neurobiology*. New York: Springer.

Seligman, M. (2006). *Learned optimism*. New York: Vintage.

Seligman, M. (2009). *Authentic happiness*. New York: Free Press.

Seung, S. (2012). *Connectome: How the brain's wiring makes us who we are*. Boston: Houghton Mifflin.

Seyle, H. (1956). *The stress of life*. New York: McGraw-Hill.

Shabris, C., & Simon, D. (2009). *The invisible gorilla: And other ways our intuitions deceive us*. New York: Broadway.

Shalev, A., Ankri, Y., Israeli-Shalev, Y., Peleg, M., Adessky, R., & Freedman, S. (2012). Prevention of posttraumatic stress disorder by early treatment: Results from the Jerusalem trauma outreach and prevention study. *Archives of General Psychiatry, 69*, 166–176.

Sharpley, C., & Guidara, D. (1993). Counselor verbal response mode usage and client-perceived rapport. *Counseling Psychology Quarterly, 6*, 131–142.

Sharpley, C., & Sagris, I. (1995). Does eye contact increase counselor-client rapport? *Counselling Psychology Quarterly, 8*, 145–155.

Shenk, D. (2010). *The genius in all of us*. New York: Doubleday.

Sherrard, P. (1973). *Predicting group leader/member interaction: The efficacy of the Ivey Taxonomy*. Unpublished doctoral dissertation, University of Massachusetts, Amherst.

Shostrom, E. (1966). *Three approaches to psychotherapy* [Film]. Santa Ana, CA: Psychological Films.

Siegel, D. (2007). *The mindful brain*. New York: Norton.

Siegel, D. (2012). *The developing mind*. New York: Guilford Press.

Singer, T., Seymour, B., O'Dougherty, J., Kaube, H., Dolan, R., & Frith, C. (2004). Empathy for pain involves the affective but not sensory components of pain. *Science, 303*, 1157–1161.

Somers, T. (2006). The sounds of silence: Brains are active in absence of sound. *Society for Neuroscience.* Retrieved from www.sfn.org/index.aspx?pagename=news_010406

Stephens, G., Silbert, L., & Hasson, U. (2010). Speaker–listener neural coupling underlies successful communication. *Proceedings of the National Academy of Sciences of the United States of America, 107*(32), 14425–14430.

Steward, R., Neil, D., Jo, H., Hill, M., & Baden, A. (1998). *White counselor trainees: Is there multicultural counseling competence without formal training?* Poster session presented at the Great Lakes Regional Conference of Division 17 of the American Psychological Association, Bloomington, IN.

Stöppler, M. (2011). Surprising health benefits of sex. *Medicine Net.* Retrieved from www.medicinenet.com /sexual_health_pictures_slideshow/article.htm

Strauch, B. (2010, January 3). How to train the aging brain. *New York Times*, p. ED10.

Sue, D. W. (2010a). *Microaggressions in everyday life: Race, gender, and sexual orientation.* New York: Wiley.

Sue, D. W. (Ed.). (2010b). *Microaggressions and marginality: Manifestation, dynamics, and impact.* New York: Wiley.

Sue, D. W., & Sue, D. (2013). *Counseling the culturally diverse: Theory and practice* (6th ed.). New York: Wiley.

Szczygieł, D., Buczny, J., & Bazińska, R. (2012). Emotion regulation and emotional information processing: The moderating effect of emotional awareness. *Personality and Individual Differences, 52*(3), 433–437.

Tamase, K. (1991). Factors which influence the response to open and closed questions: Intimacy in dyad and listener's self-disclosure. *Japanese Journal of Counseling Science, 24*, 111–122.

Tamase, K., Otsuka, Y., & Otani, T. (1990). Reflection of feeling in microcounseling. *Bulletin of Institute for Educational Research* (Nara University of Education), *26*, 55–66.

Tamase, K., Torisu, K., & Ikawa, J. (1991). Effect of the questioning sequence on the response length in an experimental interview. *Bulletin of Nara University of Education, 40*, 199–211.

Tang, Y., Lu, Q., Fan, M., Yang, Y., & Posner, M. (2012). Mechanisms of white matter changes induced by meditation. *Proceedings of the National Academy of Sciences, 109*, 10570–10574.

Tellis, G. (2012, February 7). Metaphors make brains touchy feely. *Science Now.* Retrieved from news.sciencemag.org /sciencenow/2012/02/metaphors-make-brains-touchy-fee .html?ref=hp

Torres-Rivera, E., Pyhan, L., Maddux, C., Wilbur, M., & Garrett, M. (2001). Process vs. content: Integrating personal awareness and counseling skills to meet the multicultural challenge of the twenty-first century. *Counselor Education and Supervision, 41*, 28–40.

Truax, C. (1961). *A tentative approach to the conceptualization and measurement of intensity and intimacy of interpersonal contact as a variable in psychotherapy.* Washington, DC: Eric Clearinghouse. Retrieved from www.eric.ed.gov /PDFS/ED133613.pdf

Tyler, L. (1961). *The work of the counselor* (2nd ed.). East Norwalk, CT: Appleton and Lange.

U.S. Department of Health and Human Services. (2000, December 28). *Federal Register, 65*(250), 82468. Retrieved from www.hhs.gov/ocr/privacy/hipaa/administrative /privacyrule/prdecember2000all8parts.pdf

U.S. Department of Health and Human Services. (2003, May). Summary of the HIPAA Privacy Rule. Retrieved from www.hhs.gov/ocr/privacy/hipaa/understanding /summary/privacysummary.pdf

U.S. Department of Health and Human Services. (2011, May 31). News Release: HHS Announces Proposed Changes to HIPAA Privacy Rule. Retrieved from www .hhs.gov/news/press/2011pres/05/20110531c.html

U.S. Department of Labor. (2012). *Occupational Outlook Handbook 2010–2011.* www.bls.gov/oco/ooh_index.htm

Utay, J., & Miller, M. (2006). Guided imagery as an effective therapeutic technique: A brief review of its history and efficacy research. *Journal of Instructional Psychology, 33*, 40–43.

Van der Molen, H. (1984). *Aan verlegenheid valt iets te doen: Een cursus in plaats van therapie* [How to deal with shyness: A course instead of therapy]. Deventer, Netherlands: Van Loghum Slaterus.

Van der Molen, H. (2006). Social skills training and shyness. In T. Daniels & A. Ivey (Eds.), *Microcounseling* (3rd ed.). Springfield, IL: Charles C Thomas.

Vedantam, S. (2010). *The hidden brain.* New York: Speigel & Grau.

Viamontes, G., & Beitman, B. (2007). Mapping the unconscious in the brain. *Psychiatric Annals, 37*, 243–258.

Vogel, G. (2012). Can we make our brains more plastic? *Science, 338*, 36–39.

Voss, M., Heo, S., Prakash, R., Erickson, K., Alves, H., Chaddock, L., et al. (2012, June 5). The influence of aerobic fitness on cerebral white matter integrity and cognitive function in older adults: Results of a one-year exercise intervention. *Human Brain Mapping.* doi: 10.1002/hbm.22119 [Epub ahead of print].

Voss, M., Prakash, R., Erickson, K., Basak, C., Chaddock, L., Kim, J., et al. (2010). Plasticity of brain networks in a randomized intervention trial of exercise training in older adults. *Frontiers in Aging Neuroscience, 2*, 1–17.

Wade, S., Borawski, E., Taylor, H., Drotar, D., Yeates, K., & Stancin, T. (2001). The relationship of caregiver coping to family outcomes during the initial year following pediatric traumatic injury. *Journal of Consulting and Clinical Psychology, 69*, 406–415.

Weir, K. (2011). The exercise effect. *Monitor on Psychology, 42*(11).

Wen, C., Wai, J., Tsai, M., Yank, Y., Cheng, T., Lee, M., et al. (2011). Minimum amount of physical activity for reduced mortality and extended life expectancy: A prospective cohort study. *Lancet, 378*, 1244–1253.

Weniger, G., Lange, C., Sachsee, U., & Irle, E. (2009). Reduced amygdala and hippocampus size in trauma-exposed women with borderline personality disorder and without posttraumatic stress disorder. *Journal of Psychiatry and Neuroscience, 34*, 383–388.

West, C. (2007). Racism's cognitive toll: Subtle discrimination is more taxing on the brain. Retrieved from www.eurekalert.org/pub_releases/2007-09/afps-rct091907.php

Westen, D. (2007). *The political brain.* New York: Public Affairs.

White, M., & Epston, D. (1990). *Narrative means to therapeutic ends.* New York: Norton.

Whiting, J. (2007). Authors, artists, and social constructionism: A case study of narrative supervision. *American Journal of Family Therapy, 35*, 139–150.

Woods, S., Walsh, B., Saksa, J., & McGlashan, T. (2010). The case for including Attenuated Psychotic Symptoms Syndrome in DSM-5 as a psychosis risk syndrome. *Schizophrenia Research, 123*, 199–207.

Woodzicka, J., & LaFrance, M. (2002). Real versus imagined gender harassment. *Journal of Social Issues, 1*, 15–30.

Zalaquett, C. (2008). *Las habilidades atencionales básicas: Pilares fundamentales de la comunicación efectiva* [DVD]. Alexandria, VA: Microtraining Associates/Alexander Street Press.

Zalaquett, C., Foley, P., Tillotson, K., Hof, D., & Dinsmore, J. (2008). Multicultural and social justice training for counselor education programs and colleges of education: Rewards and challenges. *Journal of Counseling and Development, 86*, 323–329.

Zalaquett, C., Fuerth, K., Stein, C., Ivey, A., & Ivey, M. (2008). Reframing the DSM from a multicultural and social justice perspective. *Journal of Counseling and Development, 86*, 364–371.

Zalaquett, C., Ivey, A., Gluckstern-Packard, N., & Ivey, M. (2008). *Las habilidades atencionales básicas: Pilares fundamentales de la comunicación efectiva.* Alexandria, VA: Microtraining Associates/Alexander Street Press.

Zur, O. (2011). *The HIPAA compliance kit.* Sonoma, CA: OZ Publications.

NAME INDEX

A

Abe, N., 120
Ahmad, T., 333
Allouche, P., 355
Alvarez-Buylla, A., 207
Ammentorp, J., 72
Amminger, G., 336
Anderson, P., 337
Andreasen, N., 452
Angelou, M., 351
Apergis-Schoute, J., 448
Arredondo, P., 41
Arroyo, D., 406
Asbell, B., 93
Aupperle, R., 420

B

Back, A. L., 72
Baden, A., 72
Baron, N., 323
Barrett, L. F., 448
Barrett, M., 301
Barrett-Lennard, G., 282
Bazińska, R., 174
Beck, A., 410, 420
Beck, J., 420
Becker, B., 9
Begley, S., 22
Beitman, B. D., 458
Bensing, J., 72, 150, 174
Berman, J., 301
Berra, Y., 91
Beswick, G., 355
Beylin, A., 207
Bickart, K. C., 448
Blanchard, K., 8, 265
Blumenfeld, W. J., 280
Bodie, G. D., 73
Brain, L., 170n
Brammer, L., 353
Bridges, M., 47
Brodhead, M. R., 430–431
Brown, L., 402
Buczny, J., 174
Butler, K., 207, 212n, 219n
Bylund, C. L., 72

C

Callanan, P., 31
Camus, A., 278
Cao, C., 337
Carkhuff, R., 74

Carlson, M. C., 333
Carlstedt, R., 6
Carstensen, L., 174
Carter, R., 57, 150, 251, 283, 455
Carver, C., 47
Centonzea, D., 420
Charness, N., 81
Chiao, J., 99
Churchill, W., 431
Cicchetti, D., 9
Colozino, L., 444
Contrada, R., 282
Corey, G., 31
Corey, M. S., 31
Coutinho, J., 458
Cox, H., 352
Craighero, L., 449
Croce, A., 93
Crocket, K., 15
Cross, W., 251, 401
Currier, J. M., 15

D

Dalai Lama, 3, 338, 426
Damasio, A., 22, 57, 168, 172, 173, 354, 461
D'Andrea, M., 18, 43, 46, 74, 402
Daniels, J., 18, 402
Daniels, T., 12, 65, 72, 120, 150
Danner, D., 173
Dautoff, R. J., 448
Davidson, R., 50, 57, 173, 461
Davis, D., 338
Davis, M., 326
Decety, J., 75, 150, 460
Dement, W., 333
de Waal, E., 278, 295
Dickerson, B. C., 448
Dickson, D., 71
Dietrich, A., 251
Dinsmore, J., 227
Donk, L., 73
Draganski, B., 99
Drury, B., 99
Dryden, W., 420
Duncan, B., 16, 73, 282
D'Zurilla, T. J., 353, 398

E

Eberhardt, J., 99
Egan, G., 74, 353
Ekaterina, L., 448
Ekman, P., 94, 99, 167

Eliot, G., 316
Ellis, A., 69, 410, 420
Epictetus, 410
Epston, D., 15
Erickson, K. I., 57, 332, 334
Ericsson, A. K., 81
Eshelman, E., 326
Everett, B., 333

F

Fadiman, J., 9
Fairchild, E. E., 280
Fan, M., 445
Farnham, S., 278
Fava, G., 452
Feltovich, P., 81
Fidacaro, G., 448
Fiedler, F., 282
Foley, P., 227
Fournier, J., 452
Fox, E., 46, 47
Frager, R., 9
Frank, M., 169
Frankl, V., 127, 267, 278, 280–281, 282, 288, 290, 397, 399–400, 410
Franklin, B., 187–188, 353, 397
Fredrickson, B., 173
Freed, P. J., 444
Freeman, A., 281
Freud, S., 335, 420, 450
Friedman, M., 326
Friesen, W., 173
Fromm, E., 235
Fuertes, J., 41
Fuerth, K. M., 45
Fukuyama, M., 277

G

Gabrelli, J., 99
Galiano, G., 254
Garcia-Preto, N., 441
Garrett, M., 227
Gawande, A., 358, 359
Gazzaniga, M., 251
Gearhart, C. C., 73
Geda, Y., 338
Gendlin, E., 73, 146
Gergen, K., 172
Gergen, M., 172
Gerson, R., 441
Gill, J., 278
Giordano, J., 441

Gluckstern, N., 66n
Gluckstern-Packard, N., 151, 207, 212n, 219n
Golby, A., 99
Goldapple, K., 444
Goldman, R., 174
Goldstone, S., 336
Gonçalves, O. F., 20, 22, 75, 207, 458
Goodwin, L. K., 456
Gottman, J., 176
Gould, E., 207
Grady, L., 219
Grawe, K., 98, 100, 251, 455, 460
Greenberg, L., 174
Greene, D., 99
Guidara, D., 99

H
Haase, R., 66
Haley, J., 398
Hall, E., 71, 94, 99
Hall, J. A., 99
Hall, R., 38
Hallowell, E., 326
Hargie, O., 71
Harris, J., 94
Haskard, K., 99
Hasson, U., 76
Hayes, J., 338
Hebb, D., 450
Heller, R., 117
Henckens, A., 21
Hendricks, M., 73
Hendricks-Gendlin, M., 146
Henretty, J. R., 306
Hermans, E., 21
Hill, C., 72, 99, 120, 174, 250
Hill, M., 72
Hillman, C. H., 57
Hindle, T., 117
Ho, B., 452
Hof, D., 227
Hoffman, M., 444, 457, 458
Hoffman, R. R., 81
Holland, J. M., 15
Hölzel, B. K., 22, 339
Hubble, M., 73, 282
Hunt, A., 420
Hunter, W., 73
Hwang, H., 169

I
Ikawa, J., 120
Imel, Z. E., 15
Irle, E., 448
Ishiyama, I., 228
Ivey, A. E., 11, 18, 43, 45, 46, 65, 66, 72, 73, 74, 80, 102, 120, 150, 151, 167, 186, 187, 207, 208, 209, 212, 219, 243n, 280, 331, 332, 353, 355, 357, 359, 458, 461
Ivey, M. B., 18, 43, 45, 46, 65, 72, 74, 102, 120, 142, 150, 151, 207, 212n, 219n, 243n, 280, 322, 353, 357, 403

J
Jackson, P., 75, 150, 460
Jacobs, R., 29, 58, 331
James, W., 397
Jameson, M., 336
Janis, I., 353
Jo, H., 72
Jobs, S., 283
Johnson, B. B., 20
Joshi, K. Y., 280

K
Kabat-Zinn, J., 338, 461
Kagan, J., 17
Kaiser, C., 99
Kandel, E., 444, 461
Kanel, K., 409
Kanso, R., 251
Karno, M. P., 321
Katz, D., 331
Keane, T., 405
Kesteren, M., 21
Kim, E., 46
King, K., 99
King, M. L., Jr., 256
King, R., 337
Knox, A., 406
Kofoed, P., 72
Kolb, B., 120
Korman, L., 174
Kramer, A. F., 57
Krugman, P., 458
Kunkler-Peck, K., 219

L
Lacan, J., 458
Lacey, S., 177
LaCombe, S., 449
LaFrance, M., 99
Lambert, V. A., 282
Lane, R., 102
Lange, C., 448
Lanier, J., 14
Lao Tse, 183
Larkin, G., 173
Lee, C., 128
Lee, S. M., 456
Levitt, H. M., 306
Lewin, K., 394, 397
Lewis, D., 452
Li, J., 282
Libert, Y., 72
Lindlahr, V., 335
Lindquist, K. A., 448
Loftus, E., 120, 251
Logothetis, N. K., 22
Lopez, R., 151
Lu, Q., 445
Lucas, L., 282
Luthar, S., 9

M
MacDonald, G., 353
MacLean, P. D., 446

Maddux, C., 227
Magnotta, V., 452
Mahoney, M., 281
Mainz, J., 72
Malahy, L. W., 99
Mann, J. J., 444
Mann, L., 353, 355
Manthei, R., 383, 384
Marci, C. D., 75
Marroun, H., 452
Marshall, J., 38
Martin, E., 444
Martin, S., 444
Marx, R., 68n, 389
Maslow, A., 9
Masoumi, A., 336
Masuda, T., 99
Matsumoto, D., 169
Mayo, C., 99
Mayr, U., 174
McBurnett, K., 80
McEwen, B., 50
McGlashan, T., 453
McGoldrick, M., 441
McGorry, P., 326, 336, 453
McGraw, T., 449
McIntosh, P., 42
McKay, M., 326
McLean, R., 278
McLuhan, M., 7
Meara, N., 69
Meichenbaum, D., 410
Merzenich, M., 461
Mikecz, R., 250
Miller, C., 66
Miller, D., 332, 334
Miller, G., 209, 409
Miller, K. I., 99
Miller, M., 330
Miller, S., 16, 73, 282
Miller, W., 355, 356
Minatrea, N., 152
Monk, G., 15
Moos, R., 227
Morrill, W., 66
Mozart, W. A., 81
Mulcahy, N., 339
Murray, W., 69
Myers, D. G., 141
Myers, J. E., 51, 56, 102, 243n, 280

N
Neil, D., 72
Neimeyer, R. A., 15
Nelson, B., 336
Nes, L., 46
Nesselroade, J., 174
Nezu, A. M., 353, 398
Nezu, C. M., 398
Nisbett, R., 99
Nitschke, J., 57
Normington, C., 66
Nwachuku, U., 72, 150

O

Obama, B., 43
O'Brien, K., 72, 99, 250
Ogbonnaya, O., 214
Oliveira-Silva, P., 22, 75
Ortega y Gasset, J., 211, 212
Ossewaarde, L., 21
Otani, T., 174
Othmer, S., 458
Otsuka, Y., 174

P

Packard, N. G., 125
Paré, J., 448
Parent, J., 207
Park, N., 46
Parsons, F., 353
Pasteur, L., 99
Pasupathi, M., 174
Paul, G., 384
Pavlov, I., 448
Pedersen, P., 17, 38, 120, 400
Peirce, C. S., 397
Pepinsky, H., 69
Perls, F., 69, 236
Peterson, C., 46
Pfeffer, J., 44
Pfiffner, L., 80
Piaget, J., 323, 448
Pierson, R., 452
Pizzagalli, D., 57
Ponterotto, J., 41
Popa, D., 448
Pope, M., 44
Pos, A., 174
Posner, M., 65, 445
Power, S., 151
Probst, R., 282
Puig, A. I., 456
Putnam, K., 57
Pyhan, L., 227

Q

Qin, S., 21

R

Rai, S., 444
Raji, C., 336
Ratey, J., 21, 57, 282, 283, 332, 333, 443, 445, 456, 459, 460
Reeves, A., 207
Restak, R., 150, 455
Reyes, C., 226
Riachle, M. E., 457
Richardson, M., 444
Rigazio-DiGilio, S., 219
Rigney, M., 128–129
Rizzolatti, G., 449
Roecklein, K., 332, 334
Rogers, C. R., 9, 15, 63, 69, 73, 131, 140, 160, 161, 186, 187, 188, 191, 201, 237, 282, 397, 398–399, 404
Rollnick, S., 355, 356
Rose, L., 338

Rothbaum, B., 405
Royall, R., 444
Roysircar, G., 41
Russell-Chapin, L., 458
Russo, A., 152

S

Sabroe, S., 72
Sachsee, U., 448
Sagris, I., 99, 251
Saksa, J., 453
Sampson, A., 458
Santiago-Rivera, A., 151
Sapolsky, R., 460
Sathian, K., 177
Sawamoto, K., 207
Scheier, M., 47
Schmid Mast, M., 99
Schwartz, J., 22
Segerstrom, S., 46
Seki, T., 207
Seligman, M., 46
Seung, S., 444
Seyle, H., 324
Shabris, C., 91
Shalev, A., 452
Shannon, J., 69
Sharpley, C., 99, 251
Shenk, D., 81, 82
Sherrard, P., 120, 456
Shors, T. J., 207
Shostrom, E., 69
Siegel, D., 92, 445, 449, 461
Silbert, L. J., 76
Simon, D., 91
Singer, T., 75
Skinner, L., 169
Snowdon, D., 173
Snyder, A., 457
Soares, J., 458
Somers, T., 70
Sousa, M., 458
Sparks, J., 16
Stein, C., 45
Stephens, G. J., 76
Steward, R. J., 72
Stewart, F., 99
Stilla, R., 177
Stöppler, M., 333
Strauch, B., 337
Sue, D., 6, 41, 43, 190, 227, 251, 257, 402
Sue, D. W., 6, 41, 43, 146, 171, 172, 190, 227, 251, 257, 300, 402
Sue, S., 300
Sutton, R., 44
Sweeney, T. J., 51, 56, 102, 243n, 280
Szczygiel, D., 174

T

Tamase, K., 120, 174
Tanapat, P., 207
Tang, Y., 445
Teilhard de Chardin, P., 444
Tellis, G., 177
Tenzin Gyatso, 420

Tillich, P., 5
Tillotson, K., 227
Titchener, E. B., 90
Toporek, R., 41
Torisu, K., 120
Torres-Rivera, E., 227
Tourish, D., 71
Truax, C., 73, 74
Tugade, M., 173
Tyler, L., 46

U

Utay, J., 330

V

van der Molen, H., 73, 80
van Marle, H., 21
Vartainian, O., 251
Vedantam, S., 171
Verheul, W., 150, 174
Viamontes, G. I., 458
Vogel, G., 445
Voss, M. W., 333

W

Wade, S., 282
Walsh, B., 453
Wampold, B. E., 15
Ward, S., 278
Waugh, C., 173
Weir, K., 331, 333
Weiten, W., 451
Wen, C. P., 332
Weniger, G., 448
West, C., 171
Westen, D., 98, 171, 267, 283
White, M., 15
Whiting, J. B., 15
Whittemore, F., 397
Wilbur, M., 227
Williams, T., 81
Winslade, J., 15
Wishaw, I., 120
Witgestein, A., 254
Woods, S., 453
Woodzicka, J., 99
Wright, C. I., 448
Wynn, K., 93

Y

Yang, Y., 445
Yung, A., 336

Z

Zalaquett, C. P., 45, 151, 186, 207, 212n, 219n, 227, 353, 410, 411
Zamora, N., 207, 208, 212, 216, 219, 239, 241, 246, 253
Zander, A., 138
Zander, B., 433
Zhang, W., 71, 94, 226, 276, 305
Ziebell, S., 452
Zucchi, J., 254
Zur, O., 200

SUBJECT INDEX

A

abortion issues, 223, 224–226
abstract/formal operational style, 102, 103
abstraction ladder, 102, 130
abstraction vs. concreteness, 102–103, 108, 110–111
abused clients, 228
accents, 68
acceptance, 140–141, 245, 255
accident survivors, 405
acculturation
    nonverbal behavior and, 100–101
    See also multicultural issues
accumulative stress, 146
acetylcholine (ACH), 454
action stage, 17, 193–194
    cognitive behavioral therapy and, 418–420
    counseling session examples of, 199–200, 378–380
    crisis counseling and, 404–405
    mediation process and, 256
active listening, 153
    anticipated results of, 140
    competencies in, 156, 158
    empathy and, 140–141
    feedback form on, 157
    neuroscience and, 150, 456
    practice exercises on, 154–156
    research evidence on, 150
    skills of, 139–140, 146–149, 153
    See also listening skills
addiction, 244–245, 450
additive empathy, 74, 141
adolescents
    accumulative stress in, 146
    questioning at-risk, 128
    See also child counseling
adrenal glands, 449
adrenaline, 449, 454
advocacy, 227–228, 388
African Americans
    accumulative stress in, 146
    nonverbal behavior in, 67, 99
    questioning as youth at risk, 128
    reflection of meaning and, 278, 280
AIDS clients, 276
alcoholics, 104
Alzheimer's disease, 336
American Academy of Child and Adolescent Psychiatry (AACAP), 31

American Association for Marriage and Family Therapy (AAMFT), 31
American Counseling Association (ACA), 31, 33, 41
American Psychological Association (APA), 31, 35, 41
American School Counselor Association (ASCA), 32
amygdala, 49, 50, 150, 173, 447, 448
amyloids, 336
anandamide, 454
anger
    basic feeling of, 167, 168
    nonverbal indicators of, 169
    words related to, 182
anterior cingulate cortex, 449, 458
antidepressants, 453
antipsychotics, 452–453
approach/avoidance hypothesis, 448
Asian culture, 150
assault victims, 405
assertiveness training, 328–329
assessment
    change process, 243–250, 255
    personal journal for, 62
    wellness, 51–56, 59
at-risk youth, 128
attending behavior, 10, 63–89
    anticipated result of, 65
    body language and, 66, 71–72
    brain imaging of, 65
    challenging situations and, 80–81
    child counseling and, 145
    competency evaluation, 88–89
    demystifying the use of, 186
    disabled people and, 67
    effective feedback on, 85
    empathy and, 73–76
    examples illustrating, 76–79
    eye contact and, 66–67
    feedback form on, 85–87
    four dimensions of, 66, 82
    key points related to, 82–83
    listening skills and, 64–66
    multicultural issues in, 67, 71–72
    neuroscience and, 81–82, 455
    observing in clients, 92–94
    practice exercises on, 83–88
    redirecting attention through, 70
    research evidence on, 72–73
    self-reflections on, 88, 89

silence and, 70
social skills and, 79–80
verbal tracking and, 66, 68–71
vocal qualities and, 66, 67–68
attention
    connective force of, 65
    example of focusing, 83
    neuroscience and, 455
    redirecting, 70
    See also selective attention
attention deficit disorder (ADD), 80
audio recording
    of counseling sessions, 200, 391
    natural style exercise using, 23
Auschwitz concentration camp, 280
Australian Aborigines, 128–129
Australian Psychological Society (APS), 32
authentic style. See natural style
automatic thoughts, 420
autonomic nervous system (ANS), 446
awareness, 41–42

B

balance sheet, 355–356
bargaining, 255
basic competence, 430
basic empathy, 74
basic listening sequence (BLS), 184–186
    anticipated result of, 184
    definition of, 184, 201
    examples of using, 185
    feedback form on, 204–205
    microskills overview, 185
    positive asset search and, 203
    practice exercises on, 202–203
BDNF (brain-derived neurotrophic factor), 332
Becoming a Person (Rogers), 161
beliefs, examining, 54, 223
bilingual clients, 151
biofeedback, 310
blind people, 67
BLS. See basic listening sequence
body awareness, 324
body language
    attending behavior and, 66, 71–72
    observation of, 100, 108
    See also nonverbal behavior
body scan, 338–339
bombardment, 127

brain
  attending behavior and, 65
  basic structures of, 445–447
  connectome studies of, 444
  default mode network of, 444, 457–458
  emotions and, 168, 171, 172–173,
    447–448
  exercise and, 332–333, 334
  hemispheres of, 57, 449–450
  HPA and TAP axes of, 49–50, 59
  importance of understanding, 48–49
  memory change and, 209
  mirror neurons in, 75
  neuroplasticity of, 21–22, 444–445
  optimism and, 47–48
  poverty and, 458
  psychotherapy and, 443–444
  stress and, 21, 48, 49, 59, 325
  terms associated with, 49
  wellness approach and, 57
  See also neuroscience
brain-based approach, 50
brain-derived neurotrophic factor (BDNF),
  332
brain-imaging techniques, 22, 444
brain stem, 446
breathing, 170
brief counseling, 421
brief solution-focused counseling, 286
British Association for Counselling and
    Psychotherapy (BACP), 32
building rapport, 190
burnout, 406, 409

C

calming the client, 404
Canadian Counselling Association (CCA),
  32, 35
"can" questions, 126
carbohydrates, 336
case conceptualization, 356–357
case management, 386, 388
CBT. *See* cognitive behavioral therapy
CCS. *See* Client Change Scale
CengageBrain.com website, 4, 27
Center for Contemplative Mind in Society,
  339
central nervous system (CNS), 446
cerebellum, 446
cerebral cortex, 446
challenge, 15, 337–338
change
  evaluation of, 243–250, 255
  five levels of, 243, 244, 255
  steps for nonviolent, 256
  strategies for maintaining, 388–390
  See also Client Change Scale
checklist for first session, 358, 359–360
Checklist Manifesto, The (Gawande), 358,
  359n
checkouts, 139, 153
  of confrontations, 259
  of influencing skills, 303
  of interpretation/reframes, 286

of paraphrases, 148
of reflecting feelings, 162–163
of summarizations, 149
child counseling
  accumulative stress and, 146
  advocacy and, 227–228, 388
  attending behavior and, 145
  example session of, 142–144
  informed consent in, 35, 145
  listening skills and, 142–145
  questions used in, 128, 131, 145
  reframing used in, 282
Chinese culture, 71
Christianity, 280
chunking, 428
circle of counseling stages, 188, 201
classifying concepts, 430
Client Change Scale (CCS), 243–250, 257
  assessing change with, 255
  counseling session example, 246–250
  diagram illustrating, 244
  interpretation/reframing and, 284
  practice exercise using, 259–260
  stages in, 243, 244, 245–246
Client Feedback Form, 24, 25
clients
  experience of, 383–384
  focus on, 224
  relapse plan for, 388–390
  stories of, 13–14, 190–191
  strengths of, 51–56
closed questions, 118, 132
  anticipated result of, 119
  bringing out specifics with, 127
  examples of using, 122–123, 185
  less verbal clients and, 130–131
  practice exercises on, 133–135
coaching, 421
codes of ethics, 31–32
coffee consumption, 337
Cognitive and Emotional Balance Sheet,
  355–356
cognitive behavioral therapy (CBT), 402,
  410–420
  action stage in, 418–420
  definition of, 410
  five-stage structure and, 201
  frame of reference in, 410
  group practice exercise on, 424
  interpretation/reframing and, 287
  key propositions of, 411
  medications vs., 453
  philosophy of, 410
  psychoeducation in, 320
  research on fMRI and, 420
  restory stage in, 412–418
  session transcript, 411–420
  stress management and, 326
  supplementary readings on, 420
cognitive challenge, 337–338
Commission on Rehabilitation Counselor
    Certification (CRCC), 32
common factors approach, 15
communication skills training, 72, 327–328

community genogram, 213, 215–223, 229
  debriefing of, 218–223
  developing your own, 215, 217–218, 231
  identifying strengths through, 218
  visual examples of, 216–217
community resources, 57
competence
  attending behavior, 88–89
  confrontation, 263–264
  empathic, 88–89
  ethical, 33, 58, 62
  feedback, 314
  focusing, 233–234
  instruction/psychoeducation, 346
  interpretation/reframing, 297–298
  listening, 156, 158
  logical consequences, 346
  microskills, 427–430
  multicultural, 41–45, 59, 62
  observational, 112–113
  questioning process, 136
  reflection of feeling, 180–181
  reflection of meaning, 297–298
  self-assessment of, 428–429
  self-disclosure, 314
  stress management and TLC, 346
  wellness, 62
competency practice exercises, 27
  on attending behavior, 83–88
  on basic listening sequence, 202–203
  on confrontation, 258–263
  on counseling theories, 422–424
  on ethics, 60
  on feedback, 312–313
  on focusing, 230–233
  on instruction/psychoeducation, 345
  on interpretation/reframing, 292–293,
    295–296
  on listening skills, 154–156
  on logical consequences, 342–345
  on multicultural competence, 60–61
  on observation skills, 107–111
  on questions, 133–135
  on reflection of feeling, 176–180
  on reflection of meaning, 290–291,
    293–295
  on self-disclosure, 312–313
  on session structure, 203–204
  on transcript analysis, 393
  on wellness, 61
compliance issues, 331
concerns vs. problems, 14
concreteness
  abstraction vs., 102–103, 108, 110–111
  child counseling and, 145
  empathy and, 121
concrete questions, 130–131
concrete/situational style, 102–103
confidentiality, 33, 58
conflict
  confrontation of, 259
  discrepancies and, 105
  identifying, 238–240
  internal vs. external, 240–241

conflict (*continued*)
    mediation process and, 255–256
    observing in clients, 93, 105, 107, 108
    value-based, 277–278
conflict resolution, 255–256
conflicts of interest, 37
conformity stage, 252, 254
confrontation, 235–264
    anticipated result of, 236
    Client Change Scale and, 243–250, 255
    competencies related to, 263–264
    conflict identification and, 238–240
    counseling session example of, 246–250
    cultural identity development and,
        251–254
    definition of empathic, 236
    evaluating effectiveness of, 243–250, 255
    feedback form on, 262
    key points about, 257
    listening and, 236, 238–240
    mediation process and, 255–256
    multicultural issues and, 237
    neuroscience and, 251, 456
    nonjudgmental attitude and, 237–238
    practice exercises on, 258–263
    research evidence on, 250–251
    self-reflection on, 264
    steps in process of, 238–243
    stuckness and, 236
    summary of, 256–257
connectome, 444, 457
consciousness, 458
consultation, 384
contact barriers, 450
contempt
    basic feeling of, 167, 168
    nonverbal indicators of, 169
    words related to, 182
content, reflection of, 147–149
contextual issues
    client wellness and, 51
    focusing on cultural and, 225–227
contracting, 193
control issues, 55
conversational distance, 71
conversational styles, 102–103
coping self, 53–54
corpus callosum, 449
corrective feedback, 85, 307, 308, 312
cortex, 447
corticosteroids, 449
cortisol, 49, 323, 449, 459
"could" questions, 123, 125, 126
counseling, 5
    brain-based approach to, 50
    client experience of, 383–384
    demystifying the process of, 186–187
    ethical foundations of, 30–31
    interviewing, psychotherapy, and, 5–6, 26
    memory change through, 207, 208–210
    national/international perspectives on,
        14–15
    neuroscience and, 20–23, 59, 187,
        443–461

paradigm shift in, 20
science and art of, 7, 26
theory-based interpretations in, 286–288
treatment plan for, 385–386, 387
Counseling and Neuroscience video, 461
counseling/coaching, 421
Counseling CourseMate website, 4, 27
counseling sessions
    action stage in, 193–194, 199–200,
        378–380
    analyzing transcripts of, 380–381, 391
    audio recording of, 200, 391
    circle of stages in, 188, 201
    competence in conducting, 205–206
    decision-making model and, 187–188,
        354
    empathic relationship stage in, 190,
        194–195, 361–363
    examples of five-stage decisional,
        194–200, 360–380
    first session checklist for, 358, 359–360
    five-stage structure for, 187–194
    goal setting stage in, 191–192, 197,
        366–370
    key points related to, 201
    note taking in, 200
    overview of stages/dimensions, 189
    planning the first session, 357–358
    practice exercise on conducting, 203–204
    restory stage in, 192–193, 198, 370–378
    story/strengths stage in, 190–191,
        195–196, 363–366
    transcribing of, 392
    video recording of, 187, 200, 391
counseling theories
    feedback form on, 423–424
    interpretations based on, 286–288
    microskills and, 27, 395–397
    portfolio of competence, 425
    practice exercises, 422–424
    self-reflection on, 425
    See also specific theories
counselor focus, 225
countertransference, 302
CourseMate website, 4, 27
creative self, 54–55
creativity and neuroscience, 251
crisis counseling, 402–409
    burnout related to, 406, 409
    first session example of, 406–409
    group practice exercise on, 424
    key methods used in, 403–405
    philosophy of, 402–403
    practice implications of, 405–406
    recommended reading on, 409
    research evidence on, 405
cross-cultural situations. *See* multicultural
    issues
cultural background, 58
cultural/environmental context (CEC),
    401–402
cultural identity
    client wellness and, 52
    confrontation process and, 251–254

stages in development of, 252
cultural identity development (CID), 401
cultural intentionality, 8, 26, 422, 432
cultural strength inventory, 129
culture
    description of, 17
    focus on, 225–227
    See also multicultural issues

D

deaf people, 67
debriefing the story, 405
decisional balance sheet, 355–356
decisional counseling, 352–356, 397–398
    definition of, 392
    emotions and feelings in, 354–356
    full-session transcript of, 360–380
    interpretation/reframing and, 286
    logical consequences and, 317–318,
        355
    modern practice and, 353–354
    overview of, 352–353, 397–398
    supplementary readings on, 398
    trait-and-factor theory and, 353
decision-making model, 187–188, 353
default mode network (DMN), 444,
    457–458
denial, 245, 255
depression, 173, 452, 453
dereflection strategy, 281–282
*Developmental Counseling and Therapy:
    Promoting Wellness Over the Lifespan*
    (Ivey, Ivey, Myers, & Sweeney), 102
*Diagnostic and Statistical Manual of Mental
    Disorders*, 7, 14, 22, 39, 453
diagnostic sessions, 7
diet and nutrition, 55, 335–337
directives, 321, 330, 342, 399, 439
disabled people, 67
discernment, 270, 278, 279–280, 400
disciplinary situations, 319
disclosure. *See* self-disclosure
discrepancies
    goal-related, 106
    identifying, 238–240, 258
    nonverbal behavior, 100, 105
    observing, 93, 100, 105–106, 107, 108
    verbal behavior, 105
discrimination
    accumulative stress and, 146
    See also prejudice; racism
disgust
    basic feeling of, 167, 168
    nonverbal indicators of, 169
    words related to, 182
dissonance stage, 252, 254
diversity
    ethics and, 41
    listening skills and, 149–151
    multiculturalism and, 38–40
    practice exercises on, 60–61
    See also multicultural issues
dopamine, 150, 452, 453
dual relationships, 37

E

eating habits, 335
eliciting meaning, 268, 270, 277, 289
emotional balance sheet, 356
emotional balancing, 356
emotional intelligence, 88
emotions
  decisional counseling and, 354–356
  expression of, 170–171
  key words related to, 151, 182
  language of, 167–168, 177, 182
  layers of, 163
  meaning and, 282
  naming/labeling, 162, 175
  neuroscience of, 168, 171, 172–173, 176, 447–448
  nonverbal indicators of, 169
  positive, 172–173
  processing of, 174
  racial issues and, 171–172
  regulation of, 50
  skill of working with, 167
  transmission of, 150
  wellness and, 54–55
  See also reflection of feeling
empathic confrontation, 236, 238–243, 256–257. *See also* confrontation
empathic listening, 142, 150
empathic relationship stage, 15, 190
  counseling session examples of, 194–195, 361–363
  mediation process and, 255–256
empathic self-disclosure, 303–305. *See also* self-disclosure
empathy, 73–76
  accurate use of, 142
  active listening and, 140–141
  anticipated result of, 73
  competency in, 88–89
  concreteness and, 121
  confrontation and, 236, 238–243
  definition of, 73, 83
  mirror neurons and, 75–76
  nonjudgmental attitude and, 237–238
  research evidence on, 75–76
  self-disclosure and, 303–305
  self-reflection on, 89
  three types of, 74, 141
  warmth and, 163
Empowering Black Males (Lee), 128
encouragement
  anticipated result of, 140
  child counseling and, 145
  listening process and, 140, 147, 153
  practice exercises on, 154–155
encouragers, 140, 142, 147, 148–149, 153, 185
endorphins, 454
enkephalins, 454
environmental focus, 225–227
epinephrine, 449, 454
essential self, 52–53
ethical witnessing, 228
ethics, 30–37

codes of, 31–32
competence and, 33, 58
confidentiality and, 33, 58
diversity and, 41
HIPAA privacy and, 33–35
informed consent and, 35–36, 58
key points related to, 58
mastery evaluation of, 62
power and, 36–37, 58
practice exercises on, 60
result of practicing, 30
sample practice contract, 36
self-reflection on, 62
social justice and, 37–38, 58
Ethics Updates website, 32
ethnicity
  emotions related to, 171–172
  wellness assessment and, 52
  See also multicultural issues
Ethnicity and Family Therapy (McGoldrick, Giordano, & Garcia-Preto), 441
eustress, 323, 324–325
evaluation
  change process, 243–250, 255
  See also assessment
exercise, 55–56, 331–333
  brain and, 332–333, 334, 445
  research on benefits of, 333
expression of emotions, 170–171
external conflicts, 240–241
eye contact
  attending behavior and, 66–67
  confrontation and, 250–251
  cultural differences in, 67, 71, 94
  observing in clients, 92, 108

F

facial expressions, 100, 147, 150. *See also* nonverbal behavior
false memories, 120, 251
family focus, 224, 226
family genogram, 218, 229, 231, 441–442
family strength inventory, 129
fear
  basic feeling of, 167, 168
  neuroscience research on, 447–448
  nonverbal indicators of, 169
  words related to, 182
feedback, 306–311
  anticipated result of, 306
  competence in providing, 313–315
  corrective, 307, 308, 312
  counseling session example of, 308–310
  definition of, 306
  feedback form on, 314
  guidelines for effective, 85, 307
  key points about, 311–312
  negative, 307, 308, 312
  neuroscience and, 310–311
  positive, 307, 308, 312
  practice exercises on, 312–313
  rewards and risks of, 311
  self-reflection on, 315

feedback forms
  on attending behavior, 85–87
  on basic listening sequence, 204–205
  on confrontation, 262
  on counseling theories, 423–424
  on feedback, 314
  on focusing, 232–233
  on instruction/psychoeducation, 344
  on interpretation/reframing, 296
  on listening skills, 157
  on logical consequences, 344
  on observation skills, 109–110
  on questioning process, 135
  on reflection of feeling, 179
  on reflection of meaning, 294
  on self-disclosure, 314
  on stress management, 344
  on therapeutic lifestyle changes, 344
feelings. *See* emotions; reflection of feeling
feminist therapy, 225, 401
first counseling session
  checklist for, 358, 359–360
  client experience of, 383
  crisis counseling example of, 406–409
  planning for, 357–358
first impressions, 383
fish consumption, 336
five-stage session structure, 187–194
  action stage in, 193–194, 199–200
  decisional counseling and, 354
  empathic relationship stage in, 190, 194–195
  example of using, 194–200
  goal setting stage in, 191–192, 197
  Ivey Taxonomy and, 437
  overview of, 187–189, 201
  practice exercise on, 203–204
  restory stage in, 192–193, 198
  story/strengths stage in, 190–191, 195–196
flexibility, 7
focusing, 211–234
  advocacy and, 227–228
  analysis of, 382
  anticipated result of, 213
  applying to challenging issues, 224–227
  community genogram and, 213, 215–223, 231
  competencies related to, 233–234
  counseling session example of, 219–223
  dimensions of, 224–227, 229
  examining beliefs through, 223
  family genogram and, 218
  feedback form on, 232–233
  international perspective on, 226
  interpretation/reframing and, 284–285
  key points about, 229–230
  multicultural issues and, 225–227
  neuroscience and, 456
  overview of using, 213–215
  practice exercises on, 230–233
  research evidence on, 227
  selective attention and, 213, 229
  self-reflection on, 234

focusing (*continued*)
   social justice and, 227–228
   summary of, 228
follow-up process, 193–194, 405
formal operational style, 102, 103
friendship, 53
frontal lobe, 445–446
functional magnetic resonance imaging
   (fMRI), 22, 311, 420, 444
future diary homework assignment, 356
future theoretical/practical integration,
   431–432

G

GABA, 449–450, 453
gender
   listening skills and, 149
   nonverbal behavior and, 99
   power differentials and, 37
   reflection of meaning and, 280
   sexual harassment and, 284, 288
   strength inventory based on, 129
   wellness assessment and, 52
gender identity, 18, 19, 52
gene expression, 283
generalization plan, 383
Genius in All of Us, The (Shenk), 81
genograms
   community, 213, 215–218, 229, 231
   family, 218, 229, 231, 441–442
genuineness, 303
Gestalt strategies, 170
giftedness, 81
gladness
   basic feeling of, 167, 168
   nonverbal indicators of, 169
   words related to, 172, 182
glutamate, 453
goals, 16, 191–192
   counseling session examples of, 197,
     366–370
   discrepancies related to, 106
   future development of values and, 432
   identifying life mission and, 278,
     279–280
   mediation process and, 256
   mutual setting of, 191–192, 197, 366–370
grilling, 127
group practice, 60
guided imagery, 330, 356
guilt, 168

H

happiness
   basic feeling of, 167, 168
   nonverbal indicators of, 169
   words related to, 172, 182
helping
   core skills in process of, 10–13
   demystifying the process of, 186–187
   ethics in process of, 31–37
   natural style of, 23–25
helping professionals, 6
hemispheres of the brain, 57, 449–450

here-and-now dimension, 225
HIPAA Compliance Kit, The (Zur), 200
HIPAA privacy, 33–35, 200
hippocampus, 171, 445, 447–448
Hispanics. *See* Latinas/Latinos
homework assignments, 193, 330
"how" questions, 126
HPA axis, 49–50, 59, 447, 448–450
human connectome project, 444
human relationships, 337
humor, sense of, 55
hyperreflection, 281
hypervigilance, 459
hypothalamus, 449

I

identifying concepts, 430
Identity acronym, 58
imagery, 330, 356
immediacy, 162, 215, 225, 303–304
incongruity
   confronting, 259
   identifying, 238–240
   observing, 93, 107, 108
Indivisible Self model, 51, 59
influencing skills, 265–266
   competency assessments, 313–315,
     345–346
   1-2-3 pattern related to, 303
   self-reflection on, 347
   See also specific skills
informed consent, 35–36, 58
   child counseling and, 35, 145
   developing a form for, 61
instructional strategies, 320–322, 341–342.
     See also psychoeducation
integrating skills. *See* skill integration
integrative awareness stage, 252, 254
integrative theories, 286
intentional competence, 27, 430
intentionality, 7–8
   cultural, 8, 422
   definition of, 8, 26
   resilience and, 9
intentional practice, 81–82
intentional wellness plan, 56
interchangeable empathy, 74, 141
internal conflicts, 240
International Union of Psychological
   Science (IUPsyS), 32
interpersonal distance, 71
interpretation/reframing, 267–270,
   283–288
   anticipated result of, 269
   caution about using, 288
   client change and, 284
   competencies in, 297–298
   definition of, 268
   feedback form on, 296
   influencing skills and, 268–269
   key points about, 290
   neuroscience and, 457
   other microskills and, 284–286
   practice exercises on, 292–293, 295–296

reflection of meaning vs., 270–273
research evidence on, 282
self-reflection on, 298
skills of, 283–284
summary consideration of, 289
theories of counseling and, 286–288
See also reflection of meaning
interviewing, 5
   counseling, psychotherapy, and, 5–6, 26
   motivational, 356, 421
introspection stage, 252, 254
issues vs. problems, 15
"I" statements, 104–105, 302
Ivey Taxonomy, 435–440
   definitions and anticipated results,
     435–440
   five-stage interview structure, 437

J

Japanese culture, 94, 99
Jewish tradition, 280–281
job satisfaction, 282
journaling process, 193

K

key words
   bilingual clients and, 151
   encouragers as, 147, 148–149
   listening for, 101, 108
   paraphrasing using, 148, 153
   reflecting feelings with, 151
knowing–doing gap, 44, 59
knowledge, 27, 42–43

L

language
   bilingual clients and, 151
   conversational styles and, 102–103
   emotional, 167–168, 177
   problem-oriented, 14
   respectful use of, 40
   See also verbal behavior
Latinas/Latinos, 67, 226, 278, 280
leading questions, 131
leisure time, 53
life mission, 278, 279–280
lifestyle changes. *See* therapeutic lifestyle
   changes
limbic system, 50, 51, 168, 447
linking, 272–273
listening skills, 138–159
   acceptance and, 140–141
   active listening, 139–140, 146–149
   attending behavior and, 64–66, 82
   basic listening sequence, 184–186
   bilingual clients and, 151
   child counseling and, 142–145
   competencies in, 156, 158
   confrontation and, 236, 238–240
   diversity and, 149–151
   empathy and, 140–142
   encouraging, 140, 147
   example of using, 142–144
   feedback form on, 157

importance of practicing, 152
key points related to, 153
logical consequences and, 318–319, 343
multicultural issues and, 149–151
neuroscience and, 150, 456
overview of, 139–140
paraphrasing, 140, 147–149
practice exercises on, 154–156
research evidence on, 150
self-reflection on, 158, 206
stress responses and, 146
summarizing, 140, 149
logical consequences, 317–320, 341
competency evaluation, 346
counseling example of, 318–319
decision making and, 317–318, 355
feedback form on, 344
guidelines for using, 319
practice exercises on, 342–345
reflection exercise on, 320
logotherapy, 280–282, 399–400
long-term memory (LTM), 209–210
love
listening as, 5
wellness and, 53
Luminosity program, 338

M

Maintaining Change Worksheet, 389–390
Man's Search for Meaning (Frankl), 267,
281, 400
Martin Luther King Jr. Center, 256
Mayo Clinic, 338
meaning
centrality of, 268, 269
elicitation of, 268, 270, 277, 289
interpretation and, 290
logotherapy related to, 399–400
neuroscience and, 282–283
wellness and, 52
See also reflection of meaning
mediation process, 255–256
medications, 452–453
meditation, 278, 338–339
memory change, 207, 208–210
metaphors and similes, 166
microaggressions, 43, 146
microcounseling model, 12
microexpressions, 169
microskills
competence with, 427–430
counseling theories and, 27, 395–397
definition of, 26
future development of, 432
hierarchy of, 10–12
Ivey Taxonomy and, 435–437
model for learning, 12, 26
multicultural training in, 186, 322–323, 431
neuroscience related to, 455–457
professional account on using, 430–431
research on, 12–13
self-assessment of, 428–429
session analysis of, 381–383
supplementary readings on, 433

See also skill integration
microskills hierarchy, 10–12
definition of, 26
personal competence with, 427–430
self-assessment form, 428–429
mindfulness meditation, 338–339
mirroring behavior, 100
mirror neurons, 75–76, 449
mixed messages, 238–240
morality, 283
motivation, 282
motivational interviewing (MI), 356, 421
movement complementarity, 100
movement dissynchrony, 100
movement harmonics, 107, 108
movement synchrony, 100, 150
multicultural competence, 41–45
action related to, 44–45
importance of, 17–18
key points about, 59
mastery evaluation of, 62
practice exercises on, 60–61
RESPECTFUL model and, 18–19
self-awareness and, 41–42
self-reflection on, 62
skills development, 43
worldview and, 42–43
multicultural counseling and therapy
(MCT), 10–11, 287, 400–402
multiculturalism, 17
diversity and, 38–40
microtraining and, 186
multicultural issues
attending behavior and, 67, 71–72
bilingual clients and, 151
confrontation and, 237
example counseling session on, 95–98
focusing on contextual and, 225–227
global overview of, 322–323
listening skills and, 149–151
microskills training and, 187, 322–323,
431
nonverbal behavior and, 93–94, 99,
100–101
observation skills and, 93–94, 100–101,
107
prejudice and, 44
questions and, 128–129, 132
reflection of feeling and, 171–172
reflection of meaning and, 278, 280
result of considering, 40
self-disclosure and, 304–305
supplementary readings on, 434
theories of counseling and, 422
verbal communication and, 94, 105
Muslims, 278, 280
mutual goal setting, 191–192, 197.
See also goals
mutuality focus, 225

N

narrative theory, 15
National Association of School Nurses
(NASN), 32

National Association of School Psychologists
(NASP), 32
National Association of Social Workers
(NASW), 32, 37
National Board of Certified Counselors
(NBCC), 421
National Career Development Association
(NCDA), 32
National Institute of Mental Health
(NIMH), 22
National Organization for Human Services
(NOHS), 32, 36, 41
Native Americans
example counseling session with, 95–98
nonverbal behavior of, 67, 94
vision quest of, 278
natural style, 23–25, 427
audio/video exercise on, 23
client feedback on, 24
future integration with, 431–432
identifying for yourself, 427
self-assessment of, 25
negative consequences, 318
negative feedback, 307, 308, 312
negative self-talk, 329
neural networks, 450–452
neurofeedback, 310–311, 458
neurogenesis, 445
neuroimaging, 444
neurons, 450–452
neuroplasticity, 22, 27, 444–445
neuroscience, 20–23, 443–461
attention and, 455
brain plasticity and, 21–22
confrontation and, 251, 456
counseling and, 20–23, 59, 187, 443–461
creativity and, 251
default mode network and, 444, 457–458
emotions and, 168, 171, 172–173, 176,
447–448
empathy and, 75–76
feedback and, 310–311
focusing and, 456
future of, 22, 460
helping process and, 187
importance of, 48–50
intentional practice and, 81–82
interpretation/reframing and, 457
listening skills and, 150
materials for further study on, 460–461
microskills related to, 455–457
nonverbal behavior and, 99, 100
observation skills and, 455
questions and, 120, 455
reflection of feeling and, 171, 172–173,
176, 456
reflection of meaning and, 282–283, 456
stress and, 21, 48, 49, 59, 325
terminology of, 49–50
web resources on, 451–452, 458, 461
See also brain
neurotransmitters, 450–454
New, the, 236, 237, 251, 284, 453
newspaper questions, 132

New Zealand Association of Counsellors (NZAC), 32
nonattention, 70
nonjudgmental attitude, 237–238
nonverbal behavior
    acculturation and, 100–101
    body language and, 66, 71–72, 100
    discrepancies and, 100, 105, 106
    emotional content and, 169
    facial expressions and, 100
    mirroring of, 100
    multiculturalism and, 93–94, 99, 100–101
    neuroscience and, 99, 100
    observation of, 92–93, 98–101, 106, 108, 170
    research on, 99
    self-evaluation of, 108
    web resources on, 93, 99
    See also verbal behavior
nonviolent change, 256
norepinephrine, 449, 454
normalizing the situation, 404
note taking, 200
nucleus accumbens, 449
nutrition, 55, 335–337

O

observation skills, 90–113
    competencies in, 112–113
    counseling session and, 92
    discrepancies and, 93, 100, 105–106, 107
    example illustrating, 95–98
    facial expressions and, 100
    feedback form on, 109–110
    group practice of, 111
    importance of, 91, 106
    internal conflict and, 93, 105, 107
    key points related to, 106–107
    multiculturalism and, 93–94, 100–101, 107
    neuroscience and, 455
    nonverbal behavior and, 92–93, 98–101, 106, 108, 170
    practice exercises on, 107–111
    reflection of feeling and, 169, 170
    result of practicing, 92
    self-reflection on, 113
    summary of, 106
    verbal behavior and, 93, 101–105, 107, 108
occipital lobe, 446
omega 3 treatments, 336
On Becoming a Person (Rogers), 399
1-2-3 pattern, 303, 311
open questions, 118, 132
    anticipated result of, 119
    beginning sessions with, 125–126
    bringing out specifics with, 127
    elaborating client's story with, 126
    examples of using, 123–125, 185
    less verbal clients and, 130–131
    practice exercises on, 133–135

oppression, 37, 459
optimism, 46–48
Orygen Youth Health program, 336
"other" statements, 104

P

pacing clients, 170
paraphrases
    anticipated result of, 140
    child counseling and, 145
    examples of using, 185
    four dimensions of, 148, 153
    interpretation/reframing and, 285
    listening process and, 140, 147–149, 153
    practice exercises on, 154–155
    reflection of feeling vs., 161, 162, 177–178
parasympathetic nervous system, 446
parietal lobe, 446
partial acceptance, 255
partial examination, 245
perception checks. See checkouts
peripheral nervous system (PNS), 446
personalizing sessions, 190
personal resources, 57
personal strength inventory, 129
personal style. See natural style
person-centered counseling, 201, 398–399
    interpretation/reframing and, 286
    key methods used in, 398–399
    philosophy of, 398
    practice exercise on, 203–204
    supplementary reading on, 399
person-in-community concept, 214
physical self, 55–56
pituitary gland, 449
planning
    first session, 357–358
    treatment, 385–386, 387
Political Brain, The (Weston), 283
political correctness (PC), 39–40
Portfolio of Competence, 27
positive asset search
    basic listening sequence and, 203
    community genogram and, 218
    dereflection and, 282
    example of using, 196
    questions for, 129–130, 132
    wellness and, 191
positive consequences, 318
positive emotions, 172–173
positive feedback, 307, 308, 312
positive guided imagery, 330
positive psychology, 46, 59, 281
positive regard, 140–141
positive stressors, 323
positron emission tomography (PET), 311, 444
Posit Science programs, 338
posttraumatic stress (PTS), 39
posttraumatic stress disorder (PTSD), 14, 39, 404, 452
posttraumatic stress injury (PTSI), 39
posttraumatic stress reaction (PTSR), 14, 39, 404

poverty, 458, 459
power
    ethics related to, 36–37, 58
    privilege related to, 42, 59
practice, 81–82
practice contract, 36
practice exercises. See competency practice exercises
pragmatism, 402–403
praise, 307
prefrontal cortex, 173
prejudice
    accumulative stress and, 146
    multiculturalism and, 44
    See also racism
privacy, 33–35
privilege, 42, 59, 280
problems
    language for describing, 14–15
    stages in solving, 187–188, 353
problem-solving counseling. See decisional counseling
Promise of Sleep, The (Dement), 333
psychiatric medications, 452–453
psychiatry, 7
psychodynamic theory, 287
psychoeducation, 320–322
    anticipated result of, 321
    competency evaluation, 346
    definition of instruction and, 321
    feedback form on, 344
    group practice session on, 345
    guidelines for using, 321–322
    key points about, 341–342
    social skills and, 79–80
    stress management and, 327–330
psychoeducational teaching competence, 27
psychotherapy
    brain and, 50, 443–444
    ethical foundations of, 30–31
    interviewing, counseling, and, 5–6, 26
    neuroscience and, 453–454
    number of sessions in, 6
punishment, 318, 319

Q

questions, 117–137
    at-risk youth and, 128
    beginning sessions with, 125–126
    bringing out specifics with, 127, 130–131
    child counseling and, 128, 131, 145
    closed, 118, 119, 122–123, 132
    competence in asking, 136
    elaborating client's story with, 126
    examples of using, 121–125, 137
    exploring emotions through, 170
    feedback form on, 135
    key points about, 132
    less verbal clients and, 130–131
    missing data and, 120–121
    multicultural issues and, 128–129, 132
    neuroscience and, 120, 455
    newspaper, 132

objections to, 119–120
open, 118, 119, 123–125, 126, 132
overview of, 118–119
potential problems with, 127
practice exercises on, 133–135
research evidence on, 120
self-reflection on, 137
stems for asking, 126, 133–134
strengths identified using, 129–130
summary about using, 131
value of, 118, 132

R

race/ethnicity
  wellness assessment and, 52
  See also multicultural issues
Racial/Cultural Identity Development (R/
  CID) model, 251–252, 254, 257, 261
racism
  accumulative stress and, 146, 459
  emotions related to, 171–172
  microaggressions of, 43, 146
Rainy Brain, Sunny Brain (Fox), 47
rapport building, 190, 361–363
rational emotive behavior therapy (REBT),
  410
rational emotive therapy (RET), 410
realistic beliefs, 54
receptor sites, 452
recognition, 245
redirecting attention, 70
referral process, 384–385
reflection of content, 147–149
reflection of feeling, 160–182
  anticipated result of, 161
  competencies in, 180–181
  definition of, 161
  empathic understanding and, 163
  examples of using, 164–166, 185
  expression of emotion and, 170–171
  feedback form on, 179
  interpretation/reframing and, 285
  key points about, 175–176
  language of emotion and, 167–168, 182
  multicultural issues and, 171–172
  neuroscience and, 171, 172–173, 176, 456
  nonverbal indicators and, 169
  observation skills and, 169, 170
  paraphrasing distinguished from, 161,
    162, 177–178
  positive emotions and, 172–173
  practice exercises on, 176–180
  research evidence on, 174–175
  self-reflection on, 181–182
  skill of working with, 167
  strategies for positive, 173–174
  summary consideration about, 175
  techniques used for, 162–163
  words related to, 167–168, 177, 182
reflection of meaning, 267–283
  anticipated result of, 269
  competencies in, 297–298
  counseling session example of, 273–276
  definition of, 268

discernment and, 270, 278, 279–280
elicitation and, 268, 270, 277, 289
feedback form on, 294
interpretation/reframing vs., 270–273
key points about, 289
life mission/goals and, 278
logotherapy and, 280–282
multicultural issues and, 278, 280
neuroscience and, 282–283, 456
practice exercises on, 290–291, 293–295
religion/spirituality and, 277, 278, 280
research evidence on, 282–283
self-reflection on, 298
seriously ill clients and, 276
summary consideration about, 288
value conflicts and, 277–278
See also interpretation/reframing
reframing. *See* interpretation/reframing
relapse prevention, 201, 388–390
relational cultural therapy (RCT), 401
relationship stage. *See* empathic relationship
  stage
Relaxation and Stress Reduction Workbook, The
  (Davis, Eshelman, & McKay), 326
religion
  contextual focus on, 225
  discernment process in, 278
  neuroscience and, 283
  privileges for dominant, 280
  reflection of meaning and, 277, 278, 280
  traumatic events and, 282
  wellness and, 52
  See also spirituality
reptilian complex, 446–447
research
  on attending behavior, 72–73
  on confrontation, 250–251
  on crisis counseling, 405
  on empathy, 75–76
  on exercise, 333
  on fMRI and CBT, 420
  on focusing, 227
  on interpretation/reframing, 282
  on listening skills, 150
  on microskills, 12–13
  on neuroscience, 20–23, 443–461
  on nonverbal behavior, 99
  on observation, 99
  on questions, 120
  on reflection of feeling, 174–175
  on reflection of meaning, 282–283
  on self-disclosure, 306
  on wellness, 56–57
resilience, 9, 26, 47
resistance, client, 321
resistance and immersion stage, 252, 254
RESPECTFUL model, 18–19, 26, 402
restatements, 140, 142, 147, 148, 149, 153
restory strategy, 16, 192–193
  cognitive behavioral therapy and, 412–418
  counseling session examples of, 198,
    370–378
  mediation process and, 256
role-plays, 35, 64–65, 88

S

sadness
  basic feeling of, 167, 168
  neuroscience research on, 444
  nonverbal indicators of, 169
  words related to, 182
safety concerns, 404
Sample Practice Contract, 36
samurai effect, 81, 432
School Social Work Association of America
  (SSWAA), 32
selective attention, 68–70
  focusing and, 213, 229
  listening and, 148, 153
  observing patterns of, 101, 108
selective serotonin reuptake inhibitors
  (SSRIs), 452
self-actualization, 9, 26
self-assessment, 60, 428–429
self-awareness, 41–42
self-care, 53
self-disclosure, 301–306
  anticipated result of, 301
  appropriateness of, 305
  competence related to, 313–315
  counselor focus and, 225
  cultural implications of, 304–305
  dangers of, 302
  definition of, 301
  feedback form on, 314
  genuineness of, 303
  immediacy of, 303–304
  key points about, 311–312
  overview of using, 301–302
  practice exercises on, 312–313
  research evidence on, 306
  rewards and risks of, 311
  self-reflection on, 315
  skill of, 302–303
  timeliness of, 304
self-in-relation concept, 38, 214, 228
self-reflection process, 62
self-regulation, 310
self-talk, 329
self-worth, 54
sentence stems, 148, 153, 162
serotonin, 452, 454
sessions. *See* counseling sessions
sexual activity, 333
sexual assault victims, 405
sexual harassment, 284, 288
sexual identity, 52
sexual orientation, 18, 19, 52
short-term memory (STM), 209–210
significant others, focus on, 224
silence, usefulness of, 70
Silent Language, The (Hall), 94, 99
similes and metaphors, 177
Six-Point Optimism Scale, 47
skill integration, 351–393
  action stage and, 378–380
  anticipated result of, 352
  case management and, 386, 388
  decisional counseling and, 352–356

skill integration (*continued*)
    empathic relationship stage and, 361–363
    full-session transcript illustrating, 360–380
    goal setting stage and, 366–370
    impact on client, 381–383
    key points related to, 392–393
    planning and, 357–358, 385–386, 387
    practice exercise on, 393
    restory stage and, 370–378
    story/strengths stage and, 363–366
    transcript analysis and, 380–381
skills vs. strategies, 318
sleep, 333–335
smiling, 94, 150
social context, 58
social emotions, 168, 448
social justice, 37–38, 58, 227–228, 388, 459–460
social relations, 337
social self, 53
social skills, 73, 79–80
social systems, 459
solution generation, 245, 255
Spanish key words, 151
Spark: The Revolutionary New Science of Exercise and the Brain (Ratey), 332
speech patterns, 67–68
spirituality
    contextual focus and, 225
    discernment process in, 278
    neuroscience and, 283
    reflection of meaning and, 277, 278, 280
    traumatic events and, 282
    wellness and, 52
    See also religion
statements
    alternative focus, 230–231
    confrontation, 260–261
    logical consequence, 342–343
    questions used as, 127
    supportive, 307
step-by-step learning, 12, 455
stereotyping, 42, 101
stories
    counseling session examples of, 195–196, 363–366
    debriefing following traumas, 405
    drawing out, 13–14, 190, 195–196
    mediation process and, 256
    narrative theory and, 15
    paraphrasing, 148
    restorying, 16, 192–193
    strengths and, 16, 190–191
strategies vs. skills, 318
strengths
    assessment of, 51–56, 59
    community genogram for identifying, 218
    counseling session examples of, 195–196, 363–366
    exploring with clients, 191, 196
    inventory of personal, 129
    mediation process and, 256
    positive psychology and, 46, 281

    stories and, 16, 190–191
    wellness based on, 45
stress, 323–325
    accumulative, 146
    assessing levels of, 458–459
    beneficial effects of, 325
    brain and, 21, 48, 49, 59, 325
    child development and, 459
    confrontation and, 251, 459
    destructive effects of, 325, 459
    good vs. bad, 323–325
    posttraumatic, 39
    self-reflection on, 324
    social justice and, 38
    wellness assessment and, 53–54
stress management, 326–330, 342
    anticipated result of, 326
    assertiveness training and, 328–329
    communication skills training and, 327–328
    competency evaluation, 346
    feedback form on, 344
    group practice session on, 345
    homework assignments for, 330
    mental health counseling and, 20
    positive guided imagery and, 330
    psychoeducational strategies for, 327–330
    therapeutic lifestyle changes and, 57, 326–327
    thought stopping and, 329
    wellness related to, 53–54
stuckness, 236
subtractive empathy, 74, 141
summarizations
    anticipated result of, 140
    child counseling and, 145
    examples of using, 185
    four dimensions of, 153
    listening process and, 140, 149, 153
    practice exercises on, 154–155
supervision, 384
supplementary readings, 433–434. *See also* web resources
supportive statements, 307
surprise
    basic feeling of, 167, 168
    nonverbal indicators of, 169
    words related to, 182
sympathetic nervous system, 446
synapses, 450
synaptic vesicles, 452
systematic emergency therapy, 405

T

talk time, 70–71
TAP axis, 49, 50, 59, 173, 447, 448–450
task positive brain (TPB), 457
teaching competence, 430
temporal lobe, 446
thalamus, 449
themes, session, 214
theories of counseling. *See* counseling theories

Theories of Counseling and Psychotherapy (Ivey, D'Andrea, & Ivey), 46
therapeutic lifestyle changes (TLC), 20, 331–340, 342
    body scan, 338–339
    client compliance with, 331, 332
    cognitive challenge, 337–338
    competency evaluation, 346
    contemplative practices, 339–340
    exercise, 331–333, 334
    feedback form on, 344
    group practice session on, 345
    mindfulness meditation, 338–339
    neuroplasticity and, 445
    nutrition, 335–337
    self-reflection on, 339
    sleep, 333–335
    social relations, 337
    stress management and, 57, 326–327
therapeutic relationship, 15
    discrepancies in, 106
    empathy in, 73
therapeutic self-disclosure (TSD), 306
thoughts
    chart for recording, 415
    observing your own, 258–259
    strategy for stopping, 329
    wellness assessment and, 54
timeliness of self-disclosure, 304
TLC. *See* therapeutic lifestyle changes
tone of voice, 68
touching behavior, 93
training as treatment, 79
trait-and-factor theory, 353
transcendence, 246, 255
transcripts of counseling sessions
    checklist for creating, 392
    full-session transcript analysis, 360–381
trauma
    accumulative stress and, 146
    burnout related to, 406
    crisis counseling and, 403
    neuroscience research on, 448
    posttraumatic responses to, 39
    reflecting on meaning after, 282
treatment plan, 385–386, 387
Tree of Contemplative Practices, 339–340
triune brain, 446–447
trust
    cross-cultural situations and, 128–129
    less verbal clients and, 130

U

unconditional positive regard, 140–141
Unheard Cry for Meaning, The (Frankl), 278

V

value conflicts, 277–278
verbal behavior
    concreteness vs. abstraction in, 102–103
    discrepancies in, 105
    "I" and "other" statements in, 104–105
    key words used in, 101

multiculturalism and, 94, 105
observing in clients, 93, 101–105, 107, 108
selective attention and, 101
self-evaluation of, 108
See also language; nonverbal behavior
verbal tracking, 66, 68–71, 101, 108
verbal underlining, 67
video recording
of counseling sessions, 187, 200, 391
natural style exercise using, 23
vision quest, 278
visual contact. *See* eye contact
vitamin D3 supplements, 336
vocal qualities, 66, 67–68, 108

W
warmth, 163
warnings, use of, 319

web resources, 4, 27
on HIPPA privacy, 34
on microskills training, 433
on mindfulness meditation, 339
on neuroscience, 451–452, 458, 461
on nonverbal communication, 93, 99
on professional ethical codes, 31–32
"we" focus, 225
wellness, 51–57
assessment of, 51–56, 59
brain health and, 57
contextual issues in, 51
exercise and, 55–56, 331–333
Indivisible Self model and, 51
intentional plan for, 56, 59
key points related to, 59
mastery evaluation of, 62
nutrition and, 55, 335–337
positive psychology and, 46, 59

practice exercises on, 61
research on, 56–57
self-reflection on, 62
strengths and, 45, 191
therapeutic lifestyle changes and, 331
"what else" questions, 121, 130, 132
"what" questions, 126
whistle-blowers, 228
"why" questions, 124, 126, 127
Wikipedia resources, 461
working alliance, 15, 73
working memory, 209–210
work issues, 55, 227, 228, 282
worldview, 42–43
"would" questions, 124, 126

Y
youth at risk, 128